Medicinal and Aromatic Plants of Oman

Medicinal plants have received an impetus recently due to the revival of attention to natural products in the form of herbal medicines, demanding convincing knowledge on their phytochemistry, status, cultivation, and utilization. Medicinal plants in the Sultanate of Oman have tremendous potential in terms of ethnomedicinal and food applications. However, these treasures of Oman have not been fully explored until recently. Several medicinal plants indigenous to Oman have been used for several centuries in the form of traditional medicine for the prevention and treatment of diseases. The first volume of this book covers information about sustainable use of selected medicinal plants, their phytochemistry, and efforts made to conserve them to ensure continued supply. This volume reveals systematic information about 104 to 151 promising medicinal plants in Oman with botanical names, common names, synonyms, local names, family, habitat and distribution, chemical constituents, pharmacological activity, uses (common uses, traditional uses, and therapeutic uses), toxicity and clinical data, patents, and their marketed products or industrial application. Each of the 151 plants has a high-resolution color photograph (where available), plant description, and location. The chapters are written by well-experienced authors in the field. A uniform chapter structure throughout this book has been designed to maintain consistency. This book is especially valuable to academicians, phytochemists, pharmacologists, herbalists, agriculturists, biotechnologists, natural product researchers, and industrialists involved in the production of herbal drugs. Salient features of this book are as follows:

- It provides common uses, traditional uses, and therapeutic uses of 132 medicinal plants.
- It covers the detailed phytochemistry of each medicinal plant.
- It offers systematic information about the geographical distribution of these plants.
- Updated information about the pharmacology, mechanisms of action, and toxicity of these plants is covered in detail.

Overall, this book provides an overview of botanical description, geographical distribution, chemical composition and uses of the medicinal plants native to the Sultanate of Oman.

Many of the plants listed in this book—particularly in Volume One—are endemic, near-endemic, or regionally endemic medicinal and aromatic species of Oman, which remain largely unexplored and are supported by limited scientific literature. Due to this scarcity of information, the authors have included relevant data from closely related species within the same genus. This approach enables readers to better understand the chemistry, taxonomy, and potential uses of the respective species in the context of their taxonomic relatives.

Medicinal and Aromatic Plants of Oman

Phytochemistry and Ethno-Pharmacological Aspects, Volume 1

Ahmed Al-Harrasi, Ahmed Al-Rawahi, Saif Amur Al Hatmi, and Saurabh Bhatia

CRC Press
Taylor & Francis Group
Boca Raton London New York

CRC Press is an imprint of the
Taylor & Francis Group, an **informa** business

Designed cover image: Author's own image

First edition published 2026
by CRC Press
2385 NW Executive Center Drive, Suite 320, Boca Raton FL 33431

and by CRC Press
4 Park Square, Milton Park, Abingdon, Oxon, OX14 4RN

CRC Press is an imprint of Taylor & Francis Group, LLC

Library of Congress Cataloging-in-Publication Data
Names: Al-Harrasi, Ahmed, author. | Al-Rawahi, Ahmed, author. | Hatmi, Saif Al, author. | Bhatia, Saurabh, author.
Title: Medicinal and aromatic plants of Oman : phytochemistry and ethno-pharmacological aspects, Volume 1 / Ahmed Al-Harrasi, Ahmed Al-Rawahi, Saif Amur Al Hatmi, and Saurabh Bhatia.
Description: First edition | Boca Raton, FL : CRC Press, 2026. | Includes bibliographical references and index.
Identifiers: LCCN 2025007675 (print) | LCCN 2025007676 (ebook) |
ISBN 9781032281469 (hardback) | ISBN 9781032281476 (paperback) | ISBN 9781003295488 (ebook)
Subjects: LCSH: Medicinal plants–Oman. | Materia medica, Vegetable–Oman. |
Ethnopharmacology–Oman. Classification: LCC QK99.O5 A44 2025 (print) | LCC QK99.O5 (ebook)
LC record available at https://lccn.loc.gov/2025007675
LC ebook record available at https://lccn.loc.gov/2025007676

ISBN: 9781032281469 (hbk)
ISBN: 9781032281476 (pbk)
ISBN: 9781003295488 (ebk)

DOI: 10.1201/9781003295488

Typeset in Times
by codeMantra

Contents

About the Authors

Ahmed Al-Harrasi completed his BSc in Chemistry at Sultan Qaboos University (Oman) in 1997. He then pursued his MSc and PhD in Organic Chemistry at the Free University of Berlin, earning both degrees in 2002 and 2005, respectively, as a DAAD fellow under the supervision of Professor Hans-Ulrich Reissig. His doctoral research focused on novel transformations of enantiopure 1,2-oxazines, and he was awarded his PhD *summa cum laude*, the highest academic distinction. Notably, he is part of the academic lineage of the renowned chemist Rolf Huisgen, as depicted in the academic family tree (Angew. Chem. Int. Ed. 2020, 59, 12228–12232).

In 2008, Prof. Al-Harrasi was awarded a Fulbright Scholarship for postdoctoral research in chemical biology, which he conducted in Professor Tadhg Begley's group at Cornell University (USA). There, he worked on the synthesis of isotopically labeled thiamin pyrophosphate. Upon completing his postdoctoral work in 2009, he returned to Oman to establish his independent research career at the University of Nizwa. He founded the Natural and Medical Sciences Research Center (NMSRC), a pioneering institution that now houses over 80 full-time researchers across 30 research laboratories and has become a national reference point in research.

Prof. Al-Harrasi's interdisciplinary research sits at the interface of chemistry and biology. He applies chemical principles to deepen our understanding of biological systems and uses biological tools to advance new chemical insights. He currently serves as Professor of Organic Chemistry and Vice Chancellor for Graduate Studies, Research, and External Relations at the University of Nizwa. His funded research portfolio exceeds $25 million, spanning multiple disciplines and engaging collaborators both nationally and internationally.

He was a member of UNESCO's International Basic Sciences Programme (IBSP) and a Fellow of the Royal Society of Chemistry. He has chaired and been invited to speak at numerous international conferences and serves as a referee for over 20 international chemistry journals. Prof. Al-Harrasi has led more than 40 externally and internally funded research projects and has published over 1,100 scientific papers, 46 book chapters, 15 books, and 20 patents. His work has garnered over 20,000 citations, and he holds an h-index of 63.

His excellence in research has earned him several national honors, including the Order of Royal Commendation from His Majesty the Sultan of Oman and the National Research Award from the Research Council of Oman. He has been consistently featured on Stanford University's global list of the top 2% of scientists from 2019 to 2024, underscoring the global impact and recognition of his work.

Prof. Al-Harrasi has always approached chemistry with a desire to move beyond the superficial and descriptive, embracing biology as an essential partner in understanding the organic chemistry of complex natural phenomena. Though rooted in fundamental science, his research is consistently translated into tangible societal benefits. Under his leadership, NMSRC has developed innovative solutions to pressing national challenges in health, agriculture, and the environment, leveraging world-class research infrastructure and forging strategic partnerships to attract competitive funding.

He leads an independent, highly interdisciplinary research team working across chemistry, biology, and biomedical sciences. His research focuses on gaining a chemical understanding of biologically relevant systems and translating this knowledge into applications for societal good. His integrated approach unites synthetic organic chemistry, catalysis, drug design and synthesis, organometallic chemistry, biosynthesis and bioengineering, medicinal chemistry, natural product chemistry, food chemistry, microbiology, molecular biology, and genomics. His laboratories employ cutting-edge techniques such as NMR spectroscopy, mass spectrometry, and X-ray diffraction for structural and dynamic characterization of a wide array of compounds.

Ahmed Al-Rawahi earned his MSc and PhD in Plant Pathology from the University of California at Berkeley in 1992 and 1995, respectively. Earlier (in 1988), he had completed, with honors, his BSc in Biological Science at North Carolina State University (Raleigh, USA). Upon his return to Oman, he became a Lecturer at Sultan Qaboos University, where he taught courses in the fields of Plant Pathology and Microbiology. He also conducted research in soil-borne pathogens, biocontrol, and disease management programs and published articles in high-impact international journals.

From December 1997 until May 2001, he had the honor of serving as the Minister of Agriculture and Fisheries, and played a major role in developing various strategies and development plans for these vital sectors of the Omani economy. He served as a member of the State Council from 2001 to 2011 and led important studies and reports for the government of Oman. He was also the Chairman of the Academic Foundation Committee for the University of Nizwa Project from 2000 to 2004, the year when he was appointed as Chancellor of the newly founded University. Throughout his tenure as Chancellor, he embarked on bringing the project to reality and creating a functioning academic institution that adheres to institutionalization, quality standards, ethics, and procedures. Under his dynamic leadership, thousands of Omani students graduated with Bachelor's and higher degrees, and the university contributed strongly to community services and produced innovative research relevant to the needs of Oman. He was promoted to the academic position of Founding Professor by a scholarly independent academic committee in December 2006. Professor Ahmed Al-Rawahi is a member of various national academic committees, including the Education Council and Research Council. He was also a member of the American Pathological Society. In November 2000, he was greatly honored to receive from HM Sultan Qaboos bin Said the Oman Medal of Merits (second degree, civil). In December 2013, the University of Wisconsin Oshkosh awarded him the Honorary Doctorate Degree in recognition of his role as "a distinguished and visionary academic leader with an international voice and perspective."

Saif Amur Al Hatmi is both a botanist and an ethnobotanist who has worked since 2006 in Oman Botanic Garden, specializing mostly in the Omani flora and their economic importance in society. He studied Environmental Biology at the Sultan Qaboos University. In 2006, he earned a Bachelor's degree in Environmental Biology and a Master's degree in Ethnobotany from the Kent University (UK) in 2012. He has led and conducted numerous scientific expeditions and research teams throughout the country to discover and document the plant heritage of Omani flora and conservation-related issues that affect the vegetation and native plant habitats.

Saurabh Bhatia is a pharmaceutical biologist with almost 16 years of research and academic experience in a multitude of transdisciplinary research related to pharmaceutical and food sciences. He earned his PhD degree (2015) in Pharmaceutical Technology from Jadavpur University, Kolkata. He has authored more than 140 articles and 15 books in the area of biopolymers and natural products. His research focus is exploring the food and pharmaceutical applications of biopolymer-based films as sustainable packaging materials and drug delivery systems.

Acknowledgments

The unexplored flora of the Sultanate of Oman holds immense medicinal and aromatic potential. However, the absence of a systematic body of literature presents a remarkable challenge for researchers seeking to investigate the therapeutic properties of these medicinal plants. This comprehensive set of data on Omani medicinal flora aims to inspire a new generation of researchers to harness these plants to address environmental and health-related issues and to add value to them.

We dedicate this book to all students, researchers, academicians, and scholars who have made sincere contributions to the field of medicinal plants—whether at the national or international level. Their efforts continue to deepen our understanding of medicinal plants.

We gratefully acknowledge the support of the University of Nizwa, Sultanate of Oman, for the generous support of this book project. Specifically, our deep appreciation goes to the Natural and Medical Sciences Research Center (NMSRC), whose state-of-the-art facilities and commitment to interdisciplinary research provided the foundation for much of the work presented here. The Center has been instrumental in advancing scientific exploration in medicinal plant studies and has played a vital role in bringing this book to fruition.

We extend our heartfelt thanks to Mr. Ahmed Hamood Al-Busaidi for his unwavering support and dedication in capturing high-quality photographs of medicinal plants from remote regions across Oman. His significant contribution to this project is deeply appreciated.

Finally, we express our sincere gratitude to the entire team at CRC Press, with special thanks to Dr. Renu Upadhyay and Ms. Jyotsna Jangra for their enthusiastic cooperation, guidance, and continued support throughout the publication process.

Prof. Ahmed Al-Harrasi
Professor of Organic Chemistry
Natural and Medical Sciences Research Center
University of Nizwa, Oman

Mr. Saif Amur Al Hatmi
Senior Botanist & Ethnobotanist
Oman Botanic Garden
Muscat, Sultanate of Oman

Prof. Ahmed Al-Rawahi
Chancellor and Professor of Plants Pathology
University of Nizwa, Oman

Dr. Saurabh Bhatia
Associate Professor
Natural and Medical Sciences Research Center
University of Nizwa, Oman

1 Genus *Abelmoschus*

1.1 *Abelmoschus esculentus* (L.) Moench

1.1.1 Synonyms

Hibiscus hispidissimus, Hibiscus esculentus and *Abelmoschus bammia*

1.1.2 Family

Malvaceae.

1.1.3 Vernacular Name

Semerhot.

1.1.4 Common Name

Lady's finger or okra.

1.1.5 Botanical Description

This plant is an annual, erected herb that reaches a height of 1 m. Leaves: alternate, palmately 3–5-lobed, 15–20 cm in length. Flowers: solitary, pale yellow with a crimson center. The fruits are cylindrical capsules that are densely covered with short, stiff hairs (Figures 1.1.1–1.1.3).

1.1.6 Geographical Distribution

It originated from the Abyssinian center. Globally, it is extensively available in tropical, subtropical, and warm temperate areas; nevertheless, its presence is more dominant in Thailand, Brazil, India, the Philippines, and West Africa. In Oman, it is mainly found in southern Oman among wet escarpment woodlands [1]. In Asia, it's widely grown and consumed in China, India, and Pakistan. *Abelmoschus esculentus* is supposed to have originated nearer to Ethiopia, where it was frequently grown by the Egyptians in the twelfth century and afterward cultivated across the Middle East and North Africa. Now, this plant is grown typically in tropical and subtropical areas across the globe and is considered a common garden food crop [2–4].

FIGURE 1.1.1 Apical part of *Abelmoschus esculentus.*

FIGURE 1.1.2 Flower of *Abelmoschus esculentus.*

FIGURE 1.1.3 Leaf shape of *Abelmoschus esculentus*

DOI: 10.1201/9781003295488-1

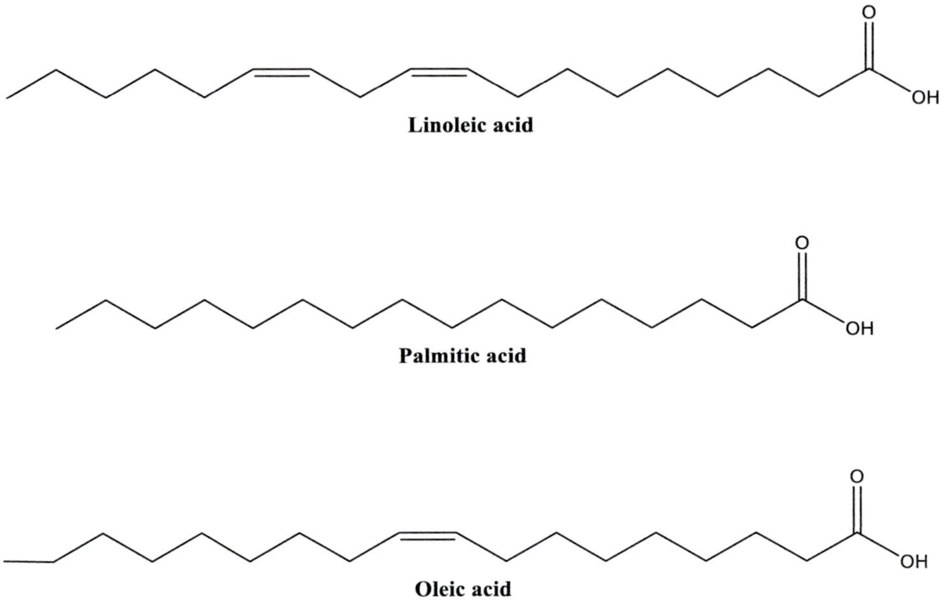

FIGURE 1.1.4 Chemical components of *Abelmoschus esculentus*.

1.1.7 Chemical Composition

Because of their nutritional value, immature pods are mainly used. However, other parts of the plant, including leaves, flowers, stems, and especially seeds, which are rich in oil and protein sources, have also been used. *A. esculentus* is an important multipurpose vegetable crop. Seeds contain protein with high lysine amounts as well as oil rich in linoleic, palmitic, and oleic acids, and polyphenolic components such as oligomeric flavonol derivatives as well as catechins (Figure 1.1.4). Roots and leaves commonly contain flavonol glycosides, while leaves are also enriched with minerals and tannins, and similarly, roots also contain carbohydrates. Seeds are enriched with high levels of fatty acids, particularly linoleic as well as palmitic acid, which correspond to more than 70% of overall fatty acids. However, the chemical composition showed considerable differences from the previous work, which suggests that the chemical profile is dependent on cultivation conditions and genotype [5]. The seeds of this plant are enriched with high-level oil containing mainly linoleic acid (Figure 1.1.4) [6].

1.1.8 Uses

A. esculentus is mainly used as a vegetable crop, usually for its immature pods. Depending upon the consumer's requirements, the harvesting stage (size of the fruit) is decided. As per the previous report [7] in Turkey, mature seeds can alternatively be used as a coffee substitute, whereas seeds, due to their high protein levels, have been reported for their use as flour in the food industry [8].

Traditionally, pods have been utilized for multiple purposes, such as aphrodisiacs, appetizers, astringents, and in the food industry. Additionally, pods have been suggested to treat urinary and gastrointestinal tract problems and gonorrhea [9]. Seeds have been evidenced for antifungal and anticancer effects, while extracts derived from the plant have also been reported for diuretic and moisturizing effects [5].

For washing and cleaning the body and scalp, people in the past would create a lather by crushing the leaves and stems of this plant with water. The stems, in particular, were known to produce a rich foam. These natural cleansers were an innovative way to maintain hygiene, though their effectiveness might not match that of modern soaps [10].

Previous reports suggested that the ethanolic fraction derived from *A. esculentus* showed significant antioxidant effects, and in the same report, the phenolic level of the plant was correlated with its antioxidant effects. Another report suggested the antioxidant effects of seed extracts derived from the *A. esculentus* plant and its correlation with phenolic content. Additionally, in the previous report, the antimicrobial effects of seed extracts from okra (*Abelmoschus esculentus*) were assessed, and findings showed that seeds demonstrated the greatest inhibition zones [11,12].

The green pods (immature) of *Abelmoschus esculentus* are often consumed as vegetables, whereas extracts derived from the pods are considered a thickening agent, mainly to increase viscosity, especially in soups and sauces [13,14]. Polysaccharides derived from this plant have several health benefits and longer shelf-lives. Thus, it is used for several applications in the food industry [15–17].

Traditionally, pods of this plant were used in food as aphrodisiacs, appetite boosters, and astringents. Moreover, pods have also been used to treat urinary disorders, dysentery, and gonorrhea [9]. It was also found that young okra pod extracts showed diuretic effects, while the seeds demonstrated fungicidal and anticancer properties [5].

Additionally, the plant and its preparations are reported for various therapeutic effects, such as antidiabetic, anticancer, antiproliferative, and proapoptotic effects; lowering bad cholesterol; antioxidant properties; pregnancy benefits; antifatigue effects; gut bacteria-friendly properties; ulcer treatment; relief and prevention of constipation; improvement of vision; antibacterial activity; antiadhesive properties; hypocholesterolemic effects; and anti-inflammatory effects [18].

1.1.9 Mechanism of Action

- Antidiabetic

  Fruits of *A. esculentus* showed antidiabetic action on alloxan-induced diabetic rodents by triggering the synthesis of glycogen in the liver and postponing glucose absorption, with a considerable glucose dialysis retardation index and high glucose adsorption capability [19].

  As per the previous report, okra powder has also shown its effect on pancreatic islets by acting on the expression of PPAR-γ and PPAR-α genes in the pancreas of streptozotocin-induced and high-fat diet rats. It was found that active components of okra powder downregulated PPAR genes in the pancreas, suggesting the potential of the components to improve β-cell impairment in diabetes via a mechanism regulated by PPAR (peroxisome proliferator-activated receptor gamma) and glucose homeostasis [20].
- Antioxidant Effects

  The polysaccharide content of okra mucilage has been reported for its antioxidant effects. It prevents damage caused by reactive oxygen species (ROS). Moreover, okra mucilage has been reported for its neutralization effect on free radicals by increasing superoxide dismutase levels [21].

  It was also found that *A. esculentus* demonstrates hypoglycemic and hypolipidemic effects. These effects could be associated with its potential in reducing the mRNA levels of proinflammatory cytokines IL-1β, IL-6, and TNF-α in THP1-derived macrophages and inducing low density lipoprotein receptor (LDLR) expression in the human hepatoma cell line Huh-7. Additionally, the sequestering ability of cholic acid was found, which could be responsible for the hypocholesterolemic activity [22].

1.1.10 Toxicity

Previous research showed that apart from arsenate, *A. esculentus* accumulated arsenite and dimethylarsinic acid mostly in the roots, with restricted transportation to shoots. Therefore, *A. esculentus* has the potential to withstand arsenic-induced stress, and this arsenic cannot be removed by boiling the pods in water [23].

1.1.11 Clinical Data

IQP-AE-103 containing dehydrated powder of okra (*Abelmoschus esculentus* (L.) Moench) pods and other ingredients has been recently investigated via a double-blind, randomized, three-armed, placebo-controlled clinical investigation (NCT03058367) to assess its potential in decreasing body weight. It was found that the formulation showed a positive effect on the lipid metabolism of subjects without any side effects [24].

A. esculentus has historically been used in the treatment of diabetes. A clinical study (double-blind, randomized clinical trial) was performed to assess the effect of *A. esculentus* fruit on glucose levels in the blood of patients with type 2 diabetes mellitus, with subsequent use of hypoglycemic agents. It was found that the treatment with *A. esculentus* fruit showed an anti-hyperglycemic effect in patients with type 2 diabetes mellitus [25].

Another investigation (randomized, double-blind clinical trial), which was performed in Iran, showed that *A. esculentus* powder consumption can elicit improvements in lipid profile, as well as glycemic markers, in type 2 diabetic patients [26].

1.1.12 Adulterants or Substituents

Since okra is widely available at an economical price, the availability of adulterants or substituents has not been reported yet. Different form of adulteration has been reported in Nigeria [27].

1.1.13 Most Cited Research

The work of Sabitha et al. [28], based on antihyperlipidemic and antidiabetic assessment of *A. esculentus* (L.) Moench, among diabetic rats induced by streptozotocin, is one of the most cited works. This work suggested that *A. esculentus* has the ability to increase body weight, reduce sugar levels, and normalize the lipid profile of diabetic rats.

iCite, Database, NIH Open Citation Collection was used from June to October 2022 to get the details of the most cited research. Filters for research papers were used by using the keywords *A. esculentus*. NIH database search showed that a viral disease known as Yellow Vein Mosaic Disease, caused by a complex consisting of the monopartite begomovirus Bhendi Yellow Vein Mosaic Virus and a small satellite DNA beta component, is considered a major limitation in the production of *A. esculentus* [29].

1.1.14 Plant Pathology

Okra is reported to be susceptible to several nematode, fungal, and viral diseases. As indicated in previous research, a viral disease known as Yellow Vein Mosaic Disease, caused by a complex consisting of the monopartite begomovirus Bhendi Yellow Vein Mosaic Virus and a small satellite DNA beta component, is considered a major limitation in the cultivation of *A. esculentus* globally [29].

1.1.15 Patent

CN104223271B (2014): A patent was filed on the basis of the development of an *A. esculentus* lactamase–fermenting beverage to offer a vitamin-, trace element–, and dietary fiber-enriched beverage with antioxidant effects.

CN104543835A (2014): This invention claims a method to prepare *A. esculentus* (L.) Moench tablet.

2 Genus *Abutilon*

2.1 *Abutilon fruticosum* Guill. & Perr.

2.1.1 Synonyms

Sida perrottetiana & Abutilon albidum subsp. fruticosum

2.1.2 Family

Malvaceae

2.1.3 Vernacular Name

Meqerqa

2.1.4 Common Name

Indian mallow, Jajaan, Texas Indian mallow, Lowaq, and Gargadan

2.1.5 Botanical Description

A grayish, perennial shrub with woolly hairs covers the entire plant. This plant contains seeds that are coated with elevated dots (papillose), leaves (ovate and petiolate; palmate venation, serrate margin), which are heart-shaped at the bottom, schizocarp fruit, and yellowish to orange flowers with long petals. This plant has a taproot system, and its leaves are deciduous (Figures 2.1.1–2.1.4) [30].

2.1.6 Geographical Distribution

This plant is prevalent in dry regions of tropical Africa. Its prevalence has also been observed in Arabia and India. In Oman it is found across the Sultanate of Oman.

2.1.7 Chemical Composition

The plant contains azulene as a main component, whereas it also contains other components, such as phytol, *n*-hexadecanoic acid, palmitic acid, omega 3 fatty acid, squalene, methyl stearate, lupeol, β-sitosterol, and neophytadiene (Figure 2.1.5). Additionally, there are other components, such as carveol, octadecanoic acid, myristic acid, tocopherol, 1-eicosanol, and linoelaidic acid, which have also been reported [31].

2.1.8 Uses

This drought-tolerant plant is used as both a medicinal and ornamental plant. Apart from these properties, the plant contains ample amounts of tough fiber that can be utilized to make cordage for ropes [32]. *A. fruticosum* has various

FIGURE 2.1.1 Green fruits of *Abutilon fruticosum*.

FIGURE 2.1.2 Apical part of flower of *Abutilon fruticosum*.

FIGURE 2.1.3 Leaves of *Abutilon fruticosum*.

DOI: 10.1201/9781003295488-2

FIGURE 2.1.4 Open flower of *Abutilon fruticosum*.

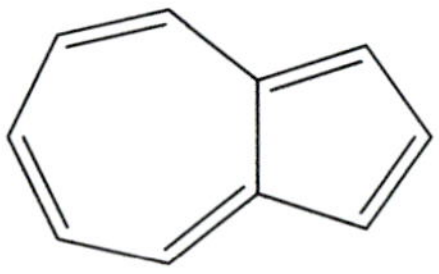

Azulene

FIGURE 2.1.5 Chemical composition of *A. fruticosum*.

medicinal effects, such as being used in the treatment of ulcers, cystitis, leprosy, piles, inflammation of bronchioles, pain in joints, and elevated levels of bilirubin [33,34].

2.1.9 Toxicity and Clinical Data

The plants contain no toxins, and toxicity has not been reported for them yet [35].

2.1.10 Adulterants or Substituents

The plant is widely available; thus, no adulterants or substitutes have been reported so far.

2.1.11 Most Cited Research

There is only one report available on National Center for Biotechnology Information (NCBI) suggesting the overall chloroplast genome, as well as the phylogenetic and evolutionary relationships of *A. fruticosum* [36].

2.1.12 Plant Pathology

There are limited reports in the literature referring to diseases of *Abutilon fruticosum*, such as the fungal disease caused by *Diaporthe phaseolorum* var. *caulivora*, which was isolated from the genus *Abutilon* [37]. Likewise, *Abutilon* mosaic virus has been reported. Nevertheless, this herb itself can become a weed in many cultivated lands.

2.2 *Abutilon bidentatum* Hochst. ex A. Rich

2.2.1 Synonyms

Sida bidentata

2.2.2 Family

Malvacea

2.2.3 Botanical Description

It's a shrubby herb with long-petiolate leaves (heart-shaped at the base, acuminate at the apex, serrate-dentate margins), yellow-colored flowers, and papillose seeds [38].

2.2.4 Geographical Distribution

The plant is available in the Indian subcontinent, from Eritrea to Niger, Namibia, and the Arabian Peninsula [39]. It is found in both Northern and Southern Oman.

2.2.5 Chemical Composition

It was found that the *Abutilon* genus contains different chemical components such as alkaloids, anthocyanins, coumarins, fatty acids, flavonoids, iridoids, megastigmanes, phenolic acids, quinones, saponins, sterols, and triterpenes [40]. In Saudi Arabia, this genus is represented by five species: *A. bidentatum*, *A. pannosum*, *A. figarianum*, *A. fruticosum*, and *A. hirtum* [41]. In 1996, Jain et al. found a new cholestane derivative in *Abutilon bidentatum* Hochst (Figure 2.2.1) [42,43].

Recently, a new oleanane-type triterpene ester, known as abubidentin A (3) (α, 3β, 30-trihydroxy-29-carboxy-olean-9(11), 12-diene-3-dotriacontanoate), along with two new components, namely stigmasta-22-ene-3-β-ol and 2-hydroxydocosanoic acid, were found from *A. bidentatum*

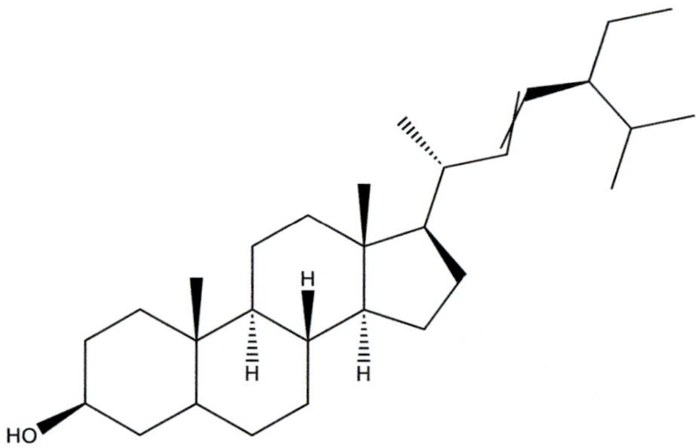

Stigmasta-22-ene-3-β-ol

FIGURE 2.2.1 Chemical components of *A. bidentatum*.

(aerial parts). Additionally, it was also found that these components have antimicrobial and antioxidant effects [43].

2.2.6 Chemical Tests

Since this genus contains flavonoids, phenolic acids, sterols, triterpenes, coumarins, alkaloids, lactones, megastigmanes, and iridoids, respective phytochemical tests can be performed [40].

2.2.7 Uses

Plants belonging to the genus *Abutilon* often contain insoluble fibers that have significant importance [44]. Medicinal plants of this genus have been reported for their health benefits in treating rheumatoid arthritis, as diuretics, and as demulcents due to the presence of high levels of mucilage [34,45].

Several therapeutic effects, including analgesic, anticancer, antidiarrheal, anti-hyperglycemic, anti-hyperlipidemic, anti-hypertensive, anti-inflammatory, antimicrobial, antioxidant, antipyretic, anti-urolithiatic, CNS activity, diuretic, hepatoprotective, immunostimulant, and wound healing effects, have been reported for plants belonging to *Abutilon* [33,46].

So far, *A. bidentatum* has been investigated for treating microbial infections and hepatoprotective effects. Different parts of this plant have been utilized for these purposes. For example, powder obtained from the roots has been used traditionally to treat diarrhea, dysentery, and colitis [47].

However, only a few reports are available related to its medicinal and phytochemical properties, such as hepatoprotective [48], antioxidant, and antibacterial [48–50] properties of *A. bidentatum*.

2.2.8 Mechanism of Action

The ethanolic extract of *A. mauritianum* considerably decreased yeast and lipopolysaccharide-induced pyrexia in rodents, possibly due to the inhibition of the release of inflammatory mediators [51]. Similarly, the ethanolic fraction obtained from *A. indicum* showed hepatoprotective activity in rodents, possibly due to the synergistic effects of the flavonoids present in the extract [52].

2.2.9 Toxicity and Clinical Data

A. bidentatum (aerial parts) aqueous methanolic fraction has not shown toxicity or mortality till 2 g/kg dose in rodents, suggesting LD_{50} of extract is more than 2 g/kg [48].

2.2.10 Most Cited Research

Recent research showed the presence of an oleanane-type triterpene ester, with two known compounds: stigmasta-22-ene-3-β-ol and 2-hydroxydocosanoic acid, isolated from the aerial parts of *A. bidentatum*. Additionally, Abubidentin A showed promising antimicrobial effects against *Staphylococcus aureus*, *Escherichia coli*, and *Pseudomonas aeruginosa* [43].

2.2.11 Patents

US9757327B2: In this invention, *Abutilon indicum* extracts have been used to impart benefits to the skin and/or improve skin conditions resulting from aging or damaged skin [53].

A novel procedure was developed to prepare the composite *Abutilon theophrasti* fiber by using basalt fiber.

CN105506805B-2015: A method for preparing a modified *Abutilon* fiber composite enhanced with basalt fiber. Another patent is based on a composition for the prevention, improvement, or treatment of inflammatory diseases, comprising an *Abutilon crispum* extract as an active ingredient. KR20210061479A: The pharmaceutical composition comprises *Abutilon crispum* extract for preventing or treating inflammatory diseases as an active ingredient.

2.2.12 Plant Pathology

Fusarium solani is an endophytic bacteria that lives in *Abutilon indicum* [54].

2.3 *Abutilon hirtum* (Lam.) Sweet

2.3.1 Synonyms

Sida hirta, *Abutilon indicum* var. *hirtum*, & *Abutilon graveolens var. hirtum*

2.3.2 Family

Malvacea

2.3.3 Common Name

Indian mallow also commonly known as Vadathuthi in India.

2.3.4 Botanical Description

A perennial herb or shrub with erect stems up to 1 m tall, covered in both glandular and non-glandular trichomes. The leaves, densely packed with both stellate and glandular hairs, possess an acute apex and a cordate base. The margins are noteworthy for their deep serrate to crenulate nature. Solitary flowers complete the picture, showcasing a color spectrum ranging from yellow to orange.

2.3.5 Geographical Distribution

A. hirtum is distributed in Tropical and Subtropical regions around the globe. Within Oman, it is found in the Southern mountains.

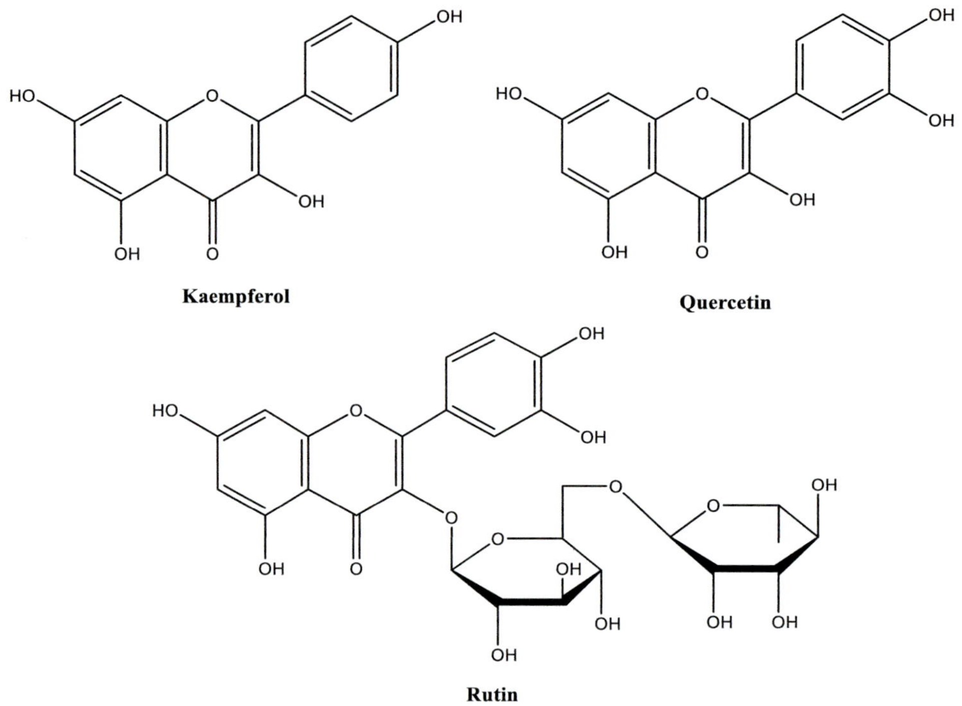

FIGURE 2.3.1 Chemical components of *A. hirtum*.

2.3.6 Chemical Composition

Phytochemical studies revealed the presence of flavonoids (kaempferol, quercetin, rutin) and phenolic acids (gallic acid, p-coumaric acid, caffeic acid) [55] (Figure 2.3.1). The presence of alkaloids in the plant has also been reported [56]. Previous investigations demonstrated the existence of chemical components such as diethyl phthalate, 4-propyl, methoxyacetic acid 3-tridecyl ester, sulfurous acid dodecyl 2-propyl ester, sulfurous acid, and butyl dodecyl ester in the chloroform extracts of leaves obtained from *A. hirtum* [57,58].

Lipoidal matter analysis of petroleum ether extracts obtained from the aerial parts of *A. pannosum*, *A. hybridum*, and *A. hirtum* showed the presence of 23 components in the unsaponifiable matter of *A. pannosum*, 19 in *A. hybridum*, and 22 compounds in *A. hirtum*. *N*-heneicosane, β-sitosterol, and n-tricosane were abundantly present in the unsaponifiable matter of the aerial parts of *A. pannosum*, *A. hybridum*, and *A. hirtum*. In the saponifiable fractions, palmitic acid (a saturated fatty acid) was identified as the major component in all three *Abutilon* species, representing 23.18% for *A. pannosum*, 23.97% for *A. hybridum*, and 19.23% for *A. hirtum*. However, the unsaturated fatty acid oleic acid was also identified as a major component in all three species (17.78%, 34.44%, and 22.13%, respectively). Thus, all species contain ample levels of sterols, unsaturated fatty acids, and saturated fatty acids [59].

2.3.7 Uses

Traditionally, several species belonging to the *Abutilon* genus have been used in the treatment of diarrhea, ulcers, inflammation, bronchitis, gonorrhea, cleaning wounds, and piles [33]. *A. hirtum* has been utilized as a poultice to relieve the pain of kidney stones in Malaysia. Sometimes, it's also blended with sticky rice (glutinous) to treat ulcers. The roots of this plant in Thailand have been used as an antipyretic and to relieve tooth pain and cough. The plant is traditionally utilized in India to treat diarrhea and is used as a demulcent and diuretic. The preparation obtained from its leaves has been utilized to treat inflammation in the bladder, ulcers, and wounds. Decoction obtained from the leaves of this plant has been utilized as a mouthwash. Fruits obtained from this plant are consumed in raw form in Kenya, whereas the leaves are fed to camels and goats. The aqueous extract of the bark has been utilized to ease discomfort in Uganda and Kenya [44,60,61]. In a few reports, this plant has also shown antioxidant, hepatoprotective, and cytotoxic effects [60,62,63].

Earlier studies showed that the extract obtained from *A. hirtum* possessed anti-inflammatory, analgesic, antipyretic, and antipyretic effects, which might be due to the presence of active phytoconstituents [64]. Fibers obtained from its bark are found to be long, fine, soft, and strong. The aqueous extract obtained from *A. hirtum* showed the highest antioxidant activity among all the extracts, whereas the ethyl acetate extract demonstrated the highest ferric-reducing antioxidant power value [60]. Similarly, the aqueous fraction of this plant obtained from leaves demonstrated considerable hepatoprotective effects in the rodent model [62].

Another study showed that the petroleum ether extract showed the highest cytotoxic effects against Ehrlich Ascites Carcinoma at 50 and 100 μg/mL [65]. However, the cytotoxic effects of the aqueous fraction of this plant against human breast cancer cell lines showed a high cell inhibition rate at a 300 μg concentration [60]. Previous studies showed the development and catalytic activity of silver nanoparticles supported on the reduced graphene oxide surface using *A. hirtum* leaf extract as a reducing as well as capping agent. It was found that the incorporation of *A. hirtum* leaf extract into an aqueous solution of silver nanoparticles and graphene oxide resulted in the fast reduction of the graphene oxide and silver ions, leading to nucleation and the development of stable silver nanoparticles on the surface of reduced graphene oxide [66].

A recent study showed that overall, ethanolic and other fractions of leaves showed antimicrobial and antiparasitic effects [67]. Previous studies also showed that different fractions of *A. hirtum* (Lam.) showed pesticidal action against *Tribolium castaneum* and *Lymnaea acuminata* through a dose-mortality assay under laboratory conditions [68].

2.3.8 Plant Pathology

Puccinia heterospora, a rust fungus isolated from *Abutilon hirtum* [69].

2.4 *Abutilon indicum* (L.) Sweet

2.4.1 Synonyms

Sida indica & *Abutilon elongatum*

2.4.2 Family

Malvaceae

2.4.3 Common Name

Indian abutilon, Indian mallow, Monkey Bush

2.4.4 Geographical Distribution

It is widely distributed from India and Southeast Asia through southern China and Taiwan to Mascarenes. Within Oman it distributed in the Northern Hajar mountains.

2.4.5 Chemical Composition

The plant contains alkaloids, flavonoids, tannins, saponins, and glycosides. Sterols [β-sitostrol, β-sitostrol-3-o-β-d-glucopyranoside, stigmasterol, 20, 23-dimethylcholesta-6, 22-dien-3β-ol, cholesterol, (24r)-5α-stigmastane-3 ,6-dione], flavonoids or flavonols (4′,6-dimethoxy kaempferol, quercetin, quercetin-3-o-β-d-glucopyranoside, querc etin-3-o-α-rhamnopyranosyl, gossypetin-7-o-β-glucoside, gossypetin-8-o-β-glucoside); flavones [luteolin, luteolin-7-o-β-glucopyranoside, chrysoeriol, chrysoerio l-7-o-β-glupyranoside, apigenin-7-O-β-glupyranoside]; anthocyanins (cyanidin-3-O-rutinoside); phenolic acid derivatives (benzoic acid, p-hydroxybenzoic acid, vanillic acid, glucovanilloylglucose, 4-hydroxyacetophenone, 4-hydroxybenzaldehyde, vanillin, syringaldehyde, methyl-4-hydroxybenzoate, eudesmic acid, gallic acid, 2,6-dihydroxy-4-methoxyacetophenone, 4-O-β-glucosylbenzoic acid, and 2,6-dihydroxy-5-methoxy-(3-c-glucopyranosyl)benzoic acid) are present in the plant. The whole plant contains p-coumaric acid, *A. hirtum*, caffeic acid, ferulic acid, 4-hydroxy-3-methoxy-e-cinnamic acid, methyl ester, methyl 4-hydroxyphenylacetate, methylcoumarate, abutilin A, fumaric acid; triterpenes [β-amyrin, β-amyrin-3-palmitate, oleanic acid, lupeol, squalene], quinones (2,6-dimethoxy-1,4-benzoquinone); coumarins [scoparone, scopoletin, 3,7-dihydroxychromen-2-one]; alkaloids and amides [aurantiamide acetate, (R)-N-(1′-methoxycarbonyl-2′-phenylethyl)-4-hydroxybenzamide, n-feruloyl tyrosine, 1-lycoperodine, 1-methoxycarbonyl-β-carboline, methyl indole-3-carboxylate, vasicine]; and ionones (3-hydroxy-β-damascone, 3-hydroxy-β-ionol). A recent study showed that the hydroalcoholic extract of *A. indicum* contains chemical components similar to paracetamol and aceclofenac, which are accountable for its analgesic activity [70]. Another study showed that syringaldehyde isolated from *A. indicum* Linn. leaves demonstrated broad-spectrum antimicrobial activity [71]. Some of the chemical components of *A. indicum* Linn. are represented in Figure 2.4.1.

2.4.6 Uses

Traditionally, it is used in the treatment of bleeding disorders, gout, ulcers, tuberculosis, and worms. It can be used as a digestive, laxative, expectorant, demulcent, astringent, analgesic, diuretic, anti-inflammatory, anthelmintic, and aphrodisiac. Decoction of this plant is also utilized for tender gums and toothache. Aqueous preparations of the leaves were locally utilized to treat ulcers. Root powder preparations are used for fever, chest infections, and urethritis. This plant is reported for several therapeutic effects, such as anti-inflammatory, analgesic, anti-arthritic, antioxidant, antihyperglycemic, hepatoprotective, anti-diarrhoeal, respiratory, gastroprotective and antiulcer, anticonvulsant activity, diuretic, anti-urolithiatic, cytotoxic, antifungal, antibacterial, anti-leishmanial, anti-stress, larvicidal, anthelmintic, anti-venom, immunostimulant, cardioprotective, abortifacient effects, aphrodisiac, wound healing, and anti-hyperlipidemic effects [40,72]. A recent study showed the cytoprotective, potent antioxidant, and DNA protective effects of the leaf extract of *A. indicum* [73]. Another recent study showed that nanoparticles synthesized using the *A. indicum* leaf extract demonstrated superior antibacterial and anticancer effects when compared to a plant extract and chemically synthesized NiO nanoparticles. Additionally, these *A. indicum* leaf extract-coated NiO nanoparticles demonstrated excellent antioxidant effects and biocompatibility with fibroblast cells [74]. Zinc oxide nanoparticles synthesized using the leaf extract of *A. indicum* showed

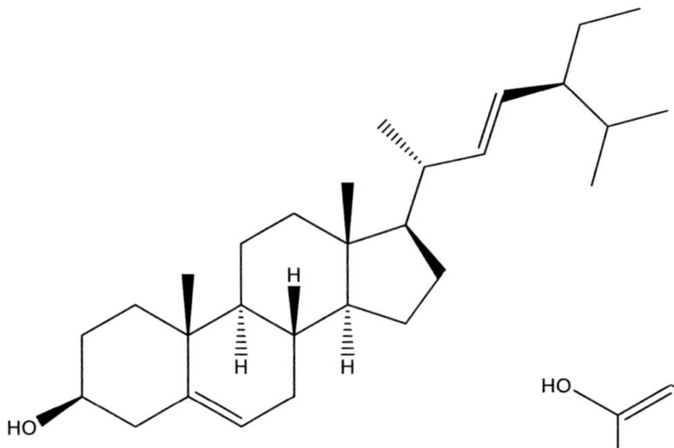

Stigmasterol

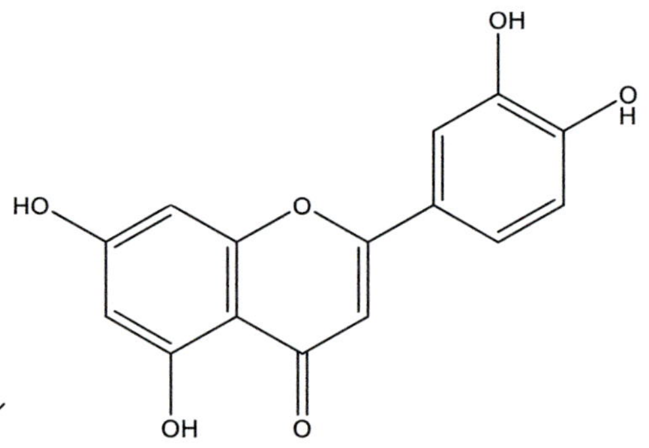

Luteolin

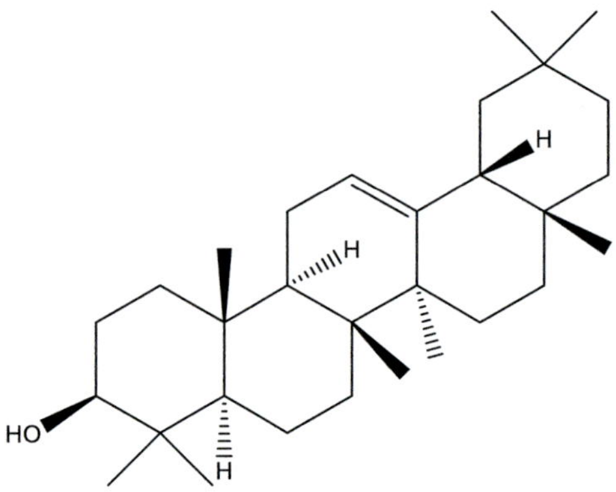

ß-amyrin

FIGURE 2.4.1 Chemical components of *A. indicum*.

potential antimicrobial effects against gram-negative bacteria [75]. Another report showed that gallic acid addition increased the apoptotic effects of *A. indicum* Linn. stem fraction in human U87 glioblastoma cells [75].

2.4.7 Mechanism of Action

The ethanolic extract of *A. indicum* showed a hepatoprotective effect in rats, possibly due to the synergistic effects of flavonoids present in the extract (luteolin, chrysoeriol, luteolin-7-O-β-glucopyranoside, chrysoeriol-7-O-β-glucopy ranoside, and quercetin-3-O-β-glucopyranoside) [52].

2.4.8 Toxicity and Clinical Data

Aqueous and ethanolic extract has not shown any toxicity and mortality at higher doses [76–79].

2.4.9 Adulterants or Substituents

It is a widely available plant.

2.4.10 Most Cited Research

Among the most cited research studies, synthesized copper oxide nanoparticles using *Abutilon indicum* showed significant antibacterial and antioxidant effects and gained more attention [80]. Another highly cited study includes the isolation of β-sitosterol as a new natural mosquito larvicidal agent from petroleum ether extract of *A. indicum* [81,82]. One more highly cited study is based on an aqueous extract of *A. indicum*, which showed hepatoprotective activity against carbon tetrachloride- and paracetamol-induced hepatotoxicity in rodents [82].

2.4.11 Patents

In this invention, the inventors developed a procedure for using *A. indicum* extracts to provide benefits to skin and/or improve skin conditions due to aging or skin damage [53].

2.5 *Abutilon mauritianum* (Jacq.) Medik

2.5.1 Synonyms

Sida mauritiana

2.5.2 Family

Malvaceae

2.5.3 Common Name

African Mallow, Mauve du pays, Lakichocho, and Makirikara in Tanzania

2.5.4 Botanical Description

A woody perennial herb reaching up to 1 m, with stems and branches covered in soft, short hairs. The leaves are heart-shaped with sharp tips and margins, featuring small tooth-like structures. The flowers are solitary, with colors ranging from yellow to orange.

2.5.5 Geographical Distribution

It thrives in the warmth of Tropical and Southern Africa, stretching across the Arabian Peninsula and reaching the volcanic islands of the Comoros. In Oman, it is distributed in Southern mountains.

2.5.6 Chemical Components

The phytochemical analysis of the *A. mauritianum* revealed that this plant contains saponins, tannins, and alkaloids [83].

2.5.7 Uses

Used as an emollient, especially internally for bronchial inflammation and externally for abdominal pain, urinary retention, etc.

The ethanolic extract of *A. mauritianum* roots showed a considerable decrease in yeast and lipopolysaccharide-induced pyrexia in rodents. A possible mechanism for this activity was the inhibition of the release of inflammatory mediators [51]. Apart from possessing antipyretic effects, *A. mauritianum* also possesses anti-inflammatory and analgesic effects [51].

A. mauritianum ethanolic extract showed high antibacterial activity against *K. pneumonia* [83]. Another report showed that a combined extract of *Funtumia africana* and *A. mauritianum* leaves showed improvement in lipid parameters and renal indices of testosterone-induced benign prostate hyperplasia in rodents. This study showed that the combined extract reduced high-density lipoprotein and increased triacylglycerol levels. Additionally, the combined extract demonstrated renoprotective potential by decreasing urea and creatinine levels [84].

Another study showed that a similar combined extract positively regulates levels of serum minerals, serum prostatic acid phosphatase activity, serum prostate-specific antigen concentrations, and enhances prostate histomorphology in benign prostatic hyperplasia-induced rats [84].

Treatment with the combined extract considerably reversed the enhanced testosterone, dihydrotestosterone, and estradiol concentrations, prostate weight, and prostate indices in comparison with the control [85].

2.5.8 Mechanism of Action

The antipyretic activity could be due to the inhibition of the release of inflammatory mediators [51,86].

2.5.9 Toxicity and Clinical Data

It was found that oral doses of *A. mauritianum* root extract (ethanolic) showed no mortality or signs of toxicity among albino rats [51].

2.5.10 Most Cited Research

The most cited research is based on the phytochemical analysis and antimicrobial assessment of *Abutilon mauritianum*, *Bacopa monnifera*, and *Datura stramonium*. Findings from this study showed that these plants contain saponins, tannins, and alkaloids, and showed inhibitory effects on *Pseudomonas aeruginosa*, *Klebsiella pneumoniae*, and *Escherichia coli* [83].

2.6 *Abutilon pannosum* (G.Forst.) Schltdl

2.6.1 Synonyms

Sida pannosa

2.6.2 Family

Malvaceae

2.6.3 Vernacular Name

Neqa

2.6.4 Common Name

Karandi, Ragged Mallow

2.6.5 Botanical Description

An erect perennial shrub reaching up to 2 m. The leaves are alternate, with silver hair shimmers on their surfaces, and have a heart shape with serrated edges. Flowers unfurl their five free petals, yellow in color. The fruit has densely packed sections, each harboring brown, hairy seeds (Figures 2.6.1–2.6.3).

2.6.6 Geographical Distribution

Globally, this plant is widely distributed throughout India, Pakistan, and Africa. In India, it is found mainly in dry and abandoned places up to an altitude of 300 m. It is also available in Yemen, Egypt, and Saudi Arabia. In Oman, it is widely available in both Southern and Northern Oman [4].

FIGURE 2.6.1 Ripe fruit of *Abutilon pannosum*.

FIGURE 2.6.3 Apical portion of *Abutilon pannosum*.

FIGURE 2.6.2 Fruiting branch of *Abutilon pannosum*.

2.6.7 Chemical Composition

Recently, HPLC-DAD (high-performance liquid chromatography with diode-array detection) investigation showed that *A. pannosum* extract contains rutin, gallic acid, and catechin, along with significant levels of phenolic components (Figure 2.6.4) [87].

A recent study revealed the phytochemistry of *A. pannosum* leaf extract collected from the Kachchh region of Gujarat, India, using liquid chromatography quadrupole time-of-flight mass spectrometry. Findings showed that phytochemicals present in methanol extracts of *A. pannosum* leaf include (-)-sedamine, oxidized dinoflagellate luciferin, and Oligomycin A [88].

Lipoidal matter analysis of petroleum ether extracts obtained from aerial parts of *A. pannosum*, *A. hybridum*,

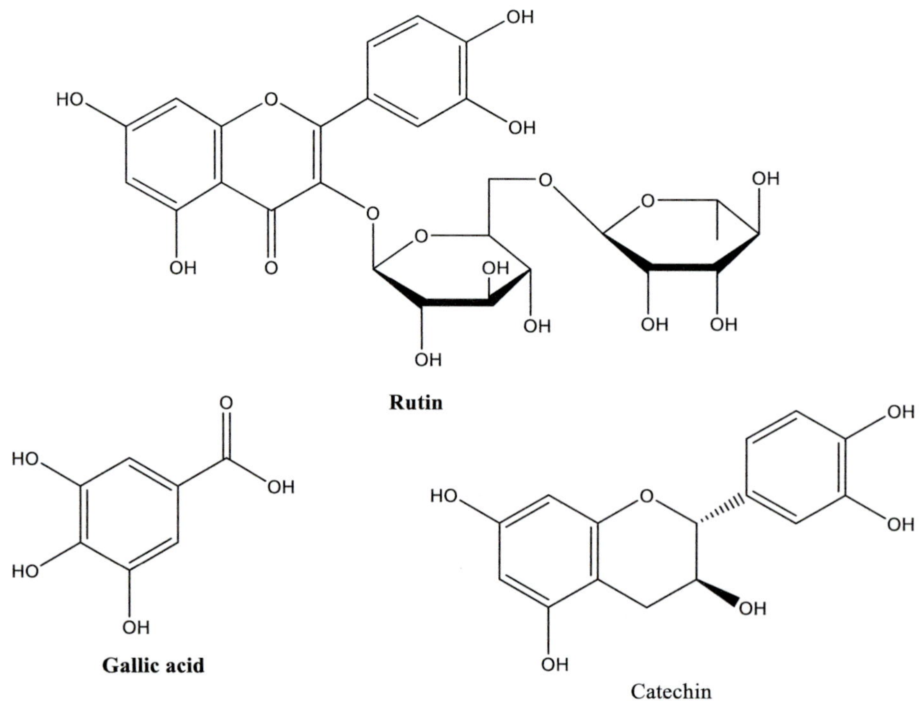

FIGURE 2.6.4 Chemical components of *A. pannosum*.

and *A. hirtum* showed the presence of 23 components in the unsaponifiable matter of *A. pannosum*, 19 in *A. hybridum*, and 22 compounds in *A. hirtum*. N-heneicosane, β-sitosterol, and n-tricosane were abundantly present in the unsaponifiable matter of the aerial parts of *A. pannosum*, *A. hybridum*, and *A. hirtum*. In the saponifiable fractions, palmitic acid (saturated fatty acid) was identified as the major component in all three *Abutilon* species, representing 23.18% for *A. pannosum*, 23.97% for *A. hybridum*, and 19.23% for *A. hirtum*. However, the unsaturated fatty acid, oleic acid, was also identified as a major component in the three species (17.78%, 34.44%, and 22.13%, respectively). Thus, all species contain ample levels of sterols and unsaturated and saturated fatty acids [59].

A previous study based on gas chromatography-mass spectrometry analysis demonstrated that the n-hexane extract of *A. pannosum* contains different types of fatty acids, such as 9, 12-octadecadienoic acid, 9-octadecenoic acid, hexadecanoic acid, octadecanoic acid, octadecatrienoic acid, 9, 12, 15-eicosanoic acid, 9-hexadecenoic acid, 11-eicosenoic acid, methyl tetradecanoate, octanoic acid, and tridecanoic acid.

Earlier study-based phytochemical analysis of seeds obtained from *A. pannosum* showed that it contains 13.4% and 15.4% oil and 23.0% and 22.2% protein. Moreover, it also contains fatty acids such as palmitic, stearic, oleic, linoleic, and malvalic acid [89].

2.6.8 Uses

Antibacterial effects of *Abutilon pannosum* (Forst.f.) extract obtained from its leaves have been examined, and it was found that all the extracts showed significant antibacterial effects when compared to the standard penicillin potassium and streptomycin sulfate against selected gram-positive and gram-negative bacteria [90].

In Pakistan, *A. pannosum* is utilized to treat lung-associated complications, diabetes, bladder inflammation, and diuretics, as well as to reduce pyrexia. Recently, the antioxidant and protective effects of *A. pannosum* against CCl_4 caused kidney and lung toxicity in rodents were examined. Results showed that rodents treated with *A. pannosum* along with CCl_4 exhibited a considerable level of recovery in comparison to the control. This suggests the protective effects of *A. pannosum* extract in CCl_4-intoxicated rats for kidney and lung damage [87].

In the same study, researchers demonstrated the antioxidant activity of the *A. pannosum* extract by showing significant activity for iron chelation, reducing potential, suppression of β-carotene oxidation, with mild potential in scavenging of nitrite, phosphomolybdenum, DPPH, and hydroxyl radicals [87]. Traditionally, it is also used in folk medicine.

2.6.9 Mechanism of Action

As stated above, *A. pannosum* demonstrated protective effects in CCl_4-intoxicated rats for kidney and lung damage. The protective effects of *A. pannosum* are possibly due to the effect of *A. pannosum* extract in reducing the increased levels of thiobarbituric acid reactive substances (TBARS), H_2O_2, nitrite, TNF-α, IL-1β, and IL-2 in lung and kidney tissues of CCl_4-treated rats [87]. Seeds of *A. pannosum* have been used as laxatives, whereas its extracts derived from flowers have been used for improving sexual performance [91].

3 Genus *Acacia* (*Senegalia* and *Vachellia*)

3.1 *Senegalia senegal* (L.) Britton

3.1.1 Synonyms

Acacia senegal

3.1.2 Family

Fabaceae

3.1.3 Vernacular Name

Tur, temmar & tulh

3.1.4 Common Name

Gum Arabic & three-thorned Acacia

3.1.5 Botanical Description

Senegalia senegal is a branched shrub that sheds its leaves annually and reaches up to a height of 15 m. The shrub is branched in a manner that causes it to divide continually, and at the top, it forms a rounded, flat crown when it matures. The stem shows variation in width, up to around 30 cm. This shrub possesses bark that is grayish-white in color and becomes darker, thinner, and scalier as it matures, although in old trees growing in the open, it may be dark, scaly, and thin. The bark displays a bright green cambium layer once it is scratched. The logging remains left in the forest after a harvest display a spotted red appearance. The shrub bears sharp, hooked thorns, enlarged at the base, and often three are present at the nodes of the branches [92]. Flowers are whitish-yellowish, fragrant, stalkless, and the shrub has small leaves with 7–25 pairs of leaflets. The fruits are flat, straight, hairy, dehiscent papery pods and carry heart-shaped seeds (Figures 3.1.1–3.1.7).

3.1.6 Geographical Distribution

This species is distributed across the tropics of Southern Africa, winding its way through the diverse ecosystems of the region, eastward through the Arabian Peninsula, and finally reaching the vibrant lands of India. Within Oman it found in Southern Oman thrives among escarpment woodlands.

3.1.7 Chemical Composition

Gum exudate from the tree includes three diverse, highly-branched carbohydrate fractions. These fractions show variation in protein content and molecular mass. The galactan present in the gum exhibits diverse biological effects. This mixture of polysaccharides and glycoproteins is broadly classified as arabinogalactan protein,

FIGURE 3.1.1 Flowering shoot of *S. senegal.*

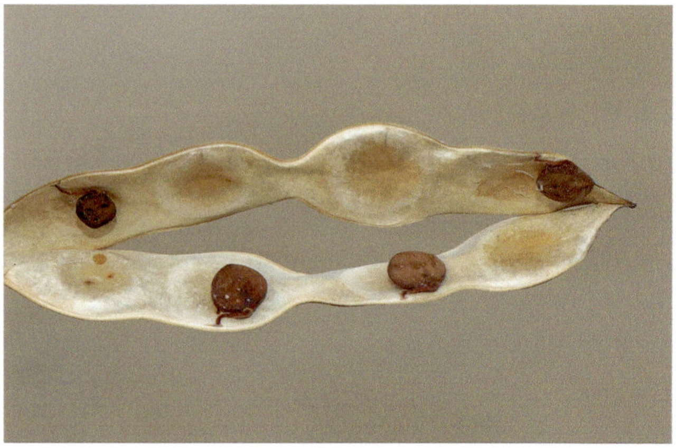

FIGURE 3.1.2 Papery walled pods of *S. senegal.*

FIGURE 3.1.3 Inflorescence of *S. senegal.*

DOI: 10.1201/9781003295488-3

FIGURE 3.1.4 Shrub with a rounded, flat-topped crown.

FIGURE 3.1.5 Immature pods of *S. senegal*.

arabinogalactan, and glycoprotein [93]. Nevertheless, the gum shows variation in chemical composition based on its geographical distribution [94,95].

3.1.8 Uses

Gum Arabic is known for its wide applications in ceramics, cosmetics, textiles, food, and pharmaceuticals [96]. This nontoxic oleogum resin is used as a stabilizer, emulsifier, and thickener for a wide range of pharmaceutical applications [97]. A recent report showed that *Senegalia senegal* gum reduced oxidative stress and genotoxicity in albino rats [98].

In Dhofar, ash is made from its bark and applied to open wounds as a disinfectant. Also, resin made from this plant was of great medicinal value; people in the past used powdered resin mixed with fat or grease to create a full-body rub. While intended to strengthen the body, act as a general tonic, keep them warm, and ward off illness [10].

FIGURE 3.1.6 Papery bark of *S. senegal*.

Bark and resin may have been used historically to create splints or casts for setting broken bones [99].

Chemical components present in this gum possess antioxidant properties and scavenge free radicals effectively [100–102]. Previous studies showed the effects of gum

FIGURE 3.1.7 Natural habit of *S. senegal.*

resin on lipid metabolism [103], kidney function [104–106], and satiety [107].

3.1.9 Mechanism of Action

As per the previous study, treatment with gum Arabic in healthy mice caused an improvement in creatinine clearance, increased renal excretion of antidiuretic hormone, Ca^{2+}, and Mg^{2+}, and reduced plasma phosphate levels and urinary excretion of phosphate and Na^{+}. Furthermore, at the molecular level, treatment with gum reduces intestinal glucose transport, angiogenin, and ß-catenin expression, which may offset the growth of colon carcinoma and inflammatory complications [108].

3.1.10 Toxicity and Clinical Data

Gum Arabic is edible in nature, is unlikely to be absorbed intact, and is somewhat fermented via intestinal microorganisms. As per the previous study, no adverse effects were reported in sub-chronic and carcinogenicity studies at the highest dose tested and there is no concern related to genotoxicity [109].

This gum is a complex polysaccharide indigestible to both humans and animals and has been considered a safe dietary fiber by the United States Food and Drug Administration (FDA) since the 1970s.

This gum was assessed for suitable daily consumption by the Joint FAO/WHO Expert Committee on Food Additives since 1969; nevertheless, people in Sudan have been using it for a long time. This gum is digested in humans as it's not degraded in the intestine but fermented in the colon to produce short-chain fatty acids. These short-chain fatty acids further possess several health benefits [94].

A recent study also showed that gum can reduce the systemic toxicity caused by CCl4 via regulation of the ROS/NF-κB signaling pathway in the rat organs [110].

3.1.11 Most Cited Research

iCite, Database, NIH Open Citation Collection was used on December 31, 2023, to get the details of the most cited research. Filters for research papers were used by using the keywords *Acacia senegal.* NIH database search revealed the three phylogenetic groups of nodA and nifH genes in *Sinorhizobium* and *Mesorhizobium* isolates from leguminous trees growing in Africa and Latin America [111].

3.1.12 Patents

Saphwan Al-Assaf, Glyn O. Phillips, Yasushi San-ei. Modified gum arabic from *Acacia senegal* EP1611159B1 2004.

- This invention changes the natural protein distribution of gum arabic, and increases arabinogalactan protein complex content.

3.1.13 Plant Pathology

This shrub is infected by a microcyclic, gall-forming rust, *Uromycladium tepperianum* (Sacc.) McAlp., in south-western Australia. This resulted in gall formation on stems, branches, phyllodes, and reproductive organs and deformity in the form of witches' brooms on twigs [112].

Another study showed that the single inoculation with arbuscular mycorrhizal fungi improves *A. senegal* seedling growth under non-sterilized field conditions. It was also found that *Senegalia senegal* develops nitrogen-fixing and phosphate-acquiring symbioses with microbes. *S. senegal* and its association with microbes not only restore soil fertility but also provide commercial products in drylands [113].

3.2 *Vachellia tortilis* (Forssk.) Galasso & Banfi

3.2.1 Synonyms

Acacia tortilis & *Mimosa tortilis*

3.2.2 Family

Fabaceae

3.2.3 Vernacular Name

Semer, horouth, & saimer

3.2.4 Common Name

Thron umbrella

3.2.5 Botanical Description

This evergreen tree or large shrub is a small to medium-sized plant that reaches up to the height of 21 m. Like *Senegalia senegal*, this tree also develops into a flat-topped or rounded-umbrella-like, spreading crown. This tree possesses gray to black or dark brown bark with young, small branches thickly covered with short, soft hair or glabrous to subglabrous. Shrubs possess red to brown hooked spines. Flowers are whitish to creamish or yellowish in color, and highly aromatic [114]. Fruit is twisted brown in color, hence called "tortilis." Leaves are compound and the leaflets are very small, glabrous to pubescent. Leaves, as well as pods, are used as fodder for bark exudate yellowish color gum resin (Figures 3.2.1–3.2.3).

3.2.6 Geographical Distribution

This tree is commonly available across arid ecosystems in Middle Eastern and African regions [115]. Moreover, this shrub is commonly available across the globe (Algeria, Asia, Africa, Pakistan, Somalia, Egypt, and India) [116]. This tree is very common throughout Oman on gravel plains, wadis, and foothills, as well as the central desert, and usually forms a distinctive woodland on the Al Batinah plain.

3.2.7 Chemical Composition

As per the previous report, essential oil from the leaves of *Vachellia tortilis* contains major components such as α-humulene, α-cadinol, nerolidol, γ-cadinene, and 2-(E)-octenal [117] (Figure 3.2.4).

The ethanolic extract obtained from the leaves of *Vachellia tortilis* contains flavan-3-ols galloyl, which showed cytotoxic and anti-inflammatory effects [118].

FIGURE 3.2.1 Immature pods of *V. tortilis*.

FIGURE 3.2.3 Bark and gum exudation of *Vachellia tortilis*.

FIGURE 3.2.2 Natural habitat of *V. tortilis*.

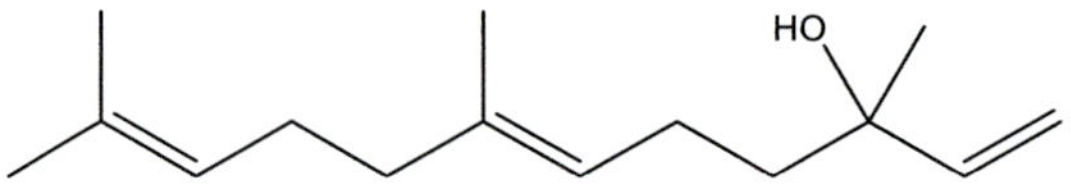

Nerolidol

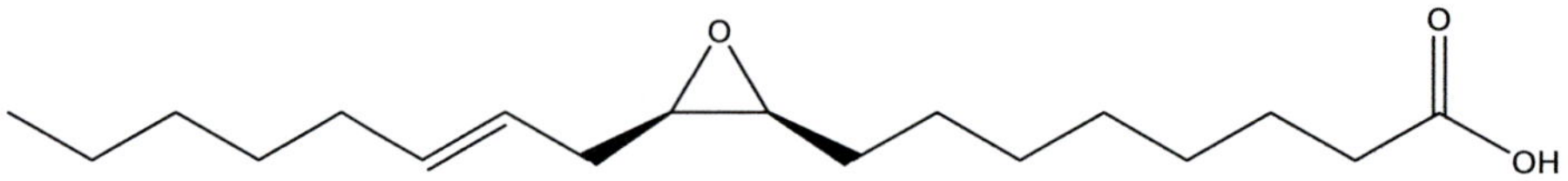

Cis-9,10-epoxyoctadec-cis-12-enoic

FIGURE 3.2.4 Active chemical components of *Vachellia tortilis.*

Vachellia tortilis has also been evidenced for the presence of linolenic acid and linoleic acid [119]. Seeds of *Vachellia tortilis* contain coronaric acid (cis-9,10-epoxyoctadec-cis-12-enoic) [120]. Plants also contain elements such as potassium, followed by calcium and phosphorous [121].

3.2.8 Uses

This shrub contains high levels of tannins [122]. In Dhofar, it was reported that the new growth of this plant was pounded into a paste, mixed with fresh milk, and boiled down. This concoction was then used as a poultice, possibly applied to soothe mastitis. This might have some antiseptic properties, but its effectiveness is questionable.

A plant belonging to this genus is known for its several therapeutic effects, such as smooth muscle relaxing effect, antimicrobial, antiasthmatic, antihyperglycemic, antiplasmodial, antiviral, antiasthmatic, and antileishmanial. Moreover, plants from this genus are used as medicine in the treatment of leukorrhea, hemoptysis, fever, diarrhea, and throat infections [121,123]

3.2.9 Mechanism of Action

This tree contains rutin and catechin, which have been reported for noteworthy inhibition of inflammatory cyclooxygenase enzymes. A polysaccharide fraction isolated from *A. tortilis* gum demonstrated α-D-glucosidase inhibitory effects [118,124].

3.2.10 Toxicity and Clinical Data

Toxicity (oral) studies showed that *A. tortilis* polysaccharide did not display toxicity, behavioral changes, and mortality up to an elevated concentration of 5,000 mg/kg [116].

3.2.11 Most Cited Research

iCite, Database, NIH Open Citation Collection was used on January 1, 2024, to get the details of the most cited research. Filters for research papers were used by using the keywords *Acacia tortilis.* NIH database search revealed the characterization of tropical tree rhizobia and description of *Mesorhizobium plurifarium* sp. nov. [125].

3.2.12 Patents

The present invention discloses a process for concurrently suppressing the cyclooxygenase COX-2 and 5-lipoxygenase enzymes using a composition containing flavans isolated from a plant belonging to the genus *Acacia.*

Qi Jia, Timothy C. Nichols, Eric E. Rhoden, Scott Waite: Isolation of a dual COX-2 and 5-lipoxygenase inhibitor from *Acacia.* WO2003082312A1. 2002.

3.2.13 Plant Pathology

This plant is highly resistant to drought stress and salt tolerance. Unfortunately, no studies have reported common diseases associated with it. However, useful microbial associations such as nitrogen-fixing bacteria and mycorrhizal associations are reported.

3.3 *Vachellia gerrardi* var. *najdensis* (Chaudhary) Ragup., Seigler, Ebinger & Maslin

3.3.1 Synonyms

Acaia gerrardii subsp. *negevensis* & *Acacia pachyceras var. najdensis*

3.3.2 Family

Fabaceae

3.3.3 Vernacular Name

Talh

3.3.4 Common Name

Red thorn

3.3.5 Botanical Description

Trees can reach up to a height of 15 m and form rather flattened or irregular crowns. The bark is rough and grayish-brown in color. Branches are hairy and possess short, straight, or sometimes hooked thorns and exude yellowish-brown gum. Leaves: compound, divided into small leaflets coated with tiny hairs. Flowers: round blooms gather in clusters, fragrant, creamish to white. Fruits: brown-colored pods with darker dots (Figures 3.3.1–3.3.9).

3.3.6 Geographical Distribution

Talh trees are well distributed in African and Middle Eastern regions [126]. This species is distributed in northern Oman mountains at middle and upper altitudes on rocky gorges among *Olea*, *Sideroxylon*, and *Dodonaea* woodlands.

3.3.7 Chemical Composition

Phytochemistry of *V. gerrardii* has not been completely explored yet. This plant reported to contain phenolic acids, including salicylic acid and flavonoids [127].

FIGURE 3.3.3 Thorns and gum exudate from *V. gerrardii* var. *najdensis*.

FIGURE 3.3.1 Inflorescence of *V. gerrardii* var. *najdensis*.

FIGURE 3.3.4 Dried gum resin of *V. gerrardii* var. *najdensis*.

FIGURE 3.3.2 Dried ripe pods with seeds of *V. gerrardii* var. *najdensis*.

FIGURE 3.3.5 Thorns and leaf arrangement of *V. gerrardii* var. *najdensis*.

FIGURE 3.3.6 Habit of *V. gerrardii* var. *najdensis.*

FIGURE 3.3.7 Immature pods of *V. gerrardii* var. *najdensis.*

FIGURE 3.3.8 Branch producing gum.

FIGURE 3.3.9 Fruiting shoot of *V. gerrardii* var. *najdensis.*

Plants belonging to the genus *Acacia* generally contain chemical components such as quercetin, kaempferol-3, robidandiol, hentriacontane, glucoside isoquercetin, dicatechin, sitosterol, betulin, β-amyrin, gallic acid, and chlorogenic acid (Figure 3.3.10) [128,129].

3.3.8 Uses

The previous finding showed that the plant is enriched with antioxidant content and thus improved lipid, serum antioxidant, and kidney function profiles in diabetic rats, suggesting its possible role in decreasing blood sugar in rats suffering from diabetic nephropathy [127]. *In vitro* antimicrobial effects of Saudi honey from *V. gerrardii* Benth. trees have been reported [130].

Additionally, the plant is known for its several benefits on ecosystems, including food for human consumption, stock pasture, traditional medicine, and honeybee forage

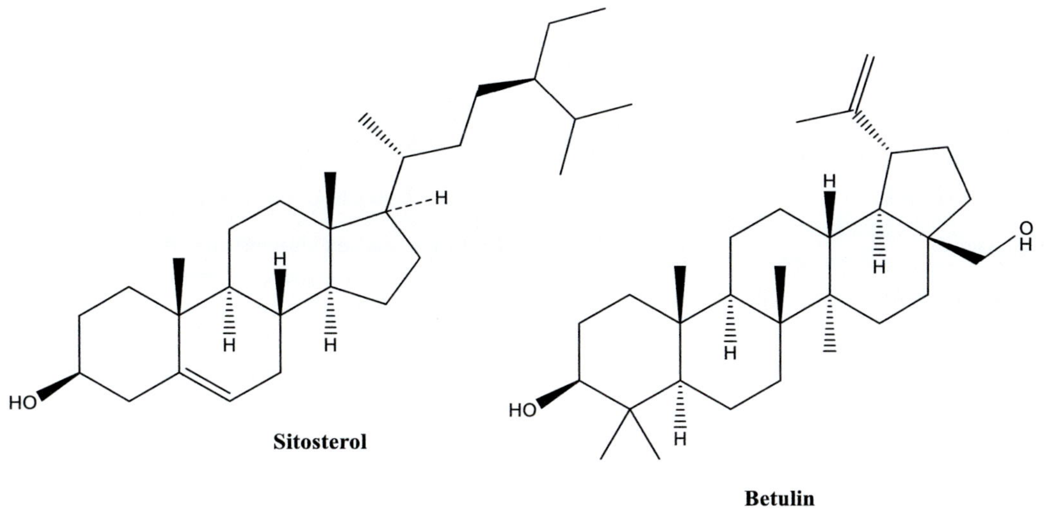

FIGURE 3.3.10 Chemical composition of *V. gerrardii.*

[131–133]. Plants from the *Vachellia* genus are considered sources of nectar and pollen for bees in the Arabian Peninsula [133,134]. The resin extracted from the stems and trunk of this plant was made into a paste and applied to the chest to treat chest pain (Al Hatmi in prep.)

3.3.9 Toxicity and Clinical Data

As plants belonging to this genus contain cyanogenic glycosides, it's important to assess the toxicity profile of this shrub. Toxic [135].

3.3.10 Most Cited Research

iCite, Database, NIH Open Citation Collection was used on January 1, 2024, to get the details of the most cited research. Filters for research papers were used by using the keywords *V. gerrardii.* NIH database search revealed the interaction between arbuscular mycorrhizal fungi and endophytic bacteria enhances plant growth of *V. gerrardii* under salt stress [136].

3.3.11 Patents

토바 실버스타인, 레이첼 루츠, 아이탄 코헨, 하야 산도리-카자즈—KR20190006985A—Composition of saponin and plant extract, 2023.

- The present invention relates generally to compositions comprising saponins and plant extracts having antimicrobial activity.

3.3.12 Plant Pathology

No reports of widespread disease affecting this wild plant in its natural habitat have been found. However, recent findings showed beneficial microbial interactions, such as endophytic bacteria and arbuscular mycorrhizal fungi participating in a tripartite mutualistic symbiosis in *V. gerrardii*, resulting in plant adaptation to salt stress tolerance [136].

4 Genus *Acalypha*

4.1 *Acalypha indica* L.

4.1.1 Synonyms

Cupamenis indica & *Ricinocarpus indicus*

4.1.2 Family

Euphorbiaceae

4.1.3 Common Name

Indian Acalypha, Indian Copperleaf & Indian nettle

4.1.4 Botanical Description

It is a vibrant annual that reaches a height of up to 75 cm and stands proudly upright throughout its season. Its leaves are arranged in an alternating pattern, resembling flattened ovals with pointed tips. The edges are adorned with tiny, sawtooth-like notches. Each leaf rests on a long, slender stalk. Flowers, formed in clusters of small, greenish flowers, appear in groups of 3–5 (Figure 4.1.1).

4.1.5 Geographical Distribution

A. indica is traditionally utilized for many purposes in many countries, especially in Africa and Asia. It's widely available in several regions of south, west, and northeast Africa. Among these regions, it is mainly found in Somalia and Ethiopia. In India, it has been used for medicinal purposes as Ayurvedic medicine [137]. *A. indica* is also available in the Arabian Gulf area, where it's consumed as food. In Oman, it is widely available in Dhofar [138].

FIGURE 4.1.1 Branch apex of *Acalypha indica*.

4.1.6 Chemical Composition

Previous studies showed that *A. indica* (aerial parts), after defatting, demonstrated the presence of triterpenoids and steroids [139]. Tannins, steroids, glycosides, alkaloids, flavonoids, and triterpenoids were found in the ethanolic extract, while chloroform extracts contained alkaloids and glycosides. It was also found that tannins and flavonoids were present in the ethyl acetate extract [140]. The plant contains cyanogenic glycosides acalyphin (0.3%), tannins, and tri-O-methyl ellagic acid as the main active chemical components (Figure 4.1.2) [141]. Regardless of the positive effects, the cyanogenic glycoside present in this plant is detrimental and lethal to humans [142]. On the other hand, this weed plant contains 20 cyanogenic agents to protect itself from herbivores [143]. Plants possess significant antioxidant activity, which might be due to the presence of phenolic components such as glucogallin, chebulagic acid, geraniin, and corilagin [144]. Antioxidant components such as gallic acid, ellagic acid, 4,4′,5,5′,6,6′-hexahydroxy diphenic acid, and kauren-18-oic acid, 16α,17-dihydroxy-ent-kauran 19-oic acid were also found in the plant [145].

The component "quebrachitol," responsible for anticancer activity and respiratory protective effects, is found in the plant leaves [146]. A steroid derivative known as stigmasterol is also found in *A. indica* [147]. *A. indica* contains ample amounts of flavonoids such as chrysin, hesperitin, and others possessing diverse biological activities [148]. Apart from non-volatile components, the plant also contains volatile oil with a strong odor, containing terpenes and terpenoids such as hexenol, decane, and trans-phytol [149]. In addition to volatile components, the plant also contains fatty acids such as n-hexadecyl ester, n-octacosanol, hexatriacontane, eicosatrienoic acid, trifluoroacetic acid, and methyl ester [150].

4.1.7 Uses

Traditionally, it's widely consumed in India as part of Ayurvedic medicine. This plant is traditionally utilized in three forms: whole plant, leaves, and roots. However, more than 60% of this plant's usage is based on leaves. *A. indica* leaves-based preparations can be used to treat diarrhea, emesis, asthma, arthritis, ulcers, and wounds, and they can act as an anthelmintic and laxative. Leaves are consumed in the form of decoction and raw form. Additionally, *A. indica* has also been utilized to relieve constipation and treat ear pain, gums and tooth-related complications, seizures, and skin-related problems. Traditionally, the whole plant is consumed in some regions of India to treat bronchitis by using its fresh juice. Additionally, whole plants have been utilized

DOI: 10.1201/9781003295488-4

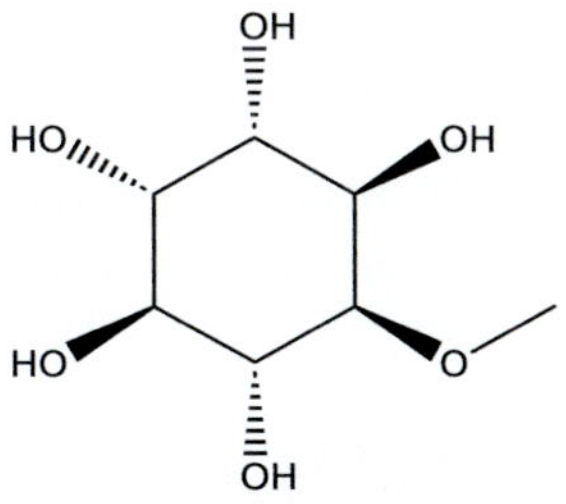

Quebrachitol

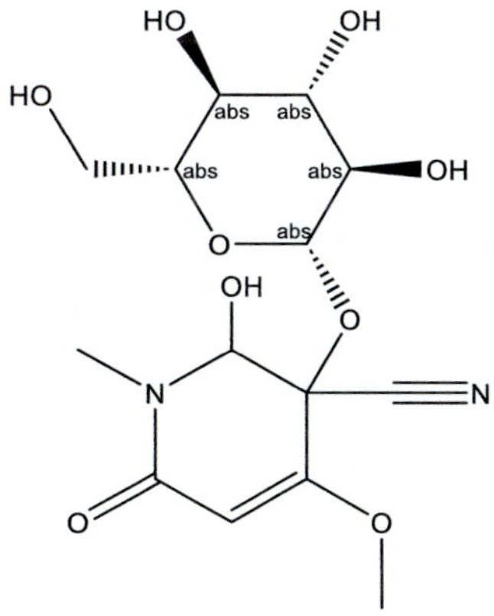

Acalyphin

FIGURE 4.1.2 Chemical components of *A. indica*.

to treat ear pain and oral complications. Besides its medicinal applications, the plant's fresh leaves are also used in the diet, especially in India and Oman [151].

In Oman, people consume *A. indica* in raw form as a medicine to treat wounds [151]. In India, some people blend this plant with others, such as *Tridax procumbens*, *Morus alba*, and *Ficus bengalensis*, to make a paste known as Pasuru, which is utilized to treat wounds [138,152]. An infusion made from the leaves of another related species, *A. fruticosa*, was reported to be used to treat insect bites [153].

4.1.8 Mechanism of Action

Extracts possess anti-venom activity, possibly because the antioxidant activity of the extract is one of the mechanisms of venom inactivation and inhibition [138,154].

4.1.9 Toxicity and Clinical Data

Since plants contain cyanogenic glycosides, their raw consumption can cause toxicity, which restricts their applications as food [148]. Symptoms presented to the end consumers were similar to cyanide poisoning [155]. After consumption by livestock, cyanogenic glycosides can be hydrolyzed by the enzyme known as β-glucosidase to yield cyanohydrin, which can further produce poisonous compounds known as HCN [143]. Thus, its excessive consumption in the form of juice or leaves decoction, without any prescription, must be restricted. Several communities are familiar with the toxicity caused by this plant [156]. A cyanogenic compound known as acalyphin, present in *A. indica*, is accountable for such toxicity. There are different types of cyanogenic compounds present in the plant, and a high proportion of cyanogenic glycosides are present in the leaves [157]. As the plant is aphrodisiac, there is the possibility of its high consumption by illiterate people. Some of the suppliers mix this plant with drugs like sildenafil to improve its effect. However, officials have now restricted the utilization of products based on this plant that are not registered [138].

4.1.10 Patents

Veterinary herbal compositions containing *A. indica* with other plants have been prepared to treat an ephemerovirus infection in cattle.

Petharajanna. WO2012131714A1. Veterinary herbal compositions and medicaments thereof 2012.

An edible product containing *Acalypha indica* has been made, which comprises a fibrous plant product and a plant extract for use as food, food supplement, medicinal, cosmetic, well-being, and nutraceutical or phytotherapeutical applications.

US20200390696A1. An edible product comprising reconstituted plant material. Philippe Ragot, Esther Pons, Bernard Mompon, Cedric Rousseau 2014.

4.1.11 Plant Pathology

Apparently, there is no data in the literature that refers to the pathology of *Acalypha indica*. However, the *Beet western yellows virus* is an important virus that affects vegetable crops and the genus *Acalypha* [158].

.... (*For additional images refer to the supplementary figures section; page no.* 299)

5 Genus *Achyranthens*

5.1 *Achyranthes aspera* L.

5.1.1 Synonyms

Centrostachys aspera & *Stachyarpagophora aspera*

5.1.2 Family

Amaranthaceae

5.1.3 Vernacular Name

Menkasa, tebaiqa & saif al jin

5.1.4 Common Name

Chaff-flower, Apaamaarga, Chirchitaa, Shikhari, Shaikharika [159]

5.1.5 Botanical Description

It is an erect, annual or perennial plant that can reach up to a height of 1–2 m. The stems are often branched from the base and stained purple [160]. The branches are striate and pubescent [161]. The leaves are petiolate, oval, thick [160,161], and pubescent on both sides [161]. The flowers are greenish white, and the seeds are reddish brown, trimmed at the top and curved at the base [159] (Figures 5.1.1–5.1.5).

5.1.6 Geographical Distribution

The plant is wild and commonly available on roadsides, and is considered a weed all over India and in the South Andaman Islands [160,162]. This herb is widely available in Baluchistan, Ceylon, Tropical Asia, Africa, Australia, and America [159]. In Oman, it is found throughout the country, colonizing disturbed areas, cultivated fields, and irrigation systems, extending to sun-baked slopes and dry wadi beds.

FIGURE 5.1.1 Fruiting shoot of *Achyranthens aspera.*

FIGURE 5.1.2 Leaves of *Achyranthens aspera.*

FIGURE 5.1.3 Branch, leaves and flowers arrangements of *Achyranthens aspera.*

FIGURE 5.1.4 Habit of *Achyranthens aspera.*

DOI: 10.1201/9781003295488-5

FIGURE 5.1.5 Flowering shoot of *Achyranthens aspera*.

5.1.7 Chemical Composition

Phytochemical investigation of seeds revealed the presence of saponins A (D-glucuronic acid) and B (β-dgalactopyranosyl ester of D-glucuronic acid) [163,164]. Additionally, the seeds also contain 10-tricosanone, 10-octacosanone, and 4-tritriacontanone [164,165]. The plants also contain alkaloids (betaine, achyranthine), amino acids, and steroids (stigmasterol) [166].

The methanolic extract of the aerial parts of *Achyranthes aspera* has been reported to contain flavonoid C-glycosides [167]. A previous study also showed the presence of triterpenoids, including betulinic acid, oleanolic acid, and ursolic acid in *A. aspera* [168].

The previous investigations revealed the presence of oleonolic acid glycosides from the seeds of *A. aspera* [159]. Hydroquinone is considered the main component [159]. Seven components including eugenol, pbenzoquinone, spathulenol, nerol, hydroquinone, α-ionone, and asarone were identified (Figure 5.1.6).

5.1.8 Uses

This herb is used to treat respiratory conditions such as cough, pneumonia, and asthma. Moreover, the herb is purgative, diuretic, laxative, antiperiodic, anti-phlegmatic, and pungent. The plant is also useful in the treatment of edema, swelling under the skin, and boils and eruptions of the skin. It was commonly used to treat skin conditions and bites, including snake bites [169]. The plant and its extracts are also used in the treatment of liver scabies, rheumatism, complaints, and other skin complications, as well as bites from wasps [159,162,170]. In Oman, it has been reported that a paste made from the roots, stems, leaves, and flowers of this plant can stop bleeding and relieve the intense pain from a scorpion sting. The effectiveness of this remedy

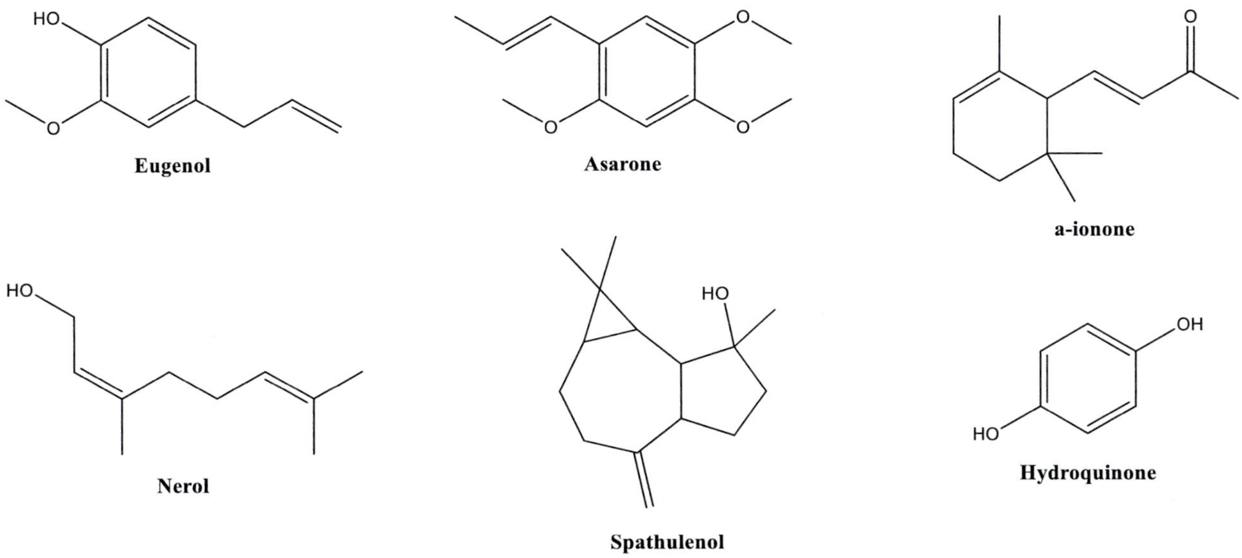

FIGURE 5.1.6 Active chemical components of *Achyranthens aspera*.

might be due to the presence of astringents, which help constrict blood vessels, or anti-inflammatory properties that reduce swelling and pain.

5.1.9 Mechanism of Action

Achyranthens aspera extract causes apoptosis in COLO-205 cells by altering the mitochondrial pathways and blocking the S phase of the cell cycle [171].

The root extract of *A. aspera* showed spermicidal activity in human and rat sperm [172]. Moreover, aqueous and methanolic extracts showed hypoglyceamic effects [173]. As methanolic extract contains alkaloid, and saponin, it has been reported to have cancer chemopreventive action [174].

5.1.10 Toxicity and Clinical Data

Recent findings showed that ethanolic leaf extracts of *A. aspera* have harmful effects on the development of rat embryos and fetuses at higher doses [175]. Plant consumption can cause hypotension and bradycardia, thus leading to dose-related transient cardiovascular toxicity [176].

The aqueous fraction of *A. aspera* showed lower piscicidal toxicity to Nile tilapia fish fingerlings compared to its molluscicidal efficacy [177]. Acute toxicity investigations revealed a toxic dose of aqueous extracts of *Achyranthes aspera* to be more than 2,000 mg/kg [166].

5.1.11 Most Cited Research

iCite, Database, NIH Open Citation Collection was used on January 1, 2024, to get the details of the most cited research. Filters for research papers were applied using the keywords *Achyranthes aspera*. The NIH database search revealed the effect of *Achyranthes aspera* on the immunity and survival of *Labeo rohita* infected with *Aeromonas hydrophila* [178].

5.1.12 Patents

Rufina Shatkina, Shlima Shatkina—US20120301573A1—Composition and method for suppressing appetite, 2012.

- This invention discloses a chocolate-based composition, including *Achyranthes aspera,* as one of the ingredients used to suppress appetite.

5.1.13 Plant Pathology

This plant has been reported to cause yellow disease, when the plant shows a characteristic yellow appearance with small leaves [179]. A severe leaf spot disease caused by *Curvularia prasadi* was reported in India post-rainy season, with severe infection causing complete plant defoliation [180]. Another aspect of this wild herb is its potential as a source of biofungicides.

6 Genus *Acridocarpus*

6.1 *Acridocarpus orientalis* A. Juss.

6.1.1 Synonyms

Anomalopterys orientalis

6.1.2 Family

Malpighiaceae

6.1.3 Vernacular Name

Qafas

6.1.4 Geographical Distribution

The species is natively distributed from Somalia northwards through the Arabian Peninsula to Iran. In Oman it is found in the foothills of the northern mountains, up to 1,500 m [181,182]. (Figures 6.1.1–6.1.4).

6.1.5 Chemical Composition

The *Acridocarpus* genus has been reported to contain several components, such as triterpenes (oleanolic acid, stigmasterol, ursolic acid, beta-sitosterol, and friedelin) and flavonoids such as quercetin, vitexin, luteolin, kaempferol, and apigenin (Figure 6.1.5) [183,184].

A previous study showed the isolation of two flavonoids known as morin and morin-3-O-β-D-glucopyranoside. Additionally, it was found that both components showed significant activity against pathogenic fungi and were phytotoxic against lettuce seeds at higher concentrations.

FIGURE 6.1.2 Unripe fruits *of Acridocarpus orientalis.*

FIGURE 6.1.3 Immature flowers *of Acridocarpus orientalis.*

FIGURE 6.1.1 Flowers of *Acridocarpus orientalis.*

FIGURE 6.1.4 Ripe fruits *of Acridocarpus orientalis.*

DOI: 10.1201/9781003295488-6

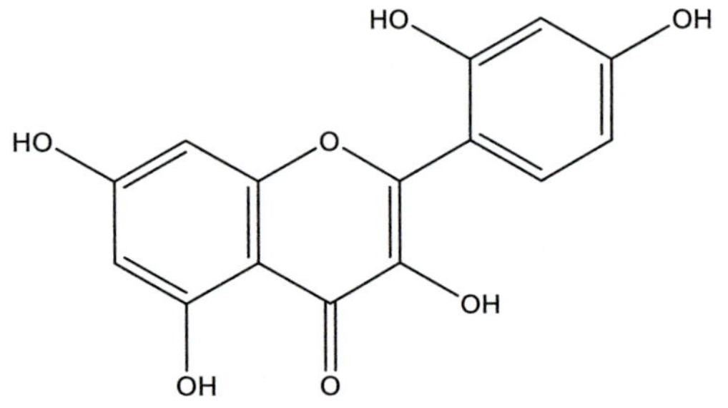

Morin

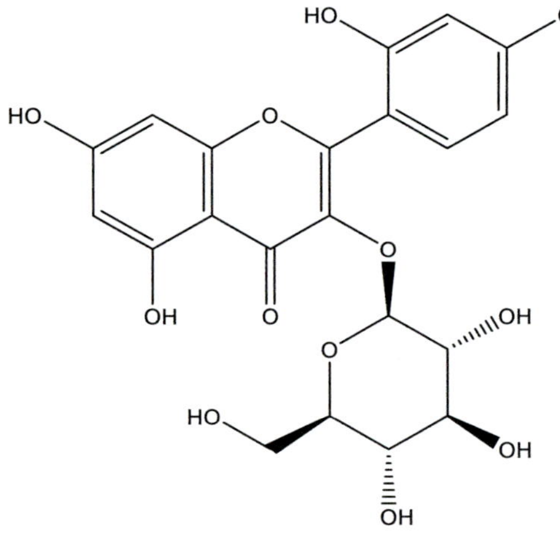

Morin-3-O-ß-D-glucopyranosid

FIGURE 6.1.5 Chemical components of plant belonging to genus *Acridocarpus*.

The anti-lipid peroxidation effects of morin-3-O-β-D-glucopyranoside were found to be more significant than *morin*. Furthermore, *morin* possesses more free radical scavenging activities than other compounds. Additionally, results obtained from anti-oxidative stress revealed that both compounds are promising antioxidant agents [185].

Earlier studies revealed that *A. orientalis* contains high polyphenolic content. In this investigation, it was found that leaves of *M. peregrina* demonstrated the highest TPC (242.2±0.13 mg/100 g), and the least amount was reported in the stem (46.6±0.14 mg/100 g) [186].

Recent investigation revealed the chemical composition of the essential oils derived from the stem and leaves of *A. orientalis* A. Juss. The key chemical volatile components present in the oil of the stem are methyl 8-pimaren-18-oate, octacosane, heptacosane, hexacosane, methyl dehydroabietate, and methyl pimar-8(14)-en-18-oate, while the leaves contain a high level of tetracosane, heptacosane, docosane, hentriacontane, heneicosane, and α-pinene. Findings obtained from the carbonic anhydrase II assay revealed that oil derived from the leaves showed more activity against carbonic anhydrase II than the stem [187].

Recent report showed the presence of β-sitosterol (steroid), β-sitosterol-3-O-β-D-glucopyranoside (one steroidal glycoside), 2,5-dimethoxy-1,4-benzoquinone (benzoquinone), 2,6-dimethoxy-1,4-benzoquinone (benzoquinone), 1-docosanol (one long-chain alcohol), betulin (lupane-type triterpenoid), betulinic acid lupane-type triterpenoid), quercetin (flavonoid), choerospondin (flavonoid), morin (flavonoid), morin-3-O-α-L-rhamnopyranoside (flavonoid), and morin-3-O-β-D-glucopyranoside (flavonoid) in the ethyl acetate fraction of *A. orientalis* [188].

A recent study also revealed that morin has considerably attenuated the cisplatin-prompted mitochondrial damage in the renal tissues of adult male Sprague-Dawley rats [189].

Recent findings revealed that morin has significantly activated *in vivo* pro-osteogenesis, resulting in osteoblast development via activation of osteoblast-associated marker genes and proteins, which suggests its important role in the regeneration of bone defects [190]. A recent study showed that the ethyl acetate fraction of leaves showed higher flavonoid levels, and all fractions showed inhibitory effects against *Aspergillus niger*, *Chitomium globosum*, and *Fusarium oxysporum*. Furthermore, an aqueous fraction of leaves demonstrated higher anti-lipid peroxidation effects. The chloroform fraction (at high dose) of the stem showed more cytotoxic effects for hepatoma and adenocarcinoma cancer cell lines [191].

6.1.6 Uses

A. orientalis is known for reducing inflammation and relieving pain. The extract obtained from the seeds and oil from *A. orientalis* can be utilized to relieve pain in the head and joints [192]. This plant also has veterinary ethnopharmacology applications, and it is used in the treatment of inflamed mammary glands of cattle in Oman [193]. In African countries, plants belonging to *Acridocarpus* have been utilized to treat gastrointestinal and skin complications [156]. Furthermore, the genus belonging to this plant has also been documented for its medicinal potential, such as antileishmanial, antimalarial, antitrypanosomal, and anti-parasitic effects [156].

Apart from its medicinal applications, the plant is also documented for its application as a source of yellow dye. Moreover, the extract obtained from the reddish hair of the young leaves can be utilized as a tanning agent [182,194,195]. The plant has also been documented to treat headaches, muscle pain, paralysis, tendon, and joint pain [193].

6.1.7 Mechanism of Action

A. orientalis extract demonstrated anti-cancer effects, perhaps due to its antioxidant effects. It was found that a compound known as morin, isolated from *A. orientalis* leaf extract, is accountable for the suppression of 4T1 cell proliferation as well as mesenchymal stem cells (MSC) expansion [196].

Another study showed anti-proliferative effects of dichloromethane extract from *A. orientalis*, demonstrating the most cytotoxic effect against both breast and cervical cancer cell lines. However, n-butanol extract from the stems of the same plant was found to be more effective against cervical cancer cells without harming normal cells. Thus, findings obtained from this study revealed that anti-proliferative components present in *A. orientalis* induce cell death in cancer cells via both apoptotic and non-apoptotic mechanisms [195].

The earlier study suggested that aqueous ethanolic crude extract of *Acridocarpus* aerial parts obtained from Al Ain and Oman showed antioxidant, anti-lipoxygenase, and anti-histone deacetylase effects, possibly because of the presence of high phenolic content [197].

Recently, hepatoprotective effects of *A. orientalis* ethanolic extract were investigated to study its potential to protect against paracetamol-induced hepatotoxicity in mice. Findings revealed that pretreatment with *A. orientalis* offers protection to mice against hepatotoxic adverse effects of paracetamol by reducing serum alanine transaminase and aspartate transaminase enzymes, with elevation of glutathione enzyme [198].

Treatment of rats with ethanolic extract obtained from an aerial plant of *A. orientalis* showed a marked reduction in blood glucose levels, with an increase in the number of insulin-positive cells and a decrease in glucagon immunoreactive cells in pancreatic islets of rodents. This treatment also raised superoxide dismutase levels, with a decrease in creatinine and bilirubin levels in the serum [199].

6.1.8 Toxicity and Clinical Data

Histopathological studies demonstrated that treatment of rodents with ethanolic extract obtained from an aerial plant of *A. orientalis* showed no toxic effect on the structure of the pancreas, liver, and kidney.

7 Genus *Adansonia*

7.1 *Adansonia digitata* L.

7.1.1 Synonyms

Baobabus digitata

7.1.2 Family

Malvaceae

7.1.3 Vernacular Name

Tabaldi & enkize

7.1.4 Common Name

Baobab, magic tree, monkey bread, pharmacist tree, and chemist tree.

7.1.5 Botanical Description

It is a giant deciduous tree having a gigantic trunk that grows up to 15 m. Leaves, simple and alternate with long petioles, are divided into 5–7 leaflets. This tree has large, overhanging white flowers and a big, egg-sculpted fruit capsule that contains many dark brown seeds (Figures 7.1.1–7.1.3).

7.1.6 Geographical Distribution

The native range of this species covers most of the tropical and subtropical regions of Africa, south of the Sahara as well as southern Arabian Peninsula (Southern Oman and Yemen). Within Oman its distribution restricted to two locations in the southern mountains.

FIGURE 7.1.1 Fruit of *Adansonia digitate.*

FIGURE 7.1.2 Branches with dry fruit of *Adansonia digitate.*

7.1.7 Chemical Composition

Owing to its unique chemical composition, this African tree possesses medicinal and nutritional value. It contains chemical components that possess anti-inflammatory and antioxidant effects [200,201]. Various phytochemical analyses showed that extracts derived from leaves contain alkaloids, anthraquinones, cardiac-active glycosides, flavonoids, phenols, reducing sugars, resins, saponins, steroids, tannins, and terpenoids [202]. In addition, extracts derived from leaves also contain high levels of mucilage carbohydrates, protein, fiber, fat, and minerals [203]. An earlier study has confirmed the presence of chemical components possessing antioxidant activity [204].

Another study confirmed the presence of high levels of fiber, minerals, and vitamins (B1, B2) [205,206]. Seeds are highly enriched with fat, protein, and energy and have high levels of aspartic and glutamic acid [207]. The fatty acid analysis confirmed the presence of palmitic acid as the main saturated acid and linoleic as well as oleic acids as the major unsaturated fatty acids. Chemical analysis revealed the presence of phenylethanoid glycosides, hydroxycinnamic acid glycosides, and iridoid glycosides [208]. It was also found that *A. digitata* contains high levels of flavonoids and phenolic acids [209]. Previous studies showed the presence of saponins, proanthocyanidins, flavonols, and phenolic acids [209]. Baobab fruit shells have been reported for their high level of phenolic component [210]. Baobab leaves have been reported for their high levels of proteins, minerals, and amino acids [211].

A. digitata fruit pulp contains high levels of flavonol as well as procyanidin glycosides. C-methylated flavone glycoside, known as tiliroside, is considered the main component [212,213] (Figure 7.1.4).

DOI: 10.1201/9781003295488-7

FIGURE 7.1.3 Habit of mature *Adansonia digitate.*

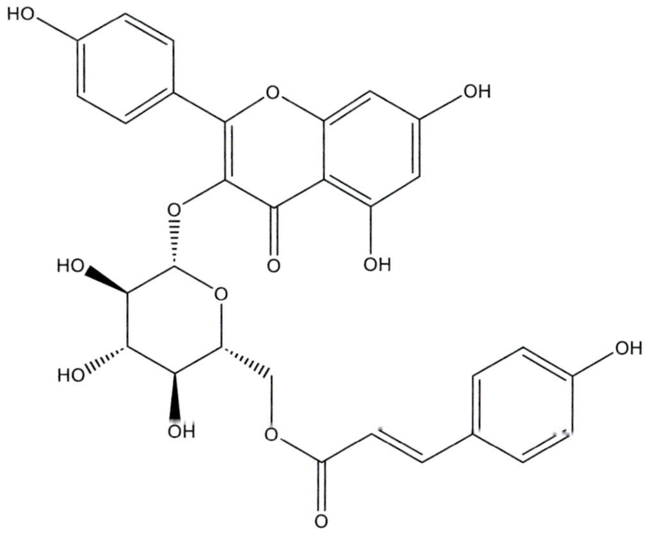

FIGURE 7.1.4 Chemical components of *Adansonia digitata.*

A recent study suggested the presence of fluorescent lipid compounds, N-acylserotonins, mainly N-lignoceryl-serotonin, in baobab (*Adansonia digitata*) seed oil, which has shown significant antioxidant activity [214].

Previous studies revealed the relation of the biochemical composition of baobab (*Adansonia digitata*) pulp, leaves, seeds, and soil types. It was found that the physicochemical features of the soil impact the biochemical composition of baobab parts [215].

In addition to flavonoids, phytosterols, amino acids, fatty acids, vitamins and minerals, and the presence of significant quantities of lysine, thiamine, calcium, and iron in seeds, this plant contains a high level of vitamin C, which is 7–10 times greater than the vitamin C content of oranges [200].

7.1.8 Uses

Parts of the tree and extracts derived from its different parts present an extensive range of economic benefits, such as food, medications, textiles, and non-wood forestry products that can be utilized for multiple purposes [216].

Fruits, leaves, seeds, and flowers are considered as the active sources of oils, milk, flavoring agents, phytochemicals, and spices. Recent research showed the medicinal potential of African baobab. Extracts derived from leaves, bark, and roots demonstrated high antiviral and antimicrobial activity. Tree bark contains an antimalarial and antipyretic alkaloidal component known as "Adansonin".

Pulp derived from the fruit of this tree is known for its stimulating effect on the activity of antioxidant enzymes and is used in the prevention of hyperlipidemia.

Baobab is known for its wide applications in food and medicine. The fruits of this tree are enriched with high vitamin C and thus can be used in food such as juice. Seeds are enriched with protein, carbohydrates, and fatty acids. Fruit pulp also contains high levels of lysine, zinc, thiamine, calcium, and iron. The dry baobab fruit contains high levels of vitamin C, potassium, calcium, and thiamine. Because of its high natural vitamin C content, baobab fruit pulp has a well-documented antioxidant capability. Due to the high level of vitamin C and other phytochemicals such as sterols, saponins, triterpenoids, flavonoids, phenolics, and other components, extracts derived from this plant possess antioxidant effects by reducing oxidative damage that can result in inflammation, cancer, aging, and other chronic complications related to health [217].

Recent results showed the antioxidant effects of *A. digitata* in reducing the testicular injury caused by cotton seed extract administration in rats. These protective effects could be due to the antioxidant effects of *A. digitata* in suppressing cotton seed extract-induced testicular damage [217].

Another recent study showed that the methanolic extract of *A. digitata* leaf possesses substantial antidiabetic and hypolipidemic effects in rats [218].

A recent study showed that enrichment with *Adansonia digitata* L. "Baobab" pulp flour as a nutritious and unconventional material up to 5% increased the nutritional value without adverse effects [219]. Due to its high ecological and socioeconomic importance and because it bears multipurpose fruit, it has been utilized traditionally and for commercial purposes more frequently as fiber, food, fodder, medicine, and for improving soil quality. The pulp derived from the fruit is highly nutritious and used as a functional food to supplement other foods. Additionally, pulp is also utilized to prepare beverages and is traded commercially by all national groups.

7.1.9 Mechanism of Action

In neuropathic pain (as an analgesic), *A. digitata* has been reported for its potential analgesic properties due to the presence of high levels of polyphenolic compounds and flavonoids. A recent report showed that *A. digitata* reduced neuropathic pain via the following mechanisms:

- suppress NF-κB to reduce inflammation
- stimulate Nrf2 to activate oxidative defense
- inhibit TRPV1 transcript to reduce pain
- suppress NR2B subunit of NMDA receptor to inhibit pain
- allow zinc-NR2A subunit interaction to suppress pain [220]

7.1.9.1 Analgesic Effects

A recent report demonstrated that aqueous bark extract of *A. digitata* exhibited a strong analgesic property [221].

7.1.9.2 Rheological Property

The rheological properties of the *A. digitata* leaf polysaccharide fraction were investigated, and it was found that it possessed strong shear thinning properties and viscoelastic behavior and was sensitive to temperature and the minimum energy required to initiate flow [222].

7.1.9.3 Antioxidant and Antidiabetic Effects

A recent study showed that baobab oil and butanolic extract from the fruit pulp protected rats against STZ-induced diabetic changes and attenuated lipid peroxidation, hypercholesterolemia, and oxidation [223].

Previous studies also revealed the antioxidant effects of *Adansonia digitata* red fiber, mainly baobab red fiber, which was 66 times higher than that of orange pulp. These red fibers showed greater antioxidant activity than pulp, leaves, fruit, and cuticle [224].

7.1.9.4 Anti-Inflammatory Effects

The previous studies demonstrated that petroleum ether and dichloromethane extracts of *A. digitata* bark presented the best anti-inflammatory effects [225].

15-Lipoxygenase inhibitory and antioxidant effects could be ascribed to the presence of high levels of phenolics and flavonoids [226].

7.1.9.5 Antimalarial Effects

Aqueous extract of the stem bark of *A. digitata* possessed the highest chemo-suppression of parasitaemia in a murine model without showing toxicity to brine shrimp (LD_{50} > 1,000 µg/mL). This could be due to the presence of alkaloids, flavonoids, sesquiterpene lactones, and saponins [227].

7.1.9.6 Glycemic Effects

In one of the reports, it was hypothesized that the phytochemicals present in baobab could be accountable for reducing the glycaemic and insulin response [228].

Another *in vivo* study based on evaluating baobab fruit extract on the effects of glycemic response showed that fruit extract, at both low and high doses, considerably reduced the glycemic response [229].

7.1.9.7 Nutraceutical Properties

Supplementation of beef patties with baobab seed extract showed an improvement in quality by enhancing lipid stability, antioxidant, and antimicrobial effects, and further improved the shelf life [213].

Furthermore, the addition of fermented baobab/acha (African grain) flour in baobab milk showed an improvement in the overall nutritional value of the diet compared to using flour alone [230].

7.1.10 Toxicity and Clinical Data

The safety and efficacy of baobab seed oil for topical application were assessed using healthy adult female Caucasian participants. The findings suggested that the application of baobab seed oil improved moisture hydration and reduced transepidermal water loss [231].

Recent research showed that the LD_{50} of all n-hexane, chloroform, ethyl acetate, and methanol extracts was found to be greater than 5,000 mg/kg, suggesting that the extract is nontoxic [232]. Baobab fruit pulp extract was also found to be nontoxic at all concentrations [233].

Another report demonstrated the LD_{50} of the methanolic extract of the stem bark of *A. digitata*, which was found to be ≥5,000 mg/kg [234].

Another acute toxicity study suggested that there were no behavioral changes, toxic signs, or death in the chicks when treated at the highest dose (3,000 mg/kg) of the extracts (obtained from the different parts of *A. digitata*) orally [235].

It was also found that the administration of cooked baobab (*Adansonia digitata* L.) seed meal on West African dwarf goats showed a normal range of serum biomarkers. Additionally, cooked baobab seed meal of up to 75% showed no toxic effects on the blood profile or performance of the goats [236].

7.1.11 Adulterants or Substituents

Due to the high demand for baobab fruit pulp powder, increases in the incidences of adulteration have been reported, such as substitutions with physically similar products like sifted cereal flours. To authenticate this cereal-based adulteration, raffinose was identified as a marker compound [237].

7.1.12 Patent

Gilles Pauly, US6274123B1, 1998 (use of an extract of the genus *Adansonia*): This invention is related to cosmetic and dermatological applications of *Adansonia digitata* (baobab) extract for topical application on the skin and/or the hair, eyelashes, and nails.

In this invention, researchers developed *Adansonia digitata* nanoparticles for their anti-cancer and antimicrobial application.

US9789146B1: Synthesis of *Adansonia digitata* nanoparticles. Manal Ahmed Gasmelseed AwadAwatif Ahmed HENDIKhalid Mustafa Osman Ortashi, 2017.

KR20140088302A: Another invention that demonstrated an antioxidant and anti-inflammatory active composition contained an extract of *Adansonia digitata*.

남궁우김희선박은희김예정. KR20140088302A. Composition Comprising Extract of *Adansonia digitata* Having Antioxidant activity and Anti-inflammatory Activity.

7.1.13 Plant Pathology

The pathogen *Oidiopsis haplophylli* has been exclusively documented on a solitary *Bombacaceae* species and *Adansonia digitata* [238]. Several fungal diseases causing leaf spots, blight, and stem cankers were reported on *A. digitate* in Kenya. These were attributed to *Pestalotia* sp., *Botryosphaeria* sp., and *Fusarium* sp. [239]. Likewise, three viral diseases were reported in Ghana [240].

8 Genus *Adesnociadium*

8.1 *Adenosciadium arabicum* (T. Anderson) H.Wolff

8.1.1 Synoyms

Ammoides arabica & *Ptychotis arabica*

8.1.2 Family

Apiaceae

8.1.3 Vernacular Name

Sefeq

8.1.4 Botanical Description

This annual plant grows up to 50cm with multiple stems arising from the base. It has alternate, pinnately lobed leaves and white flowers arranged in axillary and terminal umbels (Figures 8.1.1 and 8.1.2).

FIGURE 8.1.1 *Adenosciadium arabicum* during flowering.

FIGURE 8.1.2 Plant habit of *Adenosciadium arabicum*.

8.1.5 Geographical Distribution

This species exhibits a geographically restricted global distribution, being endemic to the eastern regions of Yemen, the southwestern portion of the Kingdom of Saudi Arabia, and Oman. Within Oman, its occurrence is primarily confined to the arid environments of the central and southern governorates.

8.1.6 Uses

This plant is used in some areas of the Central Desert of Oman, the leaves and flowers of this plant are used to treat stomach problems by preparing them as a tea, which is taken on an empty stomach [241].

DOI: 10.1201/9781003295488-8

9 Genus *Adenium*

9.1 *Adenium obesum* (Forssk.) Roem. & Schult.

9.1.1 Synonyms

Cameraria obesa & *Nerium obesum*

9.1.2 Family

Apocynacea

9.1.3 Vernacular Name

Asfed

9.1.4 Common Name

Desert rose

9.1.5 Botanical Description

The plant is a succulent shrub to a small tree with swollen pale gray and smooth bark. Leaves, alternately, often cluster together at the very tips and range from perfect ovals to playful circles with smooth edges. Flowers are bright pink with five delicate petals, each forming a perfect funnel (Figures 9.1.1–9.1.3).

9.1.6 Geographical Distribution

It is distributed in West Tropical Africa, Eastern Africa, and Arabian Peninsula. Within Oman, it is found in southern Oman. This plant has a taproot (fleshy), and leaves are arranged in a spiral manner.

FIGURE 9.1.1 Flower of *Adenium obesum*.

FIGURE 9.1.2 Paired fruits of *Adenium obesum*.

9.1.7 Chemical Composition

In 1990, pregnanes and 30 cardiac glycosides were isolated from the roots and the stems of *A. obesum*. Oleandrigenin beta-gentiobiosyl-beta-D-thevetoside was identified as the main glycoside [242].

This plant has been reported for the presence of chemical compounds including flavonoids, steroids, tannins, alkaloids, saponins, cardiac glycosides, and anthraquinones [243,244]. Rosmarinic acid was identified as an active antimicrobial compound from chloroform crude extract of *A. obesum* [245].

Recent findings reported cyanidin 3-O-[6-O-(rhamnosyl)-galactoside] and cyanidin 3-O-(galactoside) as the main and minor anthocyanins in *A. obesum* seedlings [246]. A previous study showed the presence of two antimicrobial compounds, 5,7,3′,4′-tetrahydroxy flavone 1 and 3,5,7,3′,4′,5′-hexahydroxyflavone 2 in the ethyl acetate crude extract of the stem of *A. obesum* [247].

Earlier research showed the presence of cardenolides somalin, hongheloside A, 16-acetylstrospeside, and honghelin, and the flavonol 3,3′-bis(O-methyl)quercetin in ethanolic extract of *Adenium obesum* that possess cytotoxic effects against the human epidermoid carcinoma of the nasopharynx test system (Figure 9.1.4) [248].

Other studies showed the presence of cardiac glycosides such as honghelin, obeside B, and obeside C in methanolic extract of *A. obesum*. These components demonstrated high IC50 against 3T3 cell lines with standard cycloheximide [249].

9.1.8 Uses

A. obesum is used in traditional medicines by local peoples in several regions from Saudi Arabia, Oman, and Yemen.

DOI: 10.1201/9781003295488-9

FIGURE 9.1.3 Natural habitat of *Adenium obesum*.

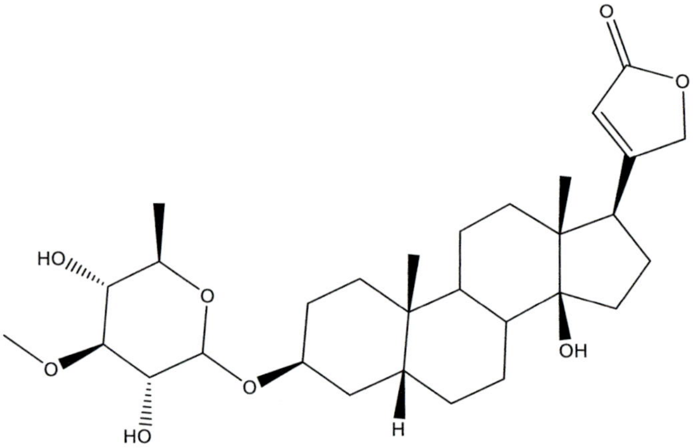

Honghelin

FIGURE 9.1.4 Chemical components of *Adenium obesum*.

A. obesum has been used by Omani people in the treatment of muscular and joint pain and skin-related complications [245]. In Oman, few plants belonging to this family have been used by the local communities as medicine. The majority of the plants belonging to this family possess medicinal properties and exudate a milky sap.

Plant preparations have been utilized by local peoples to treat sexually transmitted diseases, skin complications, wounds, and tooth decay. Additionally, plants also possess abortifacient, antiplasmodial, anti-trypanosomal, and anti-leishmanial effects [245].

The application of sap directly to swollen areas could potentially reduce inflammation and discomfort. Aching joints and sprained ligaments might have found relief through the topical use of the sieved liquid. Additionally, the plant's extract might have offered comfort to individuals suffering from joint dislocations and paralysis, although the effectiveness of this treatment remains unclear.

Ethanolic extract derived from *A. obesum* seeds extracts has also been reported for cytogenotoxic effects against breast cancer (against MCF-7 cells) [250]. Previous studies showed cytotoxic activity [251], anti-influenza virus [251], and antimicrobial effects. Also, cardiac glycosides present in *Adenium obesum* showed Molluscicidal effects against land snails [252].

Recent research showed antioxidant, anticancer, and anti-inflammatory effects of ethanolic extracts of

A. obesum by demonstrating cytotoxicity against A549 lung cancer cells and decreasing the levels of inflammatory mediators [253].

Recent findings showed that ethyl acetate fraction obtained from leaves crude extracts of *A. obesum* relatively showed better antioxidant and antibacterial effects than others [254]. Previous studies also demonstrated that all crude extracts showed potential antimicrobial effects against all pathogenic bacterial strains (Gram-positive and Gram-negative) [255].

Previous studies showed the development of biosynthesis of silver nanoparticles using aqueous leaf extract of *A. obesum*. It was found that these nanoparticles increased the level of ROS, led to the increase in DNA damage, apoptosis, and autophagy in MCF-7 cells [256].

9.1.9 Mechanism of Action

Molecular docking demonstrated anticancer effects of Δ16-3-Acetyldigitoxigenin (16-anhydro-3-acetylgitoxigenin) present in *Adenium obesum* by showing them the best binding energy scores with the majority of the targeted proteins [257].

9.1.10 Toxicity and Clinical Data

An earlier study showed that ethanolic extract of *A. obesum* stem bark was found to be toxic to the exposed fish and thus can be utilized as a natural piscicide [258].

9.1.11 Most Cited Research

In the most cited work, *A. obesum* leaf extract-loaded silver nanoparticles were fabricated, and it was found that these nanoparticles showed increased DNA damage, apoptosis, and autophagy via the generation of reactive oxygen species [256].

9.1.12 Patents

Frédéric Bonte, Marc Dumas, Jean-Christophe Archambault, US8377486B2 Cosmetic composition containing an *Adenium obesum* extract, use thereof, and method for cosmetic care including the use thereof. 2007: This invention claimed a cosmetic composition that includes *Adenium obesum* for topical applications.

PH12013000308A1, Vivian L MoyaMa Luningning L AmparadoLolito O Amparado: Another invention claimed the application of *Lantana camara* and *Adenium obesum* leaves as Organic piscicide.

9.1.13 Plant Pathology

The obligate phytoplasma pathogen *Candidatus Phytoplasma asteris* (16SrI group) is the causative agent of witches' broom disease in roses from China and has affected sesame and *Adenium obesum* in India [259]. In addition, leaf spot disease was reported on ornamental plants of *A. obesum* caused by *Pestalotiopsis microspora*. Likewise, leaf lesions caused by *Nigrospora sphaerica* were reported in Brazil [260].

10 Genus *Aerva*

10.1 *Aerva javanica* (Burm.f.) Juss. ex Schult.

10.1.1 Synonyms

Achyranthes javanica, Iresine javanica, Ouret persica var. javanica, & Illecebrum javanicum

10.1.2 Family

Amaranthaceae

10.1.3 Vernacular Name

Ra, ara, & ral

10.1.4 Common Name

Desert cotton

10.1.5 Botanical Description

Aerva javanica is an erect, perennial herb attaining a height of up to 150 cm. The plant exhibits a loosely branched, bushy growth habit. Leaves are alternate, simple, and vary from ovate to lanceolate in shape; both surfaces are densely pubescent, imparting a silvery appearance to the foliage. Inflorescences are elongated panicles bearing numerous small, white flowers (Figures 10.1.1 and 10.1.2).

10.1.6 Geographical Distribution

Aerve javanica is native to Africa and is distributed in other parts of the world, mainly in Asian and Australian countries [262]. It is a very common shrub throughout Oman.

FIGURE 10.1.1 Field photograph of *Aerve javanica.*

FIGURE 10.1.2 Inflorscense of *Aerve javanica.*

10.1.7 Chemical Composition

This plant contains alkaloids, flavonoids, tannins, sterols, saponins, and rutin [263]. Some of the phytochemicals from this plant demonstrated inhibitory action against tyrosinase [264]. It was found that *A. javanica* seeds' essential oils contain heptacosane, 3-allyl-6-methoxyphenol, pentacosane, 6,10,14-trimethyl-2-pentadecanone, nonacosane, tricosane, α-farnesene, dodecanal, and octacosane [265].

In 1990, an acylated isorhamnetin glycoside was identified from *Aerva javanica* [266]. Previous research showed that the methanolic extract of the flowers of *A. javanica* contains acylated flavone glycosides such as kaempferol and its derivatives, p-coumaric acid, caffeic acid, gallic acid, eicosanyl-trans-p-coumarate, hexadecyl ferulate, and hexacosyl ferulate. Among all, certain compounds showed weak inhibitory activity against certain enzymes (Figure 10.1.3) [267].

Previous reports showed the presence of ecdysteroids such as β-ecdysone (5), 5-β-2-deoxyintegristerone A and 24-epi-makisterone A [268]. A recent study showed the presence of several volatile compounds, of which angustione was identified as a major compound [261].

The earlier study demonstrated the presence of 3-hydroxy-4 methoxybenzaldehyde, ursolic acid, and (*E*)-N-(4-hydroxy-3-methoxyphenethyl)-3-(4-hydroxy-3-ethoxyphenyl) acrylamide. Among all the compounds, the last one showed considerable anti-ulcer effects [269].

10.1.8 Uses

Traditionally, *A. javanica* is utilized for treating wounds, cough, diarrhea, ulcers, and hyperglycemia [269]. *A. javanica* is used in Ayurvedic preparations to treat

DOI: 10.1201/9781003295488-10

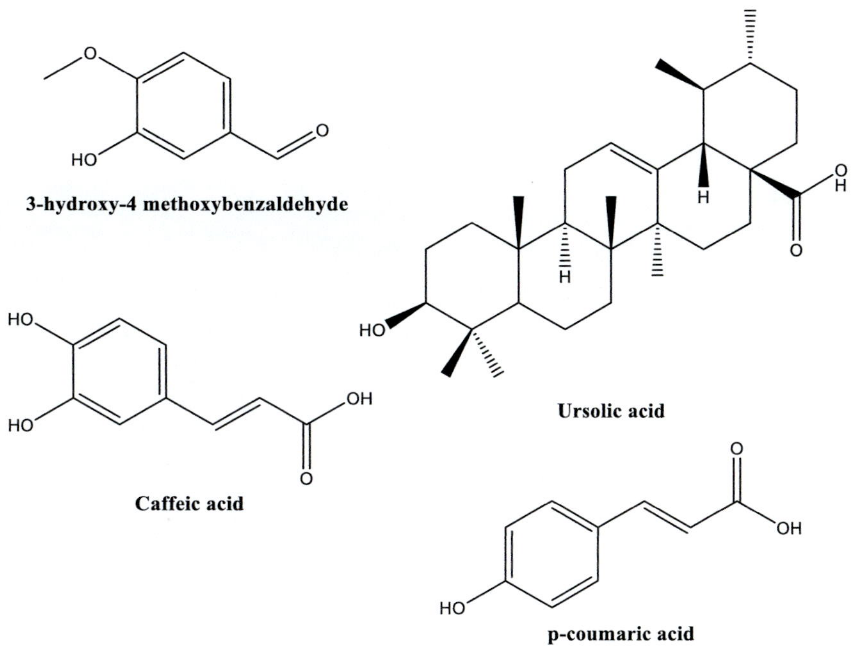

FIGURE 10.1.3 Chemical components of *Aerve javanica*.

stones in the kidney and bladder [3]. Additionally, the flowers of this plant can be used to treat rheumatism as well as kidney-related complications [182,261].

Hepatoprotective and antioxidative effects of the ethanolic extract of *A. javanica* aerial parts have been studied recently, and it was found that the extract attenuated hepatotoxicity and oxidative damage [264].

Inhibitory effects of isolated fractions derived from the methanolic extract of the flowers of *A. javanica* against the enzymes acetylcholinesterase, butyrylcholinesterase, and lipoxygenase are reported [268]. Anti-diabetic and antioxidant effects of hexane fractions were studied, and it was found that the *A. javanica* hexane fraction showed antidiabetic and antioxidant effects [261,270].

A recent study showed that *Aerva javanica* showed significant antibacterial effects against all investigated microorganisms except *Salmonella typhi*. Additionally, extracts derived from *Aerva javanica* showed poor antifungal effects against three species of *Aspergillus* and *Fusarium solani* [271].

Analgesic and anti-inflammatory effects of the aqueous methanolic extract derived from *A. javanica* have been reported [272]. Recent research showed that flavonoids and ecdysteroids present in the leaf extract inhibited the growth of granulation tissue in horses, with restoration of normal skin physiology [273].

10.1.9 Toxicity and Clinical Data

Copper oxide nanoparticles synthesized using *A. javanica* plant leaf extract can be utilized as an antimicrobial agent with negligible cytotoxicity at concentrations less than 60 µg/mL [274].

Another study showed that *A. javanica* essential oil, containing flavonoid derivatives, hydroxylated and glycosylated metabolites, showed antimicrobial, antioxidant, and cytotoxic effects. *A. javanica* extracts and essential oil demonstrated cytotoxicity in the range of 5.82%–14.47% [275]. The leaf extract of *A. javanica* demonstrated a higher inhibitory effect against human breast cancer cell lines (MCF-7) followed by other sample extracts [276].

10.1.10 Most Cited Research

Among the most cited papers, the enzyme inhibition assay of bioactive components present in *Aerva javanica* was assessed, and it was found that the ethyl acetate fraction showed considerable activity compared to other fractions [269].

10.1.11 Patents

Yitzhak Yannai. *Aerva javanica* plant named as Desert Yearning, US20030088898P1. A new and distinct variety of *Aerva javanica* called 'Desert Yearning' was identified in the Arava Valley, Israel.

10.1.12 Plant Pathology

Cercospora sp. PM018, an endophytic fungus, was isolated from *Aerva javanica* [277].

11 Genus *Aizoon*

11.1 *Aizoon canariense* L.

11.1.1 Synonyms

Aizoon canariense var. denudata & Aizoon procumbens

11.1.2 Family

Aizoaceae

11.1.3 Vernacular Name

Meshhaima & shahmet al ard

11.1.4 Common Name

Galsiekslaai

11.1.5 Geographical Distribution

This species exhibits a native distribution across a significant portion of the Old World's arid and semi-arid zones, encompassing several distinct floristic regions; Arabian Peninsula, southwest Asia, North Africa, Macaronesia and Southern Africa. In Oman it is widespread throughout the country.

11.1.6 Botanical Description

Thick-stemmed annual or short-lived perennial with zig-zagging branches. Its leaves are alternate, bright green, oval with smooth edges, and solitary and sessile flowers; yellowish inside and greenish or reddish outside with hairy exteriors (Figures 11.1.1–11.1.3).

FIGURE 11.1.1 Fruits of *Aizoon canariense*.

FIGURE 11.1.2 Habit of *Aizoon canariense*.

FIGURE 11.1.3 Fruits and flowers of *Aizoon canariense*.

11.1.7 Chemical Composition

Earlier investigation revealed the presence of alkaloids, phenolics, triterpenes, coumarins, saponins, tannins, flavonoids, and steroids in *A. canariense* [278]. Various studies reported the presence of phenolics, including quercetin [279], lauric acid, stearidonic acid, and omega-3 (Figure 11.1.4) [280].

A recent study showed the presence of behenic, tricosylic acid, and new adenine-based alkaloids in the extracts derived from aerial parts of *A. canariense*. This fraction demonstrated in vitro anticholinesterase, antimicrobial, and cytotoxic effects [281].

DOI: 10.1201/9781003295488-11

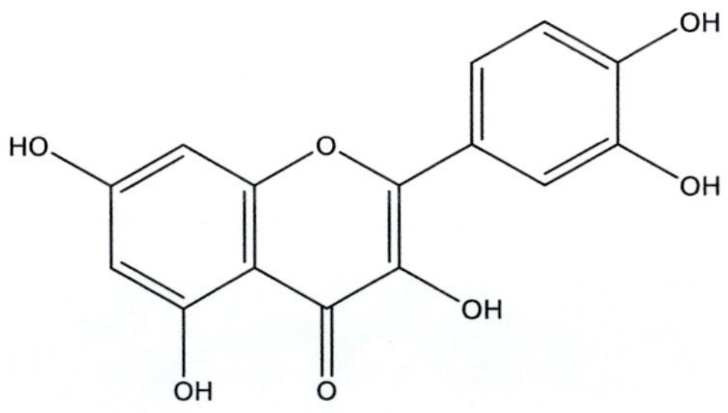

Quercetin

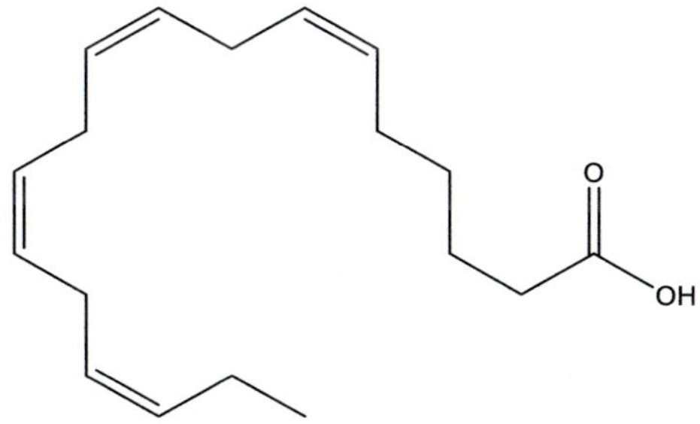

Stearidonic acid

FIGURE 11.1.4 Chemical components of *Aizoon canariense*.

11.1.8 Uses

This medicinal plant is traditionally used to treat gastrointestinal complications and is used as a hypotensive [278]. *A. canariense* has also been reported for its traditional uses, such as hypertension and flatulence [278]. *A. canariense* has been reported for its antibacterial, antioxidant, and cytotoxic effects [282] in liver cancer [278].

A recent study showed that *A. canariense* and a few others exhibited the highest antioxidant activity among the 18 plants from the Erkowit Region, Eastern Sudan [283].

Another study showed that *Aizoon canariense* L., *Asphodelus tenuifolius* Cav., and *Emex spinosus* (L.) Campdera from Bahrain possess antioxidant and antiradical effects [278]. Extracts derived from *A. canariense* have been reported to have antioxidant and radical-reducing effects [278].

11.1.9 Mechanism of Action

Aizoon canariense extract showed in vitro cytotoxic effects in human malignant melanoma cells by inducing G2/M arrest [284].

11.1.10 Toxicity and Clinical Data

Contains alkaloids. Thus, depending upon the quantity consumed, it may cause irritations of mucous membranes, cramps, and paralysis, including respiratory paralysis [285].

11.1.11 Plant Pathology

The endophytic fungus *Phoma sp.* 7210 was isolated from *Aizoon canariense* [286].

12 Genus *Alectra*

12.1 *Alectra parasitica* A. Rich.

12.1.1 Synonyms

Melasma parasiticum

12.1.2 Family

Orobanchaceae

12.1.3 Botanical Description

A. parasitica is a parasitic plant that reaches up to 30 cm in height. Its leaves are arranged oppositely along the stem. The inflorescence manifests as dense spikes containing numerous small flowers that exhibit intriguing characteristics: purple veins traverse five petals (Figures 12.1.1–12.1.3).

12.1.4 Geographical Distribution

Alectra parasitica is distributed in southern mountains in the wet escarpment woodland. In India, it is widely available [287] in various regions of Madhya Pradesh, Uttar Pradesh, Rajasthan, and Maharashtra. This plant is also native to Angola, Cameroon, Eritrea, Ethiopia, Kenya, Malawi, Oman, Saudi Arabia, Somalia, Tanzania, Uganda, Yemen, Zambia, and Zimbabwe. *A. parasitica* is an annual plant and grows primarily in the seasonally dry region.

FIGURE 12.1.1 Flowering stems of *Alectra parasitica.*

FIGURE 12.1.2 *Alectra parasitica.*

FIGURE 12.1.3 Stem and roots of *Alectra parasitica.*

DOI: 10.1201/9781003295488-12

12.1.5 Chemical Composition

Phytochemical analysis showed the presence of alkaloids, carbohydrates, sterols, glycosides, saponin, flavonoids, quinone, coumarins, and phenolic compounds [288].

Recent phytochemical investigation of *A. parasitica* subsp. *chitrakutensis* rhizome showed the presence of azafrin, rehmaionoside-C, and mussaenoside (Figure 12.1.4) [289].

So far, only azafrin and mannitol have been reported to be present in *Alectra parasitica* rhizomes [287]. A previous study based on high-performance liquid chromatography (HPLC) demonstrated the presence of azafrin in *A. parasitica* rhizomes [290].

In a recent study, glycosidic components such as azafrin, rehmaionoside-C, and mussaenoside from *Alectra parasitica* have been reported to stimulate glucose uptake in L6 myoblasts without causing toxicity [290,291].

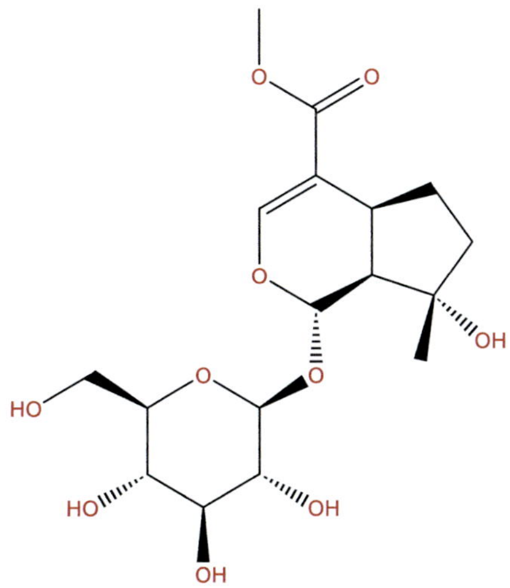

Mussaenoside

FIGURE 12.1.4 Chemical components of *Alectra parasitica*.

A recent report suggested the presence of iridoids, flavonoids, steroid glycosides, and strigolactones in the methanolic fraction of *A. parasitica* rhizome powder, which altogether possesses an anti-cancer effect on A549 [287].

12.1.6 Uses

Traditionally, *A. parasitica* had been used in the treatment of leprosy for several decades as a part of Ayurvedic medicine [292–294].

The plant comprises tremendous medicinal properties to treat many diseases such as leprosy, vitiligo, skin diseases, rheumatism, constipation, intestinal worms, fevers, swellings, paralysis, and tuberculosis. It has also been reported to treat piles, cardiovascular complications, and blood-borne non-infectious diseases [289].

In Ayurveda, *A. parasitica* has been mentioned for the treatment of disorders like rheumatism, constipation, fevers, swellings, paralysis, and skin diseases like vitiligo and leprosy [295].

12.1.7 Toxicity and Clinical Data

As it contains alkaloids, glycosides, and saponins, there is a need to study its toxicity; however, its toxicity has not been studied yet.

12.1.8 Plant Pathology

There isn't much information in the literature about the disease impact of *Alectra parasitica*, a parasitic plant that harms legume crops. However, there have been efforts to address this weed using natural enemies or biocontrol agents such as *Fusarium oxysporum* and *F. solani* [296]. Five bacterial endophytes were isolated and identified from the genus *Alectra*. The identified bacterial endophytes pertain to three genera: *Lysinibacillus*, *Peribacillus*, and *Bacillus*. Among these, *Bacillus* emerged as the dominant genus, encompassing three distinct species [297].

13 Genus *Aloe*

13.1 *Aloe dhufarensis* Lavranos

13.1.1 Family

Asphodelaceae

13.1.2 Vernacular Name

Tuf & subar

13.1.3 Botanical Description

A perennial succulent herb, reaching a height of up to 1 m. It has large, succulent leaves and possesses a grayish hue. Flowers, ranging from red to yellow, are arranged along a flowering stalk that projects vertically and can attain a height of 1 m. The mature fruit is erect, possessing three valves and a hardened exterior. Seeds: brown, winged seeds (Figures 13.1.1–13.1.3).

13.1.4 Geographical Distribution

This plant is available in the Southern Dhofar province of Oman and neighboring eastern Yemen [298]. In Oman, it is found in Southern Oman, specifically in drier areas, and is dominant on the high plateau of Jabal Qamar. It can also thrive at lower elevations on the eastern sides, on stony slopes and gravelly wadis.

13.1.5 Chemical Composition

Aloe dhufarensis Lavranos contains endophytic fungal and bacterial microorganisms, such as *Sarocladium kiliense* and *Penicillium oxalicum*. A recent study showed the presence of metabolites produced by these endophytes, such as 2-furanmethanol, dodecanoic acid, tetradecanoic acid, and 2,3-butanediol [299] (Figure 13.1.4).

FIGURE 13.1.1 Flowering stalk with red flowers of *Aloe dhufarensis.*

FIGURE 13.1.2 Natural habit of *Aloe dhufarensis.*

FIGURE 13.1.3 Immature inflorescences of *Aloe dhufarensis.*

The juice contains emodin, which acts on the intestine to induce medicinal effects. Other components, such as aloesin, homonatoloside, aloeresin A, homonataloin A and B, anthrones, phenolics, tannins, and glycosides (anthracene, coumaryl), are also present. In addition to these phytocomponents, this plant also contains vitamins [300].

Head2*A. dhufarensis* is one of the most important and useful medicinal plants in both Yemen and Oman. Traditionally, dried juice obtained from *A. dhufarensis* is used as a yellow-orange dye for pottery carving and as a cosmetic preparation to apply to the neck, arms, and legs. The juice of this plant is used to cure various diseases such as headaches, aching limbs, diabetes, and constipation. This plant is also used for digestive ailments [153,299,301].

DOI: 10.1201/9781003295488-13

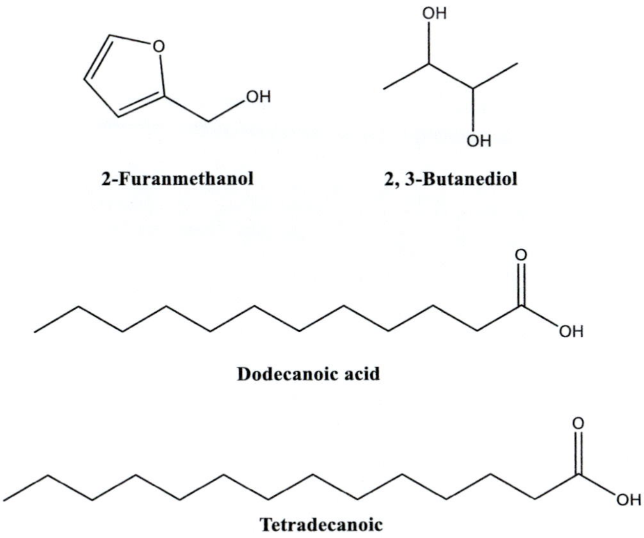

FIGURE 13.1.4 Chemical components of *Aloe dhufarensis*.

The juice derived from this plant is used as traditional medicine by the Jibali communities of Southern Oman [10]. Additionally, the plant is also used for its wound-healing as well as purgative effects [151].

A recent study showed that the leaf extracts of this plant showed significant antioxidant activity, which could be associated with high flavonoid and phenolic content. Also, the leaf extract showed considerable anti-proteinase effects [302]. Latex from this plant is used to treat sores and wounds [300].

13.1.6 Plant Pathology

Endophytic fungi, namely *S. kiliense* SQUCC-F2-2 and *P. oxalicum* SQUCC-F3-1, were isolated from the leaves of *Aerva dhufarensis* [299].

13.2 *Aloe praetermissa* T. A. McCoy & Lavranos

13.2.1 Family

Asphodelaceae

13.2.2 Vernacular Names

Sabr, Sekel

13.2.3 Botanical Description

A perennial succulent herb with dull gray-green, stemless leaves arranged in a basal rosette. The plant produces a branched inflorescence bearing red, glabrous (smooth and hairless) flowers (Figures 13.2.1–13.2.7).

FIGURE 13.2.1 Flowering stalk with flowers of *Aloe praetermissa*.

FIGURE 13.2.2 Leaves of *Aloe praetermissa*.

FIGURE 13.2.3 Dry fruits with seeds of *Aloe praetermissa*.

13.2.4 Family

Asphodelaceae

FIGURE 13.2.4 Ripe fruits of *Aloe praetermissa*.

FIGURE 13.2.5 Mature plant of *Aloe praetermissa*.

FIGURE 13.2.6 Sap droping of *Aloe praetermissa*.

13.2.5 Geographical Distribution

This medicinal plant is native to Southern Oman and is available in the coastal mountains of Southern Oman at around 1,000 m.

FIGURE 13.2.7 Cutting leaf with dropping sap of *Aloe praetermissa*.

13.2.6 Chemical Composition

Phytochemical investigation shows that species belonging to the genus *Aloe* contain numerous phytochemicals, such as carbohydrate polymers (mainly glucomannans), phenolic components, alkaloids, anthraquinones, anthrones, benzene and furan derivatives, chromones, coumarins, flavonoids, phytosterols, pyrans, and pyrones [303].

Generally, the gel obtained from plants belonging to this genus contains amino acids, anthraquinones, enzymes, flavonoids, glycoproteins, lipids, minerals, polyphenols, saccharides, and vitamins (Figure 13.2.8). Additionally, it also contains alkaloids, anthrones, benzene, chromones, coumarins, flavonoids, furans, glycoproteins, naphthalenes, and pyrones. Overall, the gel contains almost 75 different components, including vitamins, sugars, enzymes, anthraquinones, hormones, minerals, fatty acids, and others [304,305].

13.2.7 Uses

The word *Aloe* derives from the Arabic word "Allaeh," which represents a shining gel-like bitter and mucilaginous substance. This mucilaginous substance is widely

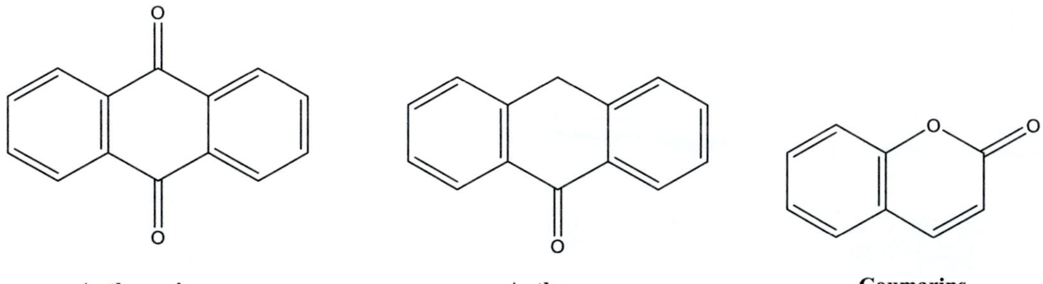

FIGURE 13.2.8 Chemical components of plant belonging to genus *Aloe*.

used for cosmetic and therapeutic applications. This plant has been used traditionally to treat skin-associated complications and is also used for therapeutic applications. Plants belonging to this genus are used to treat skin injuries (burns, cuts, insect bites, and eczemas) and complications associated with digestion. Preparations obtained from this plant have shown anti-inflammatory, antimicrobial, and wound-healing effects. The genus *Aloe* has been traditionally used to treat skin injuries (burns, cuts, insect bites, and eczemas) and digestive problems because of its anti-inflammatory, antimicrobial, and wound-healing properties.

13.2.8 Toxicity and Clinical Data

As the *Aloe* plant contains high levels of anthraquinones, its consumption may cause health complications such as diarrhea, hypokalemia, pseudomelanosis coli, renal damage, and phototoxic and hypersensitive reactions. In rodents, the administration of Aloe preparations has been reported to cause carcinogenic effects. It is also categorized by the International Agency for Research on Cancer as a potential cancer-causing agent for human beings. Plants belonging to this genus contain various components with potential benefits; however, it is the recent concern related to toxicity that may impact its utilization. The application of Aloe to the skin has caused complications such as phototoxicity, erythema, and contact dermatitis episodes. There are several reports that highlight toxicological effects, such as cytotoxicity, genotoxicity, carcinogenicity, and adverse clinical effects of *Aloe vera* whole leaf extract, gel, and latex. Utilization, phytochemistry, and many other aspects of this plant are still not fully available. Thus, we have included the general information of the plant genus [306,307].

Three distinct preparations derived from *Aloe vera* (i.e., the whole leaf extract, *Aloe vera* gel, and *Aloe vera* latex) have been used as topical and oral therapeutic remedies. However, due to the cytotoxicity, mutagenicity, and carcinogenicity of anthraquinones, it is crucial to monitor the content of these phenolic compounds in *Aloe vera* whole leaf extract and latex [308].

13.2.9 Adulterants or Substituents

Since the plant is widely available and doesn't require special conditions to grow (i.e., it can be easily cultivated in dry conditions), no adulterants or substituents have been reported.

13.2.10 Plant Pathology

The pathology of *Aloe praestermissa* has not been explored yet. However, bacterial soft rot disease of the genus *Aloe* is caused by *Pectobacterium chrysanthemi* [309].

13.3 *Aloe vera* (L.) Burm. f. synonyms: *Aloe perfoliata var. vera, Aloe chinensis* & *Aloe barbadensis*

13.3.1 Family

Asphodelaceae

FIGURE 13.3.1 Groups of of *Aloe vera* in the wild.

FIGURE 13.3.2 Inflorescence of *Aloe vera*.

FIGURE 13.3.3 Leaf section of *Aloe vera*.

13.3.2 Vernacular Name

Sekel & subar

13.3.3 Botanical Description

This medicinal plant is a succulent perennial, reaching up to 1.5 m tall. Its leaves are succulent, grayish-green, with a spine at the apex and small spines at the margins (Figures 13.3.1–13.3.3).

13.3.4 Geographical Distribution

Found in northern Oman, particularly in the foothills of the Hajar Mountains, they thrive on the rocky slopes, finding a home in the wadis (valleys) and are often seen near cultivated areas. Globally, in addition to the Arabian Peninsula, this species has also been introduced to many parts of the globe.

13.3.5 Chemical Composition

Aloe is the main source of anthraquinone glycosides. The main active component is aloin. Aloin is a mixture of various glucosides, of which barbaloin is the main component (Figure 13.3.4).

A. vera also contains aloesin, an aromatic chromone with wide applications in the cosmetic and health food industries [310].

The chemical name of barbaloin is aloe-emodin C-10 glycoside. It is hydrolyzed by ferric chloride into aloe-emodin-anthrone, aloe-emodin, and glucose. In addition to this chemical component, *Aloe* also contains

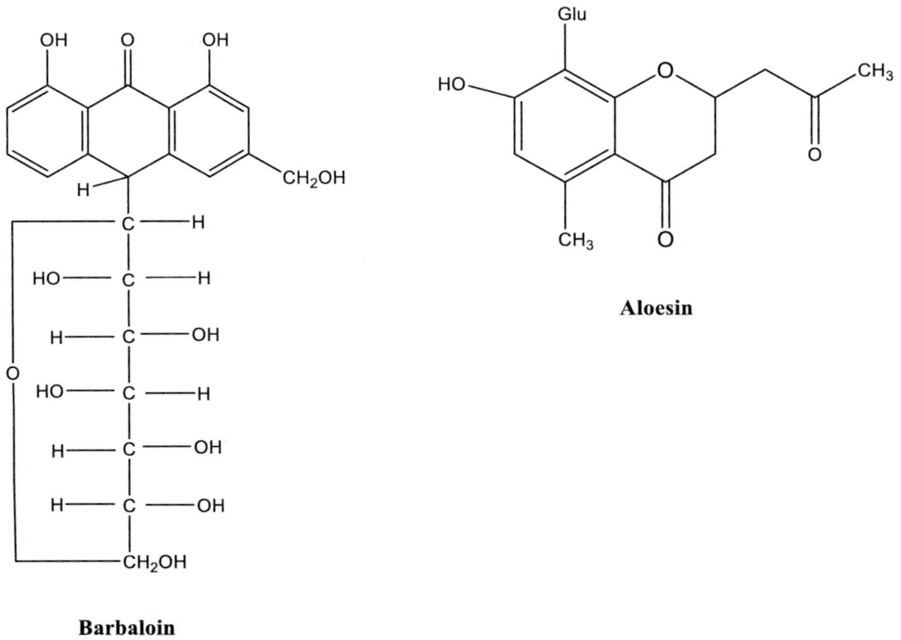

FIGURE 13.3.4 Active chemical components of *Aloe vera* (please redraw).

isobarbaloin, beta-barbaloin, aloe-emodin, and resins. The resins of *Aloe* mainly contain aloesin, which is C-glucosyl. Due to the presence of almost 75 therapeutic components, *A. vera* is considered a highly medicinally active species. Chemically, *A. vera* contains amino acids, salicylic acids, vitamins, saccharides, lignin, minerals, enzymes, saponins, and anthraquinones. Exudate derived from leaves contains anthraquinones, mainly barbaloin, which is accountable for inducing bitterness and cathartic effects. In addition, this plant also contains phytosterols, steroids, carotenoids, and terpenes. The gel derived from this plant contains high levels of polysaccharides, including acemannan.

13.3.6 Chemical Tests

Chemical tests are generally performed for the general assessment or detection of a variety of *Aloe*. A modified anthraquinone test is used to identify the presence of C-glycosides. The cupraloin test (Klunges isobarbaloin test) is used to differentiate Curacao, Cape, Sotrine, and Zanzibar aloes.

13.3.7 Uses

In Oman, this species has been used traditionally to treat various skin problems, such as burns, wounds, and irritations. The gel extracted from the leaves is also used to treat constipation.

Traditionally, *A. vera* has been used in folk medicine for almost 2,000 years. This medicinal plant has been used as traditional medicine in China, India, the West Indies, and Japan for several years. *A. vera* belongs to the genus *Aloe*, which contains almost 420 species. The Assyrian, Mediterranean, and Egyptian civilizations have used *Aloe* extensively since ancient times [311]. The first use of *Aloe* is recorded in the form of a Mesopotamian clay tablet dated to ca. 2100 BC. At the beginning of the 18th century, *A. vera* was formally registered as a skin protectant as well as a purgative by the pharmacopeia (U.S.). In the 1930s, it was clinically utilized for the treatment of skin and mucous membrane burns due to radiation-based treatment [311]. Aloes derived from *A. vera* are used as purgative.

13.3.8 Mechanism of Action

The colon is considered the main site of action. Glycosides present in *Aloe* act on the colon to exhibit strong purgative action. For this reason, it is always recommended to use carminatives while using *Aloe* oral preparations. Main chemical components such as barbaloin and isobarbaloin, present in the latex of this plant, have been reported for their cathartic and other effects. In the large intestine, barbaloin, as well as isobarbaloin, is hydrolyzed to form active components such as aloe-emodin-9-anthrone and aloe-emodin. These metabolites ultimately cause laxation via different pathways. Veracylglucan C, as well as veracylglucan B, present in *A. vera* gel, have been known for their effects in reducing inflammation [312]. On the other hand, salicylic acid, as well as other antiprostaglandin components present in non-polysaccharide gel, has been reported for its local anti-inflammatory effects via inhibiting cyclooxygenase [313].

13.3.9 Toxicity and Clinical Data

A. vera is a dietary supplement and not a regulated drug. Thus, there is no assurance related to the strength, purity, or safety of its preparations. *A. vera* is approved by the Food and Drug Administration (FDA) as a food additive for flavor. Several reports have raised safety concerns after the consumption of *A. vera*. These reports showed that treatment of rats with *A. vera* causes skin cancer and tumors, and aloin could be a possible reason behind these complications [314]. Due to the availability of inadequate information on the standardization of *A. vera* commercial formulations, it's challenging to assess its efficacy and safety. Previous reports showed that the application of *Aloe* gel on the skin is considered safe for treating moderate skin conditions, burns, wounds, and inflammation [315]. However, its use for the treatment of conditions like psoriasis, dermatitis, and surgical wounds is still not clear. Oral administration of gel for hypoglycemic response and its short-term utilization of *Aloe* latex for laxative response is considered safe. Prolonged latex utilization is considered dangerous because of the chances of electrolyte imbalance as well as dehydration [315]. Several reports suggest the importance of anthraquinone glycosides as cathartics; however, their safety concerns need to be investigated. A recent study showed that *Aloe vera* soft capsules could be considered safe before being marketed as a laxative and moistening health food. As far as toxicity is concerned, chronic oral administration of *A. vera* in rats caused toxicity to the reproductive system and substantial damage to sperm [316]. It was also found that *Aloe* latex is considered cytotoxic [317]. Several reports have shown the potential carcinogenicity and genotoxicity of anthraquinones in *Aloe vera* [306,318]. Phototoxicity of anthraquinones in *Aloe vera* has been demonstrated in animal studies [317].

13.3.10 Adulterants or Substituents

Aloe is generally obtained from Barbados or Curacao aloes (*A. barbadensis*), Cape aloes (*A. ferox*), Socotrine aloe (*Aloe perryi*), and Zanzibar aloes. Natal aloes resemble Cape aloes in microscopic characters and are thus used as substitutes. Similarly, Mocha aloes are also used as substitutes. Additionally, aloes are often adulterated with black catechu, pieces of iron, and stones. One of the most common adulterants found in *A. vera* gel, powder gel, or liquid preparations is maltodextrin [319]. Similarly, there are several additives present in *A. vera* gel liquid preparations, mainly preservatives, including benzoic acid and sorbic acid, or antioxidants, such as ascorbic acid [320].

13.3.11 Most Cited Research

iCite, Database, NIH Open Citation Collection was used from June to October 2022 to get the details of the most cited research. Filters for research papers were applied using the keywords *Aloe vera* and *Aloe barbadensis*. The most cited research work, according to an NIH database search, includes the development of biogenic gold nanotriangles and spherical silver nanoparticles via a simple protocol using *A. vera* leaf extract as the reducing agent. It was found that *A. vera* extract reduced the silver ions, resulting in the formation of silver nanoparticles with a suitable size [321].

13.3.12 Patents

- Henry H. Cobble. US3892853A, 1971: Prepared stable *Aloe vera* gel for topical application.
- Rex G. Maughan, US4446131A, 1982: Developed a controlled temperature process for the manufacturing of improved stabilized *Aloe vera*.
- 石森林蒋金珩郭菲董莉 (Stone Forest, Jiang Jinheng, Guo Fei, Dong Li). CN105363064B, 2015: Developed *Aloe barbadensis* Miller polysaccharide-based gel antiseptic for the treatment of skin burns.
- 田中美顺山田宗夫 (Mijun Tanaka, Muneo Yamada), CN103494998B, 2006: In this invention, *Aloe barbadensis* Miller extract is prepared to ameliorate hyperglycemia.
- Byung Pal Yu, Ki-Young Lee, US5939395A, 1995: identified a potent antioxidant from *Aloe barbadensis*.
- Patricia Brieva, Angelike Galdi, Natver Mehta, Jim M. Singer, Lindsay Menzer, Michell Chen. US8486463B1A, 2012: In this invention, a cosmetic composition in the form of an emulsion using *A. vera* and caprylyl salicylic acid was prepared.
- WO1998009635A9, 1996: In this invention, a glucose-rich, mannose-containing oligosaccharide was isolated from *Aloe*. This oligosaccharide inhibits the loss of skin immunocompetency caused by ultraviolet irradiation.

13.3.13 Plant Pathology

Several foliar diseases have been reported on *Aloe vera* cropped plants, including basal stem rot caused by *Fusarium* sp. Similarly, *F. proliferatum* and *F. pseudocircinatum* were identified as agents responsible for inducing leaf spots on *Aloe vera* and mango malformation disease (MMD), respectively [322]. Some wild specimens have been reported showing *Aloe* rust (*Phakopsora pachyrhizi* and *P. meibomiac*).

14 Genus *Allophylus*

14.1 *Allophylus rubifolius* (Hochst. ex A.Rich.) Engl.

14.1.1 Synonyms

Schmidelia rubifolia

14.1.2 Family

Sapindaceae

14.1.3 Vernacular Names

Zerben, Zerkim, Zirkin

14.1.4 Botanical Description

Large shrub with brown bark, reaching up to 5 m. Leaves are trifoliate, ovate with margins serrate to dentate. Flowers are greenish-yellow, whereas fruits are pale green and subglobose. Flowering and fruiting of these plants are observed between July and September (Figures 14.1.1 and 14.1.2).

14.1.5 Geographical Distribution

The plant is distributed in Southern Oman, Dhofar, on the wet escarpment mountains. It is also found in the Arabian Peninsula (Saudi Arabia, Yemen). It is widely distributed in tropical Africa [323].

FIGURE 14.1.1 Branches of *Allophylus rubifolius.*

FIGURE 14.1.2 Compound leaf of *Allophylus rubifolius.*

14.1.6 Chemical Composition

The chemical composition hasn't been revealed yet. However, other *Allophylus serratus* Kurz has been reported for its phytochemicals, such as β-sitosterol and phenacetamide [324], luteolin-7-O-β-D-glucopyranoside, rutin, apigenin-4-O-β-D- pinitol, glucoside, and quercetin [325] (Figure 14.1.3). *Allophylus laevigatus* is reported to have 11- acetoxy-4 α-methoxyeudesmane and components such as carissone and apigenin-8-C-β-rhamnopyranoside [326].

14.1.7 Uses

Allophylus species such as *A. cobbe, A. serratus, A. abyssinicus, A. africanus, A. cominia A. edulis, A. rubifolius,*

DOI: 10.1201/9781003295488-14

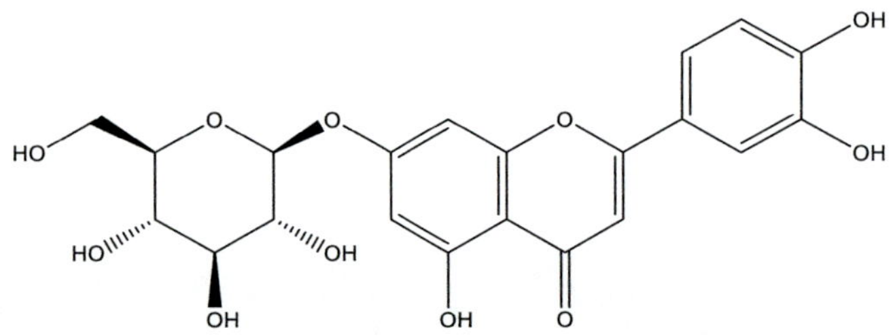

Luteolin-7-O-ß-D-glucopyranoside

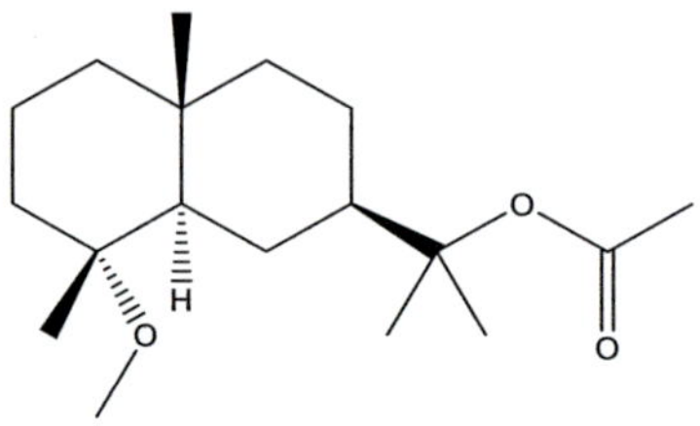

11- acetoxy-4 a-methoxyeudesmane

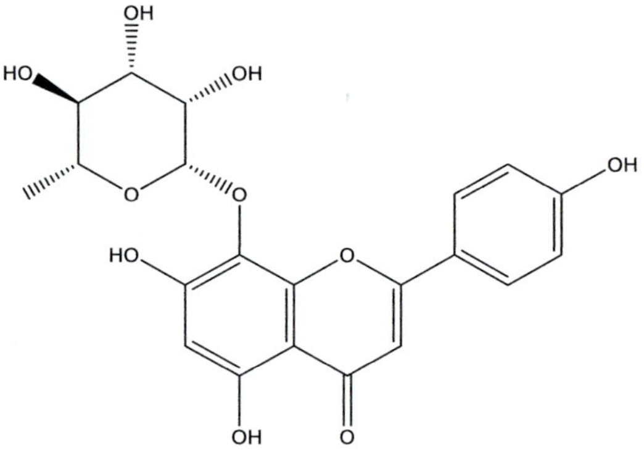

Apigenin-8-C-ß-rhamnopyranoside

FIGURE 14.1.3 Chemical components of *Allophylus serratus* and *Allophylus laevigatus.*

A. timorensis, A. occidentalis, A. racemosus, and *A. zeylanicus* have medicinal value whereas *A. cobbe, A. serratus* and *A. rubifolius* have nutritional application [327].

Traditionally, this medicinal plant is used to heal wounds. The leaves of *Allophylus rubifolius* have been used in the form of a topical paste to treat infected wounds [328]. Due to their high antioxidant activity, this plant has some non-nutritional benefits. Thus, they are frequently consumed by the aged groups in Oman.

Leaves have been used traditionally in the treatment of skin boils, ulcers, and wounds. Fruits are edible in nature and are considered a good plant source for honeybees to produce honey. Fruits of this plant are very bitter, astringent, and known for their beneficial restorative effects on human health [323].

It was found that, among several plants investigated, aqueous ethanol extracts of *Allophylus rubifolius* demonstrated the best inhibition of 2,2-Diphenyl-1-picrylhydrazyl (DPPH) radical at 89%–93%. Additionally, the ethanolic fraction derived from *A. rubifolius* showed the lowest IC50 value of 7.11 µg/ml [151].

Closely related species *A. africana* is known for its anti-inflammatory effects due to the presence of flavonoids. *A. africana* predominantly contains apigenin derivatives, luteolin derivatives, and glycosides [329].

14.1.8 Mechanism of Action

Traditional wound healing activity of this medicinal plant could be associated with radical scavenging activity.

14.1.9 Toxicity and Clinical Data

As closely related species such as *Allophylus spicatus* contain alkaloids and glycosides, it is important to assess their toxicity and clinical profile. However, as of now, there are no toxicity and clinical reports available on this plant [330].

14.1.10 Plant Pathology

No data in the literature are available on the pathology of *Allophylus rubifolius.* However, the highly potent fungal endophytes, *A. hancockii* and *P. christenseniae*, were derived through fermentation processes from the genus *Allophylus* [331].

15 Genus *Alternanthera*

15.1 *Alternanthera pungens* Kunth

15.1.1 Synonyms

Achyranthes leiantha, Alternanthera achyrantha, & Alternanthera leiantha.

15.1.2 Family

Amaranthacea

15.1.3 Common Name

Paper thorn & Khaki Weed

15.1.4 Botanical Description

Alternanthera pungens is a low-creeping perennial herb forming dense mats that hug the ground. It can reach up to 20 cm in height and spread out more. It has ovate leaves with entire margins. Flowers are inconspicuous and small, usually purple or pink, nestled in the leaf axils. Fruits are like round capsules. Flowering is observed throughout the year [332] (Figures 15.1.1–15.1.3).

In 1999, researchers isolated potexvirus from *Alternanthera pungens* (Amaranthaceae) in southern Queensland, Australia. It was found that this virus is a different species from papaya mosaic potexvirus, and its relationship with papaya mosaic potexvirus resembles that of potyvirus subgroup members [333].

Alternanthera pungens is a C4 plant, and it was found that photoactivation of phosphoenolypyruvate carboxylase (PEPC) in such plants may be mediated at least partially by the modulation of cytosolic pH [334].

FIGURE 15.1.1 Plant habit of creeping *Alternanthera pungens.*

FIGURE 15.1.2 Flowering shoot of *Alternanthera pungens.*

FIGURE 15.1.3 Inflorescence of *Alternanthera pungens.*

15.1.5 Geographical Distribution

The plant is distributed in southern and northern Oman along roadsides and compacted sand or gravel, as well as near wet areas and cultivations.

It is native to South America and also reported from tropical countries, including India [335].

15.1.6 Chemical Composition

Medicinal plants belonging to *Alternanthera* suggested the presence of flavonoids, steroids, saponins, alkaloids, triterpenoids, glycosides, and phenolic components.

DOI: 10.1201/9781003295488-15

Previous reports demonstrated the presence of low levels of phenolics in *Alternanthera pungens* (43.45 ± 14.22 µg/mg) [336].

15.1.7 Uses

The genus *Alternanthera* includes 139 species, comprising fourteen species used to treat hypertension, pain, inflammation, diabetes, cancer, microbial, and mental disorders [337]. This medicinal plant is used in folk medicine for the treatment of mouth ulcers, cough, malaria, and fever [338].

The previous study showed the effect of long-term administration of *Alternanthera pungens* herbal tea on HIV patients. It was found that a considerable decrease in plasma levels of malondialdehyde (MDA) and advanced oxidation protein end products, along with an increase in T CD4 and CD8 lymphocytes, occurred without causing hepatic and renal toxicity by the regular consumption of the herb tea [339].

Traditionally, decoction of this plant is used as a purgative, stomachic, and antispasmodic. *Zea mays* mixed with leaves of *Alternanthera pungens* and *Dysphania ambrosioides* is used as a digestive aid [340].

Earlier reports showed the immunological cross-reactivity of phosphoenolpyruvate carboxylase in leaf extracts of intermediate species of *Alternanthera*. Findings showed that anti-PEPC antiserum elevated against *A. hypochondriacus* showed high cross-reactivity with phosphoenolpyruvate carboxylase in leaf extracts of *Alternanthera pungens* [341].

It was found that crude extract of *A. pungens* demonstrated antibacterial and antioxidant effects [335]. A recent study showed novel PXR-mediated CYP3A4 activation potential of *Alternanthera pungens* in HepG2 cells. Findings obtained from this study showed that *Alternanthera pungens* was found to be the most promising inducer of CYP3A4 [342].

Recently, *Citrus aurantium*, a novel, nonedible, and waste seed oil, was used for biodiesel synthesis using recyclable zirconium oxide nanoparticles synthesized with *Alternanthera pungens* aqueous leaf extract. It was found that biodiesel production from *Citrus aurantium* waste and nonedible seed oil can be efficiently employed for the generation of renewable energy [343]. The ethanolic extract obtained from this plant is used to treat wounds, cancer, eye-related diseases, and anemia. It is also used as a diuretic [344,345].

In 2013, *A. pungens* was reported in the treatment of the investigated sexually transmitted infections [346]. Empacho (excessive food intake) is a digestive disorder and is identified as a culture-bound syndrome in Argentina. *Dysphania ambrosioides*, *Alternanthera pungens*, and *Aloysia polystachya* have been known for their potential to treat empacho [340].

15.1.8 Toxicity and Clinical Data

Toxicity and clinical data-related reports are not available for this medicinal plant. However, there are several reports available on other species belonging to the same genus. Oral administration of *Alternanthera brasiliana* and *Alternanthera bettzickiana* hydroalcoholic fraction for fourteen days in mice was studied, and it was found that *Alternanthera bettzickiana* extract showed a cytotoxic impact on Dalton's lymphoma ascites (DLA) cells [347]. Similarly, an acute toxicity study of ethanolic extract of *Alternanthera philoxeroides* in the rodent model was performed, and it was observed that no mortality, changes in body and inner organ weights, or hematological and clinical parameters were observed [348].

15.1.9 Most Cited Research

iCite, Database, NIH Open Citation Collection was used from June to October 2022 to get the details of the most cited research. Filters for research papers were used by using the keywords *Alternanthera pungens*. The most cited research work, as per an NIH database search, includes the photoactivation of phosphoenolpyruvate carboxylase in *Alternanthera pungens*.

Phosphoenolpyruvate carboxylase is considered as an important enzyme of C(4) photosynthesis. *Alternanthera pungens* is a C4 plant, and it was found that photoactivation of "phosphoenolpyruvate carboxylase" in such plants may be mediated at least partially by the modulation of cytosolic pH [334].

15.1.10 Plant Pathology

"Papaya mosaic virus" (PapMV) was isolated, and subsequent cloning and sequencing of the virus's three ends from moss rose revealed a nucleotide similarity of 94% with a putative novel viral species identified in Australia from *Alternanthera pungens*, as determined by a BLAST search [349].

16 Genus *Amaranthus*

16.1 *Amaranthus dubius* Mart. ex Thell.

16.1.1 Synonyms

Amaranthus incomptus, Amaranthus dubius var. crassespicatus, & Amaranthus dubius var. flexuosus

16.1.2 Family

Amaranthaceae

16.1.3 Common Name

Red spinach

16.1.4 Botanical Description

A. dubius is the polyploid species of the *Amaranthus*, which is commonly used as a pot herb. This annual glabrous herb is erect (rarely tall) with a branched and hairless stem and ovate or rhomboid-ovate leaves.

16.1.5 Geographical Distribution

A. dubius is an annual herb native to tropical America. This leafy vegetable is extensively grown in various regions of Asia and Africa. The phylogenetic analysis based on ten chloroplast genomes showed that *A. dubius* is sister to *A. hypochondriacus* and *A. caudatus* [350,351].

Recent findings showed that by using chloroplast phylogenomics application, researchers determined the species correlation between the different species of *Amaranthus*. It was found that *A. hybridus* possibly gave rise to several species within the Hybridus complex, specifically *Amaranthus dubius*, *Amaranthus quitensis*, *Amaranthus caudatus*, *Amaranthus cruentus*, and *Amaranthus hypochondriacus* [352].

A. dubius has been reported as a possible favorable host plant for the Phytoseiidae family containing phytoseiid mites [353]. Within Oman, it occurs in northern Oman.

16.1.6 Chemical Composition

Leucine aminotransferase was isolated and partially purified from the green leaves of *A. dubius* [354]. A recent study investigated the elemental composition of *Abelmoschus esculentus*, such as heavy and trace elements, to establish the safety profile of the plant. Findings obtained showed that plants contain high levels of K, Ca, and Fe with the least level of lead [355].

Another study suggested the elemental composition of *Amaranthus tortuosus* by inductively coupled plasma optical atomic spectroscopy. It was found that *A. tortuosus* has relatively high levels of Ag, Al, Ba, Ca, Co, Cr, Cu, Fe, K, La, Li, Mg, Mn, Mo, Na, Ni, P, S, Sr, V, Zn, and Zr [356].

An earlier study demonstrated the relationship between traffic density and metal concentrations of the vegetable weed *Amaranthus tortuosus* in Uganda. It was found that Pb concentration in *Amaranthus tortuosus* leaves decreased with increasing distance from the road edge [357].

16.1.7 Uses

A. tortuosus, also known as African leafy vegetables, has been used for a long time as food and medicine. *Amaranthus tortuosus* is the most common and highly consumed vegetable in Kenya and Tanzania, and it is affordable and available. In addition to its nutritional property, this plant is also used as medicine, and livestock feed was also reported [358]. Green leafy vegetables contain high levels of antioxidants and are suggested in the management of chronic complications [359].

Red spinach extract can also be considered as a nutritional supplement to elicit an ergogenic response by delaying the ventilatory threshold during graded exercise testing. It was found that vegetables cooked by steaming or boiling reduced vitamin C levels by up to 99% in the cooked vegetables. Thus, cooking via steaming or boiling changes the antioxidant properties of indigenous vegetables [360].

In Kenya, *Amaranthus tortuosus* has been used as folk medicine to manage obesity; however, lack of scientific validation restricts its utilization

Leaves derived from *A. tortuosus* leaves showed antidiabetic effects on alloxan-induced diabetic mice [361]. This is due to the presence of bioactive components such as cucurbitae coatin, both saponins and flavonoids, which could have been responsible for hypoglycemic effects in rats [362]. The antidiabetic activity of this plant could be due to the presence of alkaloids and flavonoids in *A. tortuosus* [363].

A recent study showed the antiobesity effects of *A. tortuosus*. It was found that the leaf extract of this plant showed significant effects on body mass index in a dose-dependent manner. This might be due to the presence of phytochemicals such as saponins, flavonoids, alkaloids, and steroids [364].

Betacyanins derived from *A. tortuosus* and *H. polyrhizus* could collectively be used to potentially decrease the risk of biofilm-associated infections caused by these bacteria on hydrophobic polymers [365].

DOI: 10.1201/9781003295488-16

A. tortuosus leaves are commonly consumed as leafy green vegetables. Additionally, apart from its nutritional properties, *A. tortuosus* showed high inhibitory effects against TNF-α release and exhibited antioxidant and anti-inflammatory effects [366,367]. Earlier studies suggested that *Amaranthus tortuosus* showed the greatest inhibition of TNF-α [368].

A recent study demonstrated the inhibitory effects of betacyanins from *Amaranthus tortuosus* on biofilm formed by *Staphylococcus aureus* and *Pseudomonas aeruginosa*. The betacyanin from *Amaranthus dubius* demonstrated better anti-biofilm activity against five *Staphylococcus aureus* strains. Additionally, the betacyanin fraction considerably reduced the hydrophobicity of *Staphylococcus aureus* and *P. aeruginosa* strains [365]. *A. tortuosus* collected from KwaZulu-Natal, South Africa, showed angiotensin I-converting enzyme (ACE) inhibitory effects [369].

16.1.8 Adulterants or Substituents

A. viridis and *Amaranthus tortuosus* are both closely related species and difficult to differentiate, they are also known in the Caribbean as callaloo [370].

16.1.9 Most Cited Research

iCite, Database, NIH Open Citation Collection was used from June to October 2022 to get the details of the most cited research. Filters for research papers were applied using the keywords *Amaranthus tortuosus*. The most cited research work, as per an NIH database search, is based on establishing a relationship between traffic density and trace metal concentrations in roadside soils, surface films, and *Amaranthus tortuosus* around Kampala City in Uganda (already discussed above) [357].

16.1.10 Patents

Benny Antony. EP3013348B1. A medicinal composition of amaranth extract origin having enriched nitrate content and a method of preparing the same. This invention revealed a procedure to obtain the medicinal composition of amaranth extract (derived from *Amaranthus caudatus*, *Amaranthus cruentus*, *Amaranthus tricolor*, *Amaranthus blitum*, *Amaranthus viridis*, *Amaranthus dubius*, *Amaranthus hypochondriacus*, and *Amaranthus hybridus*) containing a high level of nitrate content.

16.1.11 Plant Pathology

Hyphomycetes predominated within the endophytic communities of *Amaranthus tortuosus* [371]. As a crop plant, several diseases have been reported. White Blister rust, triggered by *Wilsoniana amaranthus*, along with damping-off caused by *Rhizoctonia* spp. and *Pythium* spp., have been observed. Additionally, there are necrotic lesions on leaves caused by *Colletotrichum gloeosporioides*, and wet rot caused by *Choanephora cucurbitarum*. It is also affected by root-knot nematodes caused by *Meloidogyne incognita* and Papaya leaf curl virus [372].

16.2 *Amaranthus graecizans* L.

16.2.1 Synonyms

Amaranthus angustifolius, *Amaranthus angustifolius* subsp. *Graecizans*, & *Amaranthus angustifolius* var. *graecizans*

16.2.2 Family

Amaranthaceae

16.2.3 Vernacular Name

Sender & qateefah

16.2.4 Common Name

Dadh, qutaif, shegra al santeen, shaeb, sinder, Mediterranean amaranth, tumbleweed, or short-tepalled pigweed

16.2.5 Botanical Description

A. gracias L. is an annual herb with simple, broadly ovate, and glabrous leaves, unisexual flowers, and subglobose capsules containing a single seed (Figure 16.2.1).

16.2.6 Geographical Distribution

This species exhibits a pantropical distribution, covering the tropical regions of both the Eastern and Western hemispheres. In Oman, it is common throughout the country as a weed among cultivations and in disturbed areas.

FIGURE 16.2.1 Stem with leaves and flowers of *Amaranthus graecizans*.

16.2.7 Chemical Composition

Plants contain alkaloids, flavonoids, and saponins. Methanolic fractions contain alkaloids, tannins, carbohydrates, lipids, proteins, etc. [373].

Another study showed that methanolic extracts of whole herbs contain saponins, tannins, carbohydrates, flavonoids, cardiac glycosides, sterols, lipids, and alkaloids [374].

16.2.8 Uses

Leaves of this plant are used as a vegetable throughout Africa and the Middle East [375]. Leaves obtained from this plant can be consumed in their raw form but are more often cooked or added to sauces and stews.

Leaves are used to treat snake bites and itchy rashes, mainly for soothing purposes. The leaves are used as emollients. *A. graecizans* showed antioxidant potential and thus can be utilized as a natural source of antioxidants [373].

A. graecizans is used as a folk medicine to treat various inflammatory conditions. A recent study showed the hepatoprotective effect of *A. graecizans* against rifampicin- and isoniazid-induced liver injury. Findings obtained from this study showed that *A. graecizans* methanolic extract demonstrated a considerable decrease in the raised liver enzymes and reduced the liver injury caused by isoniazid and rifampicin [376].

Similarly, another study revealed that the methanolic extract of this plant showed substantial anti-nociceptive, anti-inflammatory, and protease inhibition effects [374].

16.2.9 Mechanism of Action

It was found that the extract derived from this plant showed anti-inflammatory effects by decreasing nitric oxide and TNF-alpha synthesis in the cells of monocyte origin and decreased cytokine- or LPS-stimulated iNOS mRNA levels in both cell types [377].

16.2.10 Toxicity and Clinical Data

Plants belonging to this genus are not poisonous. However, it depends on the soil on which they are cultivated. When grown on heavy metal– or nitrogen-enriched soil, they retain nitrates and heavy metals in the leaves. Moreover, since the plant contains alkaloids, flavonoids, and saponins, it is important to assess its toxicity using a desirable model.

16.2.11 Most Cited Research

iCite, Database, and NIH Open Citation Collection were used from June to October 2022 to get the details of the most cited research. Filters for research papers were applied using the keywords *Amaranthus graecizans*. The most cited research work, as per the NIH database search, includes the uptake and release of cesium-137 by *A. graecizans*. It was found that plant 137Cs reduced after the addition of the ammonium solution, suggesting that a considerable amount of 137Cs can be released from the shoots of plants due to quick variations in soil chemistry [378].

16.2.12 Patents

Benny Antony. EP3013348B1. A medicinal composition of amaranth extract origin having enriched nitrate content and a method of preparing the same. This invention revealed a procedure to obtain the medicinal composition of amaranth extract (derived from *Amaranthus caudatus*, *Amaranthus cruentus*, *Amaranthus tricolor*, *Amaranthus blitum*, *Amaranthus viridis*, *Amaranthus dubius*, *Amaranthus hypochondriacus*, and *Amaranthus hybridus*) containing a high level of nitrate content.

16.2.13 Plant Pathology

Leaf blight in *Amaranthus graecizans* caused by *Choanephora cucurbitarum* [379].

16.3 *Amaranthus viridis* L.

16.3.1 Synonyms

Euxolus viridis, Glomeraria viridis & Pyxidium viride

16.3.2 Family

Amaranthaceae

16.3.3 Vernacular Name

Dah, quttaif, shegra tal santeen, sindar, Chauraiy

16.3.4 Common Name

Waterleaf

16.3.5 Botanical Description

A. viridis is an annual herb that grows up to 50 cm with a reddish-green and erect stem. Leaves: alternate, ovate with long petioles. Flowers: small, green or reddish, clustered in spikes (Figure 16.3.1).

16.3.6 Geographical Distribution

Amaranthus viridis is widely distributed throughout the world, grown throughout the world and has been used in folk medicine. It is widely available in northern Oman and is considered a tropical weed. It is distributed in cultivated fields and irrigation systems across Northern Oman as a weedy plant.

FIGURE 16.3.1 Inflorescence of *Amaranthus viridis.*

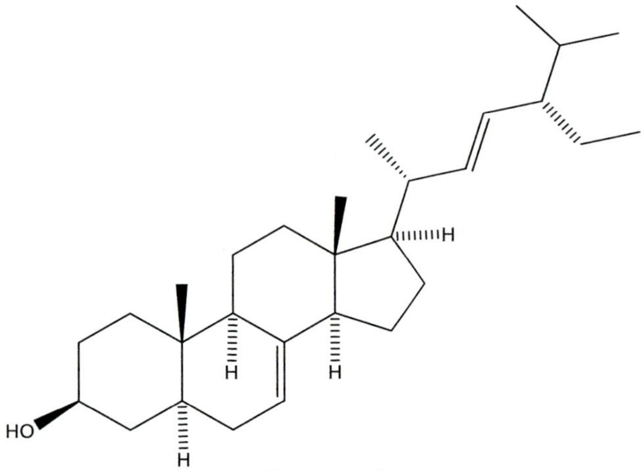

FIGURE 16.3.2 Chemical components of *Amaranthus viridis.*

16.3.7 Chemical Composition

Phytochemical analysis of the extracts of *A. viridis* revealed the presence of major phytochemical compounds, including flavonoids, alkaloids, phenolics, steroids, terpenoids, saponins, cardiac glycosides, and tannins [380]. An antiproliferative and antifungal lectin from the seeds of *Amaranthus viridis* Linn was isolated [381]. *A. viridis* L. contains quercetin, lutein, rutin, β-carotene, amasterol, squalene, spinasterol, trilinolein, polyprenol, and phytol [382].

A recent report suggested that the leafy vegetable has protein, dietary fiber, carbohydrates, Ca, K, Mg, P, S, Fe, Mn, Cu, Zn, Na, Mo, B, chlorophylls, β-cyanins, β-xanthines, betalains, β-carotene, and vitamin C. *A. viridis* had the highest nutrients, pigments, vitamins, phenolics, flavonoids, and antioxidants [383]. It was also found that this plant contains high levels of protein, fiber, minerals, and antioxidant compounds [384]. The dichloromethane extracts of *Amaranthus viridis* L. contain squalene, trilinolein, polyprenol, and phytol from the leaves; spinasterol and squalene from the stems; and spinasterol from the roots (Figure 16.3.2) [382].

Further, it was revealed that chlorogenic acid, gulonic acid, and kaempferol were found as major components responsible for the antioxidant activity (radical scavenging, lipid peroxidation inhibiting, metal chelating activity) of *A. viridis* extract. Additionally, it was found that the extract has protective effects on H_2O_2-induced DNA damage with the alkaline comet assay [380].

16.3.8 Uses

Amaranthus viridis is an important medicinal herb. Extracts derived from *A. viridis* showed antioxidant effects, antiviral, anticancer, antidiabetic, antimicrobial, and antinociceptive activity [380].

α-Amylase and pancreatic lipase inhibitory effects have been reported for polyphenol-rich aqueous extracts of *Amaranthus viridis*, suggesting its possible role in the management of diabetes and obesity [385].

An antiviral protein purified from the leaves of *Amaranthus viridis* was named amaranthin. The in vivo antiviral activity of amaranthin was confirmed in a tobacco mosaic virus (TMV) infection test [386].

Previous findings showed that methanolic extracts of *A. viridis* and *L. taraxacifolia* possess antiplasmodial activity. This can be further studied as a potential antimalarial agent [387].

A recent study also showed that *Amaranthus viridis* (Linn.) leaf could be used to treat cyclophosphamide-induced neuro-endocrine dysfunction and reproductive toxicity [388].

16.3.9 Mechanism of Action

A. viridis leaf extract has antioxidant and anti-inflammatory effects and may modulate cholesterol metabolism by inhibiting HMG-CoA reductase [389].

A previous study demonstrated the cardioprotective effect of *A. viridis* by altering C-reactive protein and glycoprotein levels [390].

A previous investigation showed the cardioprotective effects of *A. viridis* in ISO-induced oxidative damage. This could be due to an increase in endogenous antioxidants and inhibition of lipid peroxidation of the membrane [391].

16.3.10 Toxicity and Clinical Data

A recent investigation showed that *A. viridis* contamination due to contamination of water resources and food chain by persistent organic pollutants raised health risks. Therefore, in Kinshasa, proper measures must be taken to prevent this contamination by persistent organic pollutants to protect public health through direct and dietary exposure pathways [392].

A recent study showed an accumulation of Cd, Fe, Ni, and Zn in *Amaranthus viridis* collected from 11 sampling sites. Although the accumulated levels were lower, it still required routine monitoring [393].

A. viridis leaves contain unacceptable levels of toxic heavy metals and potentially dangerous pathogenic microorganisms, thus presenting significant health risks for consumers [394]. *A. viridis* releases outdoor air spora (pollen grains), which can cause airborne allergies [395].

16.3.11 Most Cited Research

iCite, Database, and NIH Open Citation Collection were used from June to October 2022 to get the details of the most cited research. Filters for research papers were applied using the keywords *Amaranthus graecizans.* The most cited research work, as per an NIH database search, includes alpha-pinene suppressing growth and inducing oxidative stress in roots. Alpha-pinene, a monoterpenoid released from several aromatic plants including forest trees, is known for its growth-inhibitory activity. It was found that alpha-pinene suppresses early root growth and induces oxidative damage in root tissue via enhanced production of ROS [396].

16.3.12 Patents

张朝凤李静任国庆 (Zhang Chaofeng, Li Jing, Ren Guoqing). CN114869996A. Application of *Amaranthus viridis* polypeptide in preparation of medicine for improving ulcerative colitis. 2022.

This invention is related to the application of *Amaranthus viridis* polypeptide or protein in the preparation of medicine or functional food for improving ulcerative colitis.

KR100266229B1. Antiviral proteins Amarandin 1 and Amarandin 2 derived from *Amaranthus viridis* and cDNAs encoding the same. 1997

This invention is related to antiviral proteins (Amarandin 1 and Amarandin 2) derived from *Amaranthus viridis* and cDNAs encoding the same proteins are provided, thereby isolating and sequencing the antiviral proteins and producing a large quantity of antiviral proteins.

16.3.13 Plant Pathology

Amaranthus viridis serves as a potential host for *Curtobacterium flaccumfaciens* pv. *flaccumfaciens* (Cff), the causal agent of bacterial wilt, is a prominent affliction affecting common beans [397]. Likewise, *A. viridis* serves as a weed host for *Meloidogyne arenaria,* while *Cercospora brachiate* has been reported to cause leaf spot disease on *A. viridi* [398].

17 Genus *Anagallis*

17.1 *Anagallis arvensis* L.

17.1.1 Synonyms

Anagallis latifolia

17.1.2 Family

Primulacae

17.1.3 Vernacular Name

Eyn a zara

17.1.4 Common Names

This plant is also known as the red chickweed, scarlet pimpernel, poor man's weather glass, red pimpernel, poor man's barometer, and shepherd's weather glass or shepherd's clock [399].

17.1.5 Botanical Description

Herbaceous annual with cylindrical, erect stems up to 30 cm high. Leaves: ovate, sessile, light green. Flowers: single, produced in the leaf axils, deep blue with five velvety petals and a vibrant red heart, adding a touch of contrast at the center. Fruits: rounded capsule forms on a slender stalk, eventually drooping as it ripens (Figures 17.1.1–17.1.3).

17.1.6 Geographical Distribution

Anagallis arvensis is widely available throughout Europe and worldwide due to its ruderal habitat. The plant is also widely distributed in Egypt, Palestine, non-tropical regions of South America, and Taiwan [400–402]. It is distributed in diverse habitats across Oman—from cultivated fields to mountain slopes up to 2,000 m.

FIGURE 17.1.1 Flowers and leaves of *Anagallis arvensis.*

FIGURE 17.1.2 Flower of *Anagallis arvensis.*

FIGURE 17.1.3 Fruit of *Anagallis arvensis.*

17.1.7 Chemical Composition

A. arvensis root actively contains cucurbitacin (a glycoside) and a triterpenoid series of saponins. The flowers of the plant have also been reported to contain spinasterol (glucoside), kaempferol, quercitin, sterol, ß-sitosterol, rutin, and stigmasterol. Additionally, chemical components such as deglucoanagallosides B, deglucoanagallosides, oleananetriterpenes, triterpene saponins, analagossides A, and anagallosides A, B, and C are present in the leaves [402– 404] (Figure 17.1.4).

17.1.8 Uses

This plant is utilized to treat liver ailments in Taiwan [405]. This herb is also used in veterinary practices in Italy. *Anagallis arvensis* is commonly used to treat conjunctivitis

DOI: 10.1201/9781003295488-17

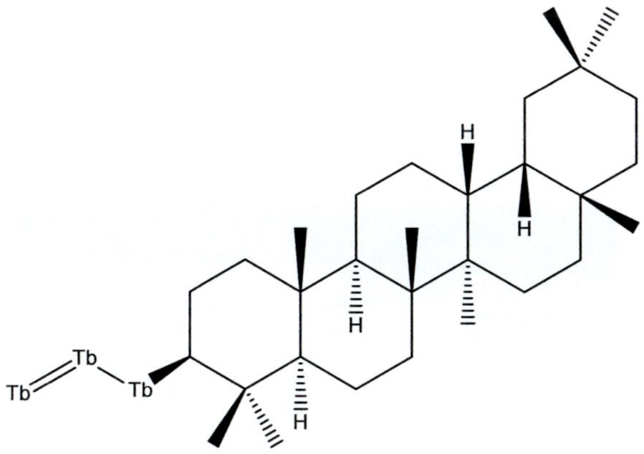

Oleananetriterbenes

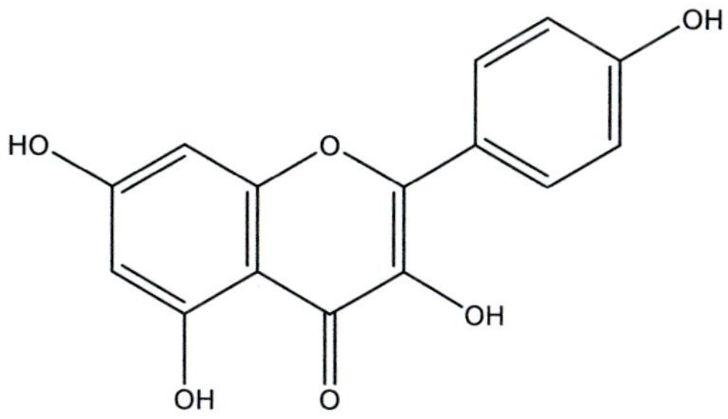

Kaempferol

FIGURE 17.1.4 Active chemical composition of *Anagallis arvensis*.

and skin complications [406]. The plant is also used for eliminating leeches from cattle's nostrils [402,407–409].

17.1.9 Mechanism of Action

Methanolic extract possesses antimicrobial and anti-inflammatory effects via inhibiting COX-1 and COX-2 [410]. Also, the methanolic extract of *A. arvensis* has been reported as potent uroprotective and hepatoprotective due to the presence of antioxidant phytochemicals in the rodent model [411].

Further, *A. arvensis* extracts were reported as a potential natural source of bioactive antioxidants, enzyme inhibitors, and anticancer agents [412]. Furthermore, the apoptotic potential of the alcoholic extract of the whole plant of *Anagallis arvensis* has been reported as a potential anticancer agent against leukemia cells [413].

Further, *A. arvensis* extract demonstrated restorative ability against CCl4-induced hepatotoxicity and could be used as a hepatocurative and hepatoprotective agent [414].

17.1.10 Toxicity and Clinical Data

At higher doses, consumption of this plant may cause toxicity [410]. The plant has also been reported for in vivo toxicity. *A. arvensis* has been reported to cause poisoning in sheep in Australia [415], and its toxicity has been confirmed by experiment [416–418].

17.1.11 Most Cited Research:

iCite, Database, NIH Open Citation Collection was used on January 1st, 2024, to get the details of the most cited research. Filters for research papers were applied using the keywords *Anagallis arvensis.* NIH database search revealed the cupric ion reducing antioxidant capacity and polyphenolic content of some herbal teas [419].

17.1.12 Plant Pathology

It was found that *Anagallis arvensis* was susceptible to both beet mild yellowing virus and an isolate of beet western yellows viruses [420]. It was also mentioned that it helps increase the resistance of chili plants to *Sclerotium rolfsii*, a fungal pathogen found in the soil that causes southern blight disease in chili and various other plants. The dried plant components were utilized alongside two rhizobacteria strains for this purpose [421].

18 Genus *Ammi*

18.1 *Ammi majus* L.

18.1.1 Synonyms

Ammi majus var. *genuinum, Apium ammi*, & *Carum majus*

18.1.2 Family

Apiaceae

18.1.3 Vernacular Name

In some regions of Oman, it is called naynia & sanoot.

18.1.4 Common Name

Bishop's weed, large bullwort, *Aatrilal*. In Arabic, it's called Hiirz al Shayateen, Hasheeshatul Arz, Khilla bariah, Rijlul Ghurab, Rijulul Tair; English: Greater Ammi, Bishop's weed, Lady lace [422]. *Aatrilal* is a Berberi word that means "Rijul ul Tair." *Aatrilal* is also called "Atrilaan." Its generic name is *Ami*.

18.1.5 Botanical Description

An annual herb grows up to 60 cm tall with vertical grooves. Leaves: arranged alternately on the stem. The lower leaves are deeply lobed or divided, and the upper leaves have sawtooth-like edges. Flowers: white, form a terminal inflorescence in an umbrella shape (Figures 18.1.1 and 18.1.2).

18.1.6 Geographical Distribution

Ammi majus is native to Egypt and extensively distributed in Europe, the Mediterranean, and West Asia. In Oman, it's widely distributed in Northern and Southern Oman.

FIGURE 18.1.1 Leaf portion of *Ammi majus*.

FIGURE 18.1.2 White lacy inflorescences of *Ammi majus*.

18.1.7 Chemical Composition

This medicinal plant mainly contains coumarins (xanthotoxin, bergapten, isopimpinellin, psoralen, imperatorin, marmesin, umbelliferone, pyrano coumarin, alloimperatorin, furanocoumarin, pyrone coumarin) and flavonoids (kaempferol, quercetin) (Figure 18.1.3). Additionally, it also contains terpenoids, proteins, essential oil constituents, etc. [423].

18.1.8 Uses

This plant has been used to treat various skin complications, such as vitiligo, for a long time. It is considered the first-line drug for the treatment of vitiligo. Traditionally, fruits obtained from this plant can be used to purify the blood and act as a diuretic as well as an emmenagogue. Extracts derived from this plant can be used to treat leprosy, urinary problems, and complications associated with digestion. Xanthotoxin, the main component of this plant, also known as methoxsalen, is utilized to increase the formation of melanin pigment in the skin when exposed to light. It's also used to treat idiopathic vitiligo and leucoderma. Additionally, this plant has been evidenced to have cytotoxic, anti-inflammatory, antibacterial, antiviral, analgesic, and other effects. Clinical investigations showed the therapeutic importance of this plant in various skin complications, such as psoriasis and vitiligo. Further, the plant has also been reported for its role in the treatment of kidney stones and urinary tract infections [424].

18.1.9 Mechanism of Action

Psoralen and its derivatives, mainly bergapten and xanthotoxin, have photosensitizing effects. These compounds induce

DOI: 10.1201/9781003295488-18

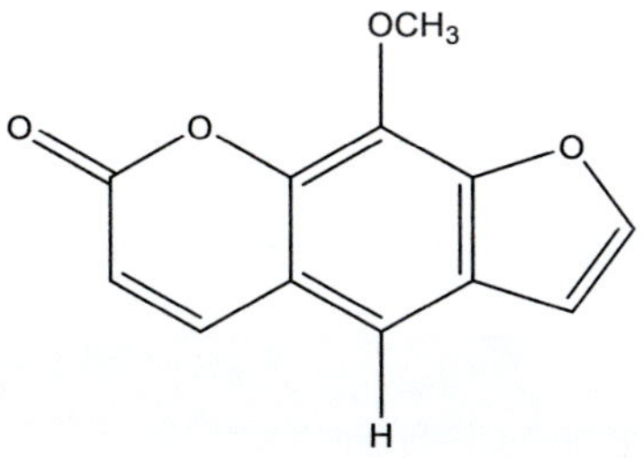

Xanthotoxin

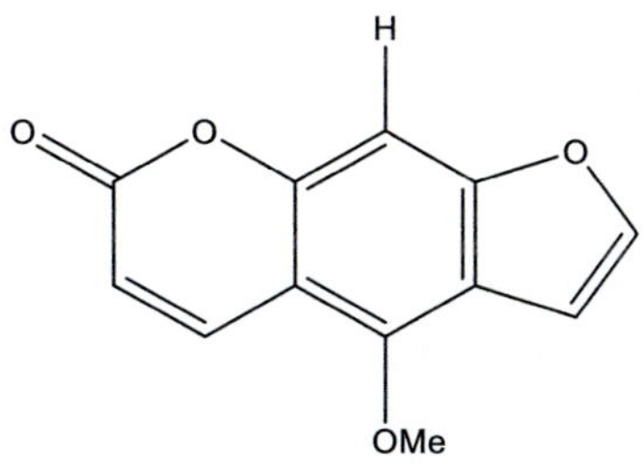

Bergapten

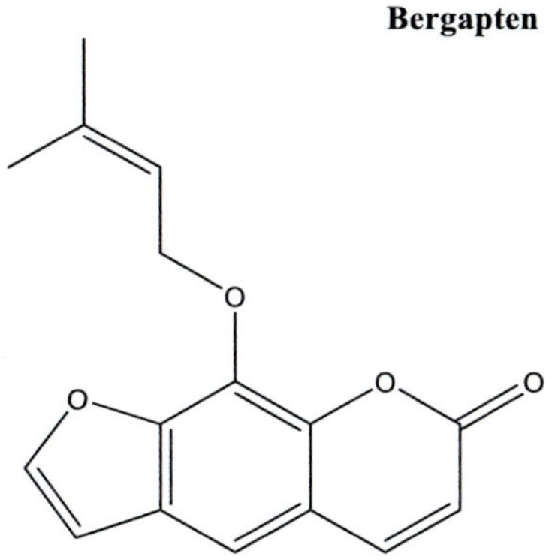

Imperatorin

FIGURE 18.1.3 Main active chemical components of *Ammi majus*.

repigmentation, reactivate the melanocytes, and release the inhibited tyrosine enzymes. This leads to a rise in the movement of active melanocytes from the normal epidermis.

18.1.10 Toxicity and Clinical Data

As per data reported by North Carolina State University, this plant has low-severity poison characteristics [425]. As the plant contains photosensitive agents, exposure to any part can cause photosensitive reactions and/or dermatitis.

18.1.11 Adulterants or Substituents

Due to its resemblance to many different crude drugs, *Aatrilal* has been a controversial crude drug. *Aatrilal* appears similar to *Anethum sowa*, *Apium graveolens*, *A. visnaga*, and *Peucedanum grande*. *Ammi majus* is sometimes confused with *A. visnaga*, as both plants have some common chemical components and pharmacological effects. However, *A. visnaga* is used in cardiac and pulmonary complications, and *Ammi majus* is used in skin complications [423].

18.1.12 Most Cited Research

iCite, Database, NIH Open Citation Collection was used from June to October 2022 to get the details of the most cited research. Filters for research papers were used by using the keywords *Ammi majus*. The most cited research work, as per an NIH database search, includes the effect of elicitors on the biosynthesis of psoralens in *Ammi majus* suspension cultures [426].

18.1.13 Patents

This invention includes pharmaceutical and cosmetic preparations containing extracts of *Ammi majus* or vasoactive agents such as visnadine, kelline, and other coumarins and flavonocoumarins [427].

18.1.14 Plant Pathology

The previous study reveals an expansive host range for *Spongospora subterranea*, with the families Malvaceae and Zygophyllaceae, along with the following species identified as novel hosts: *Malva nicaeensis*, *Chrysanthemum segetum*, *Verbesina encelioides*, and *Ammi majus* [428]. Powdery mildew caused by *Erysiphe heraclei* was reported on *Ammi majus* in Mexico [429]. Also, Bidens mottle virus and Apium virus Y were reported in Florida [430].

19 Genus *Anastatica*

19.1 *Anastatica hierochuntica* L.

19.1.1 Synonyms

Anastatica littoralis & Myagrum hierochunticum

19.1.2 Family

Brassicaceae

19.1.3 Vernacular Name

In Oman, it called makfaita & Kaf Maryam

19.1.4 Common Names

Rose of Jericho, in Arabic, is called Kaff Maryam (Mary's hand), True Rose of Jericho, or Genggam Fatimah

19.1.5 Botanical Description

An herbaceous annual with erected stems grows up to 15 cm, and the whole plant is covered with tiny star-shaped hairs. Leaves are obovate or oblong and narrow at both ends with smooth margins. Flowers: small white flowers that grow directly on the stem. The fruits curl inward when ripe, forming a tight, spherical structure that can protect and hold seeds for years. When hydrated, the structure curls upward and opens, releasing the seeds. (Figures 19.1.1–9.1.3).

19.1.6 Geographical Distribution

This desert plant is widely available in Saudi Arabia, Egypt, Jordan, Oman, Libya, Iraq, the United Arab Emirates, Iran, Kuwait, and North Africa. It is also consumed by the people of Asia and other parts of the world. In Oman found in northern part, thrives in northern gravel desert.

FIGURE 19.1.1 Natural plant habit of *Anastatica hierochuntica*.

FIGURE 19.1.2 Flowering plant of *Anastatica hierochuntica*.

FIGURE 19.1.3 Tightly curled, dry plant of *Anastatic hierochuntica*.

19.1.7 Chemical Composition

This herb contains considerable minerals, such as Mg, Ca, Mn, Fe, Cu, and Zn. Apart from its rich mineral composition, the plant also contains hierochins A, B, and C and Anastatin A and B 9 (Figure 19.1.4) [427,431].

19.1.8 Uses

Anastatic hierochuntica is used to treat metabolic and reproductive system-related disorders. Additionally, the plant is also reported as an analgesic. Furthermore, *Anastatica hierochuntica* has been reported for its possible role in the treatment of conjunctivitis, sterility, epilepsy, gastric complications, arthritis, bronchial asthma, mouth ulcers,

DOI: 10.1201/9781003295488-19

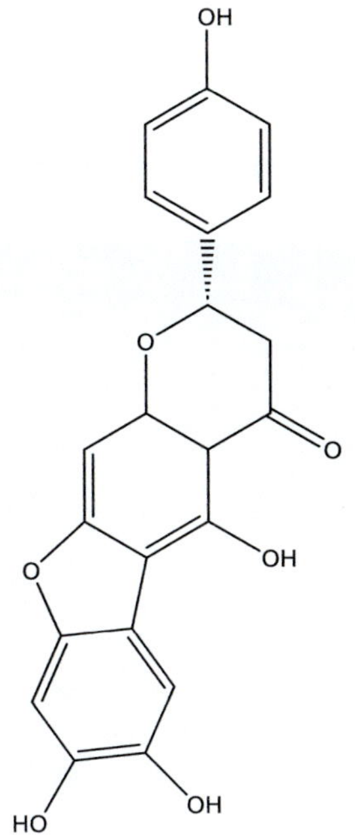

Anastantin A

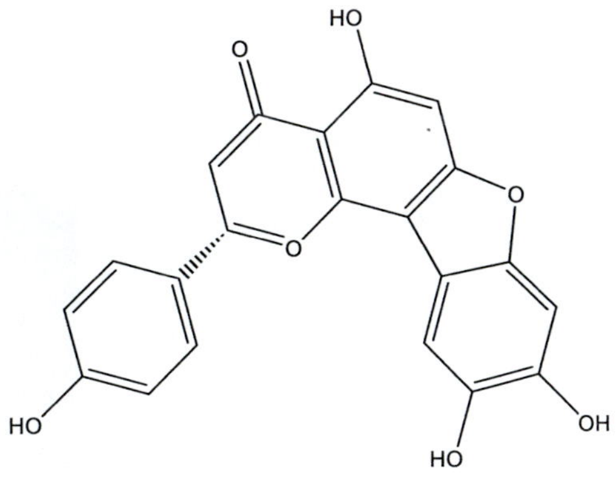

Anastantin B

FIGURE 19.1.4 Active chemical components of *Anastatic hierochuntica*.

malaria, and mental depression. Additionally, the plant has been reported for antimicrobial, anti-melanogenic, antioxidant, anti-inflammatory, antifungal, nitric oxide inhibitory effects, hypolipidemic, hypoglycemic, hepatoprotective, and gastroprotective effects [427,432].

19.1.9 Mechanism of Action

A recent report showed that aqueous and methanolic extracts of *A. hierochuntica* plant parts possess antiproliferative effects via induction of apoptosis in breast cancer MCF-7 cells [433].

19.1.10 Toxicity and Clinical Data

This herb is used as folk medicine to treat reproductive system-related problems and metabolic disorders. As plant preparation is used by pregnant women at the late stages of pregnancy to ease the labor process, there is always a concern related to its safety and toxicity profile. Results from this study showed that *A. hierochuntica* aqueous extracts have prenatal toxicity effects in Sprague-Dawley rats. Thus, this plant could be possibly unsafe for developing fetuses, particularly when consumed at the stage of organogenesis and implantation [434].

19.1.11 Most Cited Research

The methanolic extract from the whole plants of *Anastatica hierochuntica*, an Egyptian herbal medicine, was found to inhibit melanogenesis in theophylline-stimulated murine B16 melanoma 4A5 cells.

iCite, Database, NIH Open Citation Collection was used from June to October 2022 to get the details of the most cited research. Filters for research papers were applied using the keywords *Anastatica hierochuntica*. The most cited research work as per NIH database search includes *Anastatica hierochuntica* methanolic extract-based inhibition of melanogenesis in theophylline-stimulated murine B16 melanoma 4A5 cells [435].

19.1.12 Patents

심성보홍우진조진훈. (Sim Seong-bo Hong-woo Jin-jo Jin-hoon) KR20160068327A. Cosmetic composition comprising *Anastatica hierochuntica* extracts for protecting skin from stress due to harmful environment. 2014.

This invention claims a cosmetic composition having skin protection efficacy against environmental stress, containing an extract of *Anastatica hierochuntica*.

19.1.13 Plant Pathology

No diseases were reported in this herb, but two distinct fungal endophytic strains of *Micromonospora* spp., ESRAA1997 and ALAA2000, were isolated from the herbal plant *Anastatica hierochuntica* in Egypt [436].

20 Genus *Andrachne*

20.1 *Andrachne aspera* Spreng.

20.1.1 Synonyms

Andrachne gruvelii & *Andrachne aspera* var. *glandulosa*

20.1.2 Family

Phyllanthaceae

20.1.3 Vernacular Name

Arq al aswad & shagrat al aqrab

20.1.4 Common Name

Rumtotia

20.1.5 Botanical Description

A perennial herb with a woody base with solitary flowers, trigonous fruit, and orbicular leaves. The plant has hairs (trichomes) across its apical parts (Figures 20.1.1–20.1.3).

20.1.6 Geographical Distribution

This species is distributed from the Middle East to Pakistan, North Africa, southern Turkey to Sudan, and Ethiopia. In Oman, this species is found in Southern and Northern Oman, including Musandam.

20.1.7 Chemical Composition

Andrachne aspera contain andrachcinine, andrachcinidine, andrachamine, and andrachcine (Figure 20.1.4) [437].

FIGURE 20.1.1 Flower portion of *Andrachne aspera.*

FIGURE 20.1.2 Inflorescence of *Andrachne aspera.*

FIGURE 20.1.3 Leaf of *Andrachne aspera.*

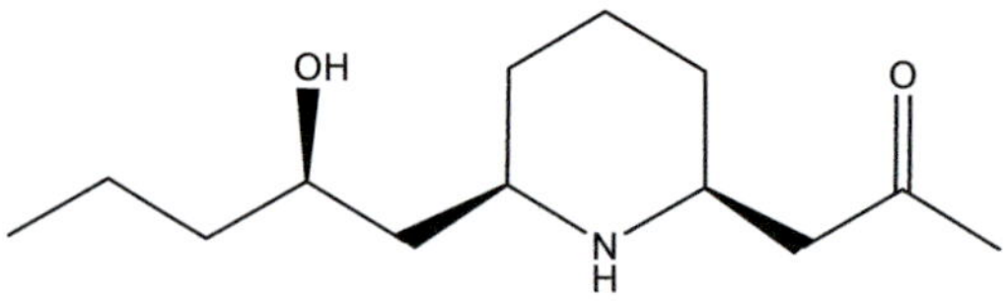

FIGURE 20.1.4 Chemical component of *Andrachne aspera* [438].

20.1.8 Uses

Fresh roots of the plant are collected and peeled off to remove the upper bark of the roots. Then, these roots are applied to the eyes as a needle (Surma Salai). This is done to remove the foreign body from the eyes as well as to relieve eye pain [437].

In Ethiopia, it's commonly known as Tekeze, which is used to treat unexplained stomach aches (megagna). The root of this plant is chewed to treat the stomach and nausea. Its consumption is not recommended during pregnancy. It's

DOI: 10.1201/9781003295488-20

used to treat gastritis, roundworm infections, liver complications, bloated stomach, malaria, and asthma [439]. The herb of this plant is used to treat cough and bronchitis [440].

20.1.9 Toxicity and Clinical Data

As it contains alkaloids, its toxicity profile needs to be assessed.

20.1.10 Most Cited Research

iCite, Database, NIH Open Citation Collection was used from June to October 2022 to get the details of the most cited research. Filters for research papers were applied using the keywords *Andrachne aspera*. The most cited research work as per the NIH database search includes the isolation of andrachcinine, andrachcinidine, andrachamine, and andrachcine from *Andrachne aspera* [438].

20.1.11 Patents

刘龙洋. CN103845492A. *Achyranthes aspera* plaster is used to treat ankle joint sprains. 2012.

- The invention claimed traditional Chinese medicine plasters containing *Achyranthes aspera* plaster for treating ankle joint sprains.

  김봉이박문여 KR101971800B1. Composition for preventing and treating cancer comprising *Achyranthes aspera* and *Melandryum firmum* rohrb. 2017.

- In this invention, the inventors claimed the anticancer properties of an *Achyranthes aspera* extract and a *Melandryum firmum* extract.

  WO2019227885A1. 徐寒梅戚微岩董瑶 (Xu Hanmei, Qi Weiyan, Dong Yao). Capsule preparation of *Achyranthes aspera* L. var. *rubrafusca* (wight) Hook polysaccharide, preparation method, and applications. 2018.

- This invention claimed a preparation method and application of a capsule from *Achyranthes aspera* polysaccharide.

  Vikas Gautam, Lipika Singhal, Satinder Kaur, Vivek Gupta, Avinash Singh, Pallab Rays P Gautam WO2018047194A1. Highly potent synergistic composition of jatyadi oil and *Achyranthes aspera* (apamarga) extract for therapeutic applications. 2017.

- In this invention, the inventors claimed the method and therapeutic applications of Jatyadi oil and *Achyranthes aspera* (Apamarga) extract.

20.1.12 Plant Pathology

The endophytic bacterium AL2–14B, isolated from *Achyranthes aspera* L. leaves, was identified as *Pseudomonas aeruginosa* through phenotypic and physiological characterization [441].

21 Genus *Argemone*

21.1 *Argemone mexicana* L.

21.1.1 Synonyms

Echtrus mexicanus, Argemone mexicana var. *typica*, & *Papaver mexicanum*

21.1.2 Family

Papaveraceae

21.1.3 Vernacular Name

Ghadrect & tashmezg

21.1.4 Common Name

Mexican poppy, Mexican prickly poppy, flowering thistle, cardo, or cardosanto

21.1.5 Botanical Description

An erect annual herb with spiny, branched stems reaching a height of about 70 cm. It is characterized by deeply lobed, spiny leaves and showy yellow flowers. Fruits are prickly, oblong, or egg-shaped (ovoid) capsules (Figures 21.1.1–21.1.3).

21.1.6 Geographical Distribution

This species is native to Central Mexico and Honduras but has been introduced and naturalized globally in many tropical and subtropical regions, including Oman. In Oman it found in Southern mountains and Musandam thrives as invasive weed.

FIGURE 21.1.1 Flower part of *Argemone mexicana*.

FIGURE 21.1.2 Adult plant of *Argemone mexicana*.

FIGURE 21.1.3 Spiny, unripe fruits of *Argemone mexicana*.

21.1.7 Chemical Composition

Argemone mexicana secretes a pale yellow color resinous material containing protopine and berberine (Figure 21.1.4). This plant contains a pale yellow color oil (non-edible), known as argemone oil. This oil chemically contains toxic alkaloids such as dihydrosanguinarine and sanguinarine (Figure 21.1.4). The whole plant contains isoquinoline alkaloids such as oxyberberine, jatrorrhizine, dehydrocorydalmine, and columbamine. Some other alkaloids have also been reported, such as coptisine, chelerythrine, argemexicaines A and B, cryptopine, and allocryptopine [442].

DOI: 10.1201/9781003295488-21

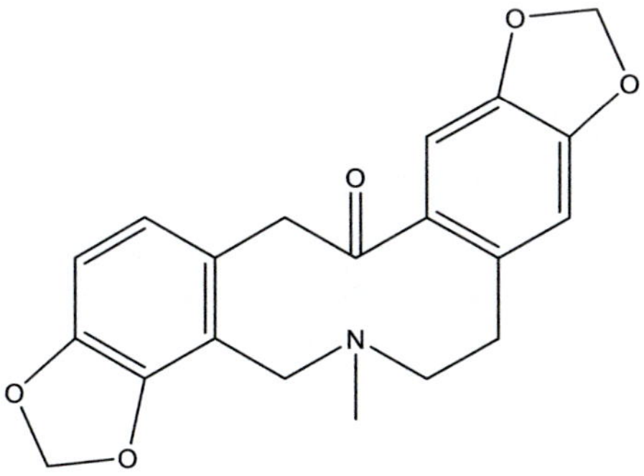

Protopine

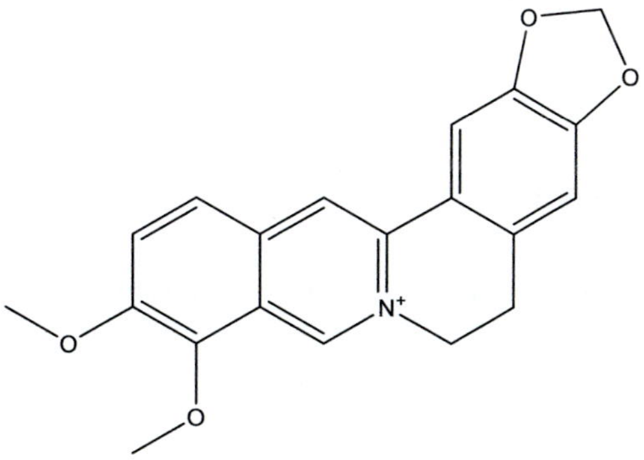

Berberine

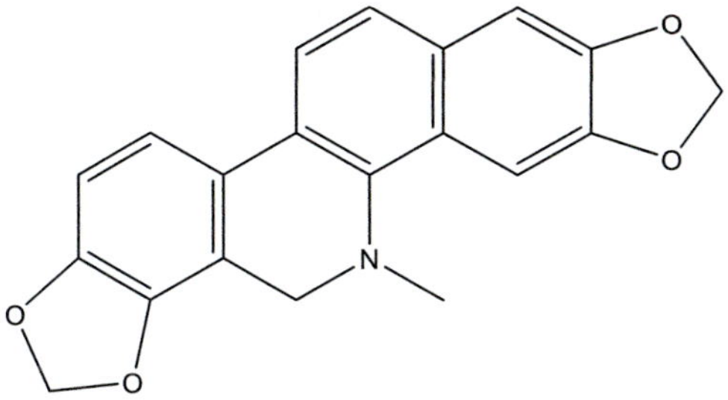

Dihydrosanguinarine

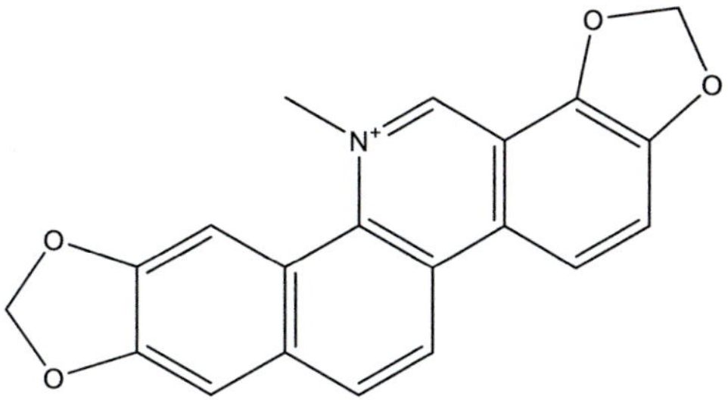

Sanguinarine

FIGURE 21.1.4 Chemical components of *Argemone mexicana*.

21.1.8 Uses

A. mexicana is considered the most effective traditional medicine for the treatment of uncomplicated falciparum malaria in Mali [443].

The plant has been reported for antidiabetic, anti-cancer, anti-HIV, CNS-related, wound healing, anti-microbial, antioxidant, anti-inflammatory, analgesic, antipyretic, hepatoprotective, anti fertility, antiallergic, nematicidal, allelopathic effect, anthelmintic, larvicidal, antifeedant action, etc. [444].

21.1.9 Mechanism of Action

Alkaloids isolated from *Argemone mexicana* have shown inhibition in SW480 colon cancer cells [445].

21.1.10 Toxicity and Clinical Data

Argemone mexicana poisoning causes epidemic dropsy, which is represented by the pathological accumulation of diluted lymph in body tissues and cavities [446,447].

Toxicity of *argemone* oil: Consumption of oil extracted from *argemone* seed can cause epidemic dropsy. Such complications have been observed in India as well as in Mauritius, Fiji Island, and South Africa. *Argemone* oil poisoning can cause vomiting, diarrhea, nausea, swelling of limbs, erythema, pitting edema, breathlessness, etc. In extreme cases, glaucoma and even death due to cardiac arrest have been encountered. The toxicity of *argemone* oil has been attributed to sanguinarine and dihydrosanguinarine [448]. Toxicity of *Argemone* seeds: Toxicity of *Argemone* seeds is also evidenced in rats [449,450].

21.1.11 Adulterants or Substituents

Its seeds are blackish brown, round, and netted. These seeds apparently have a close resemblance with mustard seeds. Due to this reason, mustard seeds are often adulterated with *Argemone mexicana* seeds either accidentally or intentionally [451].

The oil of *Argemone mexicana* seeds is pale yellow in color, is almost tasteless, and contains toxic alkaloids, namely sanguinarine and dihydrosanguinarine. *Argemone* seed oil is sometimes mixed with sunflower oil and sesame oil to increase the bulk quantity.

21.1.12 Most Cited Research

iCite, Database, NIH Open Citation Collection was used from June to October 2022 to get the details of the most cited research. Filters for research papers were used by using the keywords *Argemone mexicana*. The most cited research work, as per an NIH database search, includes a systematic chapter based on the occurrence of berberine in various plants and its role in the management of chronic cardiometabolic disorders [452].

21.1.13 Patents

Sudershan K Arora et al. WO2003057133A3. Herbal composition for treating various disorders, including psoriasis, a process for preparation thereof, and method for treatment of such disorders. 2003

- The invention revealed a method of preparation of herbal composition containing the extracts of *Argemone mexicana* and *Cuminum cyminum* to treat psoriasis and related immunological disorders.

  BR102018003523A2. Maria Gabriella Silva Araujo, Regina Célia Sales Santos Veríssimo, João Xavier De Araújo, Círia Vieira Barbosa, Edeildo Ferreira Da Silva, Thiago Mendonça De Aquino, 2018. Pharmaceutical composition for wound healing treatment containing isolated substances of leaves of the *Argemone mexicana* Linn species.
- The present invention revealed the pharmaceutical application of compositions containing berberine alkaloids from *A. mexicana*.

21.1.14 Plant Pathology

Several insect pests have been reported on this plant, but no diseases were reported. However, the endophytic fungus *Aspergillus flavus* was isolated from *Argemone mexicana* [453].

22 Genus *Anthemis*

22.1 *Anthemis odontostephana* Boiss.

22.1.1 Synonyms

Anthemis gayana, Anthemis gayana var. *hispidula*, & *Anthemis gayana* var. *multiramosa*

22.1.2 Family

Asteraceae

22.1.3 Vernacular Name

Khazmayah and ‘eyqwan

22.1.4 Botanical Description

An aromatic, small annual, grayish hairy herb with weakly spreading stems up to 15 cm. Leaves: pale green, feathery, ovate in outline, deeply segmented. Flowerhead: solitary with four to six white ray florets and a yellow disc. Fruit: achenes (Figures 22.1.1 and 22.1.2).

22.1.5 Geographical Distribution

The native range of this species is Syria to Central Asia and Pakistan, SE Arabian Peninsula, Afghanistan, Iran, Iraq, Lebanon-Syria, Pakistan, Tadzhikistan, Turkmenistan, and Uzbekistan. In Oman, this plant is found at higher elevations in the mountains of Musandam and in the Eastern Hajar Mountains.

FIGURE 22.1.1 Flower part of *Anthemis odontostephana*.

FIGURE 22.1.2 Flowering shoot of *Anthemis odontostephana*.

22.1.6 Chemical Composition

A previous study showed the presence of antheminone A, flavonoids, pectolinaringenin, and salvigenin in the flowers of *Anthemis odontostephana* (Figure 22.1.3) [454].

The main components of *Anthemis* species include natural flavonoids and essential oils. The essential oil from the flower of *A. odontostephana* contains spathulenol, hexadecanoic acid, germacrene D, 1,8-cineole, 6-methyl-5-hepten-2-one, caryophyllene oxide, β-caryophyllene, and camphor. Other than these components, the plant also contains monoterpene hydrocarbons, eleven oxygenated monoterpenes, eleven sesquiterpene hydrocarbons, nine oxygenated sesquiterpenes, and phenylpropanoids [455,456].

22.1.7 Uses

In Musandam, tea made from the flowers was used to treat gastrointestinal disorders and diabetes, and as a laxative.

In folk medicine, it has been used as an antispasmodic, anti-migraine, and hypolipidemic agent. *Anthemis odontostephana* has been recently known as an antioxidant, anti-inflammatory, and antimicrobial agent. A recent study showed hematoprotective and nephroprotective properties of *Anthemis odontostephana* on diabetic mice [457].

22.1.8 Mechanism of Action

A. odontostephana ethanol extract can regulate the biochemical parameters and inhibit hepatic damage in CCl_4-induced hepatotoxicity in mice [458].

DOI: 10.1201/9781003295488-22

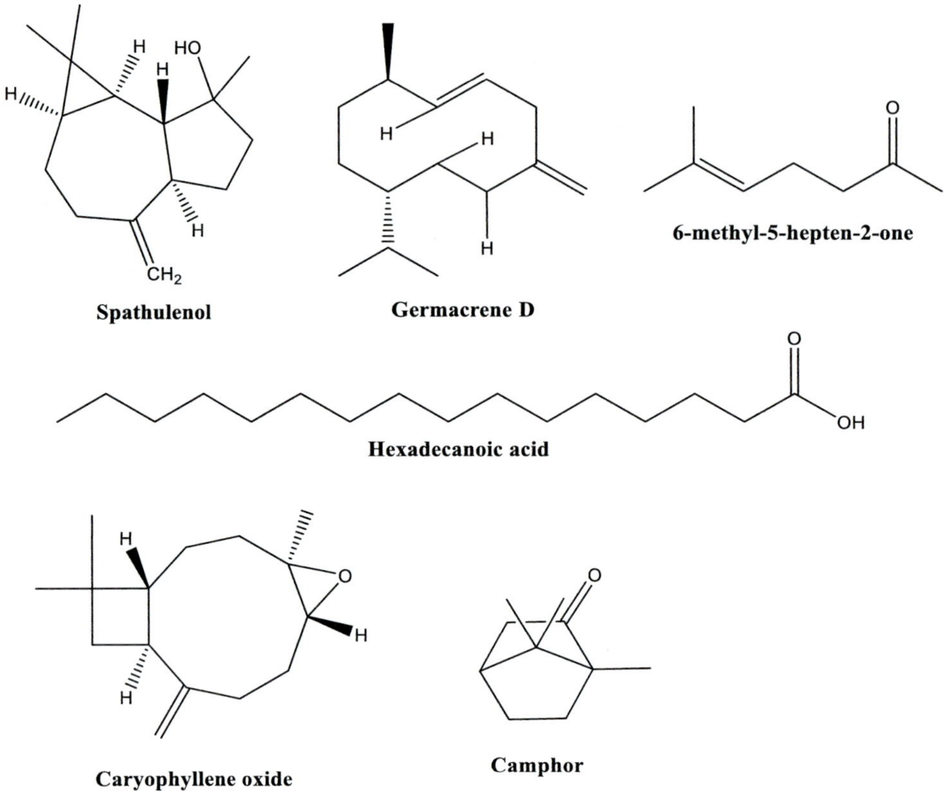

FIGURE 22.1.3 Chemical components of *Anthemis odontostephana*.

22.1.9 Toxicity and Clinical Data

No toxicity has been reported yet. However, as it contains sesquiterpenes, it is important to study its toxicity profile.

22.1.10 Most Cited Research

Based on the relevance, a search report generated from Google using *Anthemis odontostephana* as the keyword found that the most cited work includes the exploration of the chemical composition of the essential oil of the flowers of *Anthemis odontostephana* by GC-MS [455].

22.1.11 Plant Pathology

Apparently, there is no data in the literature that refers to the pathology of *Anthemis odontostephana*. However, the moth *Cnephasia longana*, which exhibits a broad host range, has been noted to feed on the flowers of *Hypochoeris*, *Anthemis*, *Aster trifolium*, *Chrysanthemum*, and *Ranunculus* [459].

23 Genus *Anticharis*

23.1 *Anticharis glandulosa* Asch.

23.1.1 Synonyms

Anticharis brevipes, Anticharis glandulosa var. *coerulea*, & *Distemon glandulosus*

23.1.2 Family

Scrophulariaceae

23.1.3 Vernucular Name

Shajrat al mutawa & remailiah

FIGURE 23.1.1 Flower of *Anticharis glandulosa*.

23.1.4 Botanical Description

It is an annual herb reaching a height of up to 30 cm and has a branched stem with hairy glands and ovate to lanceolate, sessile leaves, covered in glandular trichomes. The plant bears pink-colored, trumpet-shaped flowers. The fruits of this species are ovoid capsule (Figures 23.1.1 and 23.1.2).

23.1.5 Geographical Distribution

The native range of this species stretching from the Sahara through North and Northeast Africa, across the Arabian Peninsula and into Northwest India. Locally, it is distributed all over Oman on gravel plains and the Wadis system.

FIGURE 23.1.2 Leafy branch with glandular trichomes and flowers of *Anticharis glandulosa*.

DOI: 10.1201/9781003295488-23

24 Genus *Anogeissus* (*Terminalia*)

24.1 *Terminalia dhofarica* (A.J. Scott) Gere & Boatwr.

24.1.1 Synonyms

Anogeissus dhofarica

24.1.2 Family

Combretaceae

24.1.3 Vernacular Name

Sgot & miset

24.1.4 Botanical Description

Large tree reaching up to 10 m with broad, rounded crowns and branches extending outwards. The leaves of this tree turn to some extent a bluish-green color and shed at the beginning of November/December, and then regrow before the monsoon season. The tree bears yellowish flowers and cone-like fruits (Figures 24.1.1–24.1.3).

24.1.5 Geographical Distribution

The genus *Anogeissus* or *Terminalia* (axle wood tree, ghatti tree, button tree, and chewing stick tree) belongs to the Combretaceae and includes eight species that are distributed in Asia and Africa. *T. dhofarica* is dominant on the slopes of the Dhofar escarpment in southern Oman, forming vast, monospecific stands. The presence of this dominant tree extends beyond Oman's borders, with reports of its prevalence in the Hawf region of the Mahra Governorate in southeast Yemen.

24.1.6 Chemical Composition

Plants belonging to this genus contain flavonoid, tannin, phenolic acids, lignan, and triterpene

24.1.7 Uses

Extracts derived from this genus can be utilized to treat skin complications. Plant preparations belonging to this genus can be utilized as traditional medicine to treat several complications such as diabetes, fever, diarrhea, dysentery, tuberculosis, wound healing, skin diseases (eczema, psoriasis), and snake and scorpion venom. *T. dhofarica* showed antimicrobial activity against *Pseudomonas aeruginosa*, *Staphylococcus aureus*, and *Candida albicans* [151].

The aqueous fraction derived from this plant has been reported to clean and disinfect wounds. Leaves derived from this genus are traditionally used for skin injuries by flamed leaves. A paste formulated from the leaves of the plant is also used to treat infected wounds [328].

FIGURE 24.1.1 Tree habit of *Terminalia dhofarica*.

DOI: 10.1201/9781003295488-24

FIGURE 24.1.2 Inflorescence of *Terminalia dhofarica*.

FIGURE 24.1.3 Head-like fruit with numerous seeds at the end of the stem of *Terminalia dhofarica*.

In Oman, this plant is used to treat wounds and used as an antiseptic [151]. Additionally, it acts as a galactagogue, and wood is used in construction and other household work. Furthermore, leaves have been reported for their utilization in dyeing clothes. Also, in Oman, leaves are used as an antiseptic, cleansing agent, and to maintain personal hygiene [328].

A paste from the leaves can be used to treat sores and act as an antiseptic [151]. It was found that the TPC content in the aqueous alcoholic extract of tubers was 406 GAE mg/g dry weight [151].

The antimicrobial effects of this plant could be associated with phenolics present in the plant, which indicate potent antioxidant effects. Additionally, weak cytotoxic effects were also evidenced [151,460,461].

24.1.8 Mechanism of Action

A previous study showed that the alcoholic extract obtained from *T. dhofarica* tubers showed better results than the chloroform extract, with a high percentage inhibition in the DPPH radical scavenging assay [151].

24.1.9 Most Cited Research

iCite, Database, NIH Open Citation Collection was used from June to October 2022 to get the details of the most cited research. Filters for research papers were applied using the keywords *Anogeissus dhofarica*. The NIH database search showed the exploration of two new species from the Arabian Peninsula and Mauritania [462].

24.1.10 Patents

EP2537513A3: Use of *Anogeissus* extract for fibrillin production in the skin. Susan M. Daly, Lieve Declercq.

- This invention reveals the utilization of extract from the *Anogeissus* species to treat human skin by producing fibrillin and thus improving the elasticity of the skin.

  JP5856150B2. A composition comprising DNA repair enzyme and *Anogeous* extract. 2011.
- This invention claimed the topical application of *Anogeissus* species and their effects on DNA repair enzymes.

24.1.11 Plant Pathology

In the literature, no diseases have been reported in this wild plant. However, *Microsalliota ventricocystidiata*, a saprobic fungal mushroom, has been reported on dead leaves and wood found under *A. dhofarica* [463]

25 Genus *Apium*

25.1 *Apium graveolens* L.

25.1.1 Synonyms

Carum graveolens, Celeri graveolens, & Helosciadium graveolens

25.1.2 Family

Apiaceae

25.1.3 Vernacular Name

Muratab

25.1.4 Common Name

Ajamoda and also known as celery.

25.1.5 Botanical Description

An aromatic annual or binomial herbaceous plant that grows up to 120 cm in height. The whole plant possesses an aromatic flavor, and its leaf blades and petioles are the main edible organs. Leaves: succulent, pinnate, or trifoliate. Flowers: small, white, or greenish-white, developed on sessile compound umbels. The fruits/seeds are tiny and in an achene shape.

25.1.6 Geographical Distribution

Celery is a popular vegetable crop and is widely cultivated in Europe, East Asia, southeastern Oceania, and southern Africa. This plant is reported in northern and southern Oman [464].

25.1.7 Chemical Composition

Celery contains flavonoids, mainly comprising apigenin, kaempferol, quercetin, and luteolin (Figure 25.1.1). The apigenin content in celery is higher than that in other plants [465,466]. Apigenin and luteolin possess various therapeutic effects, such as anti-bacterial, anti-oxidation, and cardiovascular protective effects.

Seeds derived from this plant contain 2%–3% essential oil and 17%–18% fatty, pale-yellow colored oil. D-limonene (60%) and ß-selinene (10%–12%) are the main components present in the oil. Other components, such as sedanoic acid anhydride (0.5%) and sedanolide (2.5%–3.0%), are also present in the oil. The flavor of this oil is due to the presence of d-selenene, sedanolide, and sedanoic acid anhydride.

25.1.8 Uses

Traditionally, in the Arab world, celery is used as traditional medicine under the name "Karafs." As per previous records, it was also known as Udasaliyon.

As per Imam Ibn al-Qayyim, fresh celery leaves help in treating stomach, liver, renal, and menstrual problems and also act as a diuretic [468].

Traditionally, Celery is also known for its effectiveness in stimulating semen production and treating bad breath.

Seeds of this plant are used as a spice for flavor, stimulant, and carminative. Additionally, traditionally, it is also used as a nerve tonic. Furthermore, it is also used as a treatment for rheumatism. Oils obtained from the seed can be used as a flavoring agent. Fatty oil isolated from the fruit is used as a nerve stimulant and antispasmodic.

25.1.9 Mechanism of Action

The anti-inflammatory effects of this plant are due to the inhibitory effects of its component, apiin, against inducible nitric oxide synthase and nitride oxide synthesis [469].

Apiuman reduced interleukin-1β, increased interleukin-10 synthesis, and decreased neutrophil migration [470].

25.1.10 Toxicity and Clinical Data

This plant is considered safe for common use. However, it is also considered a source of allergy in Europe.

Celery components, including proteins, apigenin, and anthocyanin, have been reported to have allergic effects [471–473]. Some of the most common allergens reported from this plant are as follows:

- PR-10 (Api g 1)
- nonspecific lipid transfer protein—LTP 1 (Api g 2)
- profilin (Api g 4)
- flavoprotein (Api g 5)

The anaphylactic reaction caused by Api g 2 as well as Api g 4 has been reported [474]. This plant has been reported to cause fungal infections from *Sclerotinia sclerotiorum*, which triggers dermatitis in sensitive people. As the plant is known for its uterine stimulatory effects, it should be avoided during pregnancy [475].

25.1.11 Adulterants or Substituents

Oryza sativaa is used as an adulterant for this species.

DOI: 10.1201/9781003295488-25

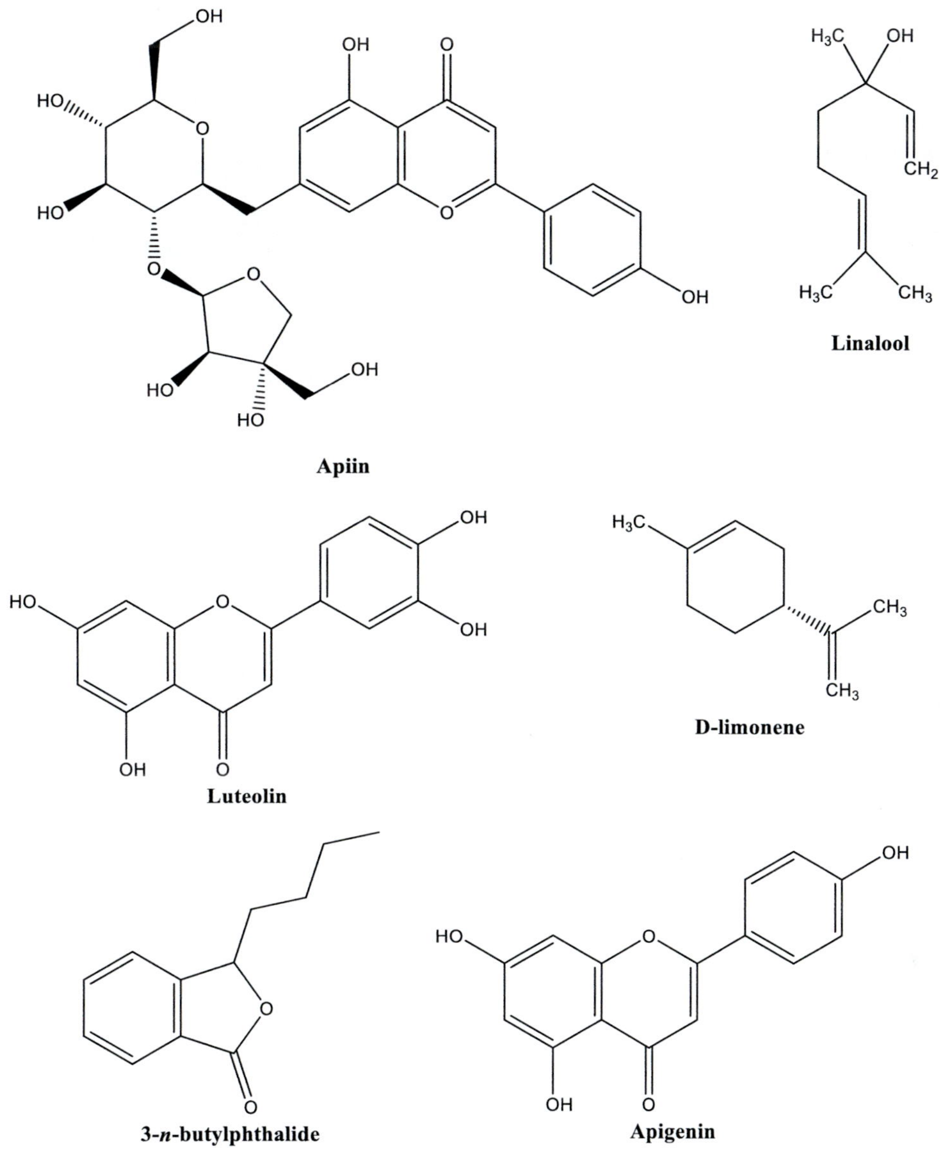

FIGURE 25.1.1 Chemical components of *Apium graveolens* [467].

25.1.12 Most Cited Research

iCite, Database, NIH Open Citation Collection was used from June to October 2022 to obtain details of the most cited research. Filters for research papers were applied using the keyword "Apium graveolens." A search in the NIH database showed that celery caused a hypersensitivity reaction.

There are several food products obtained from plants that cause immediate hypersensitivity (typeI allergy) reactions in individuals. Celery also causes a syndrome known as birch-celery syndrome, mainly due to the inhalation of tree pollen [476].

25.1.13 Patents

Hollow Celery Stick. Lawrence K. Pierce Darrell L. DUDA, US20090169687A1, 2009.

- This invention claims that the edible hollow petiole from celery plants could be used as a straw for consuming beverages or as a food product to stuff food inside the hollow celery tube.

25.1.14 Plant Pathology

Since celery is a widely cultivated crop all over the world, several pathogens attack and damage this crop, which has been widely reported. This includes bacterial, fungal, and viral diseases. Several significant plant diseases to be aware of include bacterial blight, soft rot, celery mosaic virus, and early blight, also known as Cercospora blight [477].

26 Genus *Ageratum*

26.1 *Ageratum conyzoides* L.

26.1.1 Synonyms

Carelia conyzoides & *Eupatorium conyzoides*

26.1.2 Family

Asteraceae

26.1.3 Common Name

Mentrastom, Billy goat weed, in Arabic, is called by different names such as Bergoman, Berquam, and Korink.

26.1.4 Botanical Description

It is an annual branching herb with shallow and fibrous roots, purplish to white flowers. The stems and leaves are covered with fine white hairs. Leaves are opposite and broadly ovate, and crumpled leaves produce a characteristic odor like that of male goats. Thus, it's called "goat weed" or "billy goat weed." The flower head is enclosed by two to three rows of green bracts with reddish-violet tops.

26.1.5 Geographical Distribution

This is a medicinal and aromatic tropical plant that is very common in West Africa and some parts of Asia and South America. In Oman, it is found in Northern Oman among cultivations and in Southern Oman in disturbed areas and areas affected by overgrazing, particularly areas affected by the monsoon.

26.1.6 Chemical Composition

This plant contains terpenoid, flavonoid, sterol, pyrrolon, chromene, coumarin, pyrrolizidine alkaloid, and lignin (Figure 26.1.1–26.1.4) [478]. A vast amount of literature is available on the phytochemistry of the essential oil obtained from *A. conyzoides*. The essential oil derived from this plant mainly contains precocene II, precocene I, and β-caryophyllene (Figure 26.1.1) [479].

Additionally, it also contains kaempferol, rhamnoside, quercetin, scutellarein, chromene, stigma-7-en-3-ol, sitosterol, stigmasterol, fumaric acid, caffeic acid, saponin, pyrrolidine alkaloid, ageratochromene derivatives, and alkane [480–482].

26.1.7 Uses

It's used as an ornamental plant for ceremonial uses, traditional/folklore, vegetables, and pesticides. Extracts and essential oil derived from this plant can be utilized to control *Dysdercus flavidus* and *Sitophilus zeamais*. *A. conyzoides* has been reported for its various pharmacological uses, such as analgesic, anti-inflammatory, antimicrobial, antioxidant, antimalarial, antidiarrheal, and antidiabetic effects [483].

26.1.8 Mechanism of Action

The anti-inflammatory effects of this plant have been reported. It was found that the anti-inflammatory activity of *A. conyzoides* could be due to the inhibition of p65 NF-κB and MAPK activation by its components [484].

26.1.9 Toxicity and Clinical Data

As the plant is known for nematode infections and raw consumption as a vegetable, proper cleaning has to be done before its consumption, as parasitic nematodes often interact with fungal, bacterial, and viral pathogens to cause disease. The consumption of *A. conyzoides* has been a bit

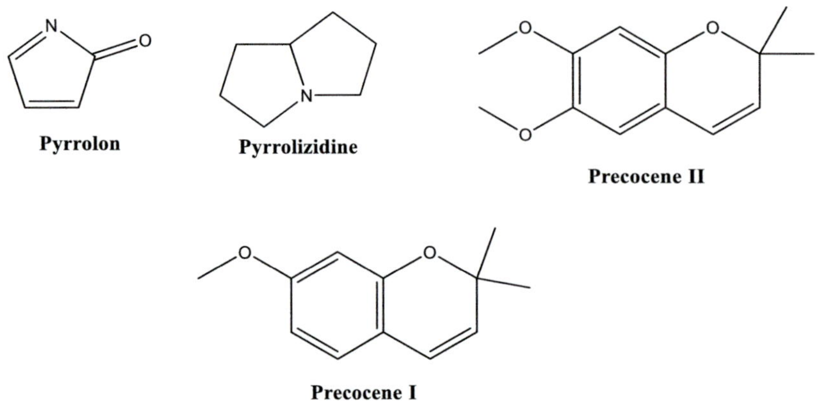

FIGURE 26.1.1 Chemical components of *Ageratum conyzoides*.

DOI: 10.1201/9781003295488-26

FIGURE 26.1.2 Flowers of *Ageratum conyzoides*.

FIGURE 26.1.3 Leaves of *Ageratum conyzoides*.

unsafe because of the presence of hepatotoxic components, mainly pyrrolizidine alkaloids, in this plant [485].

Further, it was found that the hydroalcoholic extract of *A. conyzoides* L. could be harmful to the hepatic, renal, and blood cells after using it for a long period [486]. This toxic effect could be due to the presence of pyrrolizidine alkaloids.

26.1.10 Adulterants or Substituents

A. conyzoides is considered an undesirable weed and contaminant in several harvested crops.

26.1.11 Most Cited Research

iCite, Database, NIH Open Citation Collection was used from June to October 2022 to get the details of the most cited research. Filters for research papers were applied using the keywords *Ageratum conyzoides*. An NIH database search showed that celery caused a hypersensitivity reaction. This work showed that the nanovirus-like DNA 1 component related to infected *A. conyzoides* shows no vital role in the disease and denotes a satellite-like DNA [487].

26.1.12 Plant Pathology

Several plant pathogens have been reported as causing damage to *A. conyzoides*. Since this plant itself is considered a common weed in agricultural crops, such pathogens play a magnified role in the cultivation of these crops. Among these nematodes and viruses, the begomovirus disease complex and *Ageratum yellow vein virus* are the most common examples. *A. conyzoides* is a host and is infected easily by the nematode, due to which *A. conyzoides* is always cultivated away from plants that are sensitive to nematode attack. Additionally, it's a host of pathogens and nematodes that affect crop species. This weed was also reported for an infection caused by the *Ageratum yellow vein virus*, representing prominent yellow vein symptoms. Additionally, it's also a host of the *Tomato Yellow Leaf Curl Tanzania Virus* [487–490]

FIGURE 26.1.4 Habit of *Ageratum conyzoides*.

26.1.13 Patents

章家恩方丽赵本良罗明珠全国明 (Zhang Jiaen, Fang Li, Zhao Benliang, Luo Mingzhu, National). CN103004891A. Application of alien invasive plant *Ageratum conyzoides L.* in control of golden apple snail. 2012.

- This invention reveals the application of *Ageratum conyzoides* L. dried powders to control golden apple snails.

CN102029284A. 王长方陈峰游泳王俊 (Wang Changfang, Chen Fengyun, Wang Jun). Method for remedying organochlorine pesticide-induced soil pollution by using *Ageratum conyzoides*. 2010.

- This invention reveals a reliable approach to relieving organochlorine pesticide-induced soil pollution by using *Ageratum conyzoides* L.

27 Genus *Arnebia*

27.1 *Arnebia decumbens* (Vent.) Coss. & Kralik

27.1.1 Synonyms

Arnebia decumbens var. *microcalyx, Echioides decumbens*, & *Lithospermum decumbens*

27.1.2 Family

Boraginaceae

27.1.3 Vernacular Name

Kehel, Hawareh

27.1.4 Common Name

Zicao in China, homaire in southeast Algeria

27.1.5 Botanical Description

Annual herb with erected stems grows up to 20 cm, covered with bristles and stiff hairs. Leaves: sessile, linear to lanceolate. Flowers: yellow.

27.1.6 Geographical Distribution

Algeria, Egypt, Libya, Pakistan, Afghanistan, Saudi Arabia, Morocco, Tunisia, Asia, Lebanon, Syria, Turkey, Europe, and Belgium. In Oman, this plant is only found in Musandam [491].

27.1.7 Chemical Composition

A. decumbens extract contains flavonoids [492]. A recent study showed the isolation and identification of shikonin from *Arnebia decumbens* L. and its antibacterial effects. It was found that shikonin possessed high antibacterial effects against *Pseudomonas aeruginosa*, *Escherichia coli*, *Staphylococcus aureus*, and *Klebsiella pneumoniae* [493].

Another report showed the presence of spiroethers from *Arnebia* species that are accountable for antibacterial and antitumor effects. It was found that cis-(E)- and trans-(Z)-spiroethers from different *Arnebia* species showed variations in cytotoxic effects (Figure 27.1.1) [494]. Plants belonging to the *Arnebia* genus contain shikonin derivatives containing a moiety as presented in Figure 27.1.1 [495].

Closely related species (*Arnebia benthamii*) has been reported for the presence of toxic pyrrolizidine alkaloids, which restricts the overutilization of the ethnomedicinal use of *Arnebia benthamii* [497].

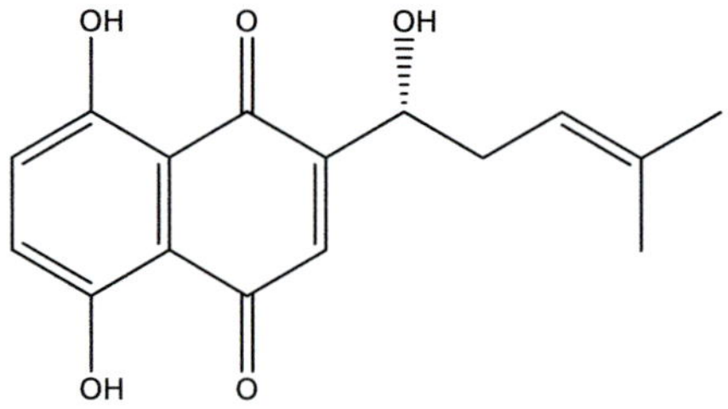

FIGURE 27.1.1 Chemical components of a plant belonging to the *Arnebia* genus [496].

Pyrrolizidine alkaloids were also found in the roots of *Arnebia euchroma* [498]. Another report showed the presence of naphthoquinones from *Arnebia densiflora*, demonstrating a cytotoxic effect via induction of apoptosis [499]. This plant contains alkannins/shikonins such as deoxyalkannin, deoxyshikonin, or arnebin-7 [500].

27.1.8 Uses

Arnebiae Radix (*A. euchroma and A. guttata*) is a traditional Chinese medicine, as mentioned in the Chinese Pharmacopoeia. This plant is usually used in traditional Chinese medicine for the treatment of inflammatory diseases [501].

The previous report suggested the antifungal effect of an extract derived from *Arnebia decumbens.* Additionally, the same extract did not cause hemolysis [502]. It was found that *A. decumbens* possesses antimicrobial and antioxidant effects [492].

27.1.9 Toxicity and Clinical Data

As its related species contain toxic pyrrolizidine alkaloids and naphthoquinones, toxicity needs to be assessed before its utilization [503].

No toxicity studies have been performed on this plant yet.

27.1.10 Most Cited Research

iCite, Database, NIH Open Citation Collection was used from June to October 2022 to get the details of the most cited research. Filters for research papers were applied using the keywords *Arnebia decumbens.* NIH database search showed antifungal activity by the herbal extract derived from *Arnebia decumbens* and other plants [502].

DOI: 10.1201/9781003295488-27

27.1.11 Patents

This invention reveals the anticancer properties of herbal extracts containing *Arnebia decumbens* extract in combination with doxorubicin and cisplatin.

- Soroush Sardari, Ghazaleh Ghavami. Method for selection of novel anticancer herbs using chemin-formatic tools. US20120009277A1. 2010

27.1.12 Plant Pathology

So far, no attention has been paid to the pathogens that cause infection in these wild plants. However, a recent report showed the use of *A. chroococcum* for the yield improvement of alkannin and shikonin derivatives in hairy roots of the genus *Arnebia* [504]. Furthermore, compounds derived from the extract of this plant show promise in fighting certain bacterial and fungal plant diseases, in addition to their medicinal properties.

27.2 *Arnebia hispidissima* (Lehm.) DC.

27.2.1 Synonyms

Dioclea hispidissima, Echiochilon hispidissimum, and *Lithospermum hispidissimum*

27.2.2 Family

Boraginaceae

27.2.3 Vernacular Name

Kehel

27.2.4 Common Name

Arabian primrose

27.2.5 Botanical Description

This herb is 2–20 cm tall, hispid with tubercle-based bristles and short hairs, with linear-lanceolate leaves and yellow subsessile flowers in spike-like cymes, heterostylous; bracts leaf-like (Figures 27.2.1–27.2.3).

27.2.6 Geographical Distribution

Arnebia hispidissima is a hispid herb that is present in northern Africa, Egypt, and northern India. In Oman it is found all over Oman [505].

27.2.7 Chemical Composition

Alkannin is a red color dye derived from the root of *A. hispidissima*. This chemical component, called alkannin, and its optical isomer shikonin, both produce a red color [506,507].

FIGURE 27.2.1 Flowers of *Arnebia hispidissima*.

FIGURE 27.2.2 Shoot with flowers of *Arnebia hispidissima*.

FIGURE 27.2.3 Plant habit in flower of *Arnebia hispidissima*.

Naphthoquinones such as 2-methyl-hexa-ienoylshikonin, shikonin, butylshikonin, and arnebifuranone were isolated from *Arnebia hispidissima* roots [508]. Naphthoquinone isolated from this plant possesses antibiotic and anticancer effects [509].

A previous study also showed the presence of flavonoid, vitexin, and naphthoquinones such as arnebin-1, arnebin-7, tiglic acid, alkannin, arnebinol, and cycloarnebin-7 (Figure 27.2.4) [509,510].

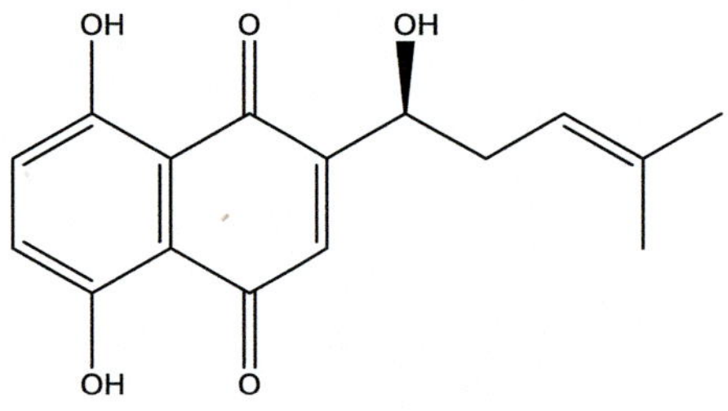

Alkannin

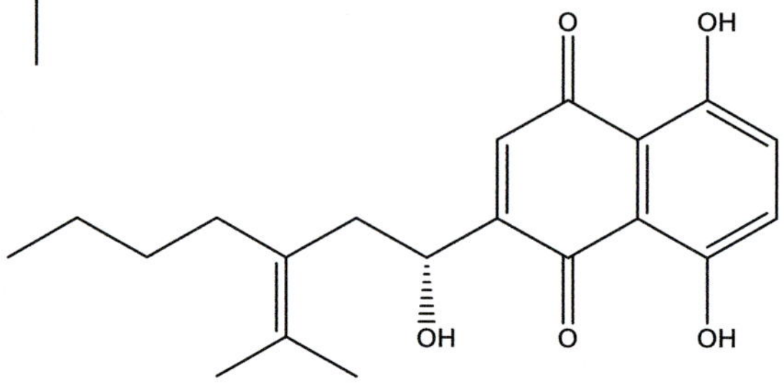

Butylshikonin

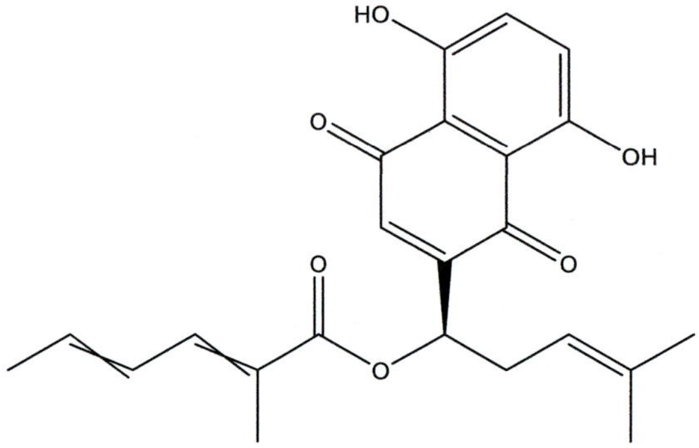

2-methyl-hexa-dienoylshikonin

FIGURE 27.2.4 Chemical components of *Arnebia hispidissima*.

Another report suggested the isolation of two unique polycyclic geranylhydroquinone-derived components, arnebacene and arnebidin. Additionally, some other phenolic components, such as arnebin-7 and vanillic acid, were also found [511].

Thus, plants contain arnebin-5, arnebin-6, arnebin-7, teracryl shikonin, arnebinone, acetyl shikonin, arnebacene, arnebidin, phenolic compounds, and geranylhydroquinone. Additionally, plants also contain triterpenoids and pyrrolizidine alkaloids [495].

After extraction from petroleum ether, the roots of this plant produced red oil. This oil, when subjected to purification via copper salt, produces a pigment as a glittering red oil-like alkannin [512]. Additionally, it also contains alkannin monoacetate and alkannin β-dimethyl acrylate [513,514]. The plant root also contains isohexenylnaphthazarin alkannin β-hydroxyisovalerate [515].

27.2.8 Uses

The plant is used as a stimulant, tonic, diuretic, and expectorant. Aqueous extracts derived from this plant are used to treat throat, fever, and cardiac complications [516].

Additionally, extracts derived from *A. hispidissima* showed antibacterial and antitumor effects, which could be due to the presence of naphthoquinones, triterpenoids, and pyrrolizidine alkaloids.

In addition, it has wide applications as a colorant in cosmetics, food, and textiles. Alkannin has applicable pharmacological effects as an antibacterial agent, granulation tissue-forming activity, and anti-ulcer activity.

27.2.9 Mechanism of Action

The previous report demonstrated the anti-inflammatory effects of shikonin derivatives derived from *A. hispidissima*. It was found that arnebinone considerably reduced carrageenan-induced paw edema and also suppressed the development of chronic arthritis in rats [495].

27.2.10 Toxicity and Clinical Data

Spiroethers present in Arnebia species have been reported to have cytotoxic effects [494].

27.2.11 Most Cited Research

iCite, Database, NIH Open Citation Collection was used from June to October 2022 to get the details of the most cited research. Filters for research papers were applied using the keywords *Arnebia hispidissima*. The NIH database search showed the anti-inflammatory effects of shikonin derivatives from *Arnebia hispidissima* in carrageenan-induced paw edema and complete Freund's adjuvant (CFA)-induced chronic arthritis in rats [495].

27.2.12 Patents

胡汛方建平 (Hu Xun, Fang Jianping). Use of alkannin in preparing medicine for treating tumor disease. CN1579378A.

- The invention reveals a new utilization of alkannin to treat tumors.

27.2.13 Plant Pathology

No diseases are reported in the literature on *A. hispidissima*. However, *Azotobacter chroococcum* (a symbiotic nitrogen-fixing bacterium) is isolated from the hairy roots of *A. hispidissima* [504].

28 Genus *Azima*

28.1 *Azima tetracantha* Lam.

28.1.1 Synonyms

Azima angustifolia, *Azima nova*, & *Azima tetracantha* var. *laxior*

28.1.2 Vernacular Name

In Oman, it is called dishi & tisi.

28.1.3 Common Names

The genus name *Azima* in Arabic means defender, and the species name in Greek, "tetracantha," means four thorns. It's also called a bee-sting bush because a wound produced by its spines is very painful. It's also called by other names in English, such as needle bush, four thorns, and stink bush. A wound, small or large, caused by one of the spines is very painful; hence, one of the common names, bee-sting bush.

28.1.4 Botanical Description

It is an evergreen, spiny shrub that grows up to 2–3 m and bears simple, oval, light green, glossy leaves with a pungent odor. Flowers: solitary, tiny, greenish to yellow color flowers and yellowish to white berries (Figures 28.1.1 and 28.1.2).

28.1.5 Geographical Distribution

It is distributed in tropical and southern Africa, the western Indian Ocean, and extends eastward across the Indian subcontinent to Myanmar. In the Arabian Peninsula, it is only found in southern mountains of Oman.

FIGURE 28.1.1 Leaf and thrones of *Azima tetracantha*.

FIGURE 28.1.2 Fruit of *Azima tetracantha*.

28.1.6 Chemical Composition

Plants contain high levels of N-methoxy-3-indolylmethyl-glucosinolate, whereas some additional components, such as indole glucosinolate, neoascorbigen, glycosides and acyl-glycosides, quercetin, isorhamnetin, rhamnetin, rhamnazin, and dimeric piperidine alkaloids, are present. The main bioactive chemical compound is friedelin, which is known for its antioxidant [517], anti-inflammatory [518], and antiulcer effects (Figure 28.1.3).

28.1.7 Uses

This plant is used as food and for various herbal medicines. It is traditionally used for several health complications, such as renal disease. Further, this shrub is reported to have various pharmacological effects, such as antimicrobial, antifungal, anti-inflammatory, analgesic, antioxidant, antipyretic, antiulcer, anticancer, anti-snake venom, diuretic, and hepato-protective activities [519–523].

The extract derived from this plant showed a nephroprotective effect in glycerol-induced acute renal failure, and this could be associated with the antioxidant activity of phytochemicals present in this plant [524].

Another study suggested that extract from *A. tetracantha* leaves inhibited the activity of toxic enzymes from *Vipera russelli* and *Bungarus caeruleus* venom. Thus, this extract can be used as a possible treatment against these poisons [525].

28.1.8 Mechanism of Action

The gastro-protective effects of friedelin from the hexane extract of leaves of *A. tetracantha* are reported. It

DOI: 10.1201/9781003295488-28

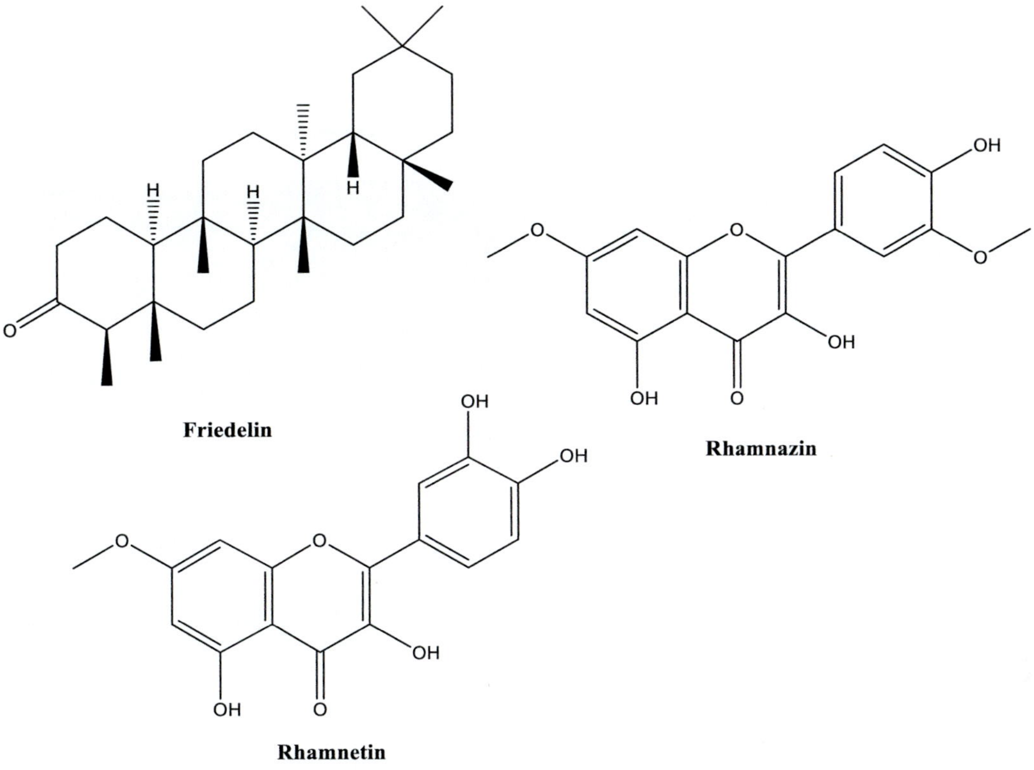

FIGURE 28.1.3 Chemical components of *Azima tetracantha*.

was found that friedelin ingestion caused a considerable reduction in vascular permeability, pro-inflammatory cytokines, inducible nitric oxide synthase (iNOS), caspase-3, and apoptosis level, suggesting the antiulcer effects of friedelin [526].

Earlier studies showed the antifungal effects of *Azima tetracantha* extracts and friedelin against *Trichophyton rubrum* and *Curvularia lunata* [519].

The anti-inflammatory and antipyretic effects of *Azima tetracantha* have also been reported, and possibly the extract also possesses antiproliferative, antioxidative, and lysosomal membrane stabilization properties [518,520].

28.1.9 Toxicity and Clinical Data

No mortality or toxicity reports were observed up to 2,000 mg/kg. Thus, the extract drug was found to be safe according to its acute oral toxicity profile [524]. However, the smoke produced from its wood is poisonous.

28.1.10 Most Cited Research

iCite, Database, NIH Open Citation Collection was used from June to October 2022 to get the details of the most cited research. Filters for research papers were applied using the keywords *Azima tetracantha*. NIH database search showed antifungal and cytotoxic effects possessed by *Azima tetracantha* [527].

28.1.11 Patents

Petharajanna. WO2012131714A4. Veterinary herbal compositions and medicaments thereof.

This invention reveals herbal preparation comprising *Azima tetracantha* with other plants for the treatment of ephemeral fever.

28.1.12 Plant Pathology

There is no data in the literature that refers to the pathology of *Azima tetracantha*. However, an endophytic fungus was isolated from *Salvadora persica* in Saudi Arabia [528].

29 Genus *Aristolochia*

29.1 *Aristolochia bracteolata* Lam.

29.1.1 Synonyms

Aristolochia abyssinica, *Aristolochia benadiriana*, & *Aristolochia benadiriana* var. *longilabia*

29.1.2 Family

Aristolochiaceae

29.1.3 Vernacular Name

Mekysah

29.1.4 Common Name

Worm killer, Dikeritimelo, in South Sudan. It's called a "worm killer" due to its supposed trypanocidal and anthelmintic effects [529].

29.1.5 Botanical Description

A. bracteolata is a small glabrous perennial that grows up to 50 cm long. Leaves: simple, broad, with wavy margins. Flowers: yellow-brown petals with a distinctive purple lip. Fruits: oblong dehiscent capsule with many endospermic seeds (Figures 29.1.1–29.1.3).

29.1.6 Geographical Distribution

Aristolochia bracteolata is extensively dispersed across tropical Asia, Africa, and South America [530] and cultivates well on lava or sandy soil, grassland, river banks, and

FIGURE 29.1.1 Ripe fruit of *Aristolochia bracteolate*.

FIGURE 29.1.2 Whole plant of *Aristolochia bracteolate*.

FIGURE 29.1.3 Flowers of *Aristolochia bracteolata*.

bushlands [531]. In Oman it found throughout the country thrives as weed in the cultivations.

DOI: 10.1201/9781003295488-29

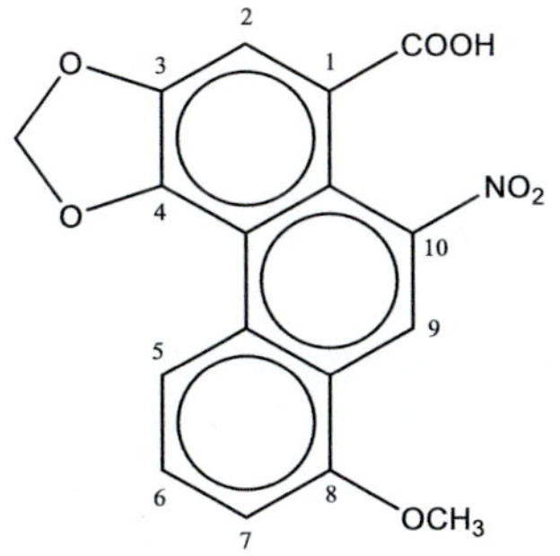

$C_{17}H_{11}NO_7$ Rel. mol. mass: 341.27

Aristolochic acid I (8-methoxy-6-nitrophenanthro [3, 4-*d*][1, 3] dioxole-5-carbosylic acid).

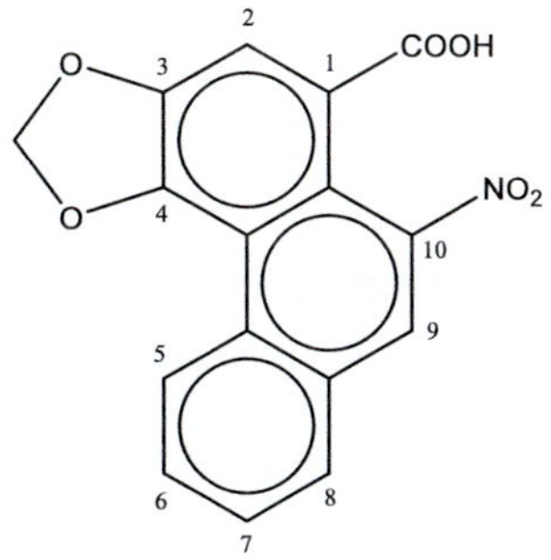

$C_{16}H_9NO_6$ Rel. mol. mass: 311.25

Aristolochic acid II (6-nitrophenanthro [3, 4-*d*][1, 3] dioxole-5-carbosylic acid).

FIGURE 29.1.4 Chemical structure of aristolochic acid I and aristolochic acid II.

29.1.7 Chemical Composition

This plant contains nephrotoxic agents, namely aristolochic acids [530] (Figure 29.1.4). The root of this plant contains acidic crystalline components. One yellow-colored needle-shaped component isolated from the roots is similar to aristolochic acid. The root and stem contain alkaloids.

29.1.8 Uses

The antiplasmodial effects of *Aristolochia bracteolata* were determined, and it was found that 100% suppression of the parasite growth at ≤ 50 µg/mL was obtained from the whole plant of *A. bracteolata*, which is used in Sudanese folk medicine [532].

Another report showed that the methanol extract of *A. bracteolata* demonstrated larval mortality (larvae of the malaria vector *Anopheles subpictus* Grassi) after 48 h [533].

In Nigeria, fresh leaves are mixed with castor oil to treat pimples, whereas flowers and roots are used to treat snakebite and scorpion stings. It is also used traditionally to treat scabies in India [534]. This plant is used in folk medicine to treat lung inflammation, gastric complications, dysentery, cancer, and snakebites [535].

Plants belonging to this family are used in Sudan as anti-snake, analgesic, and anti-scorpion agents. Additionally, plant preparations are also used to treat tumors, malaria, and fevers [536]. Leaves of this plant have been known for their anti-arthritic effects, as recorded in folk medicine literature.

The plant extract has been reported for its potential to reduce hyperuricemia in arthritic rats, perhaps because of the antioxidative activity possessed by phenolic components, including flavonoids and diterpenes. It is also known that plant leaves with excessive bitterness showed anthelmintic and anti-inflammatory effects [537]. Traditionally, *A. bracteolata* Lam. is used to treat malaria in Jubek State of South Sudan [538].

It is found that *A. bracteolata* showed anticandidal effects. The ethanol extract and essential oil possessed anticandidal effects, whereas the water extract showed no activity [539].

Recent *in-vivo* antioxidant studies showed that silver nanoparticles synthesized by using aqueous shoot extract of *Aristolochia bracteolata* showed significant free radical scavenging effects [540].

The earlier investigation demonstrated the antiallergic effects of chloroform extract of *Aristolochia bracteolata* via mast cell membrane stabilization, inhibiting NO and histamine pathways [541].

A recent study showed wound-healing effects of electrospun nanofibrous Guar gum/PVA-based scaffold matrix combined with extracts of wound-healing medicinal plants, including *Aristolochia bracteolata* [542].

Aristolochia bracteolata extracts showed protection against envenomed mice from Russell's viper and cobra venom. The computational study suggested that aristolochic acid can be responsible for these effects due to the neutralization of type I PLA2 enzyme [543].

Antimicrobial effects of extract containing aristolochic acid against *Moraxella catarrhalis* and sea urchin-derived *Bacillus* sp. have been reported. It was found that modification of aristolochic acid 1 to aristolochic acid methyl ester has abolished its antibacterial effects [544].

Another report showed that ethanol extract of *Aristolochia bracteolata* Lam. showed wound-healing effects with a significant increase in the level of superoxide dismutase and catalase in the granuloma tissue [545].

29.1.9 Mechanism of Action

Extract derived from the leaves of this plant possessed wound-healing effects via superoxide dismutase and catalase [537].

A recent study showed that methanol extract of *A. bracteolata* at 100 µg/mL has induced human gingival-derived mesenchymal stem cell differentiation into the osteogenic lineage, resulting in an increase in the expression of Runx2 and related osteogenic genes [546].

29.1.10 Toxicity and Clinical Data

As mentioned above, this plant is traditionally used to treat malaria. The recent survey showed that the majority of the users showed improvement; however, at the same time, they experienced side effects such as early-stage abortion and stomach upsets. Therefore, its utilization in pregnant women must be avoided [538].

The earlier study demonstrated 100 cases of nephropathy due to the systemic and longer-term application of Chinese snakeroot (*Aristolochia fangchi*). This has raised important concerns related to the risk of using preparations containing aristolochic acids. Many nephropathy cases have been reported after the usage of Chinese and European species of *Aristolochia* [547].

The raw consumption of plants belonging to this genus is not considered safe, as the genus *Aristolochia* has been reported for the presence of carcinogenic components known as aristolochic acid. This component is reported to cause a condition called Chinese herb nephropathy or aristolochic acid nephropathy [548,549]. This component has been reported to cause mutations, and this mutagen is more harmful than carcinogens such as UV light and tobacco smoke [550,551]. Several reports suggested that prolonged utilization of *Aristolochia fangchi* causes nephropathy [547].

In several countries, aristolochic acid and its preparations are considered toxic; however, it is still considered an important remedy in nations like India, China, Sudan, and Africa to treat menstrual disturbances, rheumatism, and infectious diseases [544].

Aristolochic acid-containing herbal dietary supplements have been discontinued worldwide, as these components are strong carcinogens. As mentioned before, aristolochic acids cause aristolochic acid nephropathy and end-stage renal failure associated with urothelial carcinoma, due to which *Aristolochia* has been banned and regulated. However, aristolochic acid-containing products are still available in the market due to the belief in the mysterious notion that the toxicity of minute carcinogens present in the plant preparation can be neutralized by the other components in the plant preparations [552].

To detect *Aristolochia* plants and related herbal products, a polymerase chain reaction (PCR) coupled with lateral flow immunochromatographic assay has recently been developed. The assay was developed specifically for the nucleotide signature in the plastid *rbcL* gene region of *Aristolochia* species [553].

29.1.11 Most Cited Research

iCite, Database, NIH Open Citation Collection was used from June to October 2022 to get the details of the most cited research. Filters for research papers were applied using the keywords *Aristolochia bracteolata*. The NIH database search includes nephropathy cases over systemic and longer-term application of *Aristolochia* species containing aristolochic acids and their associated risks [547].

29.1.12 Plant Pathology

Genera *Fusarium*, *Penicillium*, *Phialemonium*, and *Trichoderma* were identified as predominant endophytic taxa in the genus *Aristolochia* tissues [554].

30 Genus *Artemisia*

30.1 *Artemisia sieberi* Besser

30.1.1 Synonyms

Seriphidium sieberi, Artemisia glomerata & Artemisia herba-alba var. *laxiflora*

30.1.2 Vernacular Name

Ghata

30.1.3 Common Name

Desert or white wormwood & sheeh

30.1.4 Botanical Description

A robust aromatic shrublet, reaching heights of up to 40 cm. Its branches are firm and can either stand erect or spread outwards. The leaves are primarily located at the base, with lower ones have petioles, while upper leaves are almost directly attached to the stem. The flowers are bisexual, shaped like tubes, and vary in color from yellowish tones to purple hues (Figures 30.1.1 and 30.1.2).

30.1.5 Geographical Distribution

This species is native to a region stretching from the Eastern Mediterranean to western Pakistan and Tunisia to the Arabian Peninsula. In Oman, it is restricted to higher elevations of the Musandam Mountains.

FIGURE 30.1.1 Shoot system of *Artemisia sieberi*.

FIGURE 30.1.2 Habit of *Artemisia sieberi*.

30.1.6 Chemical Composition

A. sieberi alba contains a high level of herbalbin, cis-chryanthenyl acetate, hispidulin, cirsilineol, and several terpenes like monoterpenes and sesquiterpenes [555] (Figure 30.1.3). *A. sieberi* essential oil contains α, β-thujones as the main components [556].

Another study showed that *A. sieberi* showed a yield of 1.5% (w/v) oil, which contains 27 components. Out of these 27 components, camphor, 1,8-cineole, and chrysanthenone were considered as major components. It was found that the extract, as well as the oil, demonstrated the highest antioxidant effects. Moreover, the oil considerably showed a considerable reduction in the mycelium growth of *Botrytis cinerea* [557].

Two antimicrobial sesquiterpene lactones, 3R, 8R-dihydroxygermacr-4(15),9(10)-dien-6S,7S,11RH, 12,6-olide and 1R,8S-dihydroxy-11R, 13-dihydrobalchanin, with other components such as 3'-hydroxygen kwanin and 11-epiartapshin, were found in *A. sieberi*. This components showed moderate antifungal effects, whereas 3'-hydroxygenkwanin showed antimicrobial effects against *Staphylococcus aureus* and *Bacillus subtilis* [558].

A. sieberi contains spathulenol, and the essential oil obtained from this plant showed high antibacterial effects against common human pathogens [559]. In an investigation, the effect of grazing on the quality and quantity of *Artemisia sieberi* essential oil in Iran was studied, and it was found that grazing has a significant impact on the oil composition of the plant. It was found that oil derived from a plant collected from a non-grazed site contained 1,8-cineol and myrcene, while oil from a moderately grazed site contained different proportions of 1,8-cineol

DOI: 10.1201/9781003295488-30

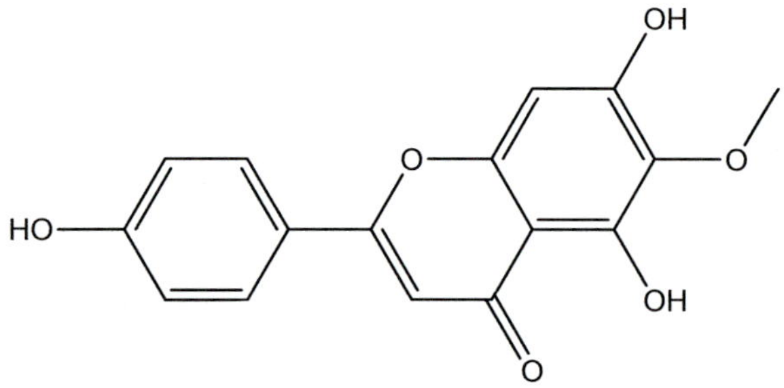

Hispidulin

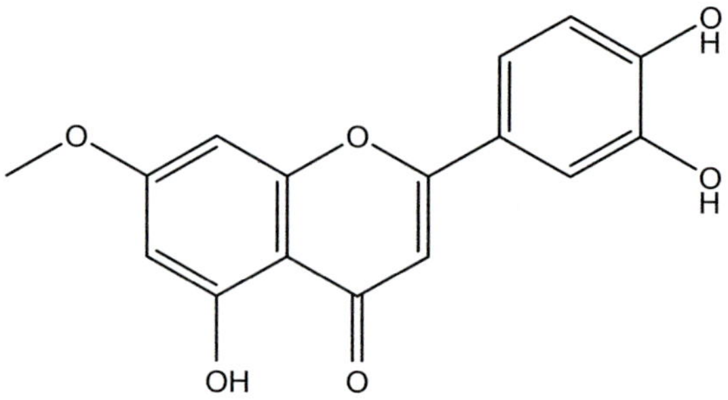

3'-hydroxygenkwanin

FIGURE 30.1.3 Chemical components of *Artemisia sieberi*.

and Eudesm-7(11)-en-4-ol. On the other side, samples of oil obtained from plants from heavily grazed sites contained 4-terpinyl acetate, davanone, and p-cymene [560].

30.1.7 Uses

Traditionally, it's used as an antiseptic, antispasmodic, and vermifuge. *Artemisia sieberi* showed the lowest antimicrobial activity against the pathogens causing oral infections [555,561]. In the Middle East, essential oil derived from *A. sieberi* is used in the treatment of infectious ulcers and as a wormicide [556]. In vitro studies showed that an aqueous extract of *A. sieberi* demonstrated protoscolicidal effects [562].

Another study investigated fumigant toxicity and sublethal effects of essential oils from *A. sieberi* against pests, *Sitotroga cerealella* Olivier. It was found that the oil showed fumigant toxicity against pests and reduced the growth, and thus can be utilized in pest management of *S. cerealella* [563].

In another study, the antifungal effects of *A. sieberi* oils were examined against different dermatophytes. It was found that the oil fraction containing higher levels of α-thujone and β-thujone demonstrated higher anti-elastase effects and thus can possibly be considered as good topical antifungal agents for the treatment of skin dermatophyte infections [564].

Similarly, the antifungal activity of silver nanoparticles using *A. sieberi* leaf extract against *A. fumigatus* was studied in vitro and *in vivo*. The findings from this study revealed that biosynthesized AS-AgNPs can be considered a novel antifungal alternative treatment against aspergillosis [565].

A recent investigation revealed the inhibitory effects of *Zataria multiflora*, *Nepeta haussknechtii*, *Artemisia sieberi*, and *Citrus aurantifolia* essential oils (EOs) in pure and zinc oxide (ZnO) nanoencapsulated formulations on the mycelial growth of *Alternaria solani*. It was found that the ZnO-ZmEO formulation can possibly be used as a biofungicide for managing and controlling tomato early blight disease [566].

Another study revealed the effects of *A. sieberi* essential oil in normal and alloxan-induced diabetic rats. Results obtained showed the oil fraction possessed antidiabetic effects in alloxan-induced diabetic rats, and thus, the oil can be used as a remedy for diabetes mellitus in humans [567].

Another study showed *A. sieberi* essential oil possessed cardioprotective, nephroprotective, and hepatoprotective activities in alloxan-induced diabetic rats [568].

In another investigation, the toxic and repellent potential of essential oil derived from this plant against *Dermanyssus gallinae*, which is considered a threat to the poultry industry worldwide, was investigated. The chemical composition of the oil analysis revealed the presence of α-thujone, β-thujone, camphor, and 1,8-cineole as main components. Findings revealed that the oil's effects are toxic and thus can be used as a repellent [569].

The earlier study demonstrated the effects of using black cumin seeds, *Artemisia* leaves, and *Camellia* L. plant extract (in the diets of broiler chicks). Results obtained revealed that black cumin seeds alone or mixed with *Artemisia* leaves improved broiler health [570]. Another study showed this plant can be used as feedstuff in the dromedary camel [571].

Anti-leishmania effects of *Artemisia sieberi* vs *Artemisinin* were investigated on *L. major* under in vitro conditions, and it was found that *A. sieberi* relatively showed higher anti-Leishmania effects than *Artemisinin* [572].

30.1.8 Mechanism of Action

A recent study showed the gastro-protective effect of *A. sieberi* oil in ethanol-induced rats. It was found that the oil showed anti-ulcerogenic effects via reducing oxidative stress and inflammatory responses associated with an ulcer. This action was mediated via numerous metabolic pathways, viz. lipid, energy, and nucleotide metabolisms [573].

In another study, the anti-inflammatory effects of this plant material containing high levels of sesquiterpene were studied. It was found that the inhibition of pro-inflammatory signaling pathways could be dependent on the type of their sesquiterpene level. Further, the plant material reduced the nitric oxide level and PGE2 production via suppression of iNOS and COX-2 expression [574].

It was revealed that *A. sieberi* contains a high level of beta-thujone, alpha-thujone, camphor, verbenol, p-mentha-1,5-dien-8-ol, davanone and showed strong fungistatic activity against some soil-borne pathogenic fungi [575].

A. sieberi has also been investigated for antimalaria effects and can be considered a potential candidate to reduce coccidial infection in chickens, as plant samples considerably reduce the number of oocysts per gram of feces in chickens [576].

30.1.9 Toxicity and Clinical Data

In vitro studies showed the antifungal effects of essential oil against saprophytic fungi, dermatophytes, *Malassezia* sp., and *Candida* sp. This has been further confirmed by six clinical studies. Clinical investigation showed *A. sieberi* essential oil (5%) lotion showed an improvement in clinical signs of fungal superficial disease with a lower recurrence rate than clotrimazole [556].

Antimalarial effects of *Artemisia sieberi* on *Plasmodium berghei* were studied. It was found that extracts derived from *A. sieberi* demonstrated antimalarial effects against murine malaria [577].

In a clinical study comparing topical *A. sieberi*, 5% lotion was used with topical clotrimazole, and 1% was used for the treatment of pityriasis versicolor. Findings obtained showed that *A. sieberi* lotion was more effective than clotrimazole lotion [578].

The anticoccidial effect of a granulated extract of *Artemisia sieberi* versus monensin in experimental broiler coccidiosis was investigated, and it was found that it improved performance attributes and considerably reduced oocysts per gram in inoculated broiler chickens [579].

A clinical investigation revealed the prophylactic and therapeutic effects of a preparation derived from *A. sieberi* extract on avian coccidiosis via reducing oocysts per gram output significantly [580].

30.1.10 Plant Pathology

The essential oil from this plant could be applied to manage *B. tabaci*, a sweet potato whitefly, which is a serious vector of several plant viruses [581]. Phytoplasma infection caused symptoms such as witches' broom and little leaves that were initially observed in *A. sieberi* grown in Iran. A previous study showed that phytoplasma infection of *A. sieberi* causes changes in its essential oil composition, especially the composition of monoterpenoids [582]. In vitro studies showed the antibacterial properties of endophytic fungi isolated from *Artemisia sieberi*.

30.1.11 Most Cited Research

iCite, Database, NIH Open Citation Collection was used from June to October 2022 to get the details of the most cited research. Filters for research papers were applied using the keywords *Artemisia sieberi*. NIH database searches include anticoccidial effects of the *Artemisia sieberi* extract in broiler chickens [576].

.... (*For additional images refer to the supplementary figures section; page no.* 299)

31 Genus *Arthrocaulon*

31.1 *Arthrocaulon macrostachyum* (Moric.) Piirainen & G.Kadereit

31.1.1 Synonyms

Arthrocnemum fruticosum var. macrostachyum, Arthrocnemum macrostachyum, & Salicornia macrostachya

31.1.2 Family

Amaranthaceae

FIGURE 31.1.1 Stems of *Arthrocaulon macrostachyum.*

31.1.3 Vernacular Name

Thaab & hamd

31.1.4 Common Name

Large-spiked glasswort

31.1.5 Botanical Description

An erected, succulent shrub that grows in masses and reaches up to a height of 100 cm. The stems are green. However, a major portion is covered with red pigments in summer due to betacyanin and anthocyanin production. This perennial shrub possesses small and scaly leaves and always grows in clumps [583–585] (Figures 31.1.1–31.1.3). Flowers are minute, produced in terminal-cylindrical spikes.

31.1.6 Geographical Distribution

A. macrostachyum is widely available in the Mediterranean, including France [586] and Portugal [587], and is also found along the delta region of the Mediterranean coast of Egypt [588]. This herb is also available in the Middle East, such as the Arabian Gulf of the UAE [589], and is found across the Arabian Sea in nations such as Pakistan. This plant is found in coastal areas throughout Oman, conquering the vast salt marsh areas known as Sabkhas and intertidal mud flats.

FIGURE 31.1.2 Dense community of *Arthrocaulon macrostachyum.*

DOI: 10.1201/9781003295488-31

FIGURE 31.1.3 Branches of *Arthrocaulon macrostachyum*.

31.1.7 Chemical Composition

This halophyte contains phenolic acids, including chlorogenic acid, protocatechuic acid, gallic acid, caffeic acid, rosmarinic acid, and p-coumaric acid. This plant also contains flavonoids such as hesperidin, rhamnetin, quercetin, and naringenin (Figure 31.1.4) [590].

In one of the previous studies, chemical components of this halophyte were divided on the basis of their biological effects, such as rhamnetin for diabetes; hesperidin, coumaric acids, chlorogenic, and salicylic for antibacterial, antiviral, and/or anti-tumoral effects; and caffeic acids, quercetin, and 4-hydroxybenzoic for antioxidant effects [591]. This halophyte is also reported for the presence of polyunsaturated fatty acids, chiefly linoleic and α-linolenic acids [592].

Additionally, ferulic acid and flavonoids like cyanidin-3-O-arabinoside, luteolin-7-glucoside, dihydroquercetin, and p-coumaroyl-glucoside were also identified in the plant [591].

31.1.8 Uses

In Tunisia, this halophyte was earlier used as an antibiotic [593] and as an alexipharmic [594,595]. Additionally, this plant was also used as a hypoglycemic agent in Ayurveda [595–598].

This halophyte has been recently used in soil desalination programs [599], as it can accumulate high levels of NaCl in its tissues. This plant has also been studied for its potential in treating degenerative disorders associated with metabolic stress, such as cancer and neurological disorders [592].

31.1.9 Toxicity and Clinical Data

The toxicity profile of this plant has not been explored yet. However, since it is edible in nature, there would be minimal acute toxicity expected [600].

No report.

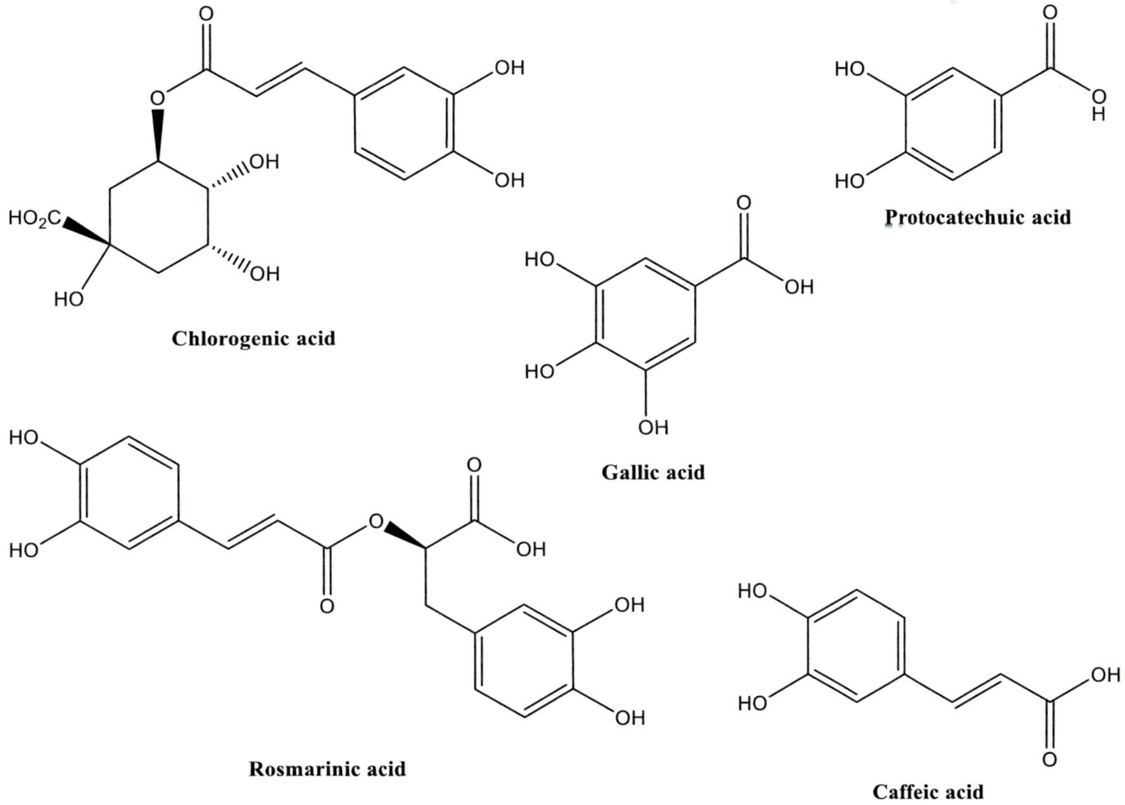

FIGURE 31.1.4 Chemical components of *Arthrocaulon macrostachyum*.

31.1.10 Most Cited Research

iCite, Database, NIH Open Citation Collection was used on January 2nd, 2024, to get the details of the most cited research. Filters for research papers were applied using the keywords *Arthrocaulon macrostachyum*. NIH database search revealed the edible halophytes and halo-tolerant species in the Apulia Region (Southeastern Italy): Biogeography, traditional food use, and potential sustainable crops [600].

31.1.11 Plant Pathology

So far, no reports related to pathogens causing damage to this wild plant are present. However, endophytic microbial diversity has been reported on this halophyte, which impacts this plant's survivability in a high-valine environment.

32 Genus *Asperugo*

32.1 *Asperugo procumbens* L.

32.1.1 Synonyms

Asperugo erecta, *Asperugo alba*, & *Asperugo vulgaris*

32.1.2 Family

Boraginaceae

32.1.3 Common Name

German-madwor or madwort

32.1.4 Botanical Description

A. procumbens is an erect herbaceous plant that grows up to 90 cm and has a slender and procumbent stems with stiff and deflexed hairs with lanceolate, subacute to obtuse, entire or slightly dentate leaves and small, blue or white flowers (Figure 32.1.1).

FIGURE 32.1.1 Flowering branch of *Asperugo procumbens*.

32.1.5 Geographical Distribution

A. procumbens is mainly present in Europe with the adjacent parts of North Africa, Siberia, and Central Asia, and it's considered to be a naturalized archaeophyte. *A. procumbens* usually occurs in anthropogenic habitats. It grows in fertile, clay, and sometimes rocky soils [601]. In Oman, it is found in the Western Hajar Mountains and Musandam.

32.1.6 Chemical Composition

A phytochemical study reveals the presence of flavonoids, phenolics, and tannins in the plant extract. The fatty acid composition has been investigated in detail. This plant is a good source of omega-3 fatty acids—stearidonic acid, α-linolenic acid, and omega-6 fatty acid—γ-linolenic acid [601] (Figure 32.1.2).

32.1.7 Uses

Traditionally, it is used as medicine in the Middle East. In Iran, plants are used to strengthen the nervous system and the heart. It's also used to treat skin and liver-related infections, asthma, and herpes. *A. procumbens* decoction is used to treat respiratory tract disorders like throat infections and chest problems. The plant is also reported for its various pharmacological effects such as antidepressant and sedative-hypnotic, antihypertensive, antioxidant, and chemopreventive effects [601]. Methanol extract of *A. procumbens* showed antibacterial effects [602].

32.1.8 Most Cited Research

iCite, Database, NIH Open Citation Collection was used from June to October 2022 to get the details of the most

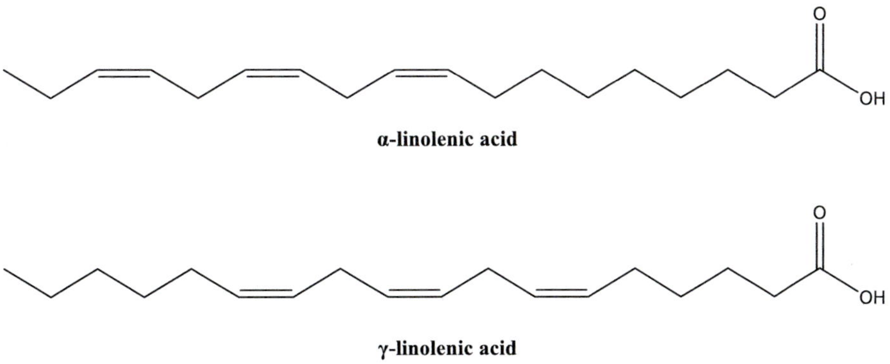

FIGURE 32.1.2 Chemical components of *Asperugo procumbens*.

DOI: 10.1201/9781003295488-32

cited research. Filters for research papers were used by using the keywords *Arthrocnemum glaucum.* NIH database search includes studies based on low antidepressant and good sedative-hypnotic activities of the *A. procumbens* hydroalcoholic extract in mice. Moreover, it was found that central benzodiazepine receptors are involved in the sedative-hypnotic effects of this plant [603].

32.1.9 Plant Pathology

No plant disease has been reported on *A. procumbens.* However, *Curvularia inaequalis*, an endophytic fungus, was isolated from *Asperugo procumbens* [604]. Other studies indicated many bacterial strains interacted with this wild weed root system.

.... (*For additional images refer to the supplementary figures section; page no.* 300)

33 Genus *Asplenium*

33.1 *Asplenium ceterach* L.

33.1.1 Synonyms

Asplenium sinuatum, Ceterach ceterach, & *Ceterach officinarum*

33.1.2 Family

Aspelinaceae

33.1.3 Common Name

Rustyback fern, Ceterach officinarum

33.1.4 Botanical Description

A desiccation-tolerant fern with tufted and leathery green fronds. The leaves are divided into many oblong leaflets with small, rounded segments.

33.1.5 Geographical Distribution

This plant is widely available in Western and Central Europe, including the Mediterranean region, and also in temperate Asia. It is often present on rocks and brick walls. This fern needs full sunlight with little humidity for its growth.-In Oman, it is found in the mountains of Musandam at higher elevations.

33.1.6 Chemical Composition

This fern has been reported for chlorogenic acid and caffeic acid (Figure 33.1.1). Additionally, it also contains enzymes such as peroxidase and polyphenol oxidase. Both polyphenolic contents, as well as antioxidant enzymes, counter the harmful effects of the reactive oxygen species produced while drying.

A previous study showed the presence of a flavanone glycoside from the fronds of *C. officinarum* [605]. Plants belonging to the *Asplenium* genus contain volatile organic components. Propionylfilicinic acid was identified in the species [606].

Previous research showed the presence of lipid derivatives as important volatile components in this plant [606]. Recent research reveals the presence of fatty acid derivatives: isomeric and decadienals, other linear aldehydes, alcohols, and related components [607]. The plant contains quercetin-3-O-β-D-gentiobioside, kaempferol-3-O-β-D-gentiobioside, quercetin-3-O-β-D-glucuronide, p-coumaroyl-4-O-β-D-glucoside, caffeic acid, chlorogenic acid, 4-vinyl-phenol-1-O-β-D-glucopyranoside, 3-(1-O-β-D-glucopyranosyl-4-hydroxyphenyl)-propionic acid, and 4-β-O-D-glucosyl-3,4-dihydroxybenzyl alcohol. Additionally, plants also contain ethyl esters of palmitic, linoleic, and α-linolenic acids [608].

HPLC analysis reveals the presence of pterosin b, catechin, quercetin, chlorogenic acid, and caffeic acid [609]. HPLC-MS/MS findings reveal the presence of flavonol hyperoside, phenolic acid, and chlorogenic acid [610].

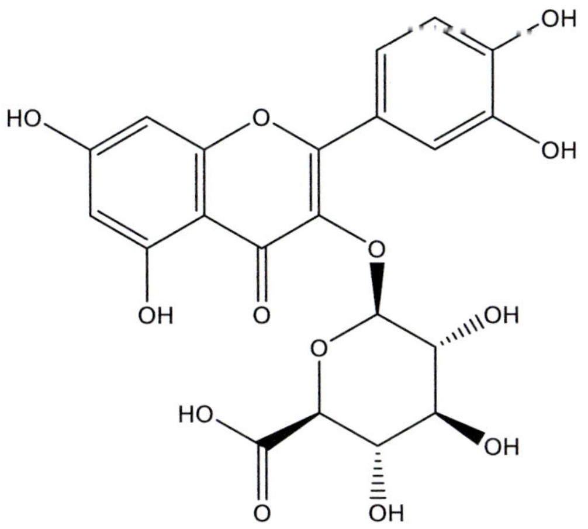

Quercetin-3-O-ß-D-glucuronide

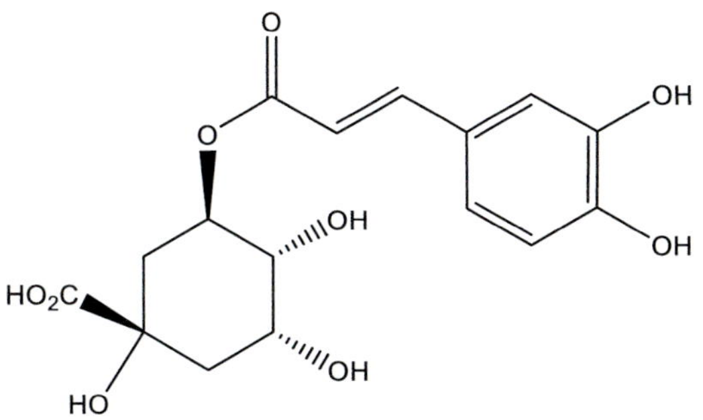

Chlorogenic acid

FIGURE 33.1.1 Chemical components of *Asplenium ceterach*.

DOI: 10.1201/9781003295488-33

33.1.7 Uses

This plant has both food and medicinal applications. *A. ceterach* is utilized to treat respiratory complications, hypertension, intestinal complications, spleen-related complaints, hemorrhoids, and kidney stone-related problems. Additionally, it is a diuretic, expectorant, laxative, etc. It is also used to treat benign prostatic hyperplasia.

Additionally, it was found that *A. ceterach* extract showed the strongest inhibitory effects against human cervical cancer as well as bacterial cells, with a low grade of cytotoxicity against mouse cell lines [611].

A. ceterach showed antioxidant effects, particularly the extract that exhibited the highest superoxide and nitric oxide radical scavenging activity, with (IC50 ¼ 145.17 and 372.03 mg/mL) [612]

Ceterach officinarum extract was found to be active against elastase, which is an important target for dermatological disorders. This could be due to the presence of flavonoids and phenolics [613].

33.1.8 Mechanism of Action

A recent study demonstrated that components of *A. ceterach* extracts behave as agonists of 5-HT2 receptors and cause contraction by stimulating the serotonergic signaling system [613].

It is assumed that *C. officinarum* aqueous extract is used to treat calcium oxalate kidney stones. Based on these data, *C. officinarum* aqueous extract could represent an attractive natural therapy for the treatment of urolithiasis [614].

33.1.9 Toxicity and Clinical Data

So far, this fern does not have any record of toxicity; however, several ferns contain cancer-causing agents and thiaminase.

33.1.10 Most Cited Research

iCite, Database, NIH Open Citation Collection was used from June to October 2022 to get the details of the most cited research. Filters for research papers were applied using the keywords *Asplenium ceterach* and *Ceterach officinarum*. NIH database searches include polyploidy, phylogeography, and Pleistocene refugia of the rock fern *A. ceterach* based on chloroplast DNA [615].

The hypotensive effect of *Ceterach officinarum*, used in Calabria (Southern Italy) as folk plant medicine, has also been reported [616].

33.1.11 Plant Pathology

Apparently, there is no data in the literature that refers to the pathology of *Asplenium ceterach*. However, *Milesia magnusiana*, a pathogenic agent responsible for inducing rust on *Asplenium* ferns—specifically targeting the black spleenwort—belongs to the genus *Asplenium* [617].

34 Genus *Asphodelus*

34.1 *Asphodelus fistulosus* L.

34.1.1 Synonyms

Asphodeloides ramosa, Ophioprason fistulosum, & *Verinea fistulosa.*

34.1.2 Family

Asphodelaceae

34.1.3 Vernacular Name

Bsulu & kwaiter

FIGURE 34.1.1 Inflorescence shoot of *Asphodelus fistulosus.*

34.1.4 Common Name

Onion weed, onion-leafed asphodel, and pink asphodel, hollow-stemmed asphodel

34.1.5 Botanical Description

It is a short-lived, tufted annual plant with a hollow stem and attains a height of up to 70cm. Its leaves are hollow, cylindrical, and elongated like onion leaves. That's why it is called onion weed. The plant has diurnal white or pinkish flowers with six petals (Figure 34.1.1).

34.1.6 Geographical Distribution

It is mainly found in warmer temperate and semi-arid regions but is also a weed of subtropical and arid environments. It is a common weed in parts of Australia, New Zealand, and Mexico. In Oman, it is found throughout the country in different habitats.

34.1.7 Chemical Composition

The earlier report showed the presence of anthraquinones such as chrysoeriol, luteolin, asphodelin, asphodelin 10′-anthron, aloesaponarin II, aloe-emodin, chrysophanol, and desoxyerythrolaccin [618] (Figure 34.1.2).

It was found that oil derived from the seed contains myristic, palmitic, stearic, and linoleic acids. Moreover, it also contains a yellow substance called fucosterol [618].

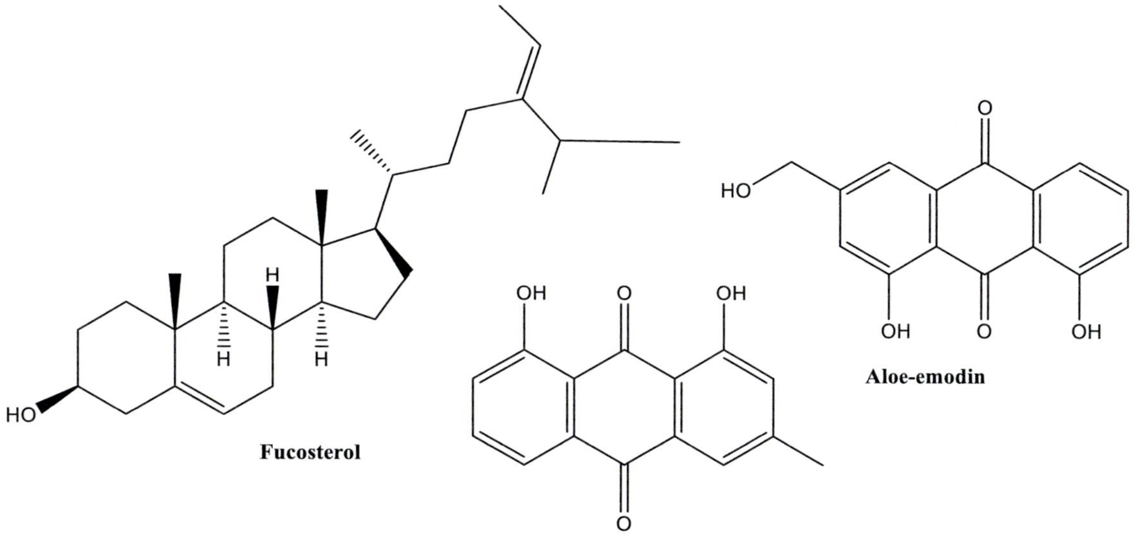

FIGURE 34.1.2 Chemical components of *A. fistulosus.*

DOI: 10.1201/9781003295488-34

The 1.8% nonsaponifiable matter contains fucosterol and a yellow, unidentified substance (Figure 34.1.2).

Chemical composition analysis showed the presence of β-sitosterol and β-amyrin in the fixed oil of seeds of *A. fistulosus* with other sterols such as stigmasterol and campesterol. The oil also contains myristic, palmitic, stearic, oleic, and linoleic acids. Seeds contain sucrose, raffinose, and stachyose [619].

34.1.8 Uses

The antibacterial effect of *A. fistulosus* aqueous and its extracts was investigated on gram-positive and gram-negative bacteria. It was found that *A. fistulosus* extracts showed antibacterial effects against both gram-positive and gram-negative bacteria [620].

34.1.9 Most Cited Research

iCite, Database, NIH Open Citation Collection was used from June to October 2022 to get the details of the most cited research. Filters for research papers were used by using the keywords *Asphodelus fistulosus*. The NIH database search includes the development of the Paris-type AM formed between *Asphodelus fistulosus* and *Glomus coronatum* grown in nurse pots [620].

34.1.10 Patents

Stavros Papadakis, 1990. WO1992001461A1. *Asphodelus* composition is used to increase white blood cell count.

- This invention reveals compositions that include *Asphodelus* to treat white blood cell deficiency in human beings or animals.

34.1.11 Plant Pathology

No plant disease has been reported on this herbaceous weed. However, McGonigle et al. [621] demonstrated the development of the Paris-type arbuscular mycorrhizas formed between *Asphodelus fistulosus* and *Glomus coronatum* grown in nurse pots [621,622]. A feature also demonstrated by onion in mycorrhizal production.

A recent study showed the effect of the natural establishment of *A. fistulosus* on microbial activity, on the composition of the fungal community, and on the mitigation of the effect of potentially toxic elements in a contaminated mine. The following findings have been observed from this study;

- *A. fistulosus* augmented the carbon level in tailings.
- *A. fistulosus* showed the highest β-glucosidase and dehydrogenase activities.
- Increased mineralization of carbon.
- *A. fistulosus* accumulated potentially toxic elements in their tissues and showed potential in the phytoremediation of lead, cadmium, and copper.
- Fungal variety was raised at the vegetated sites [623].

Another study showed that *Asphodelus fistulosus* possibly contributed to the phytostabilization and/or phytoextraction of Pb, Cd, and Zn [623].

.... (*For additional images refer to the supplementary figures section; page no.* 300)

35 Genus *Asteriscus* (*Pallenis*)

35.1 *Pallenis hierochuntica* (Michon) Greuter

35.1.1 Synonym

Asteriscus hierochunticus & *Saulcya hierichuntica*

35.1.2 Family

Asteraceae

35.1.3 Vernacular Name

Ayin al baqar

35.1.4 Common Name

Rose of Jericho

35.1.5 Botanical Description

A dwarf annual herb grows up to 10cm. Leaves form a cluster at the base of the main stem, covered with fine hairs. Flowers, sessile and yellow, form at the center (Figures 35.1.1 and 35.1.2).

35.1.6 Geographical Distribution

This species has a wide native range, stretching from south-eastern Spain and the Canary Island of Fuerteventura in the west all the way to Pakistan and the Arabian Peninsula in the east. It is found in Northern Oman, including Musandam.

35.1.7 Chemical Composition

Nerolidol derivatives, such as 6-oxo- and 6-hydroxycyclonerolidol, are considered chemical markers of the genus *Asteriscus*.

35.1.8 Plant Pathology

There is no data in the literature that refers to the pathology of *Asteriscus hierochunticus*. However, plants belonging to this family are associated with arbuscular mycorrhizal fungi [624].

FIGURE 35.1.1 *Pallenis hierochunticain in flowers.*

FIGURE 35.1.2 Habit of *Pallenis hierochuntica*.

DOI: 10.1201/9781003295488-35

36 Genus *Astragalus*

36.1 *Astragalus fasciculifolius* Boiss.

36.1.1 Synonyms

Astragalus fasciculifolius subsp. *arbusculinus* & *Tragacantha fasciculifolia*

36.1.2 Family

Fabaceae

36.1.3 Vernacular Name

Temi, anzaroot and meantres (Figures 36.1.1–36.1.5)

36.1.4 Botanical Prescription

A spiny, tough shrub with white bark, growing up to 1 meter tall, and much-branched from the base with spiny lateral branches. The leaves are compound, consisting of two to three pairs of oblong leaflets. The flowers have pink to purple petals and a calyx that is initially reddish, turning white at maturity and enclosing the developing fruit.

FIGURE 36.1.1 Flowering stems of *Astragalus fasciculifolius.*

FIGURE 36.1.2 Plant habit of *Astragalus fasciculifolius.*

FIGURE 36.1.3 Flowers and spins of *Astragalus fasciculifolius.*

FIGURE 36.1.4 Spiny branch tips of *Astragalus fasciculifolius.*

FIGURE 36.1.5 Dry gum of *Astragalus fasciculifolius.*

DOI: 10.1201/9781003295488-36

The fruits are small, two-seeded pods, completely enclosed within the persistent calyx.

36.1.5 Geographical Distribution

A. fasciculifolius is a drought-tolerant species and grows under dry climatic conditions. The native range of this plant is Iran and the Arabian Peninsula. In Oman, this species is restricted to higher elevations in the Musandam and Eastern Hajar mountains.

36.1.6 Chemical Composition

It contains glycosides, saponins, and alkaloids. It was found that *A. fasciculifolius* can prevent pain and inflammation, possibly because of the presence of a high amount of saponin [625,626]. This plant plays an important role in heart disease and inhibition of cancer cell growth [627].

36.1.7 Uses

This species produces resin from its bark. This resin holds significant medicinal value in Omani tradition. When needed, it is transformed into a powder. This powder is then mixed with eggs and a small amount of water to create a paste-like consistency. Finally, goat hair is incorporated, and all the ingredients are thoroughly combined.

The resulting plaster was used to treat bone fractures in both humans and livestock. Typically, for broken limbs, a clean cloth would be wrapped around the injured area, and then the plaster would be rubbed over the dressing repeatedly until a thick layer formed.

Another traditional use involved treating eye infections. However, due to the resin's potency, this method required great care. A tiny amount of the powdered resin would be directly applied to the eye. It is important to note that this application could be quite hot and painful.

A. fascicolifolius is used to treat problems during pregnancy. As per the literature, this plant has abortifacient effects. Additionally, this plant is used to treat joint pain, skin patches, and warts [628,629].

Recent publications demonstrated the abortifacient properties of *A. fasciculifolius* extract, as it caused a significant increase in abortion in mice by decreasing progesterone and increasing Follicle-Stimulating Hormone (FSH) and Luteinizing Hormone (LH) feedback [629].

Another study showed the inhibitory effects of *A. fasciculifolius* ethanol extract on diphenolase activity of mushroom tyrosinase when L-DOPA was used as a substrate [630].

Further, this plant also demonstrated high anti-inflammatory effects, which could be associated with high levels of saponin [631]. Moreover, topical (sarcocolla topical cream) treatment by using this plant, *Sarcocolla*, can also prove to be effective in wound healing [632].

36.1.8 Most Cited Research

iCite, Database, NIH Open Citation Collection was used from June to October 2022 to get the details of the most cited research. Filters for research papers were used by using the keywords *Astragalus fasciculifolius.* NIH database search includes the potential presence of eriophyoid mites, *Aceria fasciculifolius* sp. nov., on *Astragalus fasciculifolius* Boiss [633].

36.1.9 Patents

Calvin B. Harley, Allison C. Chin, Nancy Yuk-Yu Ip, Yung-Hou Wong, David M. Miller-Martini. Formulations containing *Astragalus* extracts and uses thereof. WO2005044179A2, 2004.

- This invention is realted to the composition containing plant extracts belonging to the same genus (*Astragalus*) and their application in impacting telomerase activity in cells.

36.1.10 Plant Pathology

No plant pathogen is reported on this wild shrub. However, some reports studied bacteria and fungi associated with *A. membranaceus* but not this species [634]. A previous study showed the potential presence of eriophyid mites, *Aceria fasciculifolius* sp. nov., on *Astragalus fasciculifolius* Boiss [633]. These phyrophagous arthropods develop an intimate relationship with their plant host and serve as potential vectors for viral plant pathogens, which ultimately affect not only this wild plant but also cultivated crops.

37 Genus *Atractylis*

37.1 *Atractylis cancellata* L.

37.1.1 Synonyms

Acarna cancellata, Carthamus cancellatus, & Cirsellium cancellatum

37.1.2 Vernacular Name

Umm droos

37.1.3 Family

Asteraceae

37.1.4 Common Name

Cage Thistle, Birdcage thistle

37.1.5 Botanical Description

A small, annual herb reaches up to 20 cm and resembles a thistle, with multiple stems branching from the base and narrow, pointed leaves (lanceolate) with prickly edges. The flowers are grouped in clusters of 3–4, with each head surrounded by prickly bracts (Figures 37.1.1–37.1.3).

37.1.6 Geographical Distribution

This species is native to a vast region, spanning from the Canary Islands in the west to western Pakistan and the Arabian Peninsula in the east. In Oman, it is found in the Western Hajar mountains and Musandam.

FIGURE 37.1.1 Prickly bracts and leaves of *Atractylis cancellata*.

FIGURE 37.1.2 Apical part of *Atractylis cancellata*.

FIGURE 37.1.3 Plant habit of *Atractylis cancellata*.

37.1.7 Chemical Composition

In addition, many studies have been reported *Atractylis* plants, contain diterpenes, triterpenes, saponins, and flavonoids [635–637]

A recent report showed the presence of the alkaloid pyrroloquinoline A, along with 15 identified components, in the petroleum ether, ethyl acetate, and n-butanol extracts of the whole plant *Atractylis cancellata* L. [638].

37.1.8 Uses

The *Atractylis* species are used in folk medicine against urinary retention, intestinal parasites, and snakebite poisoning, as well as for their anti-inflammatory and antipyretic activities [638–642].

DOI: 10.1201/9781003295488-37

Components isolated from *Atractylis cancellata* L. demonstrated antioxidant and anticholinesterase activities [638]. *Atractylis cancellata* L. is used in folk medicine for the treatment of skin disorders [643].

37.1.9 Toxicity and Clinical Data

This plant contains alkaloids; thus, there might be toxicity associated with its chemical composition. A plant belonging to this genus, such as *A.* gummifera, is identified as poisonous and is extensively present in North Africa. The juice of the rhizome is poisonous because of the presence of toxic glycosides such as atractyloside and carboxyatractyloside, which can cause severe hepatitis with fatal liver failure [644].

37.1.10 Most Cited Research

iCite, Database, NIH Open Citation Collection was used from June to October 2022 to obtain the details of the most cited research. Filters for research papers were applied using the keywords *Atractylis cancellata.* The NIH database search showed the presence of the alkaloid pyrroloquinoline A, with 15 identified components, in the petroleum ether, ethyl acetate, and n-butanol extracts of the whole plant *Atractylis cancellata* L. [638].

38 Genus *Atriplex*

38.1 *Atriplex farinosa* Forssk.

38.1.1 Synonyms

Obione farinosa

38.1.2 Family

Amaranthaceae

38.1.3 Vernacular Name

afar

38.1.4 Common Name

Saltbush, Orache, Raghal

38.1.5 Botanical Description

This low perennial shrub attains a height of up to 1–2 m, with distinctive bluish-gray oblong leaves covered in a white powdery substance. It has small pale green flowers with a pink tinge at the base, clustered in gray or red bracts. The fruit is enclosed in flat, circular bracteoles (Figures 38.1.1–38.1.3).

38.1.6 Geographical Distribution

The plants belong to the genus *Atriplex*, which is commonly known as the lagoon or sprawling saltbush. These plants are extensively present in arid and semi-arid areas, such as the Middle East.

This plant is distributed in Djibouti, Egypt, Eritrea, Gulf States, Kenya, Oman (Dhofar), Saudi Arabia, Sinai, Socotra, Somalia, Sudan, and Yemen [645,646]. In Oman it is restricted to the coastal areas of central and southern Oman.

FIGURE 38.1.1 Leaves of *Atriplex farinose*.

FIGURE 38.1.2 Flowers and fruits of *Atriplex farinose*.

FIGURE 38.1.3 *Atriplex farinose* in flowers.

38.1.7 Chemical Composition

Phytochemical investigation reveals the presence of flavonoids, alkaloids, tannins, and saponins. Additionally, this study also reveals the presence of rutin (Figure 38.1.4) [647].

38.1.8 Uses

A recent study showed the protective effect of *A. suberecta* extract against oxidative and apoptotic hepatotoxicity [647]. Plants belonging to the same genus, such as *A. halimus*, demonstrated a nephroprotective effect in the previous study [648].

DOI: 10.1201/9781003295488-38

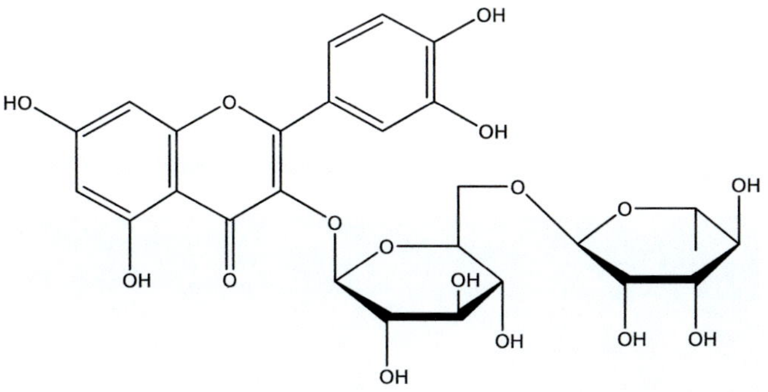

Rutin

FIGURE 38.1.4 Chemical component of plant belonging to *Atriplex genus* (*Atriplex suberecta*).

38.1.9 Toxicity and Clinical Data

Plants belonging to this genus have the tendency to accumulate high levels of cadmium, selenium, and tellurium when grown in soil contaminated with respective elements, which may cause toxicity.

38.1.10 Patents

Stephen Behr, Philippe Duret, Nathalie Gendron, Johane Guay, Bernard Lavallee, Brigitte Page. US20070122492A1: Plant extracts and dermatological uses thereof, 2004.

- This invention claims that dermatological formulations comprising one or more plant extracts, such as *Atriplex hortensis*, are capable of inhibiting one or more extracellular proteases to treat various dermatological conditions.

38.1.11 Plant Pathology

Pernospora farinosa, the downy mildew causal agent, is reported on various species of *Atriplex*. Further, an extensive report on halophytes, including *Atriplex*, indicated the presence of various fungal groups of pathogens, including rust, smuts, powdery mildew, as well as rot rots and leaf spots [649]. Other reports indicated that plants belonging to the genus accumulate selenium, which indicates phytoremediation potential and shows resistance to the beet armyworm, *Spodoptera exigua* [650].

39 Genus *Bacopa*

39.1 *Bacopa monnieri* (L.) Wettst.

39.1.1 Synonyms

Anisocalyx limnanthiflorus, Bramia monnieri, & Gratiola monnieri

39.1.2 Family

Plantaginaceae

39.1.3 Vernacular Name

Zahrat al maay

39.1.4 Botanical Description

An annual herb with multiple stems growing from the ground, covered in many small, gland-tipped hairs. The leaves are sessile and vary in shape from ovate to elliptic-lanceolate, with pointed tips and smooth edges. The flowers have a pale purple-blue hue and a tubular shape with five unequal lobes at the top (Figures 39.1.1–39.1.3).

39.1.5 Geographical Distribution

This species has a vast, pan-tropical distribution, thriving in both tropical and subtropical regions around the world. In Oman, it is found throughout the country near wet areas.

39.1.6 Chemical Composition

The phytochemical analysis reveals the presence of chemical components, mainly alkaloids such as monitoring, brahmin, nicotine, hersaponin, herpestine, and saponins A–C, triterpenoid saponins, Bacosides A1–A3 and B, acetogenins A1–A4, steroids, triterpene, and bacosine. In addition to bacopasaponins A–F, it also contains triterpenoid glycosides such as bacopasides III–V. It also contains serine and pseudojujubogenin glycoside. Additionally, it contains α-alanine, stigmastanol, β-sitosterol, glutamic acid, betulinic acid, D-mannitol, and aspartic acid [651–659]. The major chemical components are represented in Figure 39.1.4.

FIGURE 39.1.1 Flower portion of *Bacopa monnieri.*

FIGURE 39.1.2 Shoot apex of *Bacopa monnieri.*

FIGURE 39.1.3 Branch with flowers and leaves of *Bacopa monnieri.*

39.1.7 Uses

B. monnieri has been used in the form of Ayurvedic medicines for a long time, mainly to enhance brain function and improve memory. This memory-boosting plant has been a well-known part of the Indian Ayurvedic medicinal system for more than 3,000 years [661,662].

DOI: 10.1201/9781003295488-39

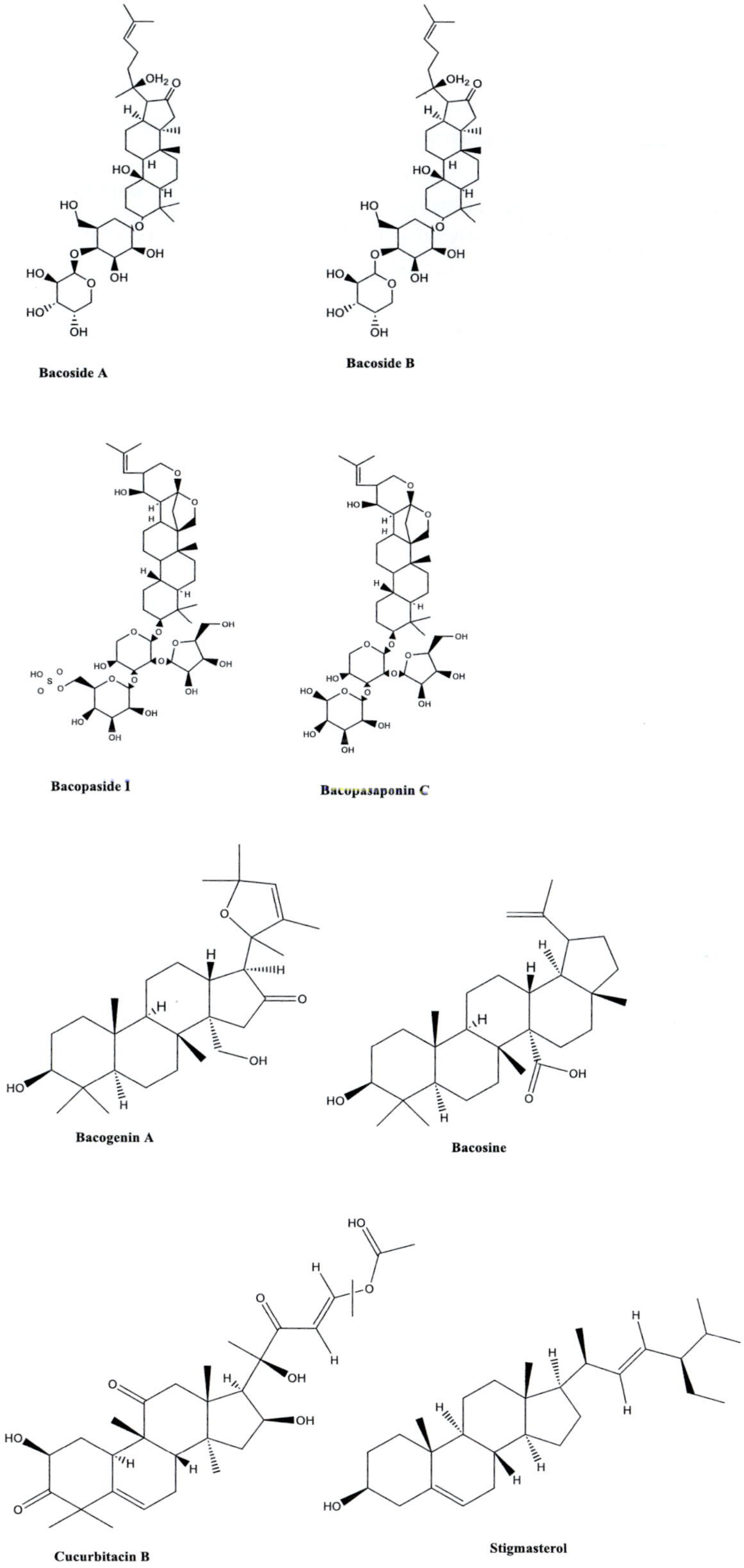

FIGURE 39.1.4 Major phytoconstituents present in extracts of *Bacopa monnieri* [660].

Traditionally, it is used to treat several complications and acts as a potent medicine, as mentioned below [663–665]:

- To treat insanity and epilepsy
- Act as a digestive aid
- Improve learning
- Improve memory and concentration
- To treat anxiety

- Skin complications
- Treat asthma

The *Bacopa* herb is a nootropic herb, i.e., it acts as a cognitive enhancer. This herb and its preparations help in repairing impaired neurons, smooth muscle relaxant effects, synaptic activity restoration, defective neuronal synthesis, and increased brain activity. Additionally, this herb exhibits various therapeutic effects, such as tranquilizer effects, hepatoprotective activities, cardiovascular, antilipidemic, adaptogenic, analgesic, and sedative effects; antidiabetic and endocrine effects; antipyretic, antimicrobial, antihypertensive, neuroprotective, and gastrointestinal effects; and antidepressant, antiepileptic, anticancer, antiarthritic, and anti-inflammatory effects [666–668].

39.1.8 Mechanism of Action

Triterpenoid saponins, mainly bacosides, present in this plant, are primarily responsible for its pharmacological effects. These components enhance the impulse transmission of nerves and also restore the physiological function of damaged neurons by increasing neuronal synthesis and kinase activity. These active chemical components are also evidenced by the synaptic activity restoration, which eventually results in an improvement in nerve impulse transmission [669].

Improvement in nerve impulse transmission further helps in cognitive functions such as improvement in learning, memory, attention span, focus, and concentration. It is also reported that *Bacopa* normalizes or balances the level of neurotransmitters by influencing their synthesis and availability [670,671]. Additionally, this plant and its preparations possess neuroprotective effects through antioxidant mechanisms and by altering neurotransmitters. Further, *Bacopa* also inhibits oxidative stress, inflammatory markers trigger microglia activation, and stimulates the PI3/Akt pathway.

39.1.9 Toxicity and Clinical Data

As far as the toxicity of this herb is concerned, it was found that the plant extract did not cause any serious undesirable effects at a dose of 5,000 mg/kg. Furthermore, the plant extract, at different doses from 30 to 1,500 mg/kg, administered for 8.8 months, did not cause any toxicity in rodents [672].

Clinical investigation revealed that *B. monnieri* decreased the rate of memory loss [673], enhanced memory performance in aged people [674], and improved cognitive performance in humans [675]. Nevertheless, the plant extract causes side effects in the gastrointestinal tract, such as augmented stool incidence, abdominal cramps, and nausea [674]. Consummation of the plant by women also causes severe liver toxicity, which was recovered after discontinuation of the consumption [672,676].

39.1.10 Most Cited Research

iCite, Database, NIH Open Citation Collection was used from June to October 2022 to get the details of the most cited research. Filters for research papers were used by using the keywords *Bacopa monnieri*. NIH database search revealed that Brahmi plants accumulated substantial amounts of Cd in different plants and tolerated high levels of Cd by producing antioxidants to reduce oxidative stress induced by cadmium [677].

39.1.11 Patents

	Title	Patent no.	Inventors	Year	Invention details
1	Herbal compositions comprising extracts of *Bacopa monnieri* and *Terminada chebula*	US11071765B2	Ganga Raju et al.	2017	Relates to synergistic compositions for improving brain health and brain function
	Uses of *Bacopa monnieri* extract	AU2013354889B2	Andrew Scholey, Con STOUGH	2013	Methods for acutely improving/ enhancing cognitive performance in a human subject comprising administration of an extract of *Bacopa monnieri*.
	The isolation and purification method of *Bacopa monnieri* total saponin extracts and its application in drugs for nervous is prepared	CN107789429A	张卫东孙青* (Zhang Weidong, Sun Qing, Sun Qing)	2016	Isolation and purification method of *Bacopa monnieri* total saponin extracts and its application in drugs for nervous is prepared

(*Continued*)

(Continued)

Title	Patent no.	Inventors	Year	Invention details
Bitterless *Bacopa monnieri* extract composition for memory improvement and a method of synthesizing the same	WO2017103831A1	Baidyanath MISHRA, Sujan GANAPATHY, Shariq Afsar	2016	Method for synthesis of non-bitter herbal composition comprising *Bacopa monnieri* extracts
Role of an herbal formulation in the prevention and management of age-related neurodegenerative disorders with special reference to senile dementia	US20120034324A1	Govind Prasad, Dubey, Aruna Agrawal, Nirupama Dubey, Shipra Dubey, Rajesh Dubey, Samathanam Mercy Deborah	2011	Herbal formulation for the prevention and management of neurodegenerative disorders
Tea composition	WO2008107296A1	Megha Sarthak Adavi, Ananthanarayana Bhima Rao Dasalukuntey, Prabha Narasimhan, Vikas Maruti Pawar, Skand Saksena, Jyoti Kumar Tiwari	2008	Composition comprising tea; and from 0.1 to 4.0 weight % *Bacopa monnieri*
Dietary supplement compositions	US11273195B2	Alexander B. Rabovsky, Stephanie Y. Nielson	2016	Dietary supplement compositions having an acetylcholinesterase inhibitor (e.g., huperzine A), a *Bacopa monnieri* extract
Composition for improving mental capabilities in mammals	US20030157201A1	Maharaj Pandita, Govind Dubey	2003	Herbal composition for improving overall mental performance in children, adults, and mentally deficient people
Process for producing enriched fractions containing up to 100% of bacopasaponins from the plant materials of *Bacopa* species	US8541381B2	Raju Gokaraju Ganga, Raju Gokaraju Rama, Subbaraju Gottumukkala Venkata, Trimurtulu Golakoti Sivaramakrishna Chillara	2005	Fraction enriched with jujubogenin and psudojujubogenin lycosides from *Bacopa* species.
Compositions for alleviating inflammation and oxidative stress in a mammal	US9265808B2	Joe Milton McCord, Paul R. Myhill	2013	Composition that increases antioxidant defense potential in a subject is disclosed. The composition contains *Bacopa monnieri* extract comprising a *Bacopa monnieri* active ingredient
A synergistic herbal composition from *Bacopa* species for management of neurodegenerative disorders and a process of preparation thereof	AU2007257480B2	Amit Agarwal	2007	Synergistic herbal composition for improving cognition, learning, and memory and a process of preparation thereof
Bacopa monnieri–based edible cookies for enhanced cognitive performance	AU2021101067A4	Kavita Narayan Gaisamudre (Sarwade), Prakash Pralhad Sarwade, Sonia Sorout, Narayan Dattatraya Totewad, Murlidhar Prabhakar Wadekar	2021	Preparation of edible cookies using Brahmi extract for enhancing cognitive abilities of the user
Cosmetic preparations containing extracts from *Phyllanthus emblica*, *Centella asiatica*, and/or *Bacopa monnieri*	US6261605B1	Shyam B. Singh-Verma	1996	Cosmetic formulations for topical application containing extracts from *Phyllanthus emblica*, *Centella asiatica*, and/or *Bacopa monnieri*, and the use of such formulations for the care of the human skin

39.1.12 Plant Pathology

There are few reports available on the association of this plant with pathogens. However, some reports indicated endophytic fungi associated with *B. monnieri* with antimicrobial potential [678]. *Bacopa monnieri* accumulates and is also found to tolerate high amounts of Cd by enhancing the level of phytochelatin synthesis and various antioxidants. This report reveals that *Bacopa monnieri* could be a suitable plant for phytoremediation [677].

Another report showed the presence of endophytic fungi in *Bacopa monnieri*. These endophyte fungi present in *Bacopa monnieri* showed cytotoxic and antimicrobial effects [679].

One more report showed the co-cultivation of endophytes, namely *Bacillus subtilis* and *Klebsiella aerogenes* with *B. monnieri* cuttings, showed improvement in the accumulation of bacoside A via modulating numerous genes of metabolic pathway [680].

Furthermore, arsenate [As(V)] reducing endophytic bacterial strains (CIMAP-A4 and BAC-7) were found in *Bacopa monnieri*. Findings from this study reveal that inoculation of arsenate [As(V)] reducing endophytic bacterial strains (BAC-7 and CIMAP-A4) in As-contaminated soil showed an increase in biomass and reduced accumulation in plant tissue. Further, one of the strains also improved arsenate reductase activity, plant nutrient uptake, and non-protein thiol synthesis [681].

A recent study showed the presence of a new strain of *Trichoderma lixii* (IIIM-B4) from *B. monnieri* with long-chain peptaibols. It was found that this strain showed antifungal properties against *Candida albicans* [682].

40 Genus *Balanites*

40.1 *Balanites aegyptiaca* (L.) Delile

40.1.1 Synonyms

Ximenia aegyptiaca & *Agialid aegyptiaca*

40.1.2 Family

Zygophyllaceae

40.1.3 Vernacular Name

Al hgleeg

40.1.4 Common Name

Desert date, Soapberry tree, or Thorn tree

40.1.5 Botanical Description

This spiny fruit-yielding multi-branched shrub attains a height of up to 10 m. The stem has yellow or green color thorns. The color of bark obtained from this tree is brownish to grey with a deep fissure. This shrub bears green asymmetric ovate leaflets and yellowish-green fragrant flowers [683] (Figures 40.1.1–40.1.3).

40.1.6 Geographical Distribution

This is native to Africa and the Middle East. It's widely available in Africa, South Asia, and the Middle East.

It is available in India, Myanmar, Iran, Jordan, Oman, Palestine, Saudi Arabia, Syria, and Yemen [684]. In Oman, it's found in the Eastern Hajar mountains.

FIGURE 40.1.1 Flowering branch of *Balanites aegyptiaca.*

FIGURE 40.1.2 Inflorescence of *Balanites aegyptiaca.*

FIGURE 40.1.3 Adult plant of *Balanites aegyptiaca.*

40.1.7 Chemical Composition

Plants contain different chemical components: saponins (spirostanol saponins, furostanol saponins, and open-chain steroidal saponins), polyphenolic components (phenolic acids, flavonoids, and coumarins), pregnane glycosides, phytosterols, alkaloids, and steroids. These components are present in different parts of the plants. Spirostanol saponins like balanitis 1–7 and deltonin are present in fruit, seeds, and root/stem bark (Figure 40.1.4). It also contains furostanol and open-chain steroidal saponins.

40.1.8 Uses

Plant materials, especially young leaves and tender shoots, are consumed as vegetables, whereas fruits and leaves are used as fodder for livestock [1]. Bark and its fibers are used for the preparation of medicines. Liquor, as well as

DOI: 10.1201/9781003295488-40

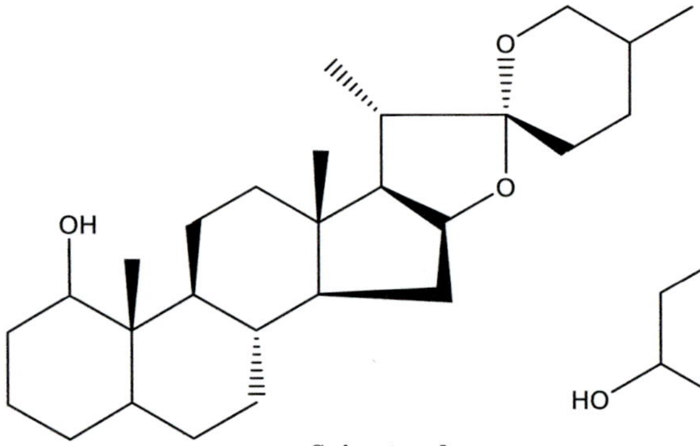

Spirostanol

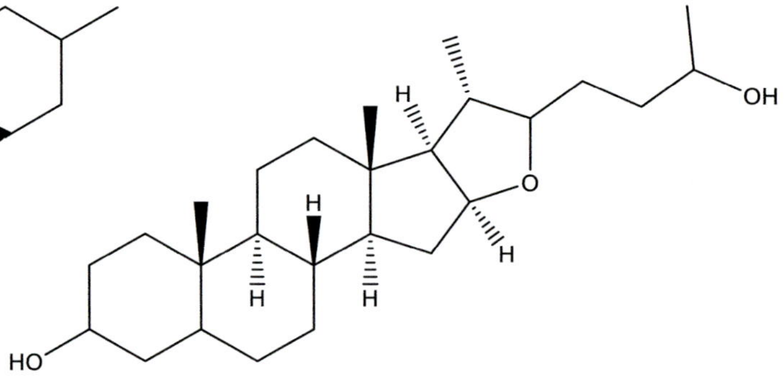

Furostanol

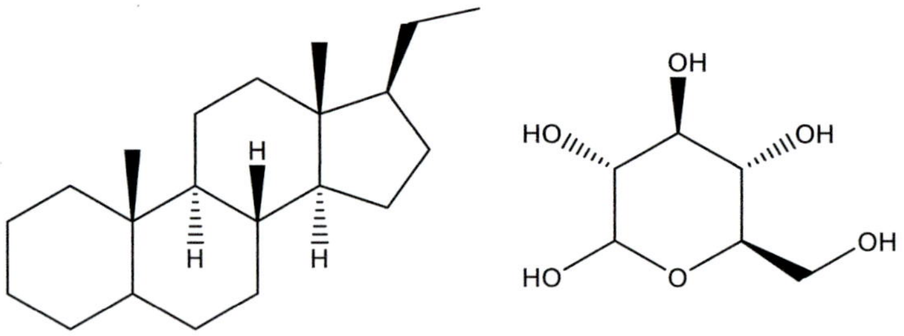

Pregnane glucosides

FIGURE 40.1.4 Chemical components of *Balanites aegyptiaca*.

beverages, are derived from fruits. Fruit of this plant was used to treat scorpion stings, whereas bark is used as a vermifuge [685].

Plant material derived from this shrub can be used for food, medicinal, biodiesel, and cattle industries. Traditionally, this shrub is used to treat asthma, wounds, epilepsy, stomach aches, fever, diarrhea, intestinal worm infection, dysentery, jaundice, hemorrhoid, syphilis, constipation, and malaria [683,686].

Further, the plant has been reported for cardioprotective, hepatoprotective, antioxidant, antimicrobial, anti-inflammatory, and antidiabetic properties and molluscicidal effects [687–693].

40.1.9 Mechanism of Action

The mechanism of action behind its anti-inflammatory effects could be due to the presence of flavonoids and saponins that exert their anti-inflammatory effects by involving the inhibition of cyclooxygenase and lipoxygenase activities [694].

40.1.10 Toxicity and Clinical Data

The susceptibility of the extract on a freshwater snail, *Biomphalaria pfeifferi*, was tested. The LC_{50} of the lethal concentrations of the extract on the fish and the snails was 1.12 and 2.0 mg/L^{-1}, respectively, while the effects of the sublethal concentrations on the growth rate and hematocrit percentage of the fish were determined. Signs of toxicosis, such as air gulping, reduction in tailfin beating rate, and overturning, were observed in the fish before death occurred. The toxicant also reduced the growth rate and hematocrit percentage of the fish.

Toxicological analysis reveals that the extracts were found to be toxic beyond 1.12 and 2.0 mg/L^{-1} on the fish and the snails, respectively. This could be due to the hemolytic or osmoregulation impairment caused by toxic components [695].

It was also found that aqueous extract administration of *Balanites aegyptiaca* bark showed a dose-dependent decrease in serum bilirubin levels in rats [696].

40.1.11 Most Cited Research

iCite, Database, NIH Open Citation Collection was used from June to October 2022 to get the details of the most cited research. Filters for research papers were applied using the keywords *Bacopa monnieri*. NIH database search revealed *Phlebotomus orientalis*, the main vector of *Leishmania donovani*, typically associated with *Balanites aegyptiaca* vegetation. Infection rates with *Leishmania* are dependent on the number of sand flies [697].

40.1.12 Patents

Zeev Wiesman *Balanites aegyptiaca* saponins and uses thereof. WO2006137069A2.

Title	Patent no.	Authors	Year	Detail
Production of biodiesel from *Balanites aegyptiaca*	US20080271364A1	Zeev Wiesman, Mordechay Herskowitz, Sarina Grinberg	2006	Production of biodiesel from *Balanites aegyptiaca* oil or crushed nuts
Synthesis of nanoparticles using *Balanites aegyptiaca*	US9889170B1	Manal Ahmed Gasmelseed Awad, Rabab Abd El Moneim Khalil El Dib, Awatif Ahmed HENDI, Shaza Mohamed Adel Al-Massarani, Khalid Mustafa Osman Ortashi	2016	Development of nanoparticles from desert date
Balanites aegyptiaca method of treatment	WO1997023234A1	Osman Abd El Moneim Hamid	1995	Reveal the effectiveness of *Balanites aegyptiaca* bark aqueous extract in treatment of both experimental obstructive jaundice in rats and infective hepatitis in humans
Effect of *Balanites aegyptiaca* on the regression of liver fibrosis and cirrhosis	WO2019221645A4	Mohamed Osman ABDELMONEM, Osman Abdelmonim ABDELMONIM, Hisham Hassan ABDELWAHAB	2019	Utilized in patients with liver fibrosis, cirrhosis, myocardial fibrosis, and lung fibrosis
Balanites aegyptiaca extracts for treatment of HIV/AIDS and leukemia	WO2001049306A1	Osman A/Monieum Hamid, Mohy Eldin Hassan A/Wahab, Zeidan Zeidan Abdu, Sakina Mohamed Idris	1999	*Balanites aegyptiaca* extracts for treatment of HIV/AIDS and leukemia

40.1.13 Plant Pathology

Several plant pathogens were reported on *Balanites aegyptiaca*, causing foliar diseases, including *Fusarium* leaf spots, *Mucor* leaf spots, *Aureobasidium* leaf rusts, and *Aspergillus* leaf rust [698].

41 Genus *Basananthe*

41.1 *Basananthe berberoides* (Chiov.) J.J. de Wilde

41.1.1 Synonyms

Carania berberoides

41.1.2 Family

Passifloraceae

41.1.3 Botanical Description

A spiny shrub with multiple stems arising from the base reaches up to 5 m. Spines range from 0.5 to 2 cm long. Leaves are small (0.2–1 cm), entire or with minute teeth, and come in various shapes (ovate, obovate, oblong). Flowers: solitary or clustered, white with hairy sepals. Fruits: oblong capsule with spiny tip contains 2–3 seeds (Figures 41.1.1 and 41.1.2).

FIGURE 41.1.1 *Basananthe berberoides* with fruit.

FIGURE 41.1.2 Natural habit of *Basananthe berberoides*.

41.1.4 Geographical Distribution

B. berberoides, found in East Ethiopia, East Somalia, Oman, and Yemen.

In 2020, *Basananthe berberoides* was initially recorded in Oman, where it was found on a gravelly wadi between Sadah and Mirbat.

41.1.5 Uses

Crushed leaves of *B. hanningtoniana* are boiled with water and administered orally to treat Darissaa (Gamoji or Zalaqa) in Ethiopia [699].

In Tanzania, the roots of closely related species (*B. sandersonii*) are chewed, and the juice is taken internally to treat snake bites [700]. Leaves of this plant are traditionally used in wound healing and as an antidote.

DOI: 10.1201/9781003295488-41

42 Genus *Berberis*

42.1 *Berberis baluchistanica* Adrendt

42.1.1 Family

Berberidaceae

42.1.2 Vernacular Name

In Oman locally known as mahlab

B. baluchistanica called "Brahvi" is known as Zarchin in Balochistan province, Pakistan [701]. It's also known as Zralag in Pashto and Korae in Balochi for diverse uses [702].

42.1.3 Botanical Description

B. baluchistanica is a spiny shrub that usually reaches up to 1–2(-3) m tall with reddish to brown spiny stems. Leaves, simple, leathery, and rigid in winter, transform into a vibrant display of bright orange-red. Flowers: bright yellow, clustered with up to ten individual blooms per cluster. Fruits: fleshy berries that morph from green to a vibrant red when unripe before finally settling into black when fully mature (Figures 42.1.1–42.1.4).

42.1.4 Geographical Distribution

Berberis baluchistanica has a restricted distribution, occurring only in Oman and Pakistan. Within Oman, it is specifically confined to the higher elevations of the Western Hajar Mountains.

42.1.5 Chemical Composition

Berberis baluchistanica has been known for the presence of berberine, and berberine-like compounds, such as pakistanine, pakistanamine, and (+)-baluchistine, have been isolated [701,703,704].

A recent study explored chemical components such as berberisinol, berberine, 8-oxoberberine, etc., that possess notable anti-leishmanial, antibacterial, antioxidant, antifungal, and anti-diabetic effects [702,705].

FIGURE 42.1.1 Flowers of *Berberis baluchistanica*.

FIGURE 42.1.2 Fruits of *Berberis baluchistanica*.

FIGURE 42.1.3 *Berberis baluchistanica* leaves and thorns.

FIGURE 42.1.4 *Berberis baluchistanica* during winter.

DOI: 10.1201/9781003295488-42

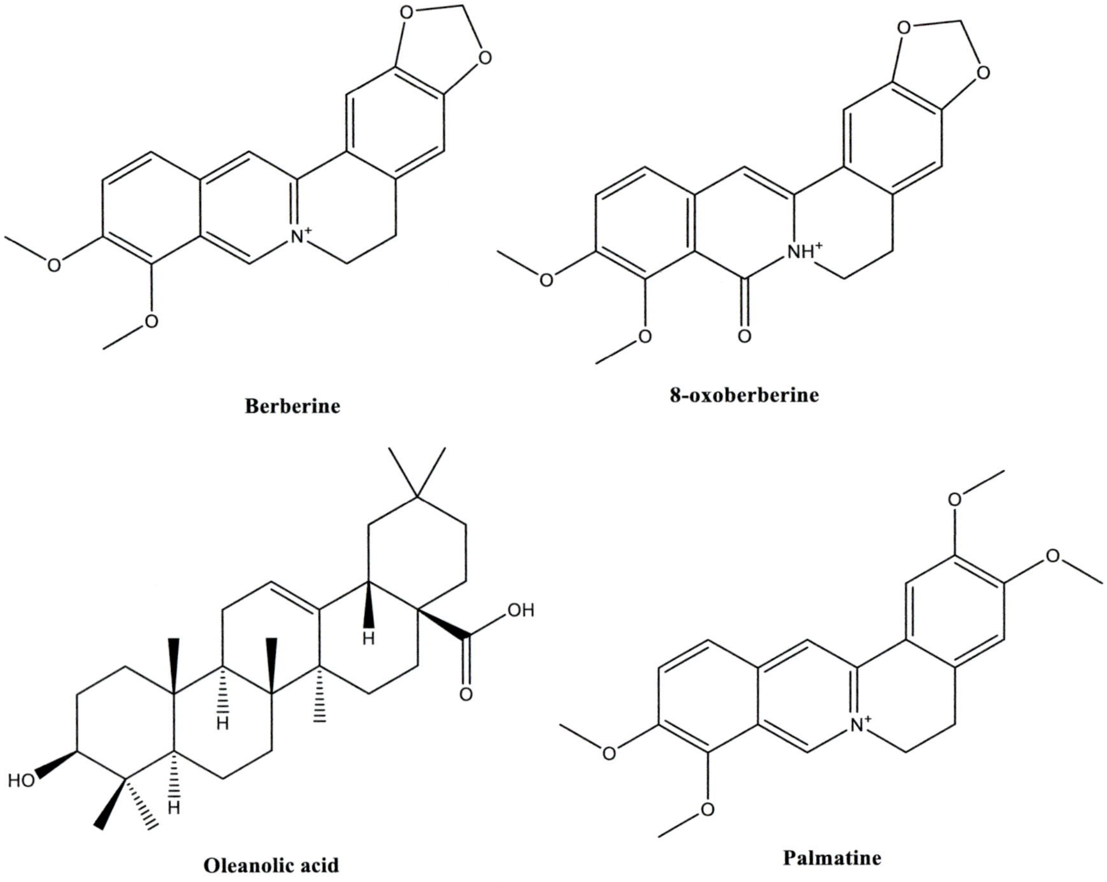

FIGURE 42.1.5 Chemical components of *Berberis baluchistanica*.

Moreover, several other chemical components such as berberisinol, gallic acid [706], 8-oxoberberine [11], palmatine [707], oleanolic acid [708], β-sitosterol [709], and berberine [710] were identified in *Berberis baluchistanica* (Figure 42.1.5).

42.1.6 Uses

Due to the presence of berberine, the plant is used in the treatment of several ailments such as wound healing, kidney-related disorders, cough, cancer, coronary heart diseases, rheumatism, eye disease, fever, gentamicin-induced renal toxicity, kidney stone removal, and other infections of human beings [701,711,712]. The plant is known for its potential role as an anti-inflammatory, anti-leishmanial, antibacterial, antifungal, and anticancer effects [701,702,712]. Additionally, plant is also used as a food preservative [713].

42.1.7 Mechanism of Action

Berberis baluchistanica shows its insulin-independent anti-diabetic effects by acting as an AMP-activated protein kinase (AMPK) activator and α-glucosidase inhibitor. This is achieved by inhibition of mitochondrial function, glycolysis stimulation, and activation of the AMPK pathway [714].

42.1.8 Toxicity and Clinical Data

This medicinal plant is considered non-toxic and used in decoction or powder form [715].

However, several toxicology effects of *Berberis vulgaris* and its active constituent, berberine, have been explored, such as cytotoxicity, genotoxicity, mutagenicity, carcinogenicity, GI upset and ulceration, immunotoxicity, phototoxicity, neurotoxicity, cardiotoxicity, and jaundice [716].

42.1.9 Most Cited Research

iCite, Database, NIH Open Citation Collection was used on January 2, 2024, to get the details of the most cited research. Filters for research papers were used by using the keywords *Berberis baluchistanica*. NIH database search revealed the endemic medicinal plants of Northern Balochistan, Pakistan, and their uses in traditional medicine [717].

42.1.10 Plant Pathology

No plant disease is reported in the wild *B. baluchistanica*. Recently, fungal endophytes were isolated from *Berberis aristata* DC. This indicates the possibility of the presence of fungal endophytes in *Berberis baluchistanica* [718].

43 Genus *Bergia*

43.1 *Bergia ammannioides* Roxb.

43.1.1 Synonyms

Lechea verticillata, Sagina ammannioides & Tillaea rubella

43.1.2 Family

Elatinaceae

43.1.3 Common Name

Water-Fire and in Hindi, it's called Tal-bhungro, Ammannia Waterwort.

43.1.4 Botanical Description

This plant is an annual erect branched shrub that attains height up to 8–35 cm. The color of the stem and branches is pinkish, and its appearance is glandular-pubescent. It bears small, white, and bisexual flowers [719] (Figures 43.1.1 and 43.1.2).

43.1.5 Geographical Distribution

In Asia, it's distributed in Afghanistan, China, India, Indonesia, Iran, Iraq, Kenya, Laos, Malaysia, Nepal, Pakistan, the Philippines, Sri Lanka, Tajikistan, Thailand, Vietnam, and Oceanic Islands. In Africa, it's available in Mozambique, Sudan, Zambia, and Australasia. *B. ammannioides* Henye ex Roth. is the one growing in Egypt. In Oman, it is recorded from Southern Mountains [4].

FIGURE 43.1.1 Stem in flowers of *Bergia ammannioides.*

FIGURE 43.1.2 Shoot system of *Bergia ammannioides.*

43.1.6 Chemical Composition

Few reports were found about the isolation of gallicin, gallic acid, lupeol, and β-sitosterol from *B. suffruticosa* (Delile) Fenzl [720]. (Figure 43.1.3).

The hexane fraction of *Bergia ammannioides* contains β-sitosterol, lupeol, cyclolaudenol, and cycloartenol, whereas the ethanol fraction contains quercetin, ellagic acid, kaempferol-3-O-α-l-rhamnoside, and quercetin-3-O-α-l-rhamnoside (Figure 43.1.3) [721].

43.1.7 Uses

Bergia species are important medicinal plants in India that are traditionally used for wound healing and applied on sores [722]. It has been traditionally used for treating skin diseases [4]. The plant has also been reported for its free radical scavenging activity [722,723]. The extract derived from *B. ammannioides* showed antibacterial, antioxidant, anti-inflammatory, and wound-healing effects [721].

Another report showed that the ethyl acetate–soluble fraction derived from this plant showed the highest antioxidant effects, which could be associated with its phenolic content [723].

43.1.8 Most Cited Research

iCite, Database, NIH Open Citation Collection was used from June to October 2022 to get the details of the most cited research. Filters for research papers were used by using the keywords *Bergia ammannioides.* NIH database search revealed the antibacterial, antioxidant, and topical anti-inflammatory activities of *Bergia ammannioides* [721].

DOI: 10.1201/9781003295488-43

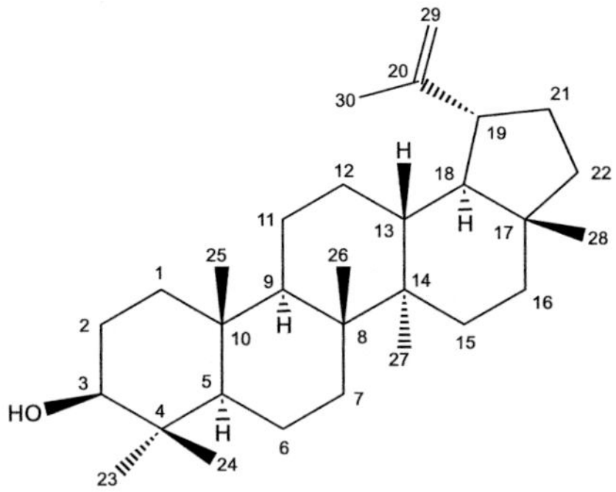

Compund 1

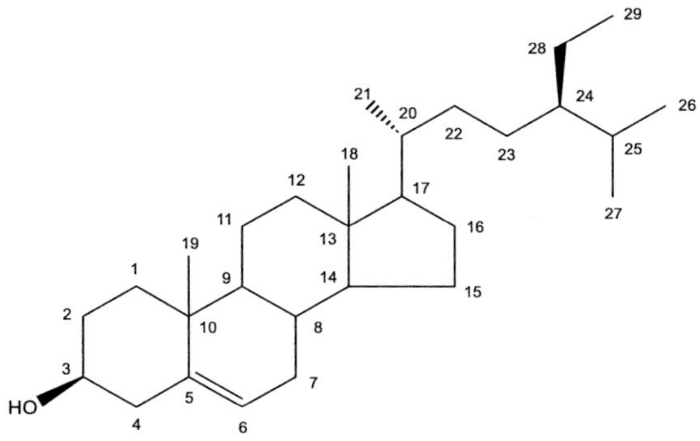

Compund 2

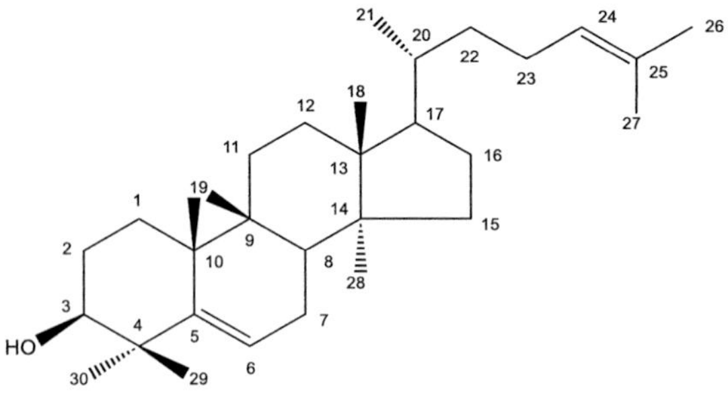

Compund 4

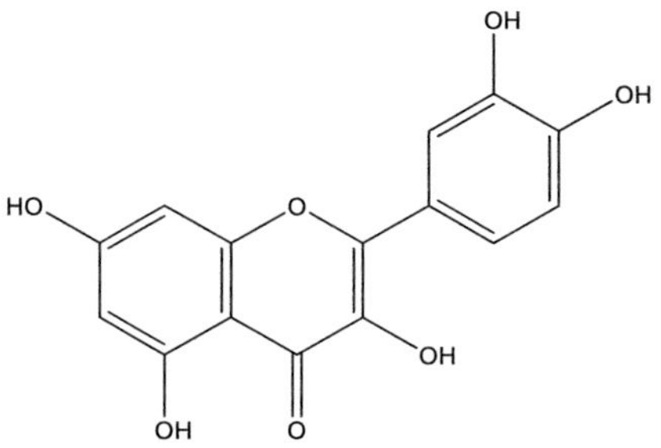

Compund 5

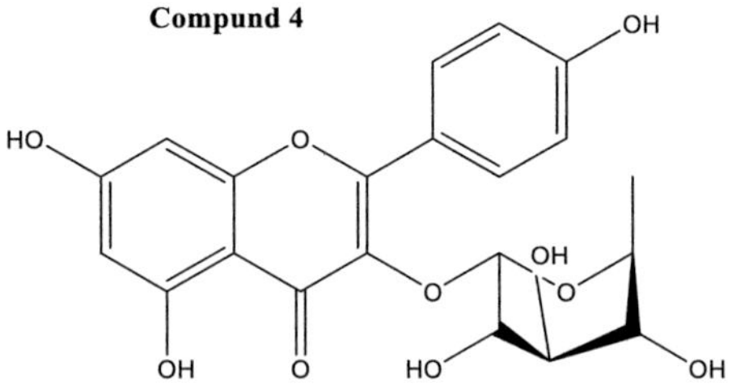

Compund 7

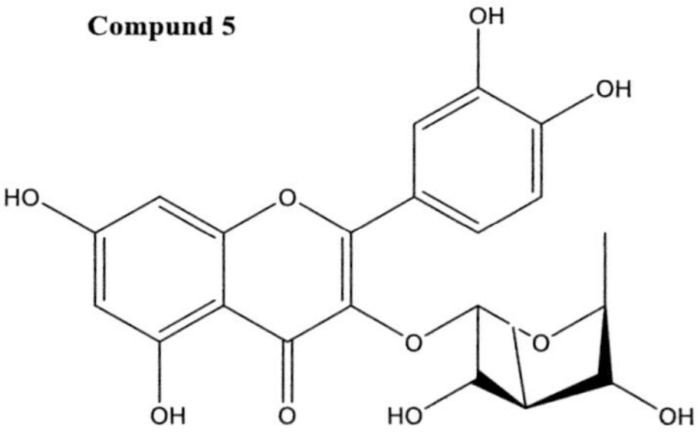

Compund 8

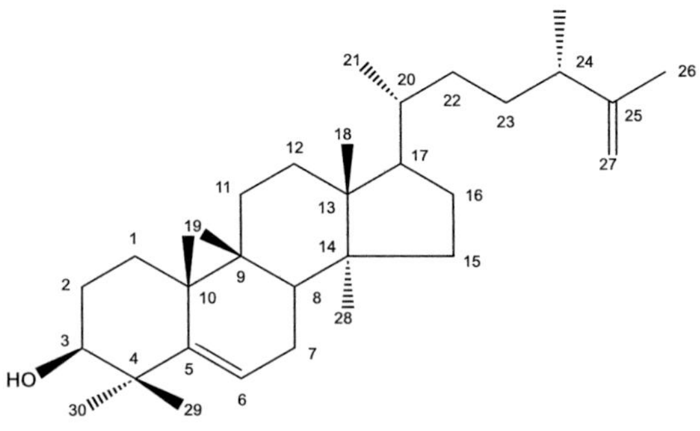

Compund 3

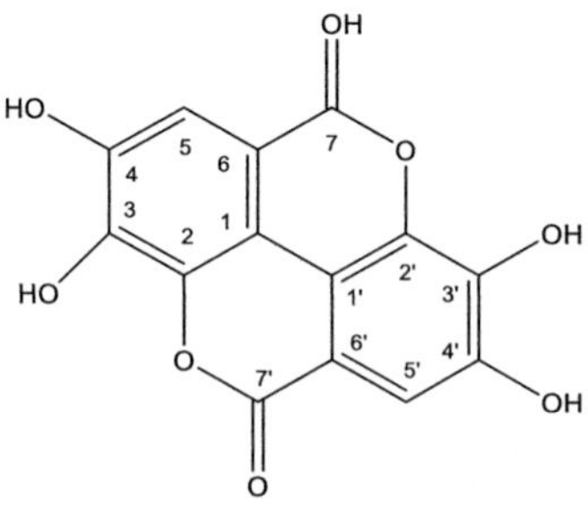

Compund 6

FIGURE 43.1.3 Structures of the isolated components: lupeol (1), b-sitosterol (2), cyclolaudenol (3), and cycloartenol (4).

43.1.9 Plant Pathology

Some species of this plant are reported as weeds in cultivated crops. Thus, many fungal, bacterial, and nematode diseases are reported. For example, *Erythricum salmonicolor* fungus causes canker, dieback, or pink disease in many plants, including various species of *Bergia* [724]. Indeed, plants belonging to this genus are susceptible to powdery mildews, root knots, and nematodes [4].

44 Genus *Bidens*

44.1 *Bidens biternata* L.

44.1.1 Synonyms

Coreopsis biternata

44.1.2 Family

Asteraceae

44.1.3 Vernacular Name

beshoukh

44.1.4 Common Name

Spanish needles, Black Jack, Jinzhan and Yinpan in China, and Phutjom, Kerrai, Kathori, and Kuro in India.

44.1.5 Botanical Description

B. biternata is an invasive weed with a quadrangular and hairy stem, trifoliate and opposite leaves, with green-colored hairy bracts at the base. This plant has yellow-colored discs and white or yellow ray florets, and bears achenes (Figures 44.1.1–44.1.3).

44.1.6 Geographical Distribution

B. biternata is native to Zimbabwe, Zambia, Zaire, Uganda, Tanzania, Taiwan, Sudan, Somalia, the Philippines, Nigeria, New Caledonia, Mozambique, Malawi, Korea, Kenya, Japan, Indonesia, Ghana, Ethiopia, Eritrea, Egypt, China, the Arabian Peninsula, Angola, and the Solomon Islands. It is naturalized in Swaziland, South Africa, Réunion, Pakistan, Namibia,

FIGURE 44.1.1 Ripe fruit of *Bidens biternata*.

FIGURE 44.1.2 Inflorescence of *Bidens biternata*.

FIGURE 44.1.3 Leaflet arrangement of *Bidens biternata*.

Lesotho, India, Botswana, Sri Lanka, and Afghanistan [725]. In Oman it is found in Southern mountains among cultivations and wet places, recently has been recorded from the Northern Oman in cultivations.

44.1.7 Chemical Composition

Plants belonging to the *Bidens* genus generally contain polyacetylenes, flavonoids, phenolics, glycosides, pheophytins, fatty acids, phytosterols, and terpenes [726,727].

Additionally, *Bidens* plants also contain sesquiterpenes, lactones, aromatic derivatives, thiophenes, carotene, coumarins (umbelliferone, scopoletin, and aesculetin), vitamin C, and C17-, C14-, C13-polyacetylenes [728,729] (Figure 44.1.4).

Plants belonging to similar genera also contain polyacetylene glucosides and caffeoyl derivatives in *Bidens parviflora* [730]. *Bidens* plants usually contain chalcone

DOI: 10.1201/9781003295488-44

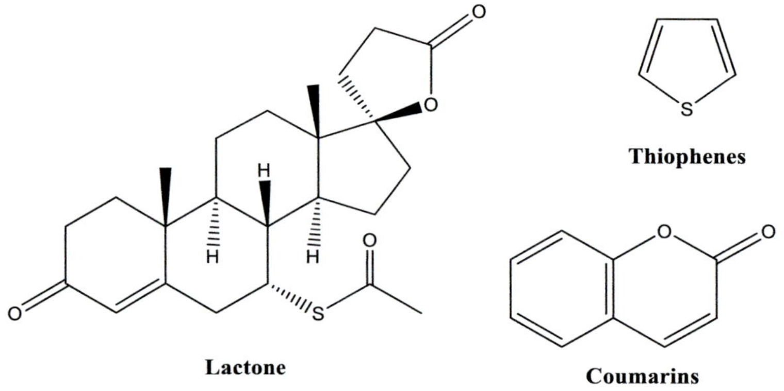

FIGURE 44.1.4 Chemical components of *Biden* plants.

glycosides, i.e., okanin. Okanin is a chalcone (1,3-diphenyl-2-propen-1-one) commonly found in this genus [731]. Further, HPLC investigation reveals the presence of compounds in *Bidens biternata* [732].

44.1.8 Uses

Traditionally, this plant is extensively utilized in both Chinese medicine and Western herbalism to treat several ailments, such as nausea, leprosy, fever, cough, and asthma [733].

B. biternata has potential benefits such as wound healing effects, anthelmintic, febrifuge, stimulant, anti-inflammatory, and diuretic properties [734]. Chinese people used its plant preparations to treat rheumatoid joint pain, blood stasis effects, sore throat, acute jaundice gastroenteritis, respiratory tract infections, snakebite, and other ailments [735].

In the Chinese-English Manual of Common-Used Herbs, under *guizhencao*, *B. biternata* along with *Bidens bipinnata* and *B. pilosa* are listed as source materials in order to clear away the superficial heat for the common cold of wind-heat type and the prevention of influenza. It also clears away heat and toxic materials for sore throat, appendicitis, snakebite, centipede bite, diarrhea, dysentery, and stomachache. Decoction of *B. bipinnata*, *B. pilosa*, and *B. biternata* with radix dichroae is considered anti-inflammatory [736].

It was also found that plants belonging to the *Bidens* genus showed substantial antimalarial effects, which could be due to the presence of acetylene compounds. The extracts of *B. biternata* contain three acetylenes that present less growth inhibition of *Plasmodium falciparum* under in vitro conditions. Thus, its antimalarial activity is not satisfactory [737].

B. biternata demonstrated higher anticancer effects than *B. bipinnata* and *B. pilosa*. This was because of the presence of anticancer components such as Tris (2,4-di-tert-butylphenyl), 4-hydroxy-2,4′-dimethoxychalc one, and 2,4-di-tert-butylphenol. Out of these components, 4-hydroxy-2,4′-dimethoxychalcone presented the highest anticancer activity [738].

Furthermore, extracts from *Bidens biternata* act as potential anticancer agents by showing significant selectivity against cancer cell lines versus normal cell lines [739].

Another study showed that *B. biternata* contains numerous nutritional factors with a low level of anti-nutritional factors and thus can be utilized to overcome the nutritional deficiency problem [740].

44.1.9 Mechanism of Action

The freeze-dried extract of *B. biternata* possesses significant antidiarrheal activity in both in *vitro* and in *vivo* models, which appears to be mediated by modulating both intestinal motility and secretory activity [741].

44.1.10 Toxicity and Clinical Data

Toxicology investigations revealed that plants belonging to the same genus, such as *B. pilosa*, are non-toxic and safe. It was found that oral administration of *B. pilosa* powdered aerial parts at 100 mg/mL is not toxic to rats at a dose limit of 2,000 mg/kg over 28 days [742]. Further, water and ethanol fractions from this plant showed insignificant toxic effects on rats *in vivo* [743]. Also, initial studies revealed no toxicity after the administration of the extract at 1 g/kg of body weight in mice [744].

44.1.11 Most Cited Research

iCite, Database, NIH Open Citation Collection was used from June to October 2022 to get the details of the most cited research. Filters for research papers were applied using the keywords *Bidens biternata*. NIH database search revealed *Bidens*' identification based on the noncoding regions of the chloroplast genome and nuclear ribosomal DNA [745].

44.1.12 Patents

The invention revealed a medicated bath bag containing *B. biternata* with other herbs for treating skin itch and a preparation method thereof.

CN105963393A. Medicated bath bag containing *Bidens biternata*, Chinese wampee leaf, *Cratoxylon ligustrinum* leaf, *Endospermum* leaf, and gardenia leaf for treating skin itch and preparation method thereof. 陈带喜莫晓茹杜睿杰汪碧涛王丽. 2016.

44.1.13 Plant Pathology

Bidens biternata is an invasive weed found in agricultural croplands. Besides its impacts on cultivated crops as a weed, it also harbors several plant pathogens. For instance, Tobacco Mosaic virus (TMV) and *Bidens Mottle virus* (BMV) are reported to infect *B. pilosa* [746]. Similarly, powdery mildew (*Golovinomyces ambrosiae*) [747] is reported on the same species. Thus, this weed can serve as an immediate source of infestation to other cultivated crops [746]. Recent investigation reveals the effects of intercropping with *Bidens* species plants on the *Ziziphus acido jujuba* seedlings' growth and accumulation of cadmium in this plant. It was found that intercropping with *Bidens* species is not appropriate for Cd-contaminated jujube orchards [748].

45 Genus *Blepharis*

45.1 *Blepharis ciliaris* (L.) B.L. Burtt

45.1.1 Synonyms

Ruellia ciliaris

45.1.2 Family

Acanthaceae

45.1.3 Vernacular Name

Kenep & kenoop

45.1.4 Common Name

Shawk al-Dhub & Eyelash plant

45.1.5 Botanical Description

Blepharis ciliaris is a subglabrous erect annual herb that attains a height up to 60 cm long. The plant has lanceolate leaves, sinuate-dentate margins, and corolla bright blue or bluish mauve (Figures 45.1.1 and 45.1.2).

45.1.6 Geographical Distribution

It is native to Iran, Pakistan, and Arabian Peninsula. In Oman it is found in Northern Oman.

45.1.7 Chemical Composition

Plants belonging to this genus, such as *B. ciliaris*, contain *Blepharis* (flavonoids), *B. sindica* contains alkaloids, whereas *B. edulis* contains phenolic acid derivatives, phytosterols, and derivatives of hydroxamic acids. These components have shown antimicrobial, anti-inflammatory, and anti-cancer effects [749]. Flavonol glycosides and blepharisides A and B were found in *Blepharis ciliaris* [750].

FIGURE 45.1.1 Inflorescence of *Blepharis ciliaris.*

FIGURE 45.1.2 Branch tips showing leaves and spines of *Blepharis ciliaris.*

Apigenin-7-O-(3″-acetyl-6″-E-p-coumaroyl-β-d-gluco pyranoside), naringenin-7-O-(3″-acetyl-6″-E-p-coumaroy l-β-d-glucopyranoside), apigenin-7-O-(6″-E-p-coumaroyl-β-d-glucopyranoside), and naringenin-7-O-(6″-E-p-coum aroyl-β-d-glucopyranoside) were isolated previously from *Blepharis ciliaris* [751].

Further, 4′-O-diglycoside of decarboxyrosmarinic acid was isolated from *Blepharis ciliaris* [752]. Another study revealed the presence of isoflavone glycoside caffeic acid ester: genistein-7-O-(6"-O-E-caffeoyl-D-glucopyranoside), methyl veratrate, methyl vanillate, protocatechuic acid, naringe nin-7-O-(3"-acetyl-6"-E-p-coumaroyl-D-glucopyranoside), nar ingenin-7-O-(6"-E-p-coumaroyl-D-glucopyranoside), apigenin-7-O-(6"-E-p-coumaroyl-D-glucopyranoside), and acteoside [753] (Figure 45.1.3).

45.1.8 Uses

B. ciliaris powdered seeds demonstrated antibacterial effects and thus can be used on boils, wounds, and sores. *B. ciliaris* seed is used to treat cough and also exhibits antibacterial, diuretic, and aphrodisiac effects [754]. Moreover, charcoal from the root is used to improve eye vision and treat sore eyes [755,756]. It is also used as food and fodder in some areas of northern Oman. The whole plant was crushed and given to livestock as fodder

DOI: 10.1201/9781003295488-45

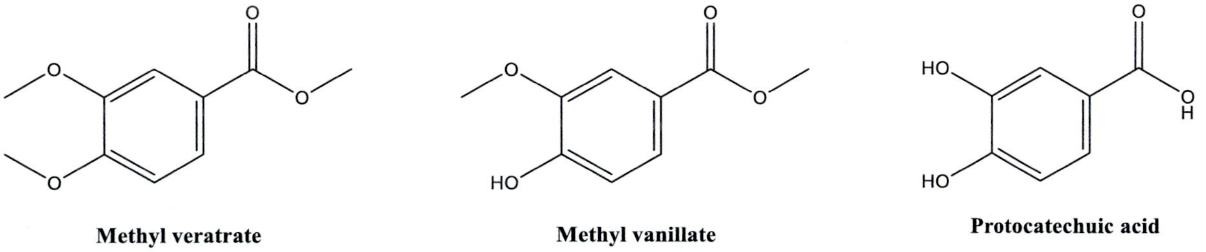

FIGURE 45.1.3 Chemical components of *Blepharis ciliaris.*

in the dry season. Its fruits were consumed by people in Musandam in the past.

45.1.9 Mechanism of Action

Apigenin present in the plant is reported for its ability to inhibit *α-glucosidase* reversibly in a noncompetitive manner [757].

45.1.10 Toxicity and Clinical Data

The previous researcher performed acute toxicity study for *B. maderaspatensis* ethanolic and chloroform extracts. It was found that extract administration showed no mortality up to 14 days [758].

45.1.11 Most Cited Research

iCite, Database, NIH Open Citation Collection was used from June to October 2022 to get the details of the most cited research. Filters for research papers were used by using the keywords *Blepharis ciliaris.* NIH database search revealed *Blepharis ciliaris* plant extract showed anti-leishmanial activity against the *L. major* amastigote stage with IC50 values less than 91 µg/mL [759].

45.1.12 Plant Pathology

No plant pathology has been reported on this wild plant (*B. ciliaris*). However, a recent study showed that *Blepharis ciliaris* showed infection caused by arbuscular mycorrhizal fungi, mainly *Claroideoglomus drummondii* and *Rhizophagus irregularis* with a rate of 98% [760].

45.2 *Blepharis linariifolia* Pers.

45.2.1 Synonyms

Acanthodium hirtum, Blepharis edulis, & *Blepharis hirta*

45.2.2 Family

Acanthaceae

45.2.3 Vernacular Name

Esbed, asker, adaan, & sebbin

FIGURE 45.2.1 Flower portion of *Blepharis linariifolia.*

FIGURE 45.2.2 Part of stem showing flowers and leaves of *Blepharis linariifolia.*

45.2.4 Botanical Description

This herbaceous perennial exhibits a thistle-like morphology characterized by prostrate or shortly ascending stems. Leaves: sessile, arranged in an opposite or seemingly whorled pattern with four leaves per whorl. Flowers: the inflorescence manifests as a spiny cone-like spike, white corolla adorned with violet veins (Figures 45.2.1 and 45.2.2).

45.2.5 Geographical Distribution

This annual or sub-shrub plant is found from the Sahara to Somalia. It is mainly present in the desert or dry shrubland biome. This plant is generally found across Indian and Arabian regions. In Oman, it is widely available in Dhofar [761].

45.2.6 Chemical Composition

As per the chemical analysis, it was found aerial parts of *B. linariifolia* contain verbascoside, vanillic acid, apigenin, and 6″-O-p-coumaroylprunin. It was found that verbascoside presented better free radical scavenging effects. Further, it was found that apigenin showed high activity against *α-glucosidase* and mushroom tyrosinase, whereas 6″-O-p-coumaroylprunin presented a significant inhibitory effect against pancreatic lipase. It's reported that plants belonging to the *Blepharis* genus contain stigmasterol tetracosanoate [762] and other flavonol glycosides and blepharisides A and B [750] from *B. ciliaris* (L.) B.L. Burtt [295,750,763].

Plants belonging to this genus also contain betaine, terniflorin, apigenin, oleanolic acid, allantoin, 2-beta-d-glucopyranosyloxy-1,4-benzoxazin-3-one, and (9s)-9-hydroxydodecanoic acid (Figure 45.2.3).

45.2.7 Uses

This edible plant has several traditional uses, such as *B. linariifolia* in Africa. African people used this plant in the treatment of wounds, urinary as well as genital infections, skin burns, and chest pain. Herbal tea prepared from this plant is sometimes recommended to treat syphilis [761]. Sudanese in Western Sudan call this plant "Begheil." Infusion derived from this plant acts as a tonic and is also utilized for the treatment of diseases caused by parasitic worms and gastric pain [764]. Additionally, aqueous extract derived from this plant is utilized to treat stones in kidney and urinary complications [765,766]. Further, plants belonging to similar genera, such as *B. maderaspatensis*, possessed anti-nociceptive and anti-inflammatory [767], wound healing [768], and gastro-protective effects [758]. Additionally, other reports showed antimicrobial, anti-cancer, and antioxidant effects of *B. maderaspatensis* [769–771].

45.2.8 Mechanism of Action

Another study reported the spasmolytic effect of the ethanolic extract of *B. edulis* mediated via Ca^{2+} channel blockade [772].

45.2.9 Toxicity and Clinical Data

The previous report demonstrated an acute toxicity study for *B. maderaspatensis*. It was found that no mortality was reported till 14 days upon administration of different

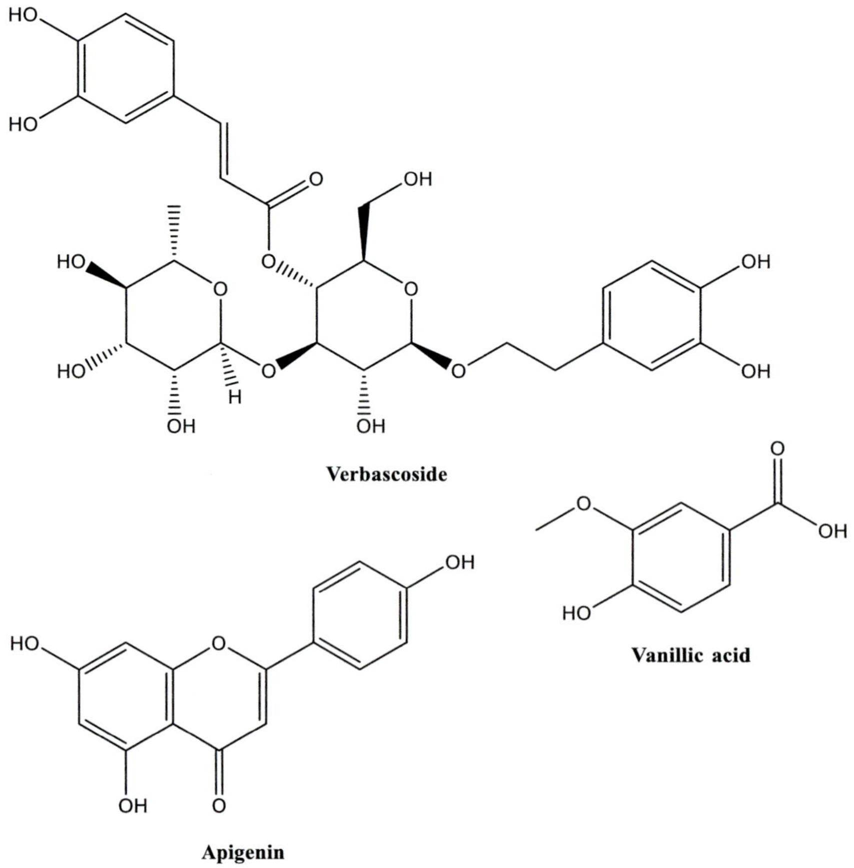

FIGURE 45.2.3 Chemical components of *Blepharis linariifolia*.

doses from 250 to 2,000 mg/kg. No toxicity was reported at 2,000 mg/kg, and thus, the oral LD_{50} was found to be more than 2,000 mg/kg in rodents [758].

45.2.10 Most Cited Research

iCite, Database, NIH Open Citation Collection was used on May 21, 2024, to get the details of the most cited research. Filters for research papers were used by using the keywords *Blepharis linariifolia*. NIH database search revealed the phenolic components present in *B. linariifolia* Pers. (Aerial Parts) and their free radical scavenging and enzyme inhibitory activities [763].

45.2.11 Plant Pathology

There is no data in the literature that refers to the pathology of *Blepharis linariifolia*. However, the genus *Blepharis* is affected by the bacterial pathogen *Xanthomonas campestris* pv. *blepharidis* (NCPPB 1757) [773].

46 Genus *Blumea*

46.1 *Blumea axillaris* (Lam.) DC.

46.1.1 Synonyms

Conyza axillaris

46.1.2 Family

Asteracea

46.1.3 Common Name

Soft Blumea

46.1.4 Botanical Description

This herb exhibits an annual to pauciennial life cycle; it is covered with glandular and softly pilose hairs that reach up to ±70 cm. The leaves are arranged in an alternate phyllotaxy and exhibit an obovate to spathulate morphology. The inflorescence represents a disc-shaped flower head structure, purple to lilac (Figures 46.1.1–46.1.4).

46.1.5 Geographical Distribution

Blumea axillaris is native to the Tropical and Subtropical Old World, with a wide distribution across tropical Africa, tropical and subtropical Asia (including Oman), and Australia. Within Oman it found in Southern Oman thrives in disturbed habitats, particularly those experiencing overgrazing.

46.1.6 Chemical Composition

The carvotanacetone derivatives, such as 4-hydroxy-7-tigloyloxycarvotanacetone and 4-hydroxy-7-ethoxycarvotanacetone, were isolated from *Blumea wightiana* [774].

Recent research showed that the acetyl derivative from monoterpene ester (4 R, 5S)-4-hydroxy-7-tigloyloxycarvotanacetone showed the highest antifungal activity against *Rhizoctonia solani* and *Aspergillus niger* [775].

Another study showed that the essential oil derived from the aerial parts of *Blumea perrottetiana* contains

FIGURE 46.1.1 Inflorescence of *Blumea axillaris*.

FIGURE 46.1.2 Leaf of *Blumea axillaris*.

FIGURE 46.1.3 Habit of *Blumea axillaris*.

FIGURE 46.1.4 Flowering shoots of *Blumea axillaris*.

DOI: 10.1201/9781003295488-46

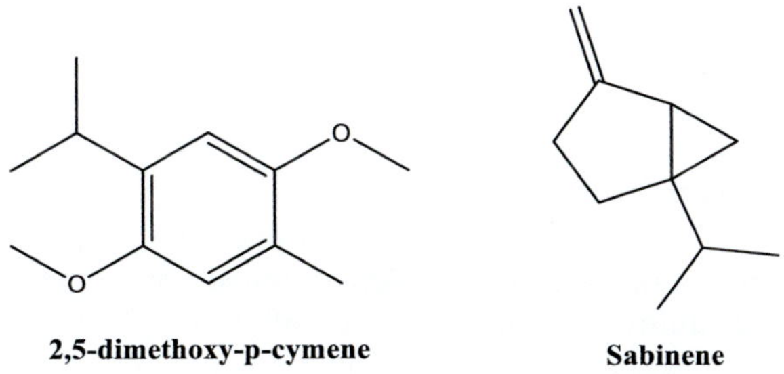

FIGURE 46.1.5 Chemical components of *Blumea perrottetiana*.

high levels of 2,5-dimethoxy-p-cymene and 1,8-cineole with lesser amounts of sabinene (8.1%), δ-cadinene, and (E)-caryophyllene (Figure 46.1.5). It was also found that the oil showed insecticidal effects against the red flour beetle, *Tribolium castaneum* [776].

46.1.7 Uses

Blumea genus plants, such as *B. lacera*, are considered aromatic weeds and thus used as insect repellents [777] and as anthelmintic [777,778].

Blumea axillaris is known for various medicinal properties, such as its role in the treatment of skin ailments, dropsy, diarrhea, external parasites, wounds, and asthma [779]. Additionally, this plant's oil and alcoholic extract also possess anti-inflammatory, antioxidant, anticancer, antibacterial, and anti-phytofungal effects [779,780] and mosquito larvicidal effects [781]. It was found that oil derived from this plant demonstrated larvicidal effects against larvae of *Culex quinquefasciatus* [781]

A recent study showed that the stem and root of *B. axillaris* were capable of synthesizing selenium nanoparticles and inhibiting multidrug-resistant pathogens of aquatic origin [782].

46.1.8 Toxicity and Clinical Data

Plants belonging to this genus, such as *Blumea lacera*, have been reported for poisoning. So, it's important to assess the toxicity of *Blumea axillaris* [783].

46.1.9 Most Cited Research

iCite, Database, NIH Open Citation Collection was used from June to October 2022 to get the details of the most cited research. Filters for research papers were used by using the keywords *Blumea axillaris*. NIH database search revealed that the acetyl derivative from monoterpene ester (4 R, 5S)-4-hydroxy-7-tigloyloxycarvotanacetone showed the highest antifungal activity against *Rhizoctonia solani* and *Aspergillus niger* [775].

46.1.10 Plant Pathology

Several pathogens are reported on this plant. For example, *Xylaria arbuscula*, *Paraphoma radicina*, *Phomopsis phaseoli*, *Sordaria fimicola*, *Aspergillus amstelodami*, and *Diaporthe eucalyptorum* were identified from *Blumea axillaris* [784].

46.2 *Blumea lacera* (Burm.f.) DC.

46.2.1 Synonyms

Blumea lacera var. *burmanni*, *Conyza lacera*, & *Placus lacerus*

46.2.2 Family

Asteraceae

46.2.3 Common Name

Anglimulli, siyalmutra, susksampatra, and kakaronda

46.2.4 Botanical Description

Perennial woody herb, branching from the base, the leaves are arranged in an alternate phyllotaxy and display linear-lanceolate morphology. The inflorescence: capitula, disc-shaped flower heads with a flat disc. Each single flower is purple to lilac (Figures 46.2.1 and 46.2.2).

46.2.5 Geographical Distribution

Blumea lacera is native to a broad area spanning Tropical and Subtropical Asia to the Southwest Pacific, including the Indian Subcontinent, Southeast Asia, East Asia, Australasia, and the Arabian Peninsula (including Oman). Within Oman it is found in southern Oman, as weeds colonize disturbed lands.

46.2.6 Chemical Composition

Chemical analysis of oil derived from *B. lacera* revealed that it contains 2,5-dimethoxy-p-cymene, β-caryophyllene, carvotanacetone, chrysanthenone, and 2,6-dimethyl phenol (Figure 46.2.3) [785]. Other reports showed that the

FIGURE 46.2.1 Cluster of Inflorescences of *Blumea lacera*.

FIGURE 46.2.2 Leaf portion of *Blumea lacera*.

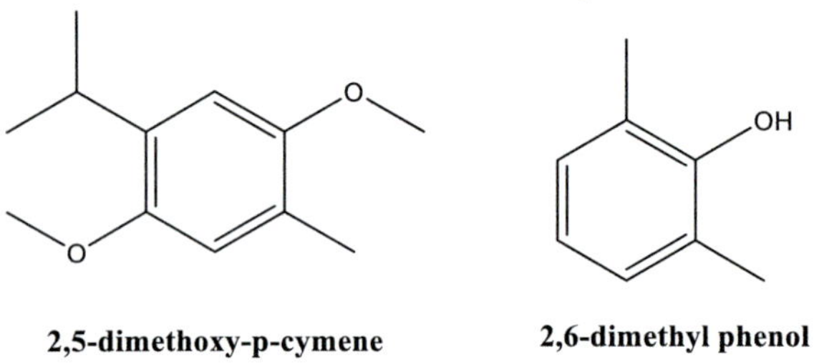

FIGURE 46.2.3 Chemical components of *B. lacera*.

oil contains (Z)-lachnophyllum ester, (Z)-lachnophyllic acid, germacrene D, (E)-β-farnesene, bicyclogermacrene, (E)-caryophyllene, and (E)-nerolidol [786].

As the plant contains steroidal glycoalkaloids, it's important to assess its toxicity. Previous research showed that steroidal glycoalkaloids isolated from *B. lacera* showed cytotoxicity [787].

46.2.7 Uses

As per Ayurveda, the plant is used as an acrid, anthelmintic, anti-inflammatory, antipyretic, astringent, bitter, deobstruent, digestive, diuretic, errhine, expectorant, febrifuge, liver tonic, ophthalmic, stimulant, styptic, and thermogenic medicine [788].

Additionally, as per previous reports, *B. lacera* possessed antidiarrheal, antifungal, antiviral, antileukemic, antibacterial, antipyretic, and cytotoxic effects [786,789,790].

The antioxidant assessment showed that chloroform and pet-ether fractions derived from *Blumea lacera* demonstrated considerable antioxidant activity. Both fractions also presented excellent dose-dependent antidiarrheal effects. Additionally, the pet-ether fraction showed considerable cytotoxicity, whereas the ethyl acetate–soluble fraction demonstrated moderate antimicrobial effects [786].

The recent investigation based on a computer-aided model revealed that thymol and cuminol showed anxiolytic and antioxidant effects, while stigmast-5-en-3β-ol and γ-sitosterol exhibited antidepressant activities [791].

It was found that (Z)-Lachnophyllum ester present in this plant possessed cytotoxic activity against MDA-MB-231, MCF-7, and 5637 human tumor cells, as well as antibacterial and antifungal activity [786].

Additionally, it was demonstrated that the extract derived from *B. lacera* leaf was effective against CCl_4-induced liver injury, perhaps because of active components such as phytol. Further, it was found that liposomal preparation containing *B. lacera* extract demonstrated improvement in bioavailability and stability of the leaf extract [792].

The essential oil derived from *B. lacera* leaves demonstrated tranquilizing, hypothermic, and analgesic effects [793]. Several reports showed that *B. lacera* showed antiviral, anti-leukemic, and cytotoxic effects against several human cancer cell lines [787,794,795].

46.2.8 Mechanism of Action

This plant is also used in the traditional system of medicine for the treatment of inflammation or irritable bowel disease. It was found that the chemical component present in *B. lacera* targets h5HTR3A to treat irritable bowel disease [785].

46.2.9 Toxicity and Clinical Data

The previous research revealed that *Blumea lacera* fresh causes plant poisoning in cattle [783].

46.2.10 Geographical Distribution

B. lacera is usually present in roadside and wasteland regions.

46.2.11 Most Cited Research

iCite, Database, NIH Open Citation Collection was used from June to October 2022 to get the details of the most cited research. Filters for research papers were applied using the keywords *Blumea lacera*. The NIH database search revealed that the methanolic extract of *Blumea lacera* showed the highest cytotoxicity against all tested cell lines [796].

46.2.12 Patents

The previous research showed that the oil (derived from *Blumea lacera*)-based preparation can be used as a mosquito repellent and insecticide.

US20040223998A1. Ganesh Natrajan Iyer, Ramesh Trimbak Sane, Sasikumar Menon: Insect repellent. 2004.

46.2.13 Plant Pathology

Apparently, there is no data in the literature that refers to the pathology of *Blumea lacera* (Burm.f.). However, the prevalence of foliar diseases caused by *Alternaria* species represents a substantial threat to the genus *Blumea* [797].

47 Genus *Boerhavia*

47.1 *Boerhavia elegans* Choisy

47.1.1 Synonyms

Boerhavia repens var. *Elegans* & *Boerhavia rubicunda*

47.1.2 Family

Nyctaginaceae

47.1.3 Vernacular Name

Keshet al rai. In Iran, it's called "Sourkho" and "Sorhmard."

47.1.3.1 Botanical Description

It is a woody perennial with red, slender, and erect stems, reaching up to 60 cm tall (Figure 47.1.1). Leaves arise in pairs on opposite sides of the stem, elongated with parallel sides and rounded ends. Flowers: tiny red. Fruit: roughly cylindrical with a wider end.

47.1.4 Geographical Distribution

B. elegans has a broad geographic range extending from Niger in West Africa to the Arabian Peninsula, and further eastward to Iran and western India.

47.1.5 Chemical Composition

The previous report suggested that bisflavan-3-ol, boerhavianane was isolated from *B. elegans* L. [798].

FIGURE 47.1.1 Flower of *Boerhavia elegans*.

47.1.6 Uses

B. elegans is used to treat various conditions in humans. It is used to purify blood and to treat menstrual cramps, intestinal infections, jaundice, renal complications, urinary complications, diabetes, and inflammatory conditions. It is also used to boost appetite and act as an expectorant, diuretic, and carminative [799,800]. Further, it's also used to treat ophthalmia, muscular pain, ophthalmic injuries, and pain in joints [801,802]. Additionally, *B. elegans* has been also reported for antimalarial and antioxidant effects [799].

Antimalarial investigation of *Boerhavia elegans* showed promising anti-plasmodial activity *in vitro* and *in vivo* without any toxicity especially dichloromethane fraction demonstrated stronger anti-plasmodial effects.

47.1.7 Most Cited Research

iCite, Database, NIH Open Citation Collection was used from June to December 2022 to get the details of the most cited research. Filters for research papers were used by using the keywords *Boerhavia elegans*. NIH database search revealed anti-plasmodial effects of *B. elegans* [800].

47.1.8 Patents

Invention based on its closely related species, *Boerhaavia diffusa*, revealed the process of isolation of bioactive eupalitin (3-O-O-D-galactopyranoside) exhibiting anti-osteoporosis effects.

Deepak MundkinajedduLila Ram MandalasRakesh MauryaSukhdev Swami Handa. Process for isolation of eupalitin from Boerhavia diffusa. 2002. US6977294B2.

47.1.9 Plant Pathology

There are no reports available on the pathology of *Boerhavia elegans* Choisy. However, *Choanephora cucurbitarum* is a fungal pathogen that is found in the genus *Boerhavia erecta* [379].

DOI: 10.1201/9781003295488-47

48 Genus *Boscia*

48.1 *Boscia arabica* Pestal.

48.1.1 Family

Capparaceae

48.1.2 Vernacular Name

In Dhofar, it is called siemr.

48.1.3 Botanical Description

This little evergreen boasts a crown as flat and dense as a perfectly groomed topiary. Its height can reach a respectable 5 m. Smooth, gray-brown bark cloaks its form, adding a touch of elegance. Leaves spiral out from short, compact branches. Each leaf is a miniature oval, its edges smooth and unbroken, adding to the plant's overall tidiness (Figures 48.1.1–48.1.3).

48.1.4 Geographical Distribution

Its native range is restricted to Oman and Yemen. Within Oman, the species is exclusively found in the Dhofar region, characteristically on the lower foothills of the escarpment.

48.1.5 Chemical Composition

Chemical analysis was carried out to determine the major chemical classes. Phytoconstituents like carbohydrates, saponins, flavonoids, triterpenes, sterols, and alkaloids were identified in the studied extracts.

Plants belonging to similar genera, such as *Boscia angustifolia*, contain proto-alkaloids such as stachydrine and 4-hydroxystachydrine (Figure 48.1.4) [803].

FIGURE 48.1.2 Intertwined branches of *Bosica Arabica*.

FIGURE 48.1.1 Adult tree of *Bosica Arabica*.

DOI: 10.1201/9781003295488-48

FIGURE 48.1.3 Leafy shoot of *Bosica Arabica*.

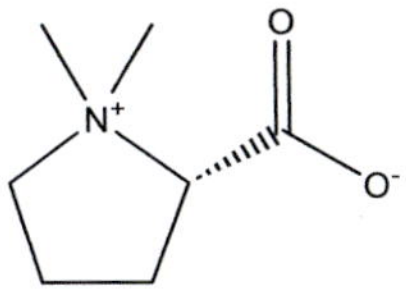

Stachydrine

FIGURE 48.1.4 Chemical component of *Boscia angustifolia*.

48.1.6 Uses

In Dhofar the tree's bark was its most crucial element. First, it was stripped, then blackened by fire, and finally pounded into a powder for storage. This powder served a vital medicinal purpose: packing deep or festering wounds to inhibit further infection, facilitate drying, and promote healing.

It is used by Yemeni people in traditional medicine for the treatment of many diseases, including kidney conditions. A recent study showed the potential antioxidant effects of *Boscia arabica* [804].

Plants belonging to this genus, such as *Boscia angustifolia*, are used for the treatment of mental illness and occasionally to combat pain and neuralgia [804].

48.1.7 Toxicity and Clinical Data

As plants belonging to this genus, such as *Boscia angustifolia*, contain proto-alkaloids, it's important to assess the toxicity of this plant.

48.1.8 Patents

This invention reveals the inhibition of intestinal absorption of sugar to treat metabolic syndrome and diabetes by using *Boscia senegalensis* aqueous extract.

Bruno Etolmar Djibrine Soudy, Youssouf Khayar Mahamoud. FR2991181A1. Composition, useful for inhibiting intestinal absorption of sugar and for treating metabolic syndrome and diabetes, and as a medicament or dietetic product against weight gain, comprises an aqueous extract of *Boscia senegalensis*. 2012.

48.1.9 Plant Pathology

Apparently, there is no data in the literature that refers to the pathology of *Boscia arabica*. However, an endophytic fungus, *Cercospora* sp. PM018, isolated from *Aerva javanica*, and an endophyte from *Catharanthus roseus*, along with two bacterial isolates from the genus *Boscia*, exhibited antibacterial activity against human pathogenic bacteria and antifungal activity against plant pathogenic fungi [805].

49 Genus *Boswellia*

49.1 *Boswellia sacra* Flueck.

49.1.1 Synonyms

Boswellia carteri, Boswellia carteri var. *undulatocrenata*, & *Boswellia undulatocrenata*

49.1.2 Family

Burseraceae

49.1.3 Vernacular Name

Luban, megerot, & mugereh

49.1.4 Common Name

Frankinscense, Olibanum tree

49.1.5 Botanical Description

Small tree up to 5 m high with a single trunk. It has papery bark with peeling characteristics. Leaves: alternate, compound, with leaflets arranged in pairs along a central rachis, clustered toward the tips of the branches. Flowers: clusters of white flowers with a sweet, honey-like fragrance. Fruits are oval-shaped capsules that start green and mature to a reddish-brown color (Figures 49.1.1–49.1.5).

49.1.5.1 Geographical Distribution

It is distributed from northern and northeastern Somalia, southern Yemen to southwestern Oman. In Dhoafr, it mainly thrives in drier mountainous areas that are outside the direct influence of the monsoon rains. Specifically, it is found on cliffs, escarpments, and wadis with a preference for north-facing slopes.

FIGURE 49.1.1 Adult tree of *Boswellia sacra*.

49.1.6 Chemical Composition

The resin derived from *B. sacra* contains oil (5%–15%), a mucus-like cluster (12%–23%), and a lipophilic part (55%–66%). Nevertheless, these components show variation among different species and grades [806]. The resin contains lipophilic parts that have high levels of therapeutically active terpenoids such as boswellic acids [807–810].

The extract derived from *B. sacra* contains different forms of boswellic acids, such as 11-keto beta boswellic acid, acetyl-11-keto beta boswellic acid, boswellic acids (α and β), and acetyl boswellic acid (α and β) [811].

The frankincense oil contains almost three hundred mono-, di-, and triterpenes. Additionally, less than one hundred thirty chemical components belonging to various classes (sterols, di, tri, and sesqui-terpenoids) have been identified from various *Boswellia* species. The most common among them are anti-inflammatory compounds named boswellic acids [812,813]. These compounds are considered

FIGURE 49.1.2 Unripe capsule of *Boswellia sacra*.

FIGURE 49.1.3 Flower of *Boswellia sacra*.

DOI: 10.1201/9781003295488-49

FIGURE 49.1.4 Ripe fruits of *Boswellia sacra*.

FIGURE 49.1.5 Leaflets of *Boswellia sacra*.

chemotaxonomic markers of the *Boswellia* genus [814]. Several investigations [812] have shown that boswellic acids and their analogues are effective phytochemicals against various human complications [813]. These phytochemicals and their respective precursors, as well as intermediates, are biosynthesized via methylerythritol 4-phosphate or mevalonic acid pathways. These molecular pathways, by using isopentenyl diphosphate and dimethylallyl diphosphate, synthesize various terpenoids [815].

49.1.7 Uses

B. sacra is known for its secretion of gum resin known as frankincense, which is obtained by incising the tree bark. This highly valuable yellowish-brown-colored oleo-gum resin shows a wide range of vital biological activities. This resin contains essential oil, which is known for its various therapeutic effects, including diuretic [816], tumor suppressive [817], anti-inflammatory [818], analgesic [819], antimicrobial, hepato-protective [820], antioxidant [821], anti-Alzheimer's [822], cardio-protective, anti-coagulant [810,823,824], and skin disorders [825–828].

Furthermore, this resin is extensively used to treat urinary as well as bronchial infections. In certain parts of Asia, this is considered a therapeutic preparation to treat menstrual pain and complications related to throat, gums, and mouth complaints. Additionally, this resinous preparation is also used to treat inflammation, hemorrhoids, sores, ulcers, and pus-filled bump skin boils. The oil derived from the resin is used as a carminative and digestive. The aqueous fraction derived from *B. sacra* oleo-gum resin, known as "Moh-Lubban," is usually used for the treatment of cough and stomach complications [811].

The oil derived from *B. sacra* has been reported to have cytotoxic effects against breast cancer cells by preventing cellular network formation and disturbing the spheroid development of breast cancer cells. Furthermore, components present in the oil suppress cell cycle regulators as well as signaling pathways [829].

A recent study showed the development of *Boswellia sacra* extract-loaded amino-functionalized mesoporous bioactive glass nanoparticles. Findings obtained from this study showed high loading capacity and release via controlled diffusion. Synthesized nanoparticles presented antibacterial effects against *S. aureus* and better cell viability against human osteoblastic-like cells [830].

49.1.8 Mechanism of Action

The resin derived from *B. sacra* is used as an anti-inflammatory, analgesic, and antimicrobial agent, which represents its immunomodulatory effects. A recent study demonstrated the effect of crude essential oil from *B. sacra* resin on the phenotype and functions of human monocyte-derived dendritic cells. It was found that the oil impacts the maturation and differentiation of dendritic cells from precursor cells by decreasing the capability of allogeneic T cells to proliferate and by improving the endocytic ability and IL-10 production [831].

49.1.9 Toxicity and Clinical Data

It was found that oral administration of *B. sacra* oleo-gum resin–derived methanolic extract in rodents for twenty-eight days did not show any symptoms of behavioral toxicity and did not affect food consumption or weight gain significantly. It was also found that the methanolic extract of *B. sacra* did not produce any significant toxicity to the liver and kidney up to doses of 100 mg/kg body weight [811].

Further, it was found that *B. sacra* extract caused hypochromic normocytic anemia among the rodents at higher doses of 500 and 1,000 mg/kg [831,832].

49.1.10 Most Cited Research

iCite, Database, NIH Open Citation Collection was used from June to December 2022 to get the details of the most cited research. Filters for research papers were used by using the keywords *Boswellia sacra*. NIH database search revealed that the essential oil derived from *B. sacra* causes cytotoxicity to breast cancer cells via suppression of cellular network formation and disruption of spheroid development of breast cancer cells [829].

49.1.11 Patents

The invention revealed compositions comprising *B. sacra* and preparations in capsule form for treating benign prostatic hyperplasia.

WO2018104877A1. Compositions comprising *Boswellia sacra* for the treatment of benign prostatic hyperplasia. Danilo RIVA, Gloria PUMILIA. 2016.

49.1.12 Plant Pathology

Like many other wild medicinal plants, no attention is paid to plant pathogens attacking *B. sacra*. This emphasizes the significance of comprehending every living and non-living factor that can harm this valuable plant and its resin. Some recent reports showed that host-associated fungal community strains, such as *Penicillium citrinum*, *Aureobasidium* sp. BSS6, and *Preussia* sp. BSL10, improved the host growth and produced extracellular enzymes (glucosidases and cellulases) and indole acetic acid (via tryptophan-dependent and independent pathways) and caused the accumulation of photosynthetic pigments [833].

50 Genus *Cadaba*

50.1 *Cadaba farinosa* Forssk.

50.1.1 Synonyms

Cadaba dubia & *Stroemia farinosa*

50.1.2 Family

Capparaceae

50.1.3 Vernacular Name

Sirmaar & khmermar

50.1.4 Common Name

Cadaba bush, herd-boy's fruit (En), Kibilazi mwitu, mvuija vumo

50.1.5 Botanical Distribution

Small bushy shrubs can grow up to 2 m, covered with white or silvery fuzz on their leaves and stems. The leaves are alternate, oval in shape, and grey-green in color. The flowers appear terminally in short racemes, each bearing four yellow petals and noticeable stamens. The fruits are farinose, cylindrical, and contain multiple seeds (Figures 50.1.1–50.1.3).

50.1.6 Geographical Distribution

Its native range encompasses the Sahara Desert to Tropical Africa, the Arabian Peninsula, and Pakistan to India. In Oman, it is found in southern and northern Oman.

FIGURE 50.1.1 Adult plant of *Cadaba farinosa.*

FIGURE 50.1.2 Inflorescence of *Cadaba farinosa.*

FIGURE 50.1.3 Interlaced branches with fruits and flowers of *Cadaba farinosa.*

50.1.7 Chemical Composition

The bark of this plant contains spermidine alkaloids cadabicine and cadabicine diacetate, whereas the leaves contain flavonols, flavonol glycosides, gallic acid, ellagic acid, p-coumaric acid, and ferulic acid, as well as the alkaloids stachydrine and 3-hydroxystachydrine (Figure 50.1.4) [834,835].

A sesquiterpene lactone, such as cadabicilone, was isolated as well. The extract derived from this plant can hydrolyze starch and thus can be used as a saccharification agent in the production of drinks.

This plant contains cadabicine, cadabicine methyl ether, cadabicine diacetate, α, β-dihydroferulic acid, capparisine, cadabicilone, aromatic acid, and kamphefrol [836,837]. Kunu drink is a common non-alcoholic drink containing *Cadaba farinosa* and other ingredients. It is widely consumed in Nigeria [838].

DOI: 10.1201/9781003295488-50

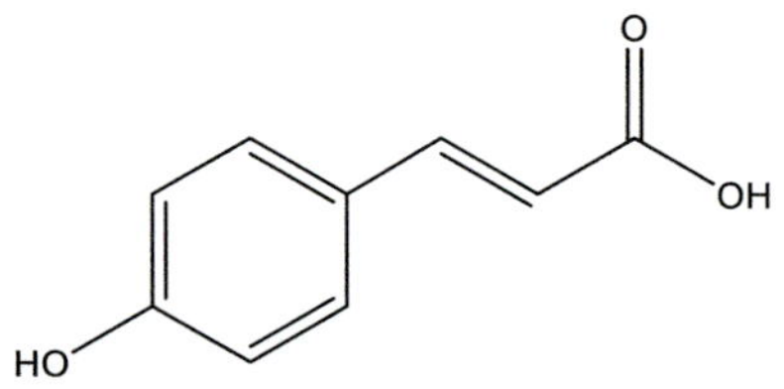

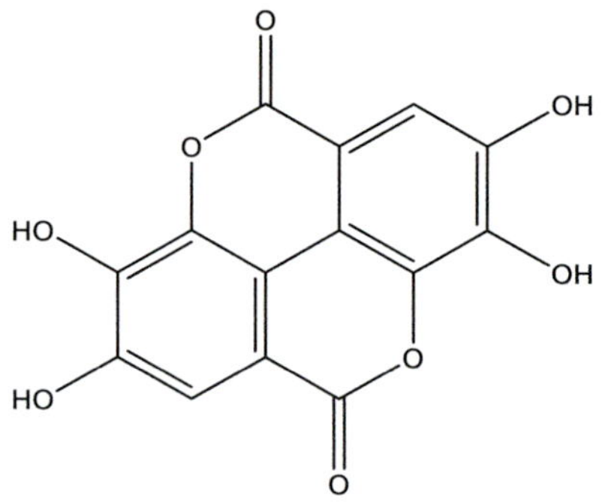

FIGURE 50.1.4 Chemical components of plants belonging to genus *Cadaba*.

The earlier report suggested that acetone/water (70%/30%) extracts of air-dried leaves of *Cadaba farinosa* showed anthelmintic effects via inhibiting larval exsheathment in a dose-dependent manner [839].

A study conducted at Jimma University in Ethiopia reported that a 93% reduction in fecal coliform was recorded after treatment with 60 mg of *C. farinosa* root powder. Recent reports reveal the bacterial removal efficiency of *Cadaba farinosa* from surface water. Findings reveal a significant difference in *E. coli* removal between *C. farinosa* seed and leaf [840,841].

This plant is used as an anti-inflammatory agent and is also used in the treatment of colic, conjunctivitis, stomach aches, and snake bites. In veterinary practice, this plant is used to treat hemorrhagic septicemia, trypanosomiasis, and anthrax in animals. In India, it is used as a hepatoprotective, whereas in Yemen, it is used to treat urinary infections and hemorrhoids. *C. farinosa* leaves are used in medicated oils to treat various ailments [842].

50.1.8 Uses

This plant has been reported to have purgative, anthelmintic, antisyphilitic, emmenagogue, aperient, stimulant, antiscorbutic, and antiphlogistic effects [837].

Additionally, the plant has also been reported for its efficiency in treating antioxidants, cancer, diabetes, and rheumatism [843].

50.1.9 Most Cited Research

iCite, Database, NIH Open Citation Collection was used from June to October 2022 to get the details of the most cited research. Filters for research papers were used by using the keywords *Cadaba farinosa*. NIH database search revealed the *in vitro* anthelmintic effect of *Cadaba farinosa* against *Haemonchus contortus* [839].

50.2 *Cadaba linearifolia* (J. Graham) M.R. Almeida.

50.2.1 Synonyms

Cadaba heterotricha & Niebuhria linearifolia

50.2.2 Family

Capparaceae

50.2.3 Varnucular Name

Ghaloot & golot

50.2.4 Botanical Distribution

This plant exhibits a shrub-like or small tree-like growth pattern, reaching a maximum height of approximately 5 m. The leaves are arranged alternately and are simple, obovate to obtuse or retuse. Flowers: white, arranged in terminal corymbose (Figure 50.2.1).

FIGURE 50.2.1 Flowering shoot of *Cadaba linearifolia*.

50.2.5 Geographical Distribution

It is distributed from northeastern Tropical Africa to Kenya, the Arabian Peninsula, and from southern Pakistan to India. In Oman, it is found across diverse landscapes, from sunbaked cliffs and gravel plains to the foothills of the northern Hajar mountains. In the southern mountains, it is found among *Acacia* and *Commiphora* woodlands.

50.2.6 Chemical Composition

Plants belonging to similar genera, such as *C. fruticosa*, an Indian plant also called "Sthalavriksha" or temple tree, contain cadabicine, cadabicine diacetate [844], capparisine, and α-β-dihydro ferulic acid [845].

Recent research showed the presence of four new kaempferol derivatives in *Cadaba rotundifolia* Forssk. Another study demonstrated the presence of kaempferol and rhamnocitrin derivatives in *Cadaba glandulosa* [846–848].

50.2.7 Uses

The name of the genus is obtained from "kadhab"—an Arabic name for *Cadaba rotundifolia* Forssk, which mentions *Cadaba farinosa* as a medicinal plant.

Plants belonging to similar genera, such as *C. fruticosa*, have several medicinal properties [849–851]. Thus, it is known as "Bahuguni" (meaning any material used for various purposes).

Other plants belonging to a similar genus, such as *C. farinosa*, have been reported for their medicinal uses, such as anthelmintic, anti-inflammatory, antisyphilitic, and anti-phlogistic [842].

The ethanol extract of the leaves of *Cadaba trifoliata* presented antibacterial effects against *Staphylococcus aureus*, *Bacillus subtilis*, *Enterobacter aerogenes*, and *Escherichia coli*, while the aqueous extract presented moderate effects against *E. coli*, *B. subtilis*, *S. aureus*, and *E. aerogenes* [852]. Sudanese people traditionally used *C. rotundifolia* to treat tumors and abscesses [846].

50.2.8 Toxicity and Clinical Data

There are no toxicity reports available for a similar plant; however, plants belonging to a similar genus, such as *Cadaba rotundifolia*, have been reported to have toxic effects on goats. The clinical signs in goats fed with *C. rotundifolia* included depression, diarrhea, frothing at the mouth, dyspnea, ataxia, loss of condition, and recumbency [853].

50.2.9 Patents

This invention reveals the hepatoprotective, hypoglycemic, or hypolipidemic agent effects of a plant from the same genus, i.e., *C. glandulosa*.

Raad MADKHALI. WO2014166588A1. *Cadaba glandulosa* leaves are used as a treatment for hepatotoxicity and hepatoviral B and C infections. 2014.

.... (*For additional images refer to the supplementary figures section; page no.* 301)

51 Genus *Cadia*

51.1 *Cadia purpurea* (*G. Piccioli*) Aiton

51.1.1 Synonyms

Panciatica purpurea & *Cadia varia*

51.1.2 Family

Fabaceae

51.1.3 Vernacular Name

Zakhamkeem

51.1.4 Botanical Description

It is a branched and leafy shrub reaching a maximum height of 1.5 m. Leaves: compound, possessing up to 50 pairs of oblong leaflets with emarginate apices. Flowers: pendulous, borne on axillary short shoots, with a corolla that is white to cream, maturing to pink or reddish. The fruit, yellow-brown in color, contains flattened, glossy, red seeds (Figures 51.1.1–51.1.4).

51.1.5 Geographical Distribution

Cadia purpurea, an unarmed shrub or small tree belonging to the Fabaceae family, subfamily Papilionoideae, exhibits a widespread distribution ranging from southern Saudi Arabia in the northern regions to southern Somalia in the southern areas. Its range extends from northern Kenya in the west to southern Oman in the east [854]. In Oman, it is only found in the southern mountains, particularly in the western part of the Dhofar Governorate on the wet escarpment.

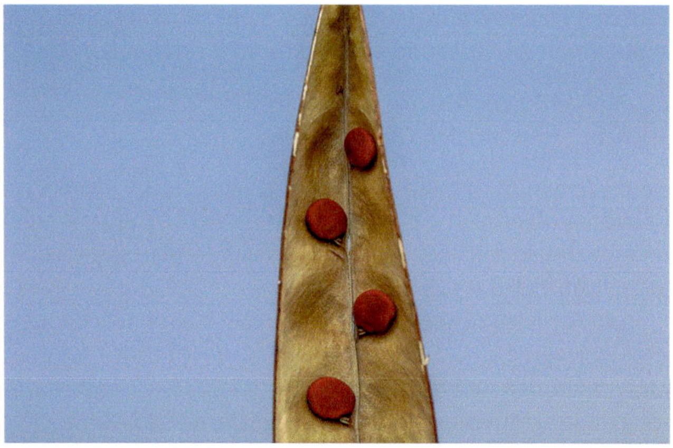

FIGURE 51.1.2 Pod with seeds of *Cadia purpurea*.

FIGURE 51.1.3 Flower portion of *Cadia purpurea*.

FIGURE 51.1.1 Field photograph of *Cadia purpurea*.

FIGURE 51.1.4 Leaflets of *Cadia purpurea*.

DOI: 10.1201/9781003295488-51

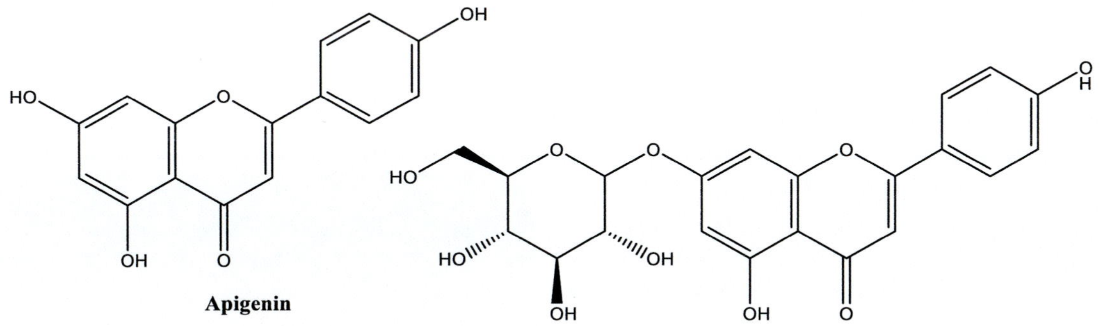

Apigenin

Apigenin 7-O-glucoside

FIGURE 51.1.5 Chemical components of *Cadia purpurea*.

51.1.6 Chemical Composition

This plant contains flavonoid components such as apigenin, apigenin 7-O-glucoside, and chrysoeriol, which were isolated from ethanol leaf extract [855].

Previous reports showed the presence of flavonoids and alkaloids as principal constituents of different parts of *C. purpurea* [856]. The previous study suggested the presence of quinolizidine alkaloids, namely alkaloids I, II, and III, from the leaves of *Cadia purpurea* [857].

Another report showed the presence of apigenin-7-O-neohesperiodoside in *Cadia purpurea* [858] (Figure 51.1.5). Other reports showed the presence of lupanin alkaloids such as l-sparteine, lupanine, α-amyrin, and cadiamine from *Cadia purpurea* [859]. Another report demonstrated the presence of a lupanine derivative known as 13-ethoxylupanine from the ethanol and chloroform extracts of *C. purpurea* [860].

Two quinolizidine alkaloids, 4 3-hydroxy-13a-O-(2′-pyrrolylcarbonyl)-lupanine (digittine) and its amino alcohol, 4 13, 13a-dihydroxylupanine, have been isolated from Ethiopian *Calpurnia aurea* ssp. *Aurea* [861].

A recent report showed the presence of five components in *Cadia purpurea*, such as tetradecahydro-17-(1-hydroxypent-4-en-2-yl)-10,13-dimethyl-1H-cyclopenta[a]-phenanthren-3-ol (1), apigenin-7-O-rhaminoglucoside (2), and 13-O-pyrrolecarboxyllupanine (3), 5-((trihydro-2-oxy-4-(hydroxymethyl)-5-(3-hydroxyphenyl)furan-3yl)methyl) benzene-1,2,3,4-tetraol, and bis(2-hydroxybutyl) phthalate. Findings showed that the tested components demonstrated dose-dependent antibacterial and antioxidant effects [862]. The leaves of *C. purpurea* were also studied for their luteolin content [863].

51.1.7 Uses

It has been reported that the Ethiopian community traditionally used the plant root to heal severe wounds [864]. Similarly, leaves obtained from this plant are used to treat skin burns in Ethiopia [865]. Furthermore, wound infection and nail inflammation treatments by using these plant leaves were also recorded in Eritrea [866]. Additionally, in Ethiopia, complications such as pyrosis, reducing gastritis, and heartburn are treated by using the nectar of *C. purpurea* [867].

51.1.8 Toxicity and Clinical Data

As the plant contains different alkaloids, it's important to assess the toxicity profile of this plant.

51.1.9 Most Cited Research

iCite, Database, NIH Open Citation Collection was used from June to October 2022 to get the details of the most cited research. Filters for research papers were used by using the keywords *Cadia purpurea*. NIH database search revealed that radial flowers of *Cadia purpurea* are dorsalized [868].

51.1.10 Patents

This invention disclosed the isolation of pharmacologically active substances from *Cadia ellisiana*.

DE2321566A1: Pharmacologically active substances and their isolation from *Cadia ellisiana*. Alfred Dr. Rer. Nat. Groebel, Ernst Prof. Dr. Med. Lindner. 1973.

52 Genus *Calligonum*

52.1 *Calligonum comosum* L'Her.

52.1.1 Synonyms

Calligonum polygonoides subsp. *comosus* & *Pallasia comosa*

52.1.2 Family

Polygonaceae

52.1.3 Vernacular Name

Art bakayer

52.1.4 Common Name

Fire Bush

52.1.5 Botanical Description

Calligonum comosum is a nearly leafless desert shrub with photosynthetic green stems and small white to pale pink flowers [869].

A rounded single and hairy carpel is a distinctive morphological characteristic of its fruit. These hairs are yellowish-green. However, they turn into brownish hairs upon ripening. *C. comosum* fruit is extremely covered by long hairs with wing-like structures [870] (Figures 52.1.1–52.1.3).

52.1.6 Geographical Distribution

This species exhibits a broad native range spanning from North Africa eastward across the Sahara Desert. It extends further into the Arabian Peninsula and continues through western Asia to Pakistan. In Oman, it is found in northern Oman and thrives in coastal and inland sand dunes.

FIGURE 52.1.1 Fruits of *Calligonum comosum*.

FIGURE 52.1.2 Plant Habit of *Calligonum comosum* in flowers.

FIGURE 52.1.3 Flowers of *Calligonum comosum*.

52.1.7 Chemical Composition

Calligonum comosum essential oil contains cuminaldehyde as the main antibacterial component. Its toxicity due to the significant level of cuminaldehyde restricts its medical application for human use [871].

A recent study showed that when compared to cuminaldehyde, the overall extract revealed similar antibacterial effects with less toxicity. Findings from this study showed the presence of bacterial species-specific phytochemicals such as cinnamyl linoleate, decanal, and linoleic acid. These phytochemicals increased the antibacterial effects of ampicillin and cuminaldehyde against *S. aureus*, including MRSA, but cinnamyl linoleate and decanal increased the effects against *E. coli* [871].

DOI: 10.1201/9781003295488-52

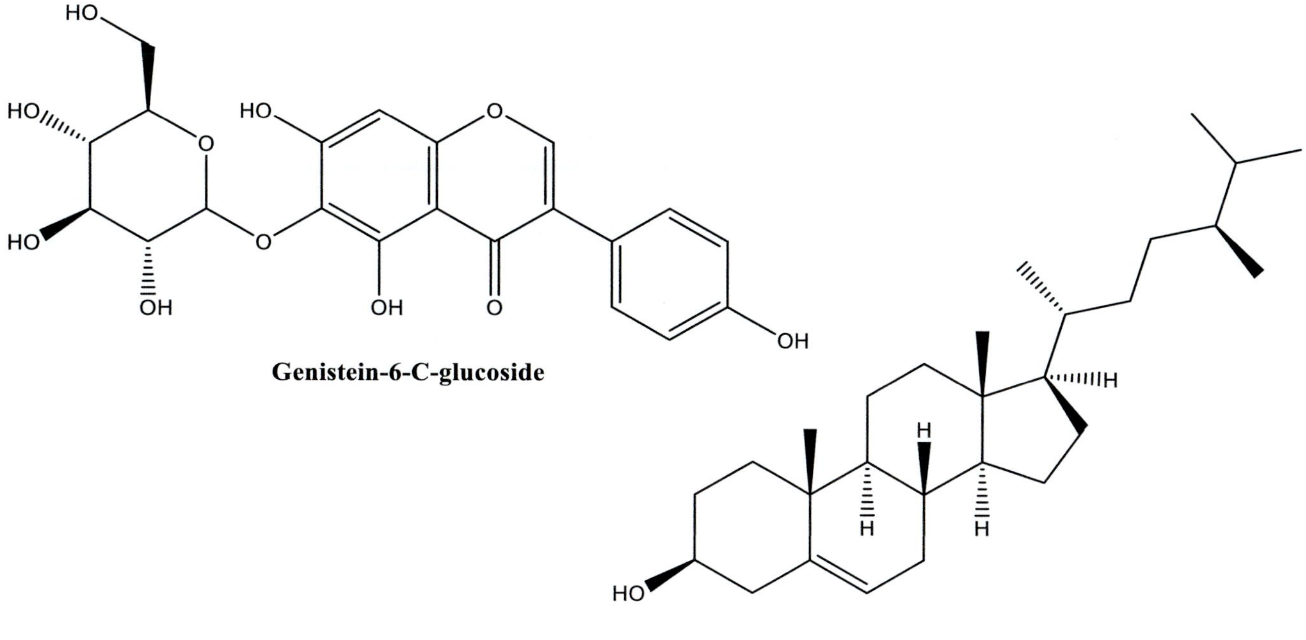

Genistein-6-C-glucoside

Campesterol

FIGURE 52.1.4 Chemical components of *Calligonum comosum*.

It was found that this plant contained high levels of nitrogen, potassium, and zinc in March, but phosphorus, calcium, iron, and copper were high in February. Furthermore, the level of magnesium, manganese, and copper was found to be high [872].

This plant contains antrakinon, kumarinat, flavonoids and alkaloids [870]. Furthermore, the plant has also been reported for catechin, isoquercetin, quercetin, kaempferol-3-O-glucuronide, and kaempferol. Besides, the roots of this plant contain soprunetin, campesterol, genistein-6-C-glucoside, and stigmasterol [873–875] (Figure 52.1.4).

In a recent study, treatment with *Calligonum* extract considerably increased the progressive motility of spermatozoa. It was concluded that treatment of spermatozoa with *Calligonum* extract slightly improved the sperm parameters due to its antioxidant activity [876].

52.1.8 Uses

A recent study showed the impact of a *Calligonum comosum* L'Hér. methanolic extract on the TNBC MDA-MB-231 cell line proliferation and related cell death mechanisms. It was found that this extract possesses anti-proliferative effects against TNBC cells via induction of apoptosis and cell growth arrest [877,878].

This plant was studied *in vitro* for its ability to inhibit the growth of *Listeria ivanovii*. This plant is used as medicine by rural people of Tunisia and is reported for its anti-inflammatory, anti-ulcer, and anticancer effects in different animal models [879,880].

Further, it has been reported that *C. comosum* extracts enhanced haloperidol-induced neuro- and hepatotoxicities among rodents [875].

52.1.9 Mechanism of Action

A recent report suggested the anti-angiogenic activity of *Calligonum comosum* ethyl acetate fraction via the following mechanisms:

- Impaired the viability of human dermal microvascular endothelial cells
- Reduced their migration, tube formation, and sprouting
- Suppressed the growth and cyst formation of developing murine endometriotic lesions
- Decreased vascularization, cell proliferation, and immune cell infiltration [881].

Another study showed hepatoprotective and chemopreventive effects of *C. comosum* via suppressing the rise of serum liver enzymes and decreased levels of oval cell proliferation induced by DEN and CCl_4. In addition, *C. comosum* extract significantly decreased the number and area of glutathione S-transferase placental form-positive preneoplastic hepatic foci induced by diethylnitrosamine and CCl_4 [882].

A recent study showed the anticancer properties of natural phenolic and flavonoid compounds contained in *Calligonum comosum*. It was found that the extract of this plant showed selective cytotoxicity to breast cancer cells due to the ROS production, resulting in apoptosis of cancer cells [883].

Other reports demonstrated the antimicrobial effects of *Calli* oil with lawsone enhanced by at least three-fold. It was found that the antimicrobial effect of *Calli* oil could be due to a benzaldehyde derivative [884].

52.1.10 Toxicity and Clinical Data

Its toxicity due to the significant level of cumin aldehyde restricts its medical application for human use [871]. Earlier studies demonstrated the presence of fungal spores of *Aspergillus* species in *Calligonum comosum* and its ability to produce aflatoxin, with induction of nephrotoxicity and hepatotoxicity in animals and even in humans [885].

A recent study showed the cytoprotective role of *C. comosum* against testicular toxicity induced by Aroclor 1260. In this study, it was found that lipoic acid was more protective than *C. comosum* extract against testicular toxicity caused by Aroclor [886].

Another report showed that systemic chemicals present in the form of toxic volatile oil containing ~50% benzaldehyde derivative (cuminaldehyde) caused strong killing effects on both mammalian cells and microbial pathogens [869].

52.1.11 Most Cited Research

The earlier research showed the presence of (+)-catechin, dehydrodicatechin A, kaempferol-3-O-rhamnopyranoside, quercitrin (quercetin-3-O-rhamnopyranoside), beta-sitosterol-3-O-glucoside, isoquercitrin (quercetin-3-O-glucopyranoside), kaempferol-3-O-glucuronide, and mequilianin (quercetin-3-O-glucuronide) in the methanolic extract of *Calligonum comosum* [879].

iCite, Database, NIH Open Citation Collection was used from June to October 2022 to get the details of the most cited research. Filters for research papers were used by using the keywords *Calligonum comosum*. NIH database search revealed the potential of cytotoxic components from *Calligonum comosum* [879].

52.1.12 Patents

Saeed Mohammad Sultan Alaryani, Humaid. GB2570508A. Preparation for treatment of wounds. 2018.

A method for the preparation of a composition comprising *Calligonum comosum* with other extracts and sodium chloride to treat various skin ailments.

52.1.13 Plant Pathology

Apparently, there is no data available in the literature on the pathogen infecting this wild plant *Calligonum*. However, a recent report indicated the isolation of Rhizobacteria from several plants, including *C. comosum*, that exhibited biocontrol inhibition of the root-rot disease of beans caused by *Rhizoctonia solani* [887].

52.2 *Calligonum crinitum* subsp. *arabicum* (Soskov) Soskov

52.2.1 Synonyms

Calligonum arabicum

52.2.2 Family

Pologanaceae

52.2.3 Vernacular Name

Arta abal

52.2.4 Common Name

Crinitum Buckthorn

52.2.5 Botanical Description

Calligonum crinitum subsp. arabicum is closely related to *Calligonum comosum*. However, it is different in terms of the color of the flower (white and pink) and fruit shape, which is quadrangular [888] (Figures 52.2.1 and 52.2.2).

FIGURE 52.2.1 Plant Habit of *Calligonum crinitum* subsp. *arabicum* in flowers.

52.2.6 Geographical Distribution

Endemic to the Arabian Peninsula: Oman, Saudi Arabia, UAE, Yemen. In Oman, it is found in the Empty Quarter sand dunes and Eastern sands.

FIGURE 52.2.2 Fruiting branch of *Calligonum crinitum* subsp. *arabicum*.

52.2.7 Chemical Composition

There is no data available on the chemical composition of this plant. However, other plants belonging to the same genus, such as *Calligonum polygonoides*, contain chemical components such as (Z, Z)-9,12-octadecadienoic acid, hexadecanoic acid, nonacosane, tetradecanoic acid, heptacosane, hentriacontane, dodecanoic acid, and pentacosane (Figure 52.2.3) [889].

Other reports suggested the presence of lauric acid, myristic acid, and palmitic acid in the oil derived from *Calligonum comosum* [890].

FIGURE 52.2.3 Fruits of *Calligonum crinitum* subsp. *arabicum*.

52.2.8 Uses

Its closely related species, i.e., *Calligonum comosum*, is used traditionally to treat menstruation and menstrual cramps [881].

52.2.9 Mechanism of Action

The ethyl acetate extract derived from its closely related species, i.e., *Calligonum comosum*, has been reported for its inhibition of angiogenesis, inflammation, and invasive tissue growth [881].

52.2.10 Toxicity and Clinical Data

Toxicity and clinical data for this plant are not available. However, plants reported to be in similar genera have toxicity concerns. Please refer to the Toxicity and Clinical Data section of *Calligonum comosum*.

53 Genus *Calendula*

53.1 *Calendula arvensis* L.

53.1.1 Synonyms

Caltha arvensis, Calendula aegyptiaca & Calendula alata

53.1.2 Family

Asteraceae

53.1.3 Vernacular Name

Honwno & henaw

53.1.4 Common Name

Marigold

53.1.5 Botanical Description

It is an annual herb ranging in height from 10 to 100 cm. Its leaves exhibit sector and secretory trichomes and are characterized by a lanceolate shape. The inflorescence comprises a solitary flower head featuring capitula in shades of yellow or orange. While this plant produces blossoms throughout the year, its peak flowering period typically spans from March to July [891] (Figures 53.1.1–53.1.4).

53.1.6 Geographical Distribution

The genus *Calendula*, primarily native to the southern European region encompassing Italy, Malta, Greece, Turkey, Portugal, and Spain, has been widely cultivated in various temperate zones globally, driven by its commercial significance [892]. This geographical distribution spans central and southern Europe, northern Africa, southwestern Asia,

FIGURE 53.1.2 Fruit portion of *Calendula arvensis.*

FIGURE 53.1.3 Flowering and fruiting shoots of *Calendula arvensis.*

FIGURE 53.1.1 Inflorescence of *Calendula arvensis.*

FIGURE 53.1.4 Fruits of *Calendula arvensis.*

DOI: 10.1201/9781003295488-53

and the Macaronesian region, representing a prevalent presence [891]. In Oman, this species is mainly found among cultivations in Northern Oman including Musandam.

53.1.7 Chemical Composition

Multiple studies have substantiated the abundance of sesquiterpene glycosides in extracts derived from the aerial components of *C. arvensis*, characterized by alloaromadendrane, eudesmane, and cubebane skeletons (Figure 53.1.5). Additionally, triterpenoid saponins like arvensoside A and B have been identified in these extracts through phytochemical analysis. Further investigations into the phytochemistry of *C. arvensis* have revealed the presence of diverse compound families, including phenolic acids and flavonoids [891].

53.1.8 Uses

Traditionally, *C. arvensis* is used as an edible alimurgic plant. Mainly, leaves and flowers were used. The flower is used as a boiling vegetable, while the leaves are used to prepare the soup. Decoction obtained from this plant is used to treat skin burns and wounds, while tea prepared from this plant has antiseptic and astringent effects. The preparations obtained from the flower are used to treat various skin inflammatory conditions [616,893,894]. The extract derived from flowers has been studied for its antimicrobial, antifungal, and antioxidant effects. Moreover, extract has been reported for their potential cytotoxic effects against breast cancer and human myeloid cells [895–897].

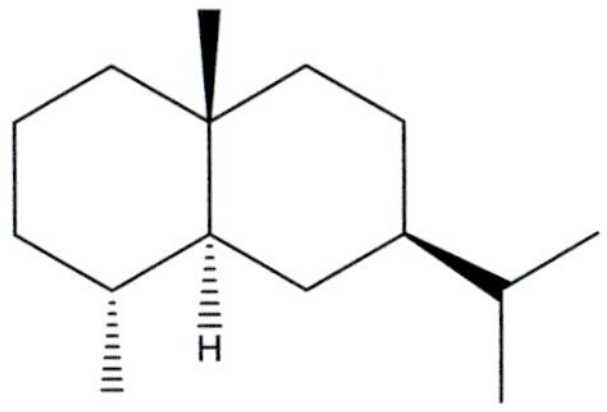

Eudesmane

FIGURE 53.1.5 Chemical component of *Calendula arvensis*.

53.1.9 Mechanism of Action

Glycosides exert their mechanism through the inhibition of cellular membrane sodium-potassium pumps, thereby elevating intracellular sodium concentrations. This elevation facilitates the accumulation of calcium within the cells. Upon subsequent release of calcium, increased myocardial contractions contribute to enhanced cardiac muscle performance [898].

53.1.10 Toxicity and Clinical Data

Calendula arvensis L is considered as non-toxic [899].

53.1.11 Most Cited Research

iCite, Database, NIH Open Citation Collection was used on November 16th, 2023, to get the details of the most cited research. Filters for research papers were used by using the keywords *Calendula arvensis* L. NIH database search revealed the efficacy of *Hypericum* and *Calendula* oils in the epithelial reconstruction of surgical wounds in childbirth with cesarean section.

53.1.12 Patents

박현호, 박현우, 이광식, 이건국 —KR102107193B1—Cosmetic composition comprising an extract of *Calendula arvensis* fermented by *Aureobasidium pullulans*, 2018.

53.1.13 Plant Pathology

Since marigold is a common annual flower that is widely planted, a wide range of pathogens are reported attacking this plant. This includes soil as well as aerial pathogens. Among the recently reported are yellow diseases caused by phytoplasma and several viral diseases. Likewise, the occurrence of *Spongospora subterranea* root infection has been identified in *Calendula arvensis* samples obtained from naturally infested agricultural fields [428].

54 Genus *Calotropis*

54.1 *Calotropis procera* (Aiton) W.T. Aiton

54.1.1 Synonyms

Asclepias procera, *Madorius procerus*, & *Calotropis gigantea* var. *procera*

54.1.2 Family

Apocynaceae

54.1.3 Vernacular Name

Shakar & ushar

54.1.4 Common Name

Country blue, Apple of sodom & Giant milkweed

54.1.5 Botanical Description

Calotropis procera is a large-branched shrub that reaches up to 5 m in height, with milky latex present within its tissues. Leaves are oppositely arranged, pale green in color, and ovate with entire margins. The leaf surface is densely covered with soft, white hairs. Flowers are arranged in clusters on stalks, either at the terminal ends of stems (terminal) or in the junctions where leaves meet the stem (axillary). Each flower possesses five petals with pointed apices, exhibiting a gradient coloration ranging from purple at the tips to pale mauve or white near the center. Fruit consists of paired capsules (Figures 51.1.1–51.1.4).

FIGURE 54.1.1 Tip of flowering shoot of *Calotropis procera.*

FIGURE 54.1.2 Phyllotaxy of *Calotropis procera.*

FIGURE 54.1.3 The adult *Calotropis procera.*

FIGURE 54.1.4 Inflorescence of *Calotropis procera.*

DOI: 10.1201/9781003295488-54

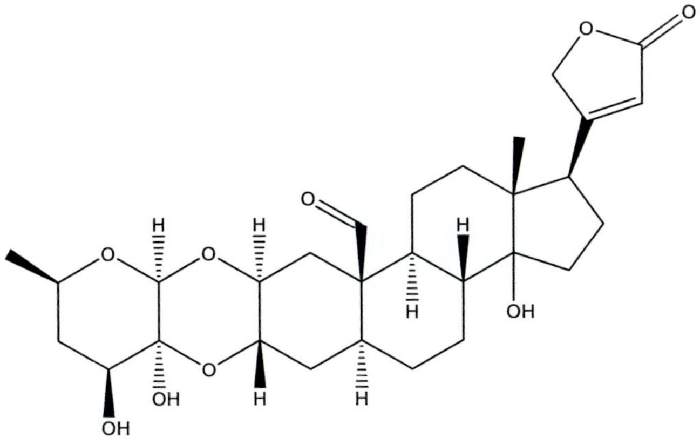

Calotropin

FIGURE 54.1.5 Chemical components of *Calotropis procera*.

54.1.6 Geographical Distribution

Calotropis procera has a wide global range of distribution encompassing Macaronesia, the entire continent of Africa, and the southeastern region of Asia. In Oman, it is found throughout the country. It is distinctive in the Dhofar region, where it is widely distributed in various habitats, including summit grasslands, and serves as an indicator of overgrazing alongside *Solanum incanum* [900].

54.1.7 Chemical Composition

Calotropis procera contains calotropin as a principal bioactive chemical component (Figure 54.1.5) [901]. It also contains metabolites like flavonoids, tannins, terpenoids, saponins, alkaloids, and steroids [902–904]. The primary phytochemical groups identified in the leaf extracts of *Calotropis procera* encompass fatty acid ethyl esters, palmitic acid esters, linoleic acids, and amino acids [905].

54.1.8 Uses

Its utilization in traditional medicinal practices extends across North Africa, the Middle East, Asia, South Asia, and Southeast Asia [906].

54.1.9 Mechanism of Action

The primary and initially established mechanism of action of cardenolides, such as calotropin, involves the inhibition of the sodium-potassium exchanger, Na^+/K^+-ATPase. This enzyme facilitates the active transport of sodium (Na^+) and potassium (K^+) ions across the cellular membrane [901].

54.1.10 Toxicity and Clinical Data

The latex derived from *Calotropis procera* is poisonous and contains the cardiac glycoside known as calotropin. This latex is employed in some regions for the production of arrow poison, where even very small quantities are reported to have lethal effects [900].

54.1.11 Most Cited Research

iCite, Database, NIH Open Citation Collection was used on September 20th, 2023 to get the details of most cited research. Filters for research papers were used by using the keywords *Calotropis procera*. The NIH database search revealed the antibacterial activity of silver and zinc nanoparticles against *Vibrio cholerae* and enterotoxic *Escherichia coli* [907].

54.1.12 Patents

Amani Shafeek Awaad, Ghada Mohamed Zain, Reham M. El-Meligy, Haya Fahd Alkanhal, Vidya Devenathadesikan Seshadri—US9533019B1—*Calotropis procera* extracts as anti-ulcerative colitis agents, 2016.

In this invention, the anti-ulcerative colitis potential of *Calotropis procera* is explored.

54.1.13 Plant Pathology

Apparently, there is no data in the literature that refers to the pathology of *Calotropis procera*. However, isolates of *C. phytoplasma trifolii* have been identified in the genus *Calotropis* (*Calotropis gigantea*) and have been previously documented [908].

55 Genus *Capparis*

55.1 *Capparis cartilaginea* Decne.

55.1.1 Synonyms

Capparis spinosa subsp. *cartilaginea*

55.1.2 Family

Capparaceae

55.1.3 Vernacular Name

Qanfar & lesafah

55.1.4 Common Name

Capper bush

55.1.5 Botanical Description

It is an evergreen shrub that sprawls across the ground or climbs with ease, often draping dramatically from rocky cliffs. Its leaves, arranged in an alternating pattern, are simple and smooth-edged, with a tough, almost leathery feel, and a pair of sharp, hooked spines at the base of each leaf. Flowers have whitish petals that are glabrous or puberulous, and 3 cm long stamens (Figures 55.1.1–55.1.5).

55.1.6 Geographical Distribution

Its native range includes Eastern Africa, western Asia to Northwest India. This area stretches from western Asia, including countries like Turkey, Syria, and Iraq, eastward to encompass the northwestern regions of India and the entire Arabian Peninsula east to the western Indian Ocean. In Oman, it is found throughout the country.

FIGURE 55.1.1 Habit of *Capparis cartilaginea.*

FIGURE 55.1.2 Unripe fruit of *Capparis cartilaginea.*

FIGURE 55.1.3 Flower of *Capparis cartilaginea.*

FIGURE 55.1.4 Phyllotaxy of *Capparis cartilaginea.*

55.1.7 Chemical Composition

Glucosinolates (isothiocyanates) are actively present in *Capparis cartilaginea* (Figure 55.1.6) [909].

DOI: 10.1201/9781003295488-55

FIGURE 55.1.5 Immature flower of *Capparis cartilaginea*.

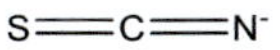

Isothiocyanates

FIGURE 55.1.6 Active chemical components of *Capparis cartilaginea*.

However, numerous phytochemical investigations of the genus *Capparis* have demonstrated the presence of terpenoids, flavonoids, alkaloids, sterols, and fatty acids in various plant parts [910].

55.1.8 Uses

Traditionally, various parts of the plant have been used in folk medicine for a range of therapeutic purposes. The juice extracted from the leaves, when diluted with water and consumed on an empty stomach each morning, is believed to aid in the elimination of intestinal worms and in the treatment of an enlarged spleen, and is also reportedly used by individuals suffering from diabetes. A small portion of the fresh root is crushed into a paste, mixed with lime and salt, and applied externally to areas of the body affected by snake bites or scorpion stings. Crushed leaves and stems are used topically to manage skin conditions such as rashes and eczema, and are also applied to swellings and wounds to promote healing. Additionally, the juice of the ripe fruit is traditionally administered as eye drops in the treatment of conjunctivitis [911].

Ethnobotanical data describe the use of *Capparis cartilaginea* leaves in tropical Africa as a laxative as well as a remedy to treat eye infections, while the roots are utilized to treat skin diseases and ulcers [909]. In traditional Arabian medicine, *Capparis* (CC) is employed for addressing inflammation, earache, headaches, wound healing, snakebite, and aiding in childbirth [912].

55.1.9 Mechanism of Action

Metabolites derived from glucosinolates exhibit diverse beneficial biological activities. Some isothiocyanates demonstrate the ability to hinder the function of the transcription regulator NF-κB, a pivotal player in stress response and inflammation pathways. Conversely, certain isothiocyanates activate NF (erythroid-derived 2)-like 2 (NFE2L2), a transcription factor that stimulates the production of Nrf2, a regulatory protein pivotal in antioxidant defenses and the activity of phase II microsomal enzymes [913–915].

Additionally, isothiocyanates exert epigenetic changes via inhibition of histone deacetylase, suppression of DNA methylation, and modulation of microRNA expression [913].

55.1.10 Toxicity and Clinical Data

According to the chemical composition of *Capparis cartilaginea*, this plant is considered to be toxic. However, the toxic effects of the genus *Capparis* when consumed by sheep, Nubian goats, and calves have not been reported [916].

55.1.11 Adulterants or Substituents

Apparently, there is no data in the literature referring to adulterants or substituents of *Capparis cartilaginea*. However, with increasing demand for *Capparis* surpassing supply and substantial economic incentives for adulteration, *Capparis erythrocarpos* faces the risk of admixture or potential replacement by closely related species [917].

55.1.12 Most Cited Research

iCite, Database, NIH Open Citation Collection was used on September 19, 2023, to get the details of the most cited research. Filters for research papers were used by using the keywords *Capparis cartilaginea*. The NIH database search revealed the chemical investigation of some *Capparis* species growing in Egypt and their antioxidant activity [918].

55.1.13 Patents

Mohammad A. Mazed—US10529003B2—Optical biomodule for detection of diseases at an early onset, 2017.

55.1.14 Plant Pathology

No data in the literature are available on the pathology of *Capparis cartilaginea*. However, the genus *Capparis* was identified as a novel host for *Sclerotium rolfsii* in the year 2006 [919].

55.2 *Capparis decidua* (Forssk.) Edgew.

55.2.1 Synonyms

Sodada decidua & *Capparis aphylla*

55.2.2 Family

Capparaceae

55.2.3 Vernacular Name

Keleel

55.2.4 Common Name

Hanbag, Margh, Sodab, and Tundub

55.2.5 Botanical Description

The plant is a densely branched, spiny shrub or tree, up to 10 m tall, with gray, rough bark and paired thorns. Its leaves are linear, and its flowers are red or pink. Berries are globose or ovoid, with hard, woody rinds, a bitter taste, and a strong odor. Seeds are reniform, covered with a grayish-white aril, and have a hard, pitted surface (Figures 55.2.1–55.2.3).

55.2.6 Geographical Distribution

This plant is distributed in northwest Africa, the Arabian Peninsula to southern Asia. In Oman, it is only restricted to a few locations along the coastal area of Al Batinah.

55.2.7 Chemical Composition

Alkaloids are the main chemical compound in *Capparis decidua*. Various phytoconstituents have been identified and isolated from different segments of *Capparis decidua*, encompassing alkaloids, glycosides, terpenoids, sterols, flavonoids, phenols, and fatty acids [920]. Chemical components such as simiarenoll, lupeol, taraxerol, and β-sitosterol are present in the aerial parts of *C. decidua* [921] (Figure 55.2.4).

FIGURE 55.2.2 Habit of *Capparis decidua*.

FIGURE 55.2.1 Grouped flowers of *Capparis decidua*.

FIGURE 55.2.3 Berry of *Capparis decidua*.

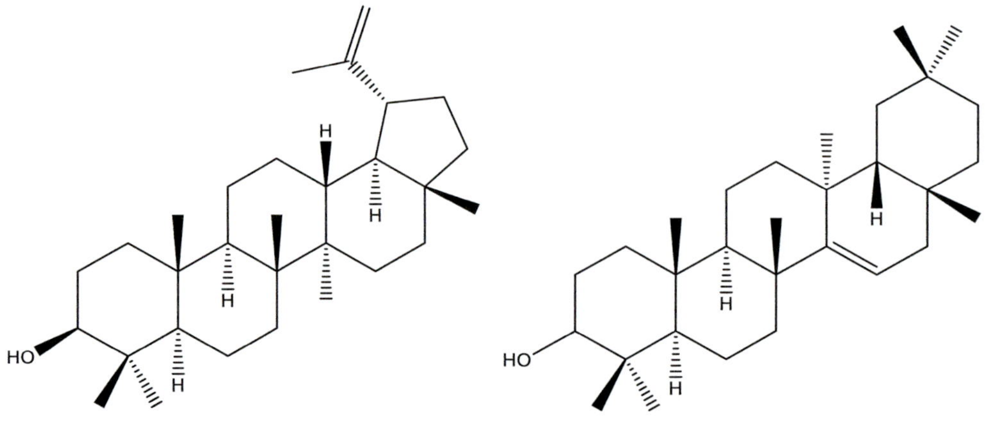

FIGURE 55.2.4 Major chemical components of *Capparis decidua*.

55.2.8 Uses

Capparis decidua, traditionally employed for its medicinal properties, has been utilized as a carminative, tonic, emmenagogue, aphrodisiac, and alexipharmic agent. It is noted for its potential to enhance appetite and its efficacy in alleviating conditions such as rheumatism, lumbago, hiccup, cough, and asthma [922].

55.2.9 Mechanism of Action

Alkaloids induce enzymatic alterations that impact physiological processes, hinder DNA synthesis and repair mechanisms through intercalation with nucleic acids, or disrupt the nervous system [923].

55.2.10 Toxicity and Clinical Data

High toxicity was observed in *Capparis decidua*, which can be attributed to the presence of volatile components found in the essential oils within its leaves and buds [924]. The plant's toxicity could be associated with its alkaloidal content.

55.2.11 Adulterants or Substituents

No data is available on adulterants or substituents of *Capparis decidua*. However, with increasing demand for *Capparis* surpassing supply and substantial economic incentives for adulteration, the genus *Capparis* faces the risk of admixture or potential replacement by closely related species [917].

55.2.12 Most Cited Research

iCite, Database, NIH Open Citation Collection was used on September 21st, 2023, to get the details of the most cited research. Filters for research papers were used by using the keywords *Capparis decidua*. The NIH database search revealed the multifarious PGPR *Serratia marcescens* CDP-13 augments induced systemic resistance and enhanced salinity tolerance of wheat (*Triticum aestivum* L.) [925].

55.2.13 Patents

US20070196528A1: Estelle Bonnet, Alain Guichon, Virginie Charton, Vincent Hubiche, Paula Lennon. Cosmetic composition based on caper flower bud extract. 2004.

- This invention reveals the procedure for the extraction of caper flower bud extract by supercritical fluid extraction.

55.2.14 Plant Pathology

No data on the pathology of *Capparis decidua* is available in the literature. However, *L. taurica* has been reported as causing infection in plants belonging to the genus *Capparis* [926].

55.3 *Capparis spinosa* L.

55.3.1 Synonyms

Capparis vulgaris & *Capparis vulgaris acutifolia*

55.3.2 Family

Capparaceae

55.3.3 Vernacular Name

Lesaf & qanfar atalab

55.3.4 Common Name

Cabbar

55.3.5 Botanical Description

Capparis spinosa is a perennial spiny shrub, typically reaching heights of 0.3–1 m. It features thorny branches, white to pinkish-white flowers, and alternate, rounded, fleshy leaves. Notably, leaf stipules have the potential to form spines known as spinose (Figures 55.3.1–55.3.4) [927].

55.3.6 Geographical Distribution

Capparis spinosa is found in regions including Pakistan, Iran, Afghanistan, Qatar, Oman, the United Arab Emirates, and Saudi Arabia [928]. In Oman, it is found throughout the country.

55.3.7 Chemical Composition

Capparis spinosa main chemical constituents are phenols and flavonoid glycosides [929]. The preliminary screening of the alcoholic extract indicated the presence of alkaloids, glycosides, carbohydrates, tannins, phenolics, flavonoids, and triterpenoids. Conversely, the aqueous extract exhibited the presence of steroids, glycosides, carbohydrates, flavonoids, and saponins [930].

FIGURE 55.3.1 Flower portion of *Capparis spinosa*.

FIGURE 55.3.2 Fruit portion of *Capparis spinosa*.

FIGURE 55.3.3 Leaf arrangements of *Capparis spinosa*.

FIGURE 55.3.4 Plant habit of *Capparis spinosa*.

Cappariloside A, stachydrin, hypoxanthine, uracil, capparine A, capparine B, flazin, guanosine, 1H-indole-3-carboxaldehyde, 4-hydroxy-1H-indole-3-carboxaldehyde, apigenin, kaempferol, thevetiaflavone, tetrahydroquinoline, kaempferol-3-glucoside, quercetin 3-O-glucoside, glucocapparin, capparispine, ginkgetin, sakuranetin, glucocapparin, and stachydrine are present in this plant [931] (Figure 55.3.5).

55.3.8 Uses

This species has a long history of medicinal use dating back to ancient times, with documented applications by both the ancient Romans and Egyptians, who employed it in the treatment of various ailments, including liver and kidney disorders as well as paralysis. It is currently regarded as an important medicinal plant across Iran, India, Pakistan, and numerous Arab countries, where it is used to treat a wide range of symptoms and diseases. In traditional Omani medicine, the leaves and root bark are carefully cleaned, crushed with water and lemon juice, and prepared as a diluted juice that is taken orally. This preparation is used to alleviate gastrointestinal conditions such as stomach pain and spleen disorders, and serves as an anthelmintic to expel internal parasites and to stop diarrhoea. The same juice is also consumed as a diuretic. Historically, decoctions made from the root bark have been used as analgesics. Additionally, the fresh roots are crushed and applied topically to treat snake bites, as well as skin conditions including rashes and localized swellings. Both the fruits and flowers of the plant are edible and have traditionally been consumed in a manner similar to Capparis cartilaginea [911].

The entire plant of *Capparis spinosa* was historically used to treat rheumatism. Roots served as diuretics, astringents, and tonics. Bitterroot bark acted as an appetizer, was astringent, and was anti-diarrheic, and it was used for conditions like hemorrhoids. Bark treated gout, rheumatism, and chest ailments and functioned as an expectorant. Stem and root bark infusions acted as anti-diarrheic and febrifuges [932].

55.3.9 Mechanism of Action

One of the significant underlying mechanisms of action of dietary flavonoids and related polyphenols is linked to their capacity to inhibit oxidative stress and its associated downstream responses, including inflammatory diseases [933].

55.3.10 Toxicity and Clinical Data

Apparently, there is no data in the literature about the toxic effects of *Capparis spinosa*. However, it contains alkaloids and glycosides, so it might have some toxic effects.

55.3.11 Adulterants or Substituents

The substitution of flower buds from other *Capparis* species is evidenced up to a certain level. In Europe, genuine capers are occasionally adulterated with young flower buds of *Caltha palustris* L. (Ranunculaceae) and *Ficaria verna* Huds [934].

55.3.12 Most Cited Research

iCite, Database, NIH Open Citation Collection was used on September 21, 2023, to get the details of the most cited research. Filters for research papers were used by using

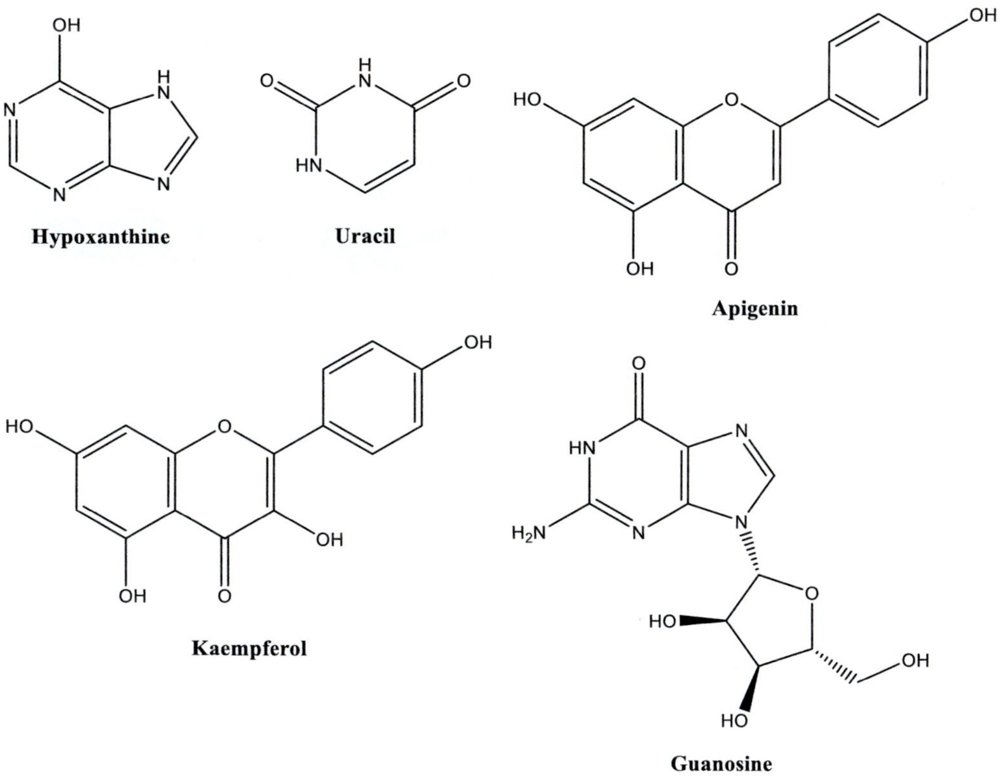

FIGURE 55.3.5 Chemical components of *Capparis spinosa*.

the keywords *Capparis spinosa*. The NIH database search revealed the antidiabetic properties of *Capparis spinosa* L. and its components [935].

55.3.13 Patents

US20070196528A1: Estelle Bonnet, Alain Guichon, Virginie Charton, Vincent Hubiche, Paula Lennon. Cosmetic composition based on caper flower bud extract. 2004.

- This invention reveals the procedure for the extraction of caper flower bud extract by supercritical fluid extraction.

55.3.14 Plant Pathology

Several viral diseases are reported on *Capparis spinosa*, including Cucumber Mosaic virus (CMV) and Eggplant Mottled Dwarf virus. Recently, a virus known as Caper latent virus (CapLV) attacks the leaves of *Capparis spinosa* and causes asymptomatic clearing, yellowing veins, and curling of leaves [936]. In addition, fungal diseases such as *Fusarium* spp., *Sclerotium rolfsii*, and *A. capparis* are well documented [919].

56 Genus *Caralluma* (*Desmidorchis*)

56.1 *Desmidorchis adenensis* (*Deflers*) Meve & Liede

56.1.1 Synonyms

Caralluma adenensis & *Caralluma rauhii*

56.1.2 Botanical Description

The robust, erect shrub grows to a diameter of 0.5 m, featuring a light gray-green stem measuring up to 0.4 m in length and 25 mm in diameter. Its apical inflorescence is globular, typically carrying about 20 flowers with pedicels 10 mm long and 2 mm thick [937]. It has star-shaped flowers that are black, purple, yellow, or red in color (Figures 56.1.1–56.1.3).

56.1.2.1 Family

Apocynaceae (formerly: Asclepeadaceae)

FIGURE 56.1.1 Flowers of *Desmidorchis adenensis*.

FIGURE 56.1.2 Immature inflorescence of *Desmidorchis adenensis*.

FIGURE 56.1.3 Overall growth form of *Desmidorchis adenensis*.

56.1.3 Geographical Distribution

This species is only distributed in Yemen and Oman. In Oman, *Desmidorchis adenensis* exhibits an extremely localized distribution. It is found on the limestone slopes of Wadi Afal, where it coexists within a community of *Acacia senegal* and *Boswellia sacra*. Apparently, the flowers in this locality exhibit a high degree of uniformity [937].

56.1.4 Chemical Composition

Desmidorchis adenensis species is thought to exert its effects through the action of active components known as pregnane glycosides [934]. The biological effects of this plant are primarily associated with the polyphenolic constituents, apparently flavonoids and phenolic acids. These compounds exhibit antioxidant properties, combating reactive oxygen species [938]. Other species, such as *Desmidorchis flava*, have been reported to contain pregnane glycosides desmiflavasides A and B [939].

56.1.5 Uses

Desmidorchis adenensis species have been used as anti-diabetic [940] and also possess anti-ulcer properties. The phytochemistry of the *Desmidorchis adenensis* is characterized by numerous pregnane glycosides, some of which have demonstrated anti-tumor activity [941].

56.1.6 Mechanism of Action

This pregnane glycoside may lower blood glucose levels by reducing the absorption of glucose in the intestines and/or by stimulating the secretion of insulin from the pancreas [942].

DOI: 10.1201/9781003295488-56

56.1.7 Toxicity and Clinical Data

According to the chemical composition of *Desmidorchis adenensis*, this plant can be considered toxic due to the availability of glycosides. However, some mild side effects, like stomach upset, intestinal gas, constipation, and stomach pain, may occur [940].

56.1.8 Adulterants or Substituents

So far, no data is available on the adulteration of *Desmidorchis adenensis*. The *Desmidorchis adenensis* genus comprises over 260 species, and the potential for adulteration among these species underscores the importance of developing analytical methods to ensure product authenticity and safety [943].

56.1.9 Most Cited Research

iCite, Database, NIH Open Citation Collection was used on September 25th, 2023, to get the details of the most cited research. Filters for research papers were used by using the keywords "Desmidorchis." The NIH database search revealed the evaluation of the *in vitro* antiplasmodial, antileishmanial, and antitrypanosomal activity of medicinal plants used in Saudi and Yemeni traditional medicine [944].

56.1.10 Patents

Ramaswamy Rajendran, Kamala Rajendran—US7976880B2—Pregnane glycoside compositions and *Caralluma* extract products and uses thereof, 2005.

- This invention is related to the treatment and management of obesity and obesity-related symptoms/problems/disorders, and other disorders using the *Caralluma* extracts and its pregnane glycoside composition.

56.2 *Desmidorchis arabica* (N.E.Br.) Meve & Liede

56.2.1 Synonyms

Caralluma arabica, Crenulluma arabica & Ceropegia arabensis

56.2.2 Family

Apocynaceae (formely: Asclepeadaceae)

56.2.3 Vernacular Name

Khansoor and dji

56.2.4 Botanical Description

It is an erect perennial with succulent stems that reach up to 40 cm. It features purplish or brownish-gray stems with angular tubercles. Its apical inflorescence is globular, bearing 20–40 reddish-purple to almost black flowers with a carrion-like scent. Pedicels are 2–5 mm long, and the circular corona ranges from red to purple. Fruits: paired, cylindrical, smooth, up to 8 cm (Figures 56.2.1 and 56.2.2).

56.2.5 Geographical Distribution

Globally, *Desmidorchis arabica* is distributed across the Arabian Peninsula. Within Oman, it is found in the northern Hajar Mountains, including the Musandam Peninsula.

56.2.6 Chemical Composition

Desmidorchis species are thought to exert their effects through the action of active biomolecules known as pregnane glycosides, such as arabincosides A–D [934]. The attributed healthy protective effects are primarily associated with the polyphenolic constituents, apparently flavonoids and phenolic acids. These compounds exhibit antioxidant properties, combating reactive oxygen species [938].

56.2.7 Uses

This plant holds a special place in local traditions. New shoots, prized for their milder bitterness, are consumed

FIGURE 56.2.1 Inflorescence of *Desmidorchis arabica*.

FIGURE 56.2.2 Fruits of *Desmidorchis arabica*.

raw or incorporated into salads with ingredients like onion, lettuce, lemon, and spices. This practice is believed to promote overall health and address ailments like high blood pressure, liver issues, intestinal worms, and even diabetes. Similarly, the sap is applied topically to wounds and burns to aid healing.

56.2.8 Mechanism of Action

This pregnane glycoside may lower blood glucose levels by reducing the absorption of glucose in the intestines and/or by stimulating the secretion of insulin from the pancreas [942].

56.2.9 Toxicity and Clinical Data

According to the chemical composition of *Desmidorchis arabica*, this plant can be considered toxic due to the availability of glycosides. *Desmidorchis* extract exhibits a notable level of toxicity, potentially leading to adverse effects on the liver, spleen, and kidneys [945].

56.2.10 Most Cited Research

iCite, Database, NIH Open Citation Collection was used on September 23, 2023, to get the details of the most cited research. Filters for research papers were used by using the keywords *Desmidorchis arabica*. The NIH database search revealed the effects of *Acacia arabica* and *Caralluma edulis* on blood glucose levels of normal and alloxan diabetic rabbits [946].

56.2.11 Patents

Ramaswamy Rajendran, Kamala Rajendran—US7976880B2—Pregnane glycoside compositions and *Caralluma* extract products and uses thereof, 2005.

56.2.12 Plant Pathology

No plant pathology has been reported on this wild plant. However, potent endophytes such as *Paecilomyces variotii*, *Penicillium purpurogenum*, and *Epicoccum nigrum* were found in *Caralluma acutangula*, which can help in the secretion of plant growth-promoting enzymes [947].

56.3 *Desmidorchis flava* (N.E.Br.) Meve & Liede

56.3.1 Synonyms

Caralluma flava, Ceropegia flava & Crenulluma flava

56.3.2 Family

Apocynaceae

56.3.3 Vernacular Name

Diji, zoja, izdot, duru, & el-kebeh

56.3.4 Botanical Description

D. flava is a small succulent herb, growing up to 30 cm tall and 50 cm wide. It has grayish-green stems with a reddish color, which are erect and four-angled with white callus scars. Its flowers are yellow, about 0.8–1.0 cm in diameter, with five-lobed calyxes and bright yellow to greenish-yellow corollas (Figures 56.3.1–56.3.5) [948].

56.3.5 Geographical Distribution

D. flava is a succulent herb that occurs in the wild across various regions of the Sultanate of Oman, Yemen, and the United Arab Emirates [949].

56.3.6 Chemical Composition

The active constituents in this plant comprise pregnane glycosides and triterpenes [151]. A recent report showed the presence of new pregnane glycosides, namely

FIGURE 56.3.1 Natural habit of *Desmidorchis flava*.

FIGURE 56.3.2 Inflorescence of *Desmidorchis D. flava*.

FIGURE 56.3.3 Flowering shoot of *Desmidorchis flava*.

FIGURE 56.3.4 Fruits of *Desmidorchis flava*.

desmiflavaside A, desmiflavaside B, desmiflavaside C, desmiflavaside D, nizwaside, desflavaside A, desflavaside B, desflavaside C, and desflavaside D [950].

56.3.7 Uses

D. flava serves the same uses as *D. arabica*.

56.3.8 Mechanism of Action

This pregnane glycoside may lower blood glucose levels by reducing the absorption of glucose in the intestines and/or by stimulating the secretion of insulin from the pancreas [942].

FIGURE 56.3.5 Inflorescence of *Desmidorchis flava*.

56.3.9 Toxicity and Clinical Data

Acute toxicity testing of the metabolic extract of *D. flava*, at doses of up to 3 g/kg, has not revealed any toxicological effects [951].

56.3.10 Most Cited Research

iCite, Database, NIH Open Citation Collection was used on September 24, 2023, to get the details of the most cited research. Filters for research papers were used by using the keywords *D. flava*. The NIH database search revealed *nizwaside*: a new anticancer pregnane glycoside from the sap of *Desmidorchis flava* [952].

56.3.11 Patents

Ramaswamy Rajendran, Kamala Rajendran—US7976880B2—Pregnane glycoside compositions and *Caralluma* extract products and uses thereof, 2005.

56.3.12 Plant Pathology

The *Caralluma* species in Oman appear to experience minimal impact from pathogens, pests, or domestic animals. Grazing has not been observed, with only occasional signs of nibbling on *D. flava* noted in the eastern Hajar region [937].

56.4 *Desmidorchis penicillata* (Deflers) Plowes

56.4.1 Synonyms

Caralluma penicillate, Ceropegia penicillata & Boucerosia penicillata

56.4.2 Family

Apocynaceae

56.4.3 Vernacular Name

Diji faresi & diji shitanb

56.4.4 Common Name

Paint Brush Caralluma

56.4.5 Botanical Description

Desmidorchis penicillata is a leafless, succulent shrub with pale green to grey-green, four-angled stems.. Numerous flowers are arranged in terminal cymes, and at the end of the corolla lobes, there is a dense cluster of mobile, purple-red hairs resembling a brush-like tuft (Figures 56.4.1–56.4.3) [953].

56.4.6 Geographical Distribution

The origin and habitat of *Desmidorchis penicillata* encompass regions including Somalia, Sudan, Djibouti, Eritrea, Ethiopia, and the Arabian Peninsula, specifically Saudi Arabia, Yemen, and Oman [953]. Within Oman, it is found in the mountainous regions of Northern Oman, including Musandam.

FIGURE 56.4.1 Overall growth of *Desmidorchis penicillata*.

FIGURE 56.4.2 Inflorescence of *Desmidorchis penicillata*.

FIGURE 56.4.3 Paired fruits of *Desmidorchis penicillata*.

56.4.7 Chemical Composition

The pregnane glycosides, specifically penicilloside E, have been isolated from *C. penicillata* [944]. Another study revealed the presence of C-15 oxypregnane glycosides, penicillosides A–C, in the chloroform fraction of the defatted ethanol extract from the aerial parts of *Caralluma penicillata* [954].

56.4.8 Uses

D. penicillata has a history of use in Yemeni traditional medicine, where it has been employed for the treatment of peptic ulcers and as an antihyperglycemic agent. Additionally, in certain populations, it has been used to treat snake and scorpion bites [955].

56.4.9 Mechanism of Action

The hypoglycemic action of russelioside B (1) is thought to involve two potential mechanisms. This pregnane glycoside may lower blood glucose levels by reducing the absorption of glucose in the intestines and/or by stimulating the secretion of insulin from the pancreas [942].

56.4.10 Toxicity and Clinical Data

The toxicity of *D. penicillata* was reported by Hamadan in 1980. It resulted in the death of 50 sheep within a month due to their consumption of the plant in the Aseer region of the Kingdom of Saudi Arabia [954].

56.4.11 Most Cited Research

iCite, Database, NIH Open Citation Collection was used on September 25, 2023, to get the details of the most cited research. Filters for research papers were used by using the keywords *Desmidorchis*. The NIH database search revealed the evaluation of the in vitro antiplasmodial, antileishmanial, and antitrypanosomal activity of medicinal plants used in Saudi and Yemeni traditional medicine [944].

57 Genus *Cardiospermum*

57.1 *Cardiospermum halicacabum* L.

57.1.1 Synonyms

Corindum halicacabum

57.1.2 Family

Sapindaceae

57.1.3 Common Name

Cardiospermum halicacabum is also known as heart pea, love-in-a-puff, heart seed, heart vine, [956], and Kanphuti [957]. In Tamil, it is called Mudakkathan [958]. It's commonly called Balloon Vine & Black Winter Cherry.

FIGURE 57.1.2 Natural habit of *Cardiospermum halicacabum*.

57.1.4 Botanical Description

Cardiospermum halicacabum is an annual or perennial twiner herb covered with hairs and almost glabrous or with slender branches, liming by means of tendrillar hooks. *Cardiospermum halicacabum* possesses a pyriform papery capsule wrangled at angles. These capsules are first greenish lamp-shaped and then become brownish, somewhat hairy, three-way inside, and usually comprise three seeds. These seeds are tightly connected to a central membrane. Seeds are black in color and have bulky white heart-developed skin [959]. Leaves are glabrous, bi-ternate, and ovate-lanceolate [960,961]. *Cardiospermum halicacabum* possesses small white to pinkish unisexual flowers (Figures 57.1.1–57.1.5).

FIGURE 57.1.3 Phyllotaxy of *Cardiospermum halicacabum*.

FIGURE 57.1.1 Fruits of *Cardiospermum halicacabum*.

FIGURE 57.1.4 Flowering shoot of *Cardiospermum halicacabum*.

DOI: 10.1201/9781003295488-57

FIGURE 57.1.5 Leaves arrangements of *Cardiospermum halicacabum*.

57.1.5 Geographical Distribution

Cardiospermum halicacabum is widely distributed in tropical and subtropical areas of the world, such as America, Africa, Bangladesh, India, and Pakistan [958]. This plant species is found in northern Oman, particularly in the foothills and coastal plains of the northern mountains. It thrives in cultivated and irrigated fields.

57.1.6 Chemical Composition

Cardiospermum halicacabum has been reported for the presence of arachidic acid, apigenin, apigenin-7-O-glucuronide, chrysoeriol-7-O-glucuronide, and luteolin-7-O-glucuronide (Figure 57.1.6) [962]. This herb also contains cardiospermin, a cyanogenic glucoside that is responsible for the anxiolytic activity of *C. halicacabum*. This herb has been reported for the presence of fatty acids such as palmitic acid, oleic acid, stearic acid, linoleic acid, and eicosenoic acid in seed oil [9].

The chemical components such as apigenin, protocatechuic acid, protocatechualdehyde, hentriacontanol, calycosin, rutin, and quercetin were identified in *Cardiospermum halicacabum* [963].

Another report revealed the presence of quercetin 3-O-β-D-glucoside, quercetin 3-O-α-L-rhamnoside, quercetin, kaempferol 3-O-α-L-rhamnoside, kaempferol, apigenin 7-O-β-D-glucuronide butyl ester,

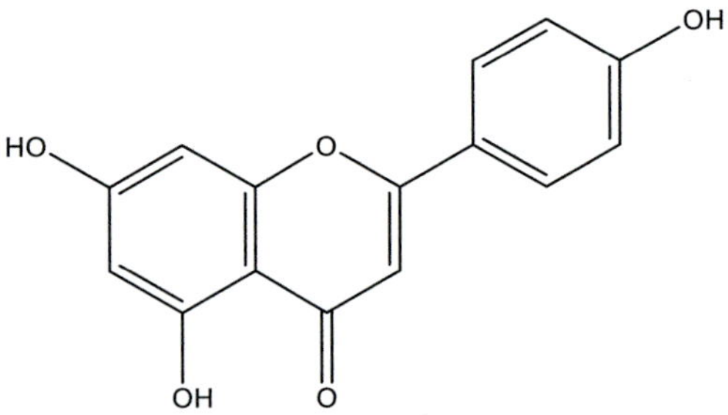

Apigenin

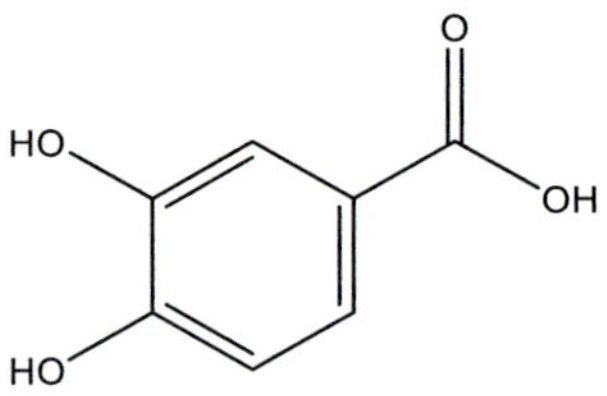

Protocatechuic acid

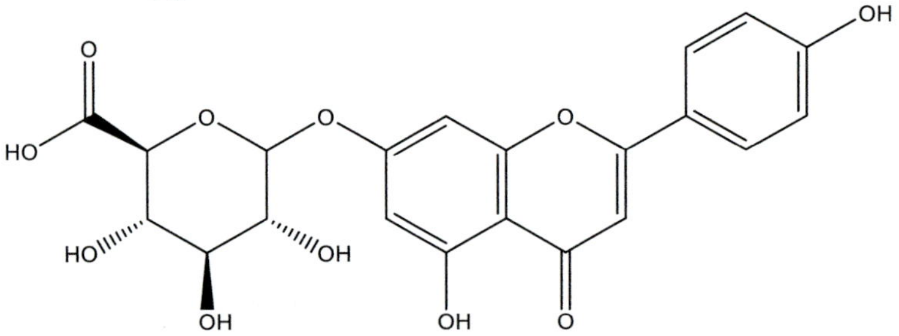

Apigenin-7-O-glucuronide

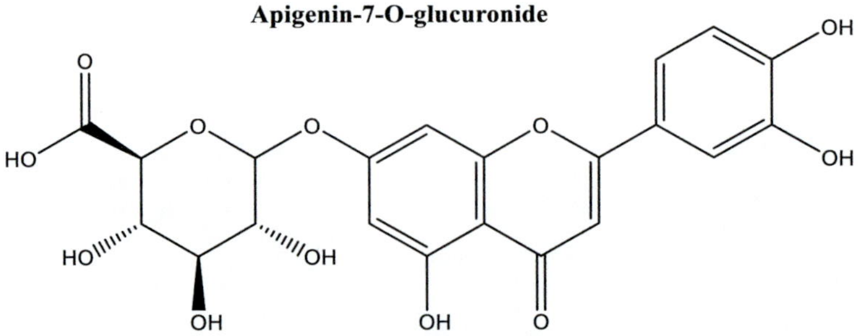

Luteolin-7-O-glucuronide

FIGURE 57.1.6 Chemical components of *Cardiospermum halicacabum*.

apigenin 7-O-β-D-glucuronide methyl ester, apigenin, and luteolin in the aerial parts of *Cardiospermum halicacabum* [964].

57.1.7 Uses

Traditionally, this herb has been used in Chinese medicine for treating rheumatism, lower back pain, nerve disorders, and demulcent [965,966]. Plant preparation demonstrated antioxidant and anti-inflammatory effects [967]. The plant shows a sedative effect on the central nervous system [960].

57.1.8 Mechanism of Action

The ethanolic extract of *Cardiospermum halicacabum* demonstrated anti-inflammatory effects by suppressing TNF-α and NO and due to the decrease of the malondialdehyde (MDA) level in the edema paw via enhancing the effects of CAT, SOD, and GPx in the liver [967].

57.1.9 Toxicity and Clinical Data

Clinical studies have been performed to study the role of *Cardiospermum halicacabum* in atopic dermatitis. It was found that this herb possesses cortisone-like anti-inflammatory effects by stimulating phospholipase A2 while maintaining the stability of cell membranes [968].

Furthermore, the toxic potential of *Cardiospermum halicacabum* is principally due to the hydrocyanic acid releasing cyano lipids [969].

57.1.10 Most Cited Research

iCite, Database, NIH Open Citation Collection was used on January 2nd, 2024, to get the details of the most cited research. Filters for research papers were used by using the keywords *Cardiospermum halicacabum*. NIH database search revealed the *in vitro* and *in vivo* antimalarial activity of *Cardiospermum halicacabum* L. and *Momordica foetida* Schumch. et Thonn [970].

58 Genus *Chenopodium*

58.1 *Chenopodiastrum murale* (L.) S.Fuentes, Uotila & Borsch

58.1.1 Synonyms

Anserina muralis, *Atriplex muralis* & *Chenopodium murale*

58.1.2 Family

Amaranthaceae

58.1.3 Vernacular Name

Labada & mahrqa

58.1.4 Common Name

Goose foot

58.1.5 Botanical Description

C. murale is an erect annual up to 60cm in height. It is slightly mealy, and its leaves are dark green, 2–8×1–6cm broad, rhombic-oblong to rhombic-ovate, angular sides, lobed, acutely toothed, base wedge-shaped, and stalked long or short. The tiny, black, lustrous edible seeds of the weed can endure in the soil for up to seven years. Seeds are sharply keeled, glossy, minutely pitted, dotted, and horizontal [971,972].

Flowers are in slender spikes, creating slack, dense axillary clusters. The fruit of *C. murale* has a membranous pericarp, which is hardly dissociable from the seeds (Figure 58.1.1). Flowers show paniculate inflorescence. Flowers bisexual or female, 1–1.5mm diam. Perianth segments 5, fused near the base. Stamens 5 or 0. Seed horizontal, 1–1.5mm diam., glossy, minutely pitted, shed in surrounding perianth (Figures 58.1.1 and 58.1.2) [973].

FIGURE 58.1.1 Flowering and fruiting shoot of *Chenopodium murale*.

FIGURE 58.1.2 Leaves of *Chenopodium murale*.

58.1.6 Geographical Distribution

This plant is a weed and is considered a pest in agroecosystems, causing harm to the growth of other plants. The plant is indigenous to Europe and available in several parts of Asian and northern African countries. Moreover, this weed is widely distributed across widespread worldwide, particularly in tropical and subtropical areas [974]. This plant is a notorious weed in Oman, readily colonizing disturbed and irrigated lands. Its presence is common across the country, with an altitudinal range of 0–2,300m.

58.1.7 Chemical Composition

The aerial parts of *C. murale* contain flavonol glycosides, aglycones, and coumarin. The aerial part of this plant contains kaempferol, its 3,7-dirhamnoside, 7-rhamnoside, 3-rhamnoside 7-glucoside, scopoletin, quercetin, and herbacetin (Figure 58.1.3) [975].

Furthermore, *C. murale* leaf extract essential oil has been reported for the presence of α-terpinene, (Z)-ascaridole, and cis-ascaridole [976].

58.1.8 Uses

Traditionally, the plant was used as folk medicine as anthelmintic, stomachic, diaphoretic, abortifacient, antispasmodic, emmenagogue, and in the treatment of amenorrhea, migraine, asthma, and catarrh.

The plant preparation was used in rural Egypt to treat oral ulcers in newborn children. These effects could be due to the presence of active chemical components such as kaempferol glycosides and aglycone [977]. *C. murale* has also been

DOI: 10.1201/9781003295488-58

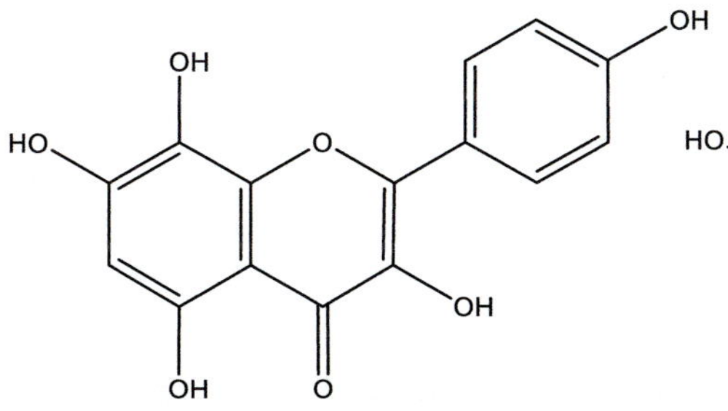

Herbacetin

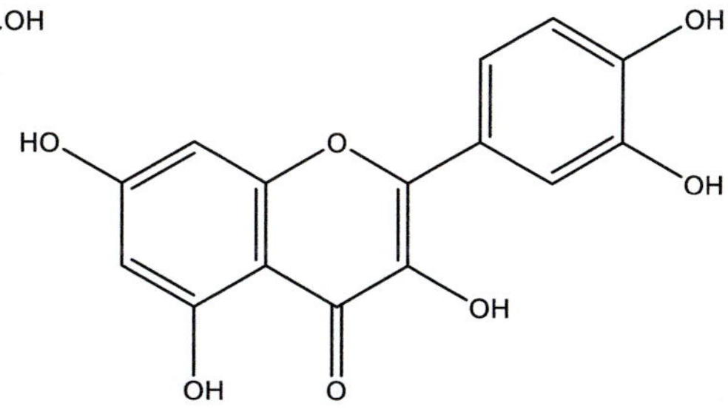

Quercetin

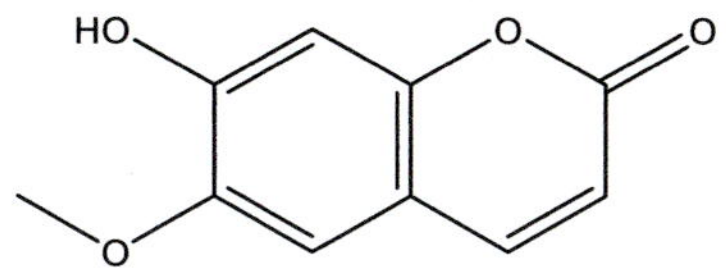

Scopoletin

FIGURE 58.1.3 Chemical composition of *Chenopodium murale*.

reported to treat hair loss, digestive issues, anxiety, depression, and cough. *C. murale* also showed antihypertensive effects due to the presence of flavonoids [971].

58.1.9 Mechanism of Action

The extract from this plant has been identified as having the potential for healing wounds. Research has indicated that this extract boosts the activity of certain genes like EGF, PDGF, and TGF-β1. This increase in gene activity encourages the growth and movement of keratinocytes to the wounded areas, ultimately aiding in the healing process and regeneration of tissue [978].

58.1.10 Toxicity and Clinical Data

Due to the somewhat edible nature of leaves and seeds, their consumption in small quantities may not cause harm. As the plant contains alkaloids and saponins, its toxicological profile must be investigated [977,979].

It must also be noted that plants belonging to the genus, such as *C. ambrosioides*, have been reported to cause slight hepatotoxic lesions in rats [980].

The leaves and seeds of all members of this genus are more or less edible. However, many of the species in this genus contain saponins, though usually in quantities too small to do any harm. Although toxic, saponins are poorly absorbed by the body, and most pass straight through without any problem [979].

58.1.11 Most Cited Research

iCite, Database, NIH Open Citation Collection was used on January 2, 2024, to get the details of the most cited research. Filters for research papers were used by using the keywords *Chenopodium murale*. NIH database search revealed the screening of Yemeni medicinal plants for antibacterial and cytotoxic activities [981].

58.1.12 Patents

Helmut Kraus, Bernd Sievernich, Mariano Etcheverry, Richard R. Evans, Ryan Louis NIELSON, Cyrill Zagar, Rex A. Liebl, Johannes Weimer—EP3319432B1—Herbicidal composition comprising cinmethylin and pendimethalin, 2021. In this invention herbicidal composition comprising cinmethylin and pendimethalin was prepared.

58.1.13 Plant Pathology

Various diseases have been documented in this area, such as *Ascochyta caulina*, *Cercospora chenopodii*, *Colletotrichum nigrum*, and *Pseudomonas syringae*. Nonetheless, it is also utilized for the large-scale cultivation of numerous plant viral pathogens. Beet leafhopper has complex host-plant requirements; *Chenopodium murale* is one of the weeds that allows the population of leafhoppers to increase rapidly [982]. In 1925, Carsner's study presented that a culture of beet curly top virus (BCTV) could be changed by transmission to *C. murale* as the similar isolate, when transmitted to beet produced only mild symptoms [983].

58.2 *Chenopodium album* L.

58.2.1 Synonyms

Botrys albus, Chenopodium album var. commune & Atriplex alba

58.2.2 Family

Amaranthaceae

58.2.3 Common Name

In Hindi, *C. album* is known as Bathua sag (Hindi), [984]. In English, it's called as Bacon Weed and Goosefoot.

58.2.4 Botanical Description

This plant is mainly identified by its goosefoot-shaped leaves. The leaves are alternate and varied in appearance [985]. This is cultivated in many areas; however, it is mainly considered a weed. It often reaches up to a height of 10–150 cm; however, it becomes recumbent after flowering. Leaves are toothed and appear waxy-coated, unwettable, and mealy in appearance, with a whitish covering on the bottom (Figures 58.2.1 and 58.2.2) [985].

58.2.5 Geographical Distribution

This weed is widely distributed in the temperate areas. It possibly originated in Europe, although its exact native range is unclear. *Chenopodium album* found in the Asian belt is different from the one found in European regions [985,986]. It thrives in the wadis and plains of Northern Oman, particularly in moist, open, and disturbed ground, including irrigated areas. Often considered a weed, it flourishes at low altitudes (0–200 m).

58.2.6 Chemical Composition

The leaf essential oil of *Chenopodium album* L. contains p-cymene, ascaridole, pinane-2-ol, α-pinene, β-pinene, and α-terpineol [987] (Figure 58.2.3). This plant has also been reported for the presence of hesperidin and rutin. The main fatty acids, such as myristic acid and cis-10-pentadecanoic acid, were also identified in *Chenopodium album* [988].

58.2.7 Uses

Traditionally, *C. album* is used as a laxative, anthelmintic, diuretic, digestive, cardiotonic, and carminative. The tender shoots are consumed in raw form as salad or with curd [989]. *C. album* is also used for the treatment of intestinal ulcers, spleen enlargement, hepatic disorders, and burns [984,990].

58.2.8 Toxicity and Clinical Data

A recent study showed that *C. album* showed maximum Mn (161.91 ± 0.9 mg/kg) accumulation potential followed by Cu (126.6 ± 1.0 mg/kg) [991]. It's also found that the seeds of this plant contain saponins, whereas leaves contain oxalates, nitrates and sulfates [992]. Thus, its high raw consumption must be avoided.

58.2.9 Most Cited Research

iCite, Database, NIH Open Citation Collection was used on January 2, 2024, to get the details of the most cited research. Filters for research papers were used by using the keywords *Chenopodium album*. NIH database search revealed the pore size of Non-Graminaceous plant cell walls is rapidly decreased by Borate Ester Cross-Linking of the Pectic Polysaccharide Rhamnogalacturonan II [993].

FIGURE 58.2.1 Leaves of *Chenopodium album*.

FIGURE 58.2.2 Stem and leaves of *Chenopodium album*.

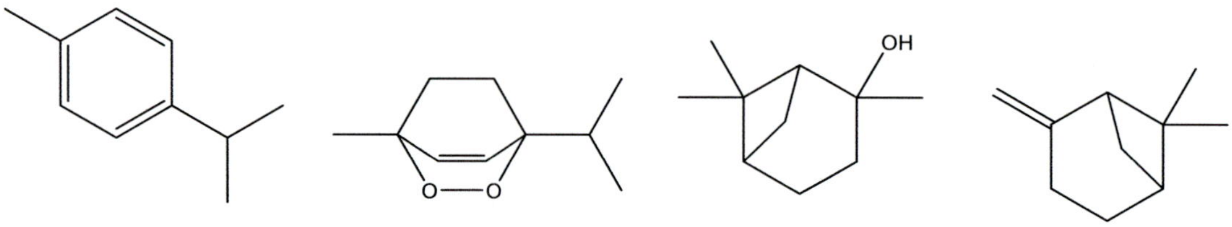

p-cymene **Ascaridole** **Pinane-2-ol** **ß-pinene**

FIGURE 58.2.3 Chemical composition of *Chenopodium album*.

58.2.10 Patents

孙锋, 何孔泉, 阮飞 —CN104758326A—*Chenopodium album* linn aqueous extract and application thereof, 2015.

- This invention reveals a *Chenopodium album* linn aqueous extract and an application thereof.

58.2.11 Plant Pathology

The previous report showed that *Atriplex patula*, *Chenopodium album*, and *Portulaca oleracea* were susceptible to beet yellows virus [420].

Chenopodium quinoa has been reported for the presence of infection caused by bona fide plant mitovirus [994].

59 Genus *Carduus*

59.1 *Carduus pycnocephalus* L.

59.1.1 Synonyms

Cnicus pycnocephalus, Carduus tenuiflorus var. pycnocephalus & Onopyxus pycnocephalus

59.1.2 Family

Asteraceae

59.1.3 Vernacular Name

Shuk al ghul & shuk al gamal

59.1.4 Common Name

Plymouth thistle, Shore thistle, and slender thistle

59.1.5 Botanical Description

Erect, spiny annual thistle reaching up to 70 cm in height, with spiny stems arising from the base. Leaves: Spathulate to lance-shaped, with lobed margins and spines along the edges. Flowerheads: Solitary flowerheads borne terminally on the stems, composed of numerous pink to purple florets (individual flowers) (Figures 59.1.1 and 59.1.2).

59.1.6 Geographical Distribution

These species have a wide distribution in the central, eastern, and northern regions of Saudi Arabia [995]. In Oman, it is distributed in Musandam.

FIGURE 59.1.1 Inflorescence of *Carduus pycnocephalus*.

FIGURE 59.1.2 Flowering shoots of *Carduus pycnocephalus*.

59.1.7 Chemical Composition

Phytochemical investigations on various *Carduus* species have identified several classes of chemical constituents, including lignans, polyacetylenes, flavonoids, coumarin, alkaloids, sterols, and triterpenes [996]. However, a literature survey on *C. pycnocephalus* indicated that only flavonoids have been reported [997]. In one of the studies, 7 flavonoidal components such as apigenin, kaempferol-3-*O*-β-D-glucoside, kaempferol-3-O-α-L-rhamnoside, kaempferol-7-methoxy -3-O-α-L-rhamnoside, diosmetin-7-*O*-β-D-xylosyl-(1‴→6″)-β-D-glucopyranoside, diosmetin-7-*O*-α-L-arabinosyl (1‴→6″)-β-D-glucopyranoside, kaempferol-3-*O*-α-L-rham nosyl-(1‴→2″)-α-L-rhamnoside, lupeol, β-sitosterol, and β-sitosterol-3-*O*-β-D-glucoside were isolated from the aerial parts of *Carduus pycnocephalus* L (Figure 59.1.3) [998].

59.1.8 Uses

In Chinese folk medicine, plants from the genus *Carduus* have been employed for treating a variety of human diseases, including colds, stomach aches, and rheumatism [999].

DOI: 10.1201/9781003295488-59

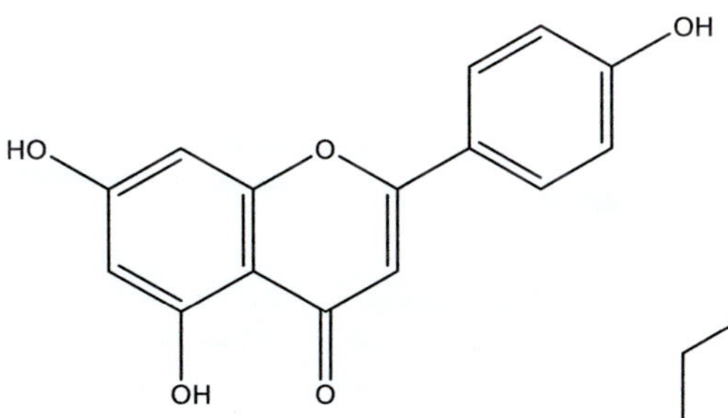

Apigenin

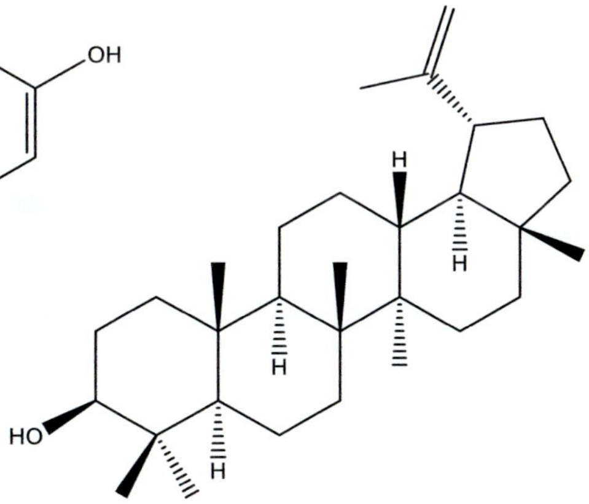

Lupeol

FIGURE 59.1.3 Primary chemical components of *Carduus pycnocephalus*.

59.1.9 Mechanism of Action

One of the significant underlying mechanisms of action of dietary flavonoids and related polyphenols in plants is linked to their ability to inhibit oxidative stress and its associated downstream responses, including inflammatory diseases [933].

59.1.10 Toxicity and Clinical Data

While certain *Carduus* species are recognized for accumulating nitrates in potentially toxic levels, there is no evidence to suggest that *C. pycnocephalus* has been associated with toxic weed-related concerns [1000].

59.1.11 Most Cited Research

iCite, Database, NIH Open Citation Collection was used on September 25th, 2023, to get the details of the most cited research. Filters for research papers were used by using the keywords *Carduus pycnocephalus*. The NIH database search revealed that mycorrhizal densities decline in association with nonnative plants and contribute to plant invasion [1001].

59.1.12 Patents

Klaus Becker, Cepta Duffy, Ellen Hoffmann, Riccardo Losa, Fergus Mould, Stefan Muetzel, Lopez Secundino, Natascha Selje, Robert Wallace. US20080008774A1. Use of plants, plant extracts, and nature-identical components from plants affects rumen fermentation and improves the energy and protein retention of ruminants. 2004.

- This plant was used as an additive in the composition to improve the energy or protein retention of rumen fermentation or to decrease lactic acidosis.

59.1.13 Plant Pathology

In 2008, the initial occurrence of rust due to *Puccinia carduorum* was documented on *Carduus pycnocephalus* [1002].

60 Genus *Carissa*

60.1 *Carissa spinarum* L.

60.1.1 Synonyms

Carissa edulis, Antura edulis & Antura hadiensis

60.1.2 Family

Apocynaceae

60.1.3 Vernacular Name

Folk

60.1.4 Common Name

Simple-spined num-num, Enkeldoringnoemnoem [1003]

60.1.5 Botanical Description

It is a fast-growing, branched, spiny, evergreen shrub that exhibits a scrambling growth habit, reaching heights of up to 5 m. It possesses a milky sap and forms a dense canopy [1004]. This plant produces edible berries that are dark red to purplish black when ripe (Figures 60.1.1 and 60.1.2).

60.1.6 Geographical Distribution

The geographical distribution of the species is evidenced in various regions, including Botswana, Cambodia, Cameroon, Eritrea, Ethiopia, Ghana, Guinea, Japan, Kenya, Myanmar, Namibia, Nigeria, Saudi Arabia, Senegal, South Africa, Sudan, Tanzania, Thailand, Uganda, Vietnam, and Yemen [1003]. In Oman, it is distributed in the escarpment woodlands of Southern mountains.

FIGURE 60.1.2 Fruits and leaves of *Carissa spinarum*.

FIGURE 60.1.1 Branch tip of *Carissa spinarum*.

60.1.7 Chemical Composition

Phytochemical screenings of *Carissa spinarum* stems have revealed the presence of tannins, flavonoids, cardiac glycosides, coumarins, steroids, terpenes, and sesquiterpenes [1005]. In contrast, the roots contain alkaloids [1006]. Tannins are actively present in *Carissa edulis* [1007].

Ethyl acetate and butanol extracts of the aerial parts of *Carissa spinarum* contain 3-O-acetyl chlorogenic acid along with four known flavonol glucosides including kaempferol 3-O-β-d glucopyranoside, quercetin-3-O-β-d glucopyranoside, rhamnetin-3-O-β-d glucopyranoside, and isorhamnetin-3-O-β-d-glucopyranoside, isorhamnetin-3-O-β-d-glucopyranoside-(2″→1‴)-rhamnopyranoside, caredulis, 1-{1-[2-(2 hydroxypropoxy) propoxy] propan-2-yloxy} propan-2-ol (VII), and (+) butyl-O-α-l-rhamnoside [1008].

60.1.8 Uses

The leaves are employed to alleviate toothache, while the roots are used to address conditions such as inflammation, rheumatism, chest pains, and infections, including gonorrhea, syphilis, rabies, malaria, and typhoid fever [1004]. Also, the plant is reported for its role in treating hypertension [1009].

60.1.9 Mechanism of Action

The mechanism of action is based on tannins. Tannins have the ability to cross the bacterial cell wall and penetrate the inner membrane, resulting in the disruption of the cell's metabolic processes, ultimately leading to its death [1010].

Leaf extract of *Carissa spinarum* caused marked endothelium-dependent relaxations and suppressed contractile responses by stimulating the endothelial formation of nitric oxide. The polyphenolic components of the butanolic fraction are stored specially in the endothelium of the arterial wall, resulting in antihypertension effects which support its traditional use in hypertension [1009].

DOI: 10.1201/9781003295488-60

60.1.10 Toxicity and Clinical Data

C. edulis toxicity is found to be low. This plant hasn't been reported for any acute toxicity [1011].

60.1.11 Adulterants or Substituents

The therapeutic applications of plants within the Apocynaceae family make them susceptible to adulteration, leading to concerns regarding adulteration. Common instances of adulteration include the substitution of *A. boonei* with *H. floribunda* and the mixing of *R. canescens* roots with *R. serpentine* root [1012].

60.1.12 Most Cited Research

iCite, Database, NIH Open Citation Collection was used on September 25, 2023, to get the details of the most cited research. Filters for research papers were used by using the keywords *Carissa spinarum*. NIH database search revealed the anti-viral activity of the extracts of a Kenyan medicinal plant *Carissa spinarum* against herpes simplex virus [1013].

60.1.13 Patents

Alternate sources of (-)-secoisolariciresinol from *Stereospermum personatum* other than *Carissa spinarum* are revealed in this invention.

CN1536990A. Pharmaceutical composition comprising (-)-secoisolariciresinol. 2001.

60.1.14 Plant Pathology

Water mold fungi such as *Pythium* cause "damping-off," the death of seedlings of a plant belonging to the genus *Carissa* [1014].

61 Genus *Caudanthera*

61.1 *Caudanthera edulis* (Edgew.) Meve & Liede

61.1.1 Synonyms

Caralluma edulis, Ceropegia edulis & Boucerosia stocksiana

61.1.2 Family

Apocynaceae

61.1.3 Vernacular Name

Daghabees

61.1.4 Common Name

Rush Milkweed

61.1.5 Botanical Description

It is a succulent herb, typically reaching up to 15–45 cm tall. It features compact, nearly round branches with four grooves. Its leaves, if present, are little, measuring 6–13 mm in length, linear in shape, and pointed. Flowers: Terminal clusters; small, bell-shaped; greenish-yellow to pale green with five lobes (Figures 61.1.1–61.1.4) [1015].

61.1.6 Geographical Distribution

The native distribution of this species extends from Niger in West Africa across northeastern tropical Africa, encompassing countries such as Sudan, Ethiopia, and Eritrea, and stretches through the Arabian Peninsula, including Saudi Arabia, Yemen, and Oman, reaching as far east as northwestern India, particularly regions such as Rajasthan and

FIGURE 61.1.1 Shoot system of *Caudanthera edulis*.

FIGURE 61.1.2 Flowers of *Caudanthera edulis*.

FIGURE 61.1.3 Flowers of *Caudanthera edulis*.

FIGURE 61.1.4 Flowering shoot of *Caudanthera edulis*.

DOI: 10.1201/9781003295488-61

Gujarat. Within Oman, It is found in southern, Central and northern Oman.

61.1.7 Uses

This plant is traditionally used as medicine for several medical conditions, including hypertension, Alzheimer's disease, rheumatism, gastric ailments, and leprosy. Traditionally, the stem of this plant is boiled in water to produce an extract which can be used for the management of diabetes [1016]. This plant is also reported for its potent, strong antioxidant, analgesic activity and anti-hyperlipidemic effects [1016].

62 Genus *Carthamus*

62.1 *Carthamus tinctorius* L.

62.1.1 Family

Asteraceae

62.1.2 Synonyms

Calcitrapa tinctoria, Carduus tinctorius & Centaurea carthamus

62.1.3 Vernacular Name

Shuran

62.1.4 Common Name

Safflower or Saffron

62.1.5 Botanical Description

Safflower is an extensively branched, herbaceous plant resembling thistles, exhibiting an annual or winter annual growth pattern. The plant bears globular flower heads, known as capitula, which commonly produce striking yellow, orange, or red flowers (Figures 62.1.1–62.1.3) [1017].

62.1.6 Geographical Distribution

Safflower, historically cultivated for dye extraction for textiles and food in southern and central Asia, as well as the Mediterranean, is now primarily grown as a dryland oilseed crop [1018]. In Oman, it is distributed in the northern region in cultivation fields.

FIGURE 62.1.1 Capitulum of *Carthamus tinctorius.*

FIGURE 62.1.2 Collected red florets of *Carthamus tinctorius.*

FIGURE 62.1.3 Natural habit of *Carthamus tinctorius.*

62.1.7 Chemical Composition

The primary constituents found in the flower of *Carthamus tinctorius* (safflower) include carthamin (flavonoid) and safflower yellow [1019]. Regarding the phytochemistry of this plant, several studies have identified active constituents, including flavonoids, phenylethanoid glycosides, coumarins, fatty acids, and steroids, isolated from various parts of the plant [1020].

Carthamidin, isocarthamidin, hydroxysafflor yellow A, safflor yellow A, safflamin C, and luteolin are the main chemical components of this plant [1019] (Figure 62.1.4).

62.1.8 Uses

Consequently, this plant is used for numerous culinary and textile applications. Safflower is mainly cultivated for orange-red dye (carthamin) extracted from its flower.

DOI: 10.1201/9781003295488-62

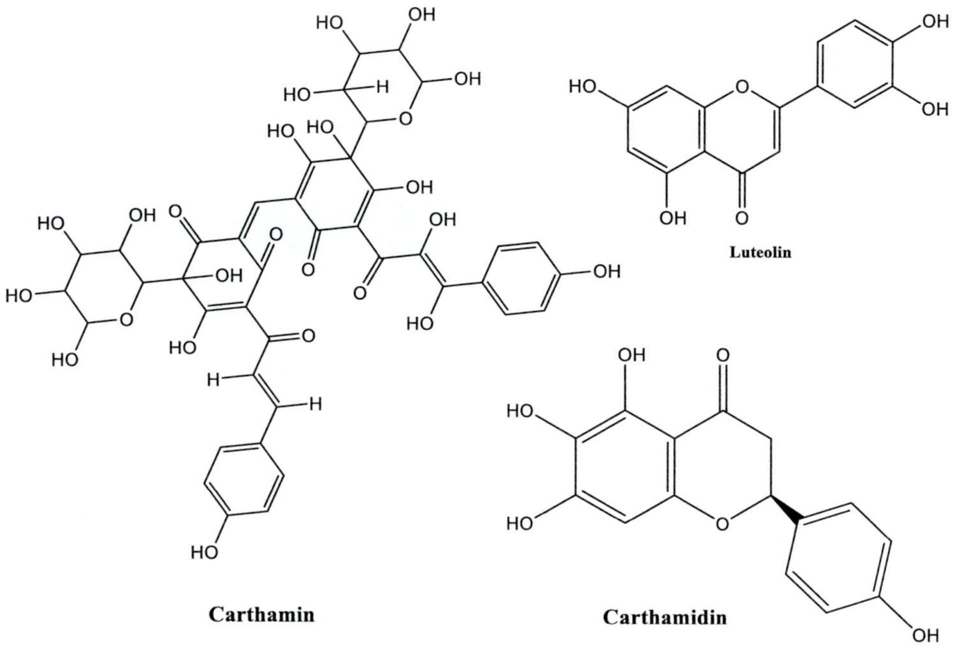

FIGURE 62.1.4 Major chemical components of *Carthamus tinctorius*.

In addition, plant is also utilized for medicine, to extract oil and its floret, seed, stem, leaf and stalk can also be utilized [1021].

Recently, this plant has been reported for its antioxidant, analgesic, anti-inflammatory, and antidiabetic effects. This plant is traditionally used as a purgative, analgesic, antipyretic, and antidote for poisoning. Additionally, it is also used to treat painful menstrual problems, post-partum hemorrhage, and osteoporosis [1019].

62.1.9 Mechanism of Action

Carthami floss, referring to the safflower flower, exhibits a natural calcium channel-blocking activity, which could be a contributing factor to its antiulcerogenic (anti-ulcer) effect. This activity involves the modulation of calcium channels, which play a role in various physiological processes, including muscle contraction and neurotransmission, and may have implications for treating conditions such as ulcers [1022].

62.1.10 Toxicity and Clinical Data

The extract of *Carthamus tinctorius* L. has demonstrated toxic effects on mouse testicular tissue, which raised a concern about its use in reproductive complications [1023]. There are several reports suggesting potential toxicological effects associated with this plant [1024].

62.1.11 Adulterants or Substituents

There are several cost-effective adulterants used for safflower (*Carthamus tinctorius*), mainly safflower (*Carthamus tinctorius*) parts, which are about 80–100 times cheaper than saffron. [1025].

62.1.12 Most Cited Research

iCite, Database, NIH Open Citation Collection was used on September 25, 2023, to get the details of the most cited research. Filters for research papers were used by using the keywords *Carthamus tinctorius* L. NIH database search revealed natural deep eutectic solvents as a new extraction media for phenolic metabolites in *Carthamus tinctorius* L. [1026].

62.1.13 Patents

Robert Maru—US10426749B2—Safflower oil emulsion as a dietary supplement and preparation thereof, 2016. This invention demonstrates stable emulsion made up of safflower oil emulsion with an application as a dietary supplement, and improves metabolism to support a healthy weight management program.

62.1.14 Plant Pathology

Safflower rust, a fungal disease of safflower caused by *Puccinia carthami*, has been recognized as a significant condition in countries such as Australia, Italy, and Oman [1027].

63 Genus *Centaurea*

63.1 *Centaurea pseudosinaica* Czerep.

63.1.1 Family

Asteraceae

63.1.2 Vernacular Name

Merar

63.1.3 Common Name

Knapweeds

63.1.4 Botanical Description

It is an annual or biennial herb. The branching stems arise from the base, reaching up to 40 cm and are covered in short, stiff hairs. Leaves: lanceolate or linear-lanceolate in shape, pinnately divided with narrow segments and undulated margins. Inflorescence: disc-shaped capitula (flower heads) positioned at the ends of the branches. Flowers: bright yellow (Figures 63.1.1 and 63.1.2).

63.1.5 Geographical Distribution

It is distributed in Iraq, Iran and Arabian Peninsula. In Oman, it is distributed in northern region including Musandam.

63.1.6 Chemical Composition

Centaurea species produce essential oils that contain several chemical components [1028]. Plants belonging to this genus contain acetylenes, flavonoids, and terpenoids, including sesquiterpene lactones of the guaianolide and germacranolide types [1029]. The sesquiterpene lactones, such as 17-epichlorohyssopifolin A and chlorjanerin, and a monoterpene lactone, loliolide, were isolated from *Centaurea pseudosinaica* (Figure 63.1.3). These sesquiterpenes are cytotoxic in nature [1029].

FIGURE 63.1.1 Inflorescence of *Centaurea pseudosinaica*.

FIGURE 63.1.2 Overall growth of *Centaurea pseudosinaica*.

63.1.7 Uses

Plants belonging to this genus are used in various folk medicinal applications, serving as diuretics, antifebrile agents, antimalarials, mild astringents, digestive aids, cytotoxic substances, phytotoxic agents, antineoplastics, and emmenagogues [1025]. The antiproliferative, antibacterial, and antifungal effects of the compounds isolated from *Centaurea pseudosinaica* and its whole extract have been reported [1029].

63.1.8 Mechanism of Action

Terpenoids present in the plant possess inhibitory effects on the production of inflammatory cytokines [1030].

63.1.9 Toxicity and Clinical Data

The cytotoxic sesquiterpene lactones, such as 17-epichlorohyssopifolin A and chlorjanerin, can cause toxicity. However, plants belonging to this genus are considered non-toxic [1029,1031].

63.1.10 Most Cited Research

iCite, Database, NIH Open Citation Collection was used on September 25, 2023, to gather details of the most cited research. Filters for research papers were used by using the

DOI: 10.1201/9781003295488-63

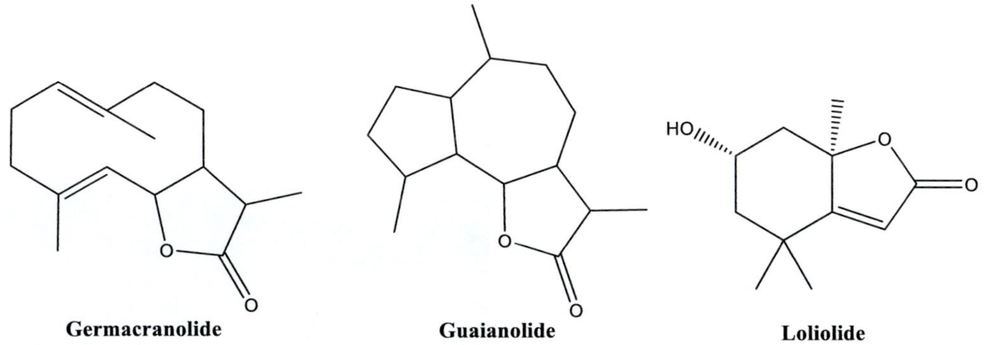

FIGURE 63.1.3 Active chemical components of *Centaurea pseudosinaica*.

keywords *Centaurea pseudosinaica*. The NIH database search revealed antimicrobial, antioxidant, and cytotoxic activities, as well as phytochemical screening of some Yemeni medicinal plants [1032].

63.1.11 Patents

Torsten Grothe, Ernst Roemer, Philipp Wabnitz—US20200055835A1—Use of tricyclic sesquiterpene lactones in the treatment of obesity and related diseases and non-therapeutic treatable conditions, 2019. In this invention, tricyclic sequiterpene lactones or known ingredients of plants were utilized in the treatment of obesity and related diseases and non-therapeutic treatable conditions.

63.1.12 Plant Pathology

Ditylenchus dipsaci is a plant pathogenic nematode that infects plants belonging to the genus *Centaurea* [1033].

64 Genus *Centaurium*

64.1 *Centaurium pulchellum* (Sw.) Hayek ex Hand.-Mazz., Stadlm., Janch. & Faltis

64.1.1 Synonyms

Chironia centaurium var. *pulchella, Chironia pulchella & Erythraea pulchella*

64.1.1 Family

Gentianaceae

64.1.2 Common Name

Lesser Centaury, Branched Centaury

64.1.3 Botanical Description

Lesser centaury (*Centaurium pulchellum*) is an erect annual herb that typically grows up to 10 cm. The leaves are opposite, simple, and sessile, with a narrow, lanceolate shape. The plant produces small, star-shaped flowers that are typically pink to pale rose in color, arranged in loose terminal clusters (Figures 64.1.1 and 64.1.2) [1034].

64.1.4 Geographical Distribution

This plant is distributed in the southern temperate regions of Europe, with a primary presence in the southern portion of the British Isles, particularly along the southern coast of England [1034]. In Oman, it is distributed in the northern region.

FIGURE 64.1.1 *Centaurium pulchellum* in flower.

FIGURE 64.1.2 Flower of *Centaurium pulchellum.*

64.1.5 Chemical Composition

Centaurium species often contain bitter secoiridoid glycosides, xanthones, and β-monoglycosides [1035] (Figure 64.1.3). Erythricin and a known erythrocentaurin were isolated from the whole plant of *Centaurium pulchellum* [1036].

64.1.6 Uses

The common name 'centaury' is derived from its historical use in traditional medicine, particularly for alleviating stomach, kidney, menstrual, and gout-related discomforts [1037]. Antibacterial and antifungal effects due to the presence of secoiridoid glycosides have been reported [1038].

64.1.7 Toxicity and Clinical Data

Centaurium pulchellum, commonly known as lesser centaury, is not known for any significant toxicity [1039].

64.1.8 Patents

Philippe Ragot, Esther Pons, Bernard Mompon, Cedric Rousseau—US20200390696A1—Edible Product Comprising Reconstituted Plant Material, 2020.

DOI: 10.1201/9781003295488-64

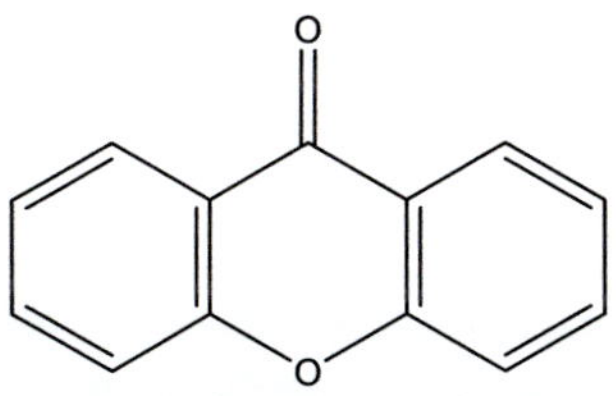

Xanthones

FIGURE 64.1.3 Active chemical components of plant belonging to *Centaurium* genus.

64.1.9 Plant Pathology

According to the literature, there is no evidence of pathological actions on *Centaurium pulchellum*. *Peronospora erythraeae* is a fungus that causes infection in plants belonging to the genus *Centaurium* [617].

65 Genus *Ceratonia*

65.1 *Ceratonia oreothauma* Hillc. & al.

65.1.1 Synonyms

Ceratonia oreothauma subsp. *Oreothauma*

65.1.2 Family

Fabaceae

65.1.3 Vernacular Name

Tiw

65.1.4 Botanical Description

Ceratonia oreothauma is a small evergreen tree characterized by gnarled branches and rough bark that reaches up to a height of approximately 8 m. The leaves of this plant are singly pinnate and consist of up to 20 leaflets. Pods are formed in clusters, fleshy, with a thick pericarp (outer wall), and are reddish-brown in color when ripe (Figures 65.1.1–65.1.3).

65.1.5 Geographical Distribution

The native range of this tree includes Somalia, Yemen, and Oman. In Oman, it is restricted to the eastern Hajar Mountains, specifically in the range of Jebel Bani Jaber.

65.1.6 Cultivation and Collection

Bees promote the pollination of the carob tree (*Ceratonia oreothauma*) [1040].

FIGURE 65.1.1 Branches of *Ceratonia oreothauma*.

FIGURE 65.1.2 Part of fruiting branch of *Ceratonia oreothauma*.

FIGURE 65.1.3 Ripe pods of *Ceratonia oreothauma*.

65.1.7 Chemical Composition

Previous research on the genus *Ceratonia* has identified various compounds within the plant, including polyphenols, flavonoids, carbohydrates, minerals, and proteins [1041]. Chemical components such as quercetin, catechin, tannic acid, quercetin-3-O-α-L-rhamnoside, apigenin flavone, chrysin aglycones, genistein, leucoanthocyanins, epigallocatechin gallate, rutin, trans-ferulic acid, vanillin, fraxidin, resorcinol, alizarin, myricetin, naringenin, kaempferol, cinnamic acid, pyrogallol, myricetin, coumaric acid, gallic acids, etc., are reported in *Ceratonia siliqua* (Figure 65.1.4) [1041].

65.1.8 Uses

Ceratonia oreothauma, also known as the carob tree, holds economic and culinary significance. Seeds of this plant,

DOI: 10.1201/9781003295488-65

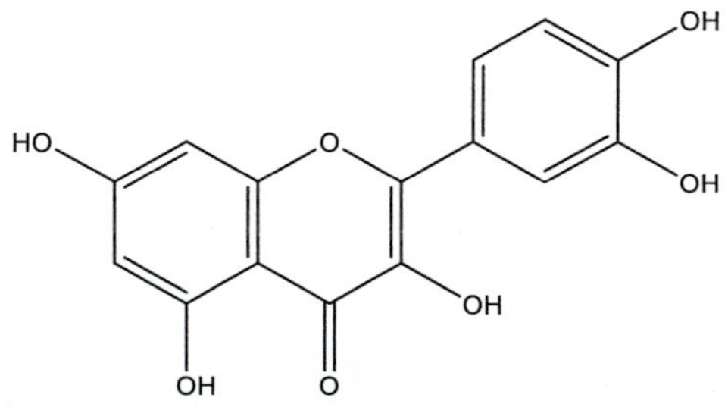

Quercetin

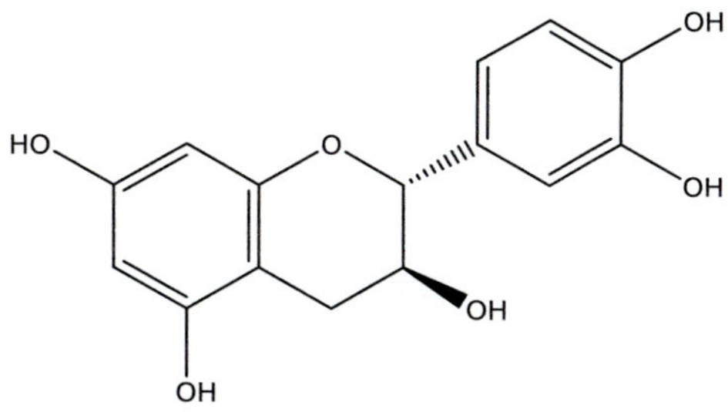

Catechin

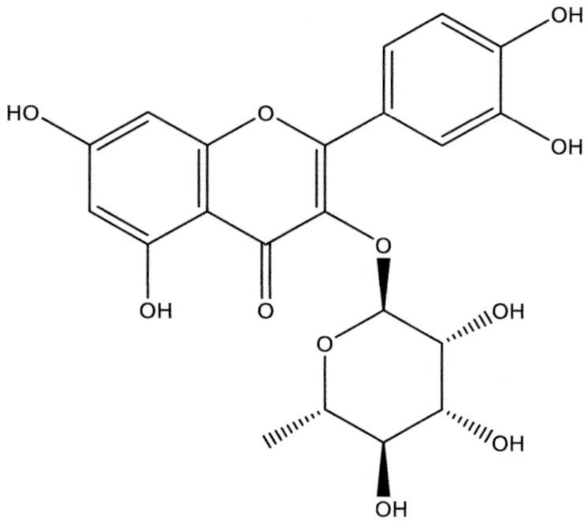

Quercetin-3-O-a-L-rhamnoside

FIGURE 65.1.4 Chemical components of *Ceratonia silique.*

referred to as carob, serve as a food source for humans and livestock [1042].

65.1.9 Mechanism of Action

Polyphenols have been reported to exhibit anti-inflammatory properties via diverse mechanistic pathways [1043].

65.1.10 Toxicity and Clinical Data

Pollen derived from male carob (genus *Ceratonia*) trees can elicit severe allergic reactions in individuals [1044].

65.1.11 Patents

Grethe Stoa Birketvedt: US20160310552A1: Grethe Stoa Birketvedt—US—Northern White Kidney Bean Extract and *Ceratonia siliqua* Extract in Combination with Green Tea Extract in the Treatment of Excess Weight and Obesity, 2016. This invention relates to compositions containing *Ceratonia siliqua* for the treatment excessive weight and obesity.

65.1.12 Plant Pathology

No data is available on the pathology of *Ceratonia oreothauma*. However, there have recently been reports of *Ceratonia* blight affecting carob trees in Libya. This blight is attributed to the fungus *Alternaria alternata* [1045].

66 Genus *Chlorophytum*

66.1 *Chlorophytum laxum* R.Br.

66.1.1 Synonyms

Chlorophytum laxiflorum & Phalangium laxum

66.1.2 Family

Asparagaceae

66.1.3 Common Name

Bichetii grass, Siam lily, false lily turf, wheat plant [1046]

66.1.4 Botanical Description

This plant is a perennial herb typically reaching a height of 10–60 cm, characterized by a rosette of elongated, slender leaves measuring 15–75 cm in length and 0.5–2 cm in width. The flowers are small, white, and borne on long racemes. The fruit is a rounded, three-angled capsule, which contains numerous small, black seeds (Figures 66.1.1 and 66.1.2) [1047].

66.1.5 Geographical Distribution

This species is native to northeastern tropical Africa, the Arabian Peninsula, India, and extends to western Malesia and northern Australia [1048]. In Oman, it is distributed in Southern mountains in areas influenced by monsoon.

FIGURE 66.1.1 Plant habit of *Chlorophytum laxum*.

FIGURE 66.1.2 Fruits of *Chlorophytum laxum*.

66.1.6 Chemical Composition

Several active components have been found in this plant, such as steroids, phenolics, terpenoids, triterpenoids, triterpene, saponins, sugars, proteins, and gallo-tannins [1049]. *Chlorophytum* species are reported to exhibit a notable abundance of bioactive saponins [1050]. A diterpene glycoside, such as chlorophytoside A, was also isolated from *Chlorophytum laxum*. Moreover, the chemical structure has been revealed as (10S)-6α-hydroxylabda-8,13-dien-15,16-olide 3R-O-β-D-glucopyranoside [1051].

66.1.7 Chemical Tests

Froth formation test could be used for detecting the presence of saponins [1052].

66.1.8 Uses

Ethnobotanically, the tubers of this species have been utilized by the Soliga tribes for the treatment of piles and as an astringent, a practice that continues. Unfortunately, the species is facing a rapid decline in its natural habitat due to the excessive harvesting of its tubers for medicinal purposes [1053]. A recent study showed that plants belonging to a similar genus, such as *Chlorophytum comosum*, demonstrated antiproliferative effects on human carcinoma cell lines [1054].

66.1.9 Toxicity and Clinical Data

There is no data on the toxicity of *Chlorophytum laxum* in the literature. Plants of the genus *Chlorophytum* are not

DOI: 10.1201/9781003295488-66

considered poisonous, and there is no known record of toxicity associated with them [1055]. A recent study showed the significant antiproliferative effect possessed by the plant extract on A549 and MCF-7 cancer cell lines [1054].

66.1.10 Most Cited Research

iCite, Database, NIH Open Citation Collection was used on September 27, 2023, to get the details of the most cited research. Filters for research papers were used by using the keywords *Chlorophytum laxum*. The NIH database search revealed the effects of biochar-immobilized bacteria on the phytoremediation of cadmium-polluted soil [1056].

66.1.11 Patents

Zhongshan Traditional Chinese Medical Hospital—CN1965992A—Use of *Chlorophytum laxum* R. Br. and method for preparing the formulation thereof, 2006.

66.1.12 Plant Pathology

No data is available on the pathology of *Chlorophytum laxum*. A fungus known as *Colletotrichum capsici* has been reported to cause leaf blight in the genus *Chlorophytum* [1057].

67 Genus *Chrozophora*

67.1 *Chrozophora oblongifolia* (Del.) A. Juss. Ex Spring.

67.1.1 Synonyms

Croton oblongifolius

67.1.2 Family

Euphorbiaceae

67.1.3 Vernacular Name

Mesheryah and tanoob

67.1.4 Botanical Description

This plant is a perennial, typically growing as an erect shrub, subshrub, or woody herb, reaching a height of up to 1 m. It is characterized by inflorescences that can measure up to 1.5 cm in length. The plant has grayish-green leaves and rounded trilobate fruit (Figures 67.1.1 and 67.1.2).

67.1.5 Geographical Distribution

The native range of *Chrozophora oblongifolia* extends from the Sinai Peninsula and northeastern Tropical Africa, through Syria and the Arabian Peninsula, to Pakistan and northwestern India [1058]. In Oman it is commonly found throughout the country and thrives in a variety of habitats.

FIGURE 67.1.1 Developing flowers of *Chrozophora oblongifolia*.

FIGURE 67.1.2 Fruits of *Chrozophora oblongifolia*.

67.1.6 Chemical Composition

Chrozophora oblongifolia contains terpenes, sterols, phenylpropanoid glycosides, xanthones, chromones, and flavonoids [1059]. Chemical components such as apeginin-7-O-β-D-glucoside, quercetin-3-O-β-D-glucuronic acid, quercetin-3-O-β-D-galacturonic acid, rutin, and 1,3, 6-trigalloyl glucose have been reported [1060] (Figure 67.1.3). Another study revealed the presence of 1-octacosanol, lupeol, para-hydroxybenzoic acid and methyl gallate in aerial parts of *Chrozophora oblongifolia* [1061].

67.1.7 Uses

Historical records suggest this plant species may have been used medicinally. According to these reports, a paste made from fresh, green fruits was applied to wounds to promote healing. It also recorded that the fruits of this species were once employed in the production of a natural blue pigment. This dye served as an alternative to synthetic blue dyes for staining the skin, particularly for facial and hand tattoos prevalent among Bedouin women.

C. oblongifolia is known for its antimicrobial and antioxidant properties. The whole plant is employed to improve wound healing, while seeds and leaves are used as a laxative and depurative agent, as well as for treating skin diseases. In Saudi Arabia, the plant is also utilized to purify the blood and address jaundice [1062]. Phenolics present in this plant inhibit α-glucosidases and advanced glycation end products and thus can be used in the management of diabetes and its complications [1060].

DOI: 10.1201/9781003295488-67

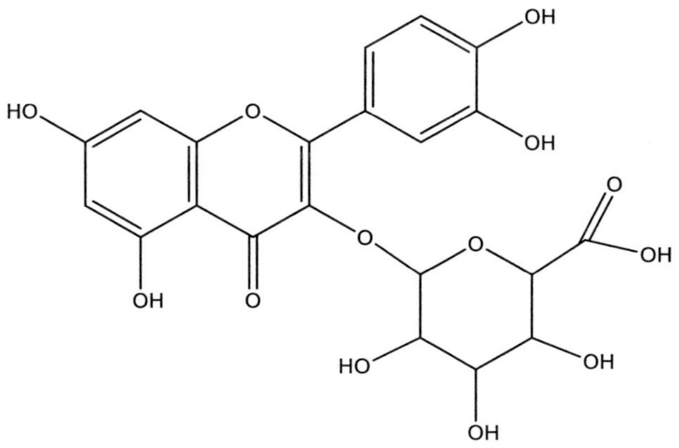

Quercetin-3-O-ß-D-glucuronic acid

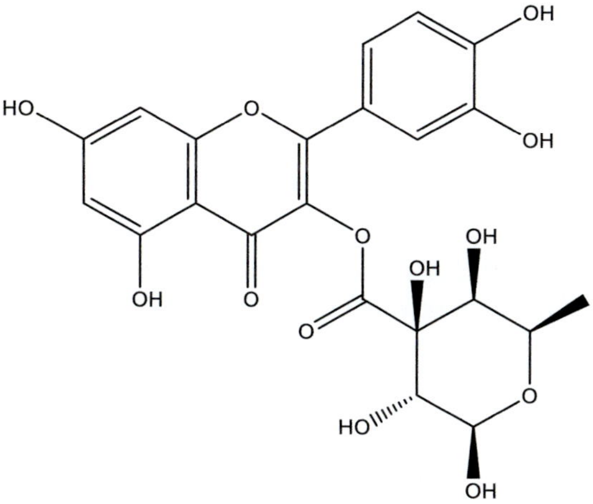

Quercetin-3-O-ß-D-galacturonic acid

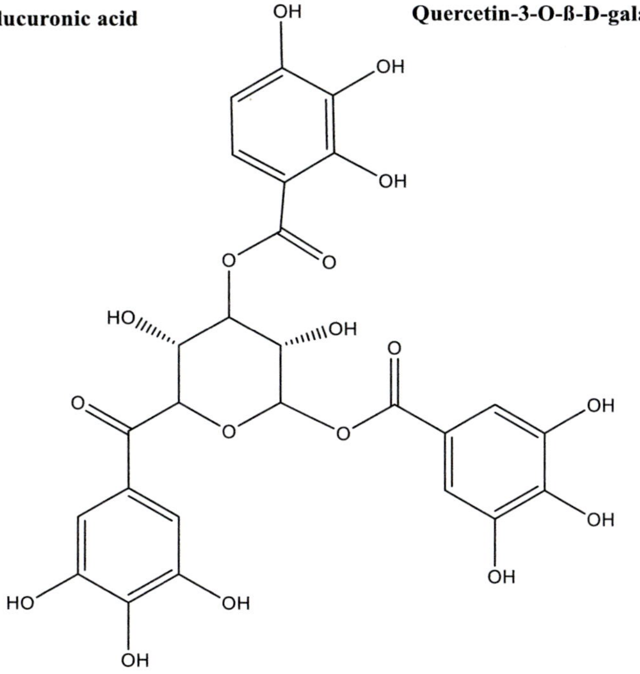

1,3,6-trigalloyl glucose

FIGURE 67.1.3 Primary chemical components of *Chrozophora oblongifolia*.

67.1.8 Mechanism of Action

One of the significant underlying mechanisms of action of dietary flavonoids and related polyphenols is linked to their ability to inhibit oxidative stress and its associated downstream responses, including inflammatory diseases [933].

The binding affinity of 1,3,6-trigalloyl glucose present in *Chrozophora oblongifolia* toward human intestinal maltase-glucoamylase α-glucosidases was recently investigated via molecular docking and dynamics studies. It was found that 1,3,6-trigalloyl glucose showed stable binding with maltase-glucoamylase α-glucosidases [1060].

67.1.9 Toxicity and Clinical Data

Apparently, there is no data on the toxicity of *Chrozophora oblongifolia* in the literature. Genus *Chrozophora* has limited medicinal uses and is toxic to livestock. As a result, its significance is primarily local and limited in scope [1063].

Acute toxicity reveals that plants belonging to the same genus, such as *Chrozophora tinctoria* (ethyl acetate and dichloromethane fractions), were found to be toxic in a dose-dependent manner and cause emesis at 0.5 g or higher and lethargy and mortality from 3 to 5 g upward [1064].

67.1.10 Adulterants or Substituents

Chrozophora tinctoria is sometimes misunderstood as *Chrozophora oblongifolia* [1065].

67.1.11 Most Cited Research

iCite, Database, NIH Open Citation Collection was used on September 27, 2023, to get the details of the most cited research. Filters for research papers were used by using the keywords *Chrozophora oblongifolia*. NIH database search revealed an assessment of selected Yemeni medicinal plants for their in vitro antimicrobial, anticancer, and antioxidant activities [1066].

67.1.12 Patents

Thumplasseril V. John, Ali SHAHBANDI, Aaron Jun KAMIMURA, Judith FLORES, Xiaodong Li; WO2025024478A1; Compositions for the modulation of taste, 2024. In this invention composition containing plant from the similar genus, *Chrozophora tinctoria* has been used with other ingredients to modulate the taste.

67.1.13 Plant Pathology

Spider mites, small insects, and extract sap from plant leaves, resulting in the formation of small yellow or white lesions on the foliage [1067].

68 Genus *Cichorium*

68.1 *Cichorium intybus* L.

68.1.1 Synonyms

Cichorium intybus var. *sylvestre, Cichorium rigidum* & *Cichorium sylvestre*

68.1.2 Family

Asteraceae

68.1.3 Vernacular Name

Ain alward and handaba

68.1.4 Common Name

Shikoryah, Hidaba, Hindaba bariah, & chicory

68.1.5 Botanical Description

Erect, woody perennial herb reaching up to 1 m in height. The plant features numerous branched stems that are hollow and round in cross-section. When crushed, any part of the plant exudes a white latex. Flowers: bright light blue flowers adorn the plant (Figures 68.1.1–68.1.3).

68.1.6 Geographical Distribution

Chicory is native to Europe and was introduced into the United States in the late 19th century. It is cultivated extensively in several European countries, including the Netherlands, Belgium, France, and Germany. Additionally, chicory is grown to some extent in North America, where it has become established as both a cultivated crop and a wild plant [1068]. In Oman, it is distributed in Musandam

FIGURE 68.1.1 Branches of *Cichorium intybus*.

FIGURE 68.1.2 Flowering branches of *Cichorium intybus*.

FIGURE 68.1.3 Flower of *Cichorium intybus*.

among agricultural terraces, and in some cultivated areas of the western Hajar Mountains.

68.1.7 Chemical Composition

The primary fraction of the plant consists of aliphatic compounds and their derivatives, while terpenoids represent minor constituents [1069]. Chicory root contains a variety of phytochemicals, including inulin, which is a starch-like polysaccharide. Other compounds found in chicory root include coumarins, flavonoids, sesquiterpene lactones (lactucin, 8-deoxylactucin, lactucopicrin, and 11ß-dihydro-derivatives), tannins, alkaloids, vitamins, minerals, and volatile oils [1070] (Figure 68.1.4). The other chemical components such as hydroxycinnamic acid derivatives, chlorogenic acids, flavonoids, benzo-isochromenes (cichorins A, B, and C), cichoriin (esculetin-7-glucoside),

DOI: 10.1201/9781003295488-68

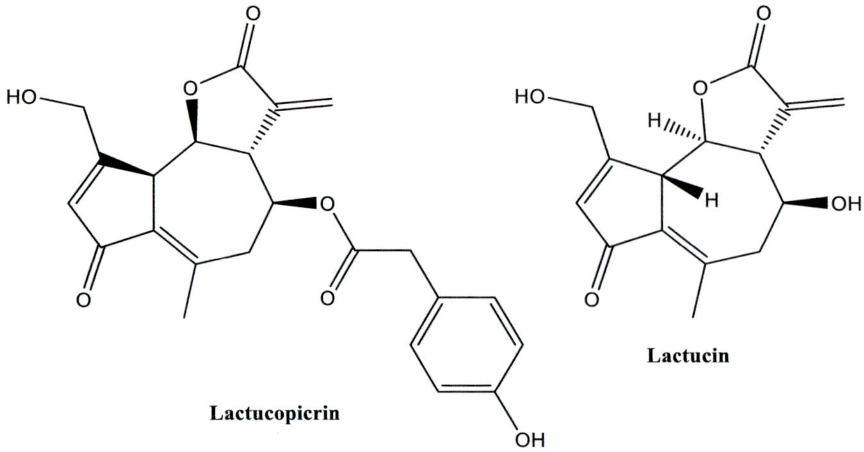

FIGURE 68.1.4 Chemical components of *Cichorium intybus*.

and (+)-germacrene have also been reported in the plant [1071].

68.1.8 Uses

Historically, chicory was cultivated by the ancient Egyptians for various purposes, including medicinal use, as a coffee substitute, and as a vegetable crop. It was also occasionally utilized as animal forage. In the 1970s, it was found that the root of *Cichorium intybus* contains a high amount of inulin, up to 40%, which has a minimal impact on blood sugar levels and is considered suitable for individuals with diabetes [1072].

This species is reputed to possess a wide range of medicinal properties and has been traditionally used in the regions where it naturally occurs, particularly across Eurasia and Africa. It has been employed in folk and traditional medicinefor the treatment of various ailments, including wounds, diabetes, headaches, fever, malaria, liver disorders, stomach discomfort, indigestion, and diarrhoea. Additionally, it is valued as a general health tonic, used to promote overall well-being [911].

68.1.9 Mechanism of Action

Terpenoids present in this plant exert inhibitory effects on the production of inflammatory cytokines [1030].

68.1.10 Toxicity and Clinical Data

Cichorium intybus has a long history of safe use, and no health hazards or side effects are known when it is administered at recommended therapeutic dosages. It is considered as safe for human use [1073].

68.1.11 Adulterants or Substituents

Campanula rotundifolia (Bluebell) and *Linum lewisii* (Prairie Flax) are the substitutes to *Cichorium intybus* [1074].

68.1.12 Most Cited Research

iCite, Database, NIH Open Citation Collection was used on September 28, 2023, to get the details of the most cited research. Filters for research papers were used by using the keywords *Cichorium intybus*. NIH database search revealed dynamics of human gut microbiota and short-chain fatty acids in response to dietary interventions with three fermentable fibers [1075].

68.1.13 Patents

Jean Paufique—US9044490B2—Active ingredient obtained from *Cichorium intybus* for acting on the barrier function of the skin that is like that of vitamin D, 2013.

68.1.14 Plant Pathology

Cichorium intybus is susceptible to various diseases, including bacterial diseases such as leaf spots caused by *Pseudomonas marginalis*, and similar diseases caused by other *Pseudomonas* species and *Pectobacterium atrosepticum* [1076].

69 Genus *Cissus*

69.1 *Cissus quadrangularis* L.

69.1.1 Synonyms

Saelanthus quadragonus & *Vitis quadrangularis*

69.1.2 Family

Vitaceae

69.1.3 Vernacular Name

aterey

69.1.4 Common Name

Adamant creeper, Bone setter

69.1.5 Botanical Description

Cissus quadrangularis is a climber herb characterized by branches with a quadrangular cross-section. The leaves are toothed and trilobed, measuring 2–5 cm in width, and they appear at the nodes. The plant has a thick quadrangular succulent stem where each node has a tendril emerging from the opposite side of the node (Figures 69.1.1–69.1.4).

69.1.6 Geographical Distribution

This species is distributed in Tropical and Southern Africa, including countries such as Mozambique, Tanzania, and South Africa, and the island of Madagascar. Its distribution continues through the Arabian Peninsula, and extends eastward across the Indian Subcontinent, including India,

FIGURE 69.1.2 Stem section *Cissus quadrangularis*.

FIGURE 69.1.3 Fruit of *Cissus quadrangularis*.

FIGURE 69.1.1 Stem and leaves of *Cissus quadrangularis*.

FIGURE 69.1.4 Shooting stem of *Cissus quadrangularis*.

DOI: 10.1201/9781003295488-69

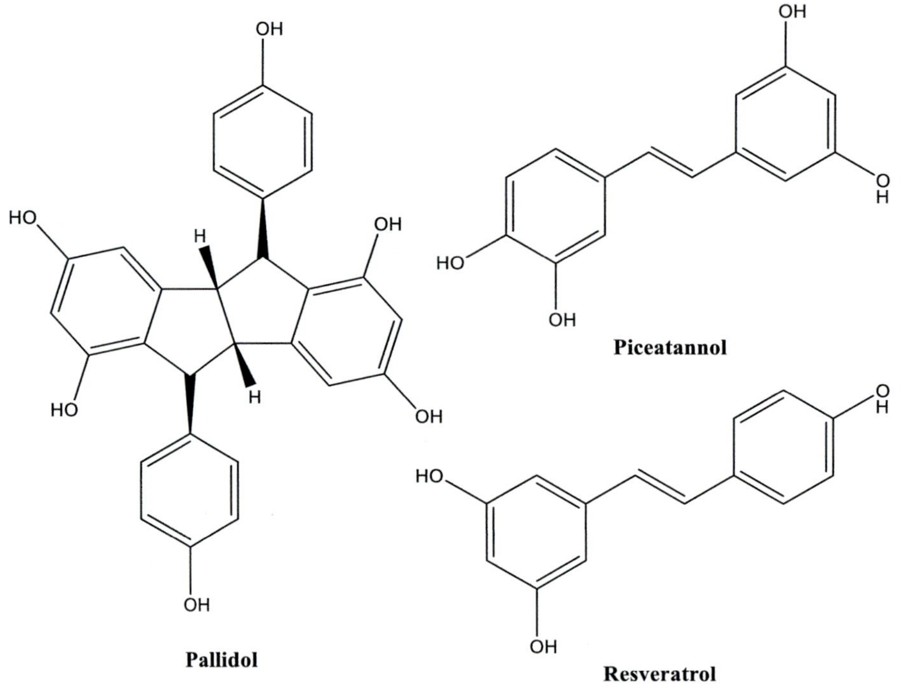

FIGURE 69.1.5 Major chemical components of *Cissus quadrangularis*.

Pakistan, Sri Lanka, and Bangladesh, reaching as far as Myanmar in Southeast Asia. In Oman, the species occurs in the wet escarpment woodlands of the southern Oman.

69.1.7 Chemical Composition

Phytochemical studies of *Cissus quadrangularis* revealed the presence of diverse constituents, including flavonoids, triterpenoids, vitamin C, stilbene derivatives, and other compounds such as resveratrol, piceatannol, pallidol, perthenocissin, and phytosterols [1077] (Figure 69.1.5). The stem parts of plant contain A and β-amyrins, β-sitosterol, ketosetosterol, phenols, tannins, taraxeryl acetate, methyl tritiacontanoic acid, and taraxeroliso-pentadecanoic acid [1078].

69.1.8 Uses

Cissus has a wide range of applications, such as promoting the healing of fractured bone and tissue repair, preventing osteoporosis, aiding in weight loss, regulating blood sugar levels, treating digestive disorders and menstrual irregularities, and addressing various other traditional medicinal uses [1079].

According to traditional practices, this plant has been used to treat skin infections and mastitis in livestock.

69.1.9 Mechanism of Action

One of the significant underlying mechanisms of action of dietary flavonoids and related polyphenols is linked to their capacity to inhibit oxidative stress and its associated downstream responses, including inflammatory diseases [933].

69.1.10 Toxicity and Clinical Data

Cissus quadrangularis is considered as safe. No toxic effect has been observed [1077].

69.1.11 Most Cited Research

iCite, Database, NIH Open Citation Collection was used on September 29, 2023, to get the details of the most cited research. Filters for research papers were used by using the keywords *Cissus quadrangularis*. NIH database search revealed a systematic review of the efficacy and safety of herbal medicines used in the treatment of obesity [1080].

69.1.12 Patents

Jacob Geissler—US7582316B1—Ketone-steroid extracts from the *Cissus quadrangularis* plant and methods thereof, 2006. This invention is related to extracts of the *Cissus quadrangularis* for preparing a ketone-steroid extract to promote the healing of connective tissue.

69.1.13 Plant Pathology

No literature is available regarding the pathology of *Cissus quadrangularis*. However, *Glomerella cingulate* is a fungal pathogen that attacks the genus *Cissus* [1081].

70 Genus *Cistanche*

70.1 *Cistanche tubulosa* (Schenk) Wight ex Hook.f.

70.1.1 Synonyms

Phelypaea tubulosa

70.1.2 Family

Orobanchaceae

70.1.3 Vernacular Name

Sharbooth & baisail

70.1.4 Common Name

Haluk, Danum, and Tarfas

70.1.5 Botanical Description

This perennial parasitic herb, typically 15–40 cm tall with a purplish tinge, displays an erect growth form featuring lanceolate leaves densely covered by 1–4 cm long scales. The plant produces dense, cylindrical spikes of yellow flowers with hairy staminal filaments and anthers. Its ovoid-oblong capsules contain numerous dark-coloured, pitted seeds, each measuring 1 mm in length (Figures 70.1.1 and 70.1.2) [1082].

70.1.6 Geographical Distribution

This species is native to a broad region encompassing North Africa and Southwest Asia. Within Oman it is found in southern, northern, and central Oman.

FIGURE 70.1.1 Inflorescence of *Cistanche tubulosa.*

FIGURE 70.1.2 Whole plant of *Cistanche tubulosa.*

70.1.7 Chemical Composition

The main active chemical compound in *Cistanche tubulosa* is phenylethanoid glycosides [1083]. *Cistanche tubulosa* has also been found to contain a variety of compounds, including phenylethanoid glycosides, syringalide A 3-O-α-L-rhamnopyranoside, campneoside I, iridoids, iridoid glycosides, phenylethyl oligosaccharides, terpenes, and lignans [1082]. The plant also contains a high level of polysaccharides that possess the following therapeutic properties:

- ease cognitive dysfunction by regulating antioxidant and anti-inflammatory processes in rodents
- enhance echinacoside absorption *in vivo* and affect the gut microbiota
- induce melanogenesis in melanocytes
- protect PC12 cells against OGD/RP-induced injury
- reduce oxidative stress [1084].

70.1.8 Uses

Cistanches herba exhibits pharmacological activities suitable for the potential treatment of neurodegenerative disorders [1085], including Alzheimer's disease [1086]. Parkinson's disease [1087], and vascular dementia [1088], as reported in prior studies. The earlier reports showed that

DOI: 10.1201/9781003295488-70

ethanolic extract of *Cistanche tubulosa* (a component of a traditional Chinese herbal formula) can regulate the gut microbial composition in rats [1084].

70.1.9 Mechanism of Action

The aqueous extract from *Cistanche tubulosa*, a traditional Chinese herbal formula, has been reported to play a role in protecting the human intestine and modulating the dramatic changes in the metabolic pathways of the gut microbiome induced by cefixime. The plant also contains a high level of polysaccharides that possess the following therapeutic properties

- ease cognitive dysfunction by regulating antioxidant and anti-inflammatory processes in rodents
- enhance echinacoside absorption in vivo and affect the gut microbiota
- induce melanogenesis in melanocytes
- protect PC12 cells against OGD/RP-induced injury
- reduce oxidative stress [1084].

70.1.10 Toxicity and Clinical Data

Cistanches herba has been a well-regarded traditional medicine in China for over a thousand years. Preparations of Cistanches herba typically result in mild adverse clinical symptoms, such as nausea, vomiting, abdominal pain, and dizziness [1083].

70.1.11 Adulterants or Substituents

Identifying adulterants in *Cistanches* Herba presents a challenge using existing analytical chemistry methods due to its striking resemblance to *C. salsa* [1089] and *C. sinensis* [1090].

70.1.12 Most Cited Research

iCite, Database, NIH Open Citation Collection was used on October 5, 2023, to get the details of the most cited research. Filters for research papers were used by using the keywords *Cistanche tubulosa*. NIH database search revealed anti-ageing active ingredients from herbs and nutraceuticals used in traditional Chinese medicine: pharmacological mechanisms and implications for drug discovery [1091].

70.1.13 Patents

Chao-Jih Wang, Ai-Ling YEH— US9931367B2—Uses of *Cistanche tubulosa* extract and isoacteoside in protecting muscle, 2017. The present invention relates to the uses of a *Cistanche tubulosa* extract and isoacteoside in protecting muscles.

71 Genus *Citrullus*

71.1 *Citrullus colocynthis* (L.) Schrad.

71.1.1 Synonyms

Cucumis colocynthis & *Citrullus colocynthoides*

71.1.2 Family

Cucurbitaceae

71.1.3 Vernacular Name

Handal & sheru

71.1.4 Botanical Description

Citrullus colocynthis is a trailing perennial that grows up to 3 m with alternate leaves, gray-green, and triangular with entire margins, deeply lobed and have tendrils emerging from the leaf nodes for gripping. Flowers: solitary appear at the leaf axils (where the leaf meets the stem). Fruit: large, round, and smooth, turning from green with yellow stripes to yellow when ripe. Seeds embedded in a bitter white pulp (Figures 71.1.1–71.1.3).

71.1.5 Geographical Distribution

Citrullus colocynthis is commonly found in sandy lands across various regions, including Northwest India, Punjab, Sind, Central India, Southern India, and the Coromandel Coast. It is also indigenous to Arabia, West Asia, Tropical Africa, and the Mediterranean region [1092]. It is distributed throughout Oman.

FIGURE 71.1.1 Ripe fruit of *Citrullus colocynthis*.

FIGURE 71.1.2 Flower of *Citrullus colocynthis*.

71.1.6 Chemical Composition

The ethanolic extract of *Citrullus colocynthis* contains various chemical components, including alkaloids, glycosides, and flavonoids, which exhibit potent antibacterial properties [1093]. The primary active constituents in *Citrullus colocynthis* are cucurbitacin, flavonoids, alkaloids, and phenolic acids [1094]. Natural cucurbitacins are triterpenoid chemicals famous for their bitter taste and toxicity. Plant contain glycosides such as elaterin (cucurbitacin E), elatericin B (cucurbitacin I), and dihydroelatericin B (cucurbitacin L) (Figure 71.1.4) [1095].

The cucurbitane-type triterpene glycoside and its aglycon, Cucurbitacin E 2-O-D-glucopyranoside, and cucurbitacin E, exhibited anti-allergic effects [1096]. Because of their cytotoxic properties, cucurbitacins play an important role in drug discovery [1097]. Plants also contain high levels of flavonoids such as kaempferol, quercetin, myricetin, catechin, gallic acid, vanillic acid, p-hydroxybenzoic acid, p-coumaric acid, caffeic acid, sinapic acid, chlorogenic acid, and ferulic acid [1098].

71.1.7 Chemical Tests

A previously reported method for confirming the presence of cucurbitacins in extracts and fractions involves combining the sample with triphenyltetrazolium chloride.

DOI: 10.1201/9781003295488-71

FIGURE 71.1.3 Natural habit of *Citrullus colocynthis*.

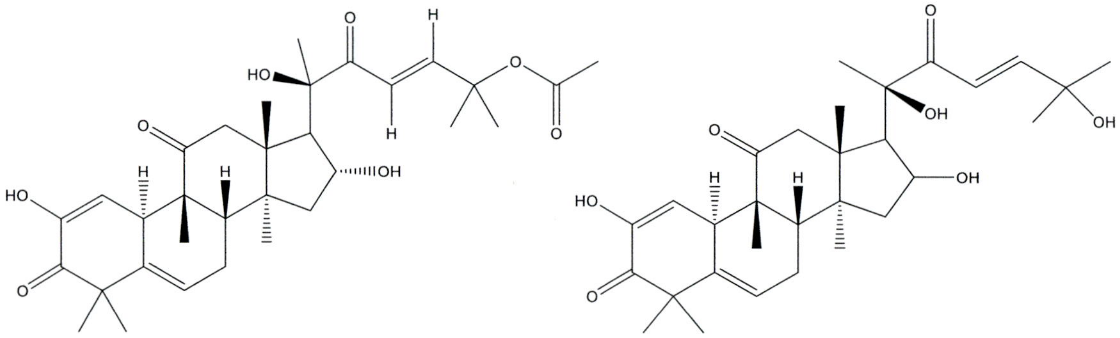

FIGURE 71.1.4 Primary chemical components of *Citrullus colocynthis*.

The formation of a red precipitate known as formazin signifies the presence of cucurbitacins [1095,1099].

71.1.8 Uses

Citrullus colocynthis root was historically employed for treating breast and joint inflammation, as well as external applications for ophthalmia and uterine pains. Additionally, a paste of the fruit and root, when mixed with water, was used to alleviate boils and pimples, and the root paste was applied to reduce abdominal swelling in children [1100].

Plants exhibit a spectrum of pharmacological activities, including antioxidative, antibacterial, anti-cancerous, anti-inflammatory, analgesic, gastrointestinal tract modulation, reproductive influence, protective effects, antimicrobial properties, antidiabetic attributes, hypolipidemic impacts, antineoplastic potential, profibrinolytic action, anti-allergic properties, pesticidal traits, and immune-stimulatory responses [1098].

71.1.9 Mechanism of Action

Cucurbitacin mechanisms of action encompass anti-proliferative effects, suppression of migration and invasion, pro-apoptotic induction, and promotion of cell cycle arrest [1101].

71.1.10 Toxicity and Clinical Data

Citrullus colocynthis has been shown to induce diarrhea and cause intestinal damage in experimental studies involving treated animals. This suggests that certain compounds or components within the plant may have a laxative or irritant effect on the gastrointestinal system, potentially leading to these adverse effects [1102].

71.1.11 Adulterants or Substituents

In the mountainous regions of northern India, the fruits of *Cucumis trigonus* and *Cucumis pseudo colocynthis* are commonly found and are sometimes used to adulterate colocynth sold in the market [1103].

71.1.12 Most Cited Research

iCite, Database, NIH Open Citation Collection was used on September 30, 2023, to get the details of the most cited research. Filters for research papers were used by using the keywords *Citrullus colocynthis*. NIH database search revealed an overview of antidiabetic medicinal plants having insulin mimetic property [1104].

71.1.13 Patents

Alberto Sirizzotti, Daniel BELLON DONA, Richard Bernard Berentsen, Daniel JIMENEZ FERNANDEZ—US20210000047A1—Cgmmv-resistant *Citrullus* plants, 2018. The present invention relates to *Citrullus colocynthis* plant cells and *Citrullus colocynthis* plants resistant to CGMMV (Cucumber Green Mottle Mosaic Virus).

71.1.14 Plant Pathology

Citrullus colocynthis displayed mild mosaic symptoms caused by zucchini yellow mosaic virus [1105]. Zucchini yellow mosaic, caused by the Zucchini yel-low mosaic virus, is an economically harmful disease of imperative cucurbit crops [1106].

72 Genus *Citrus*

72.1 *Citrus aurantiifolia* (Christm.)

72.1.1 Synonyms

Citrus medica f. aurantiifolium & *Limonia* × *aurantiifolia*

72.1.2 Family

Rutaceae

72.1.3 Vernacular Name

Omani lime, lemon, & lumi

72.1.4 Common Name

Lime, Key lime, Egyptian Lime, Indian Lime & Mexican Lime

72.1.5 Botanical Description

A small, much-branched tree that can grow up to 5 m tall and is known for its characteristic short, sharp spines. Leaves are leathery and can be elliptic or oblong-ovate. Flowers: white. Fruit: globose to subglobose (round to slightly oblong) with a very acidic pulp (Figures 72.1.1–72.1.4).

72.1.6 Geographical Distribution

Citrus aurantiifolia is native to Southeast Asia and widespread in tropical and subtropical regions. Oman has been recognized as a pivotal transit region for the dissemination of lime (*C. aurantiifolia*) to Africa and the New World [1107]. Within Oman, is primarily found as a cultivated crop throughout Oman, thriving in various agricultural regions across the country.

FIGURE 72.1.1 Flower of *Citrus aurantiifolia*.

FIGURE 72.1.2 Fruit of *Citrus aurantiifolia*.

FIGURE 72.1.3 Phyllotaxy of *Citrus aurantiifolia*.

FIGURE 72.1.4 Apical fruiting portion of *Citrus aurantiifolia*.

DOI: 10.1201/9781003295488-72

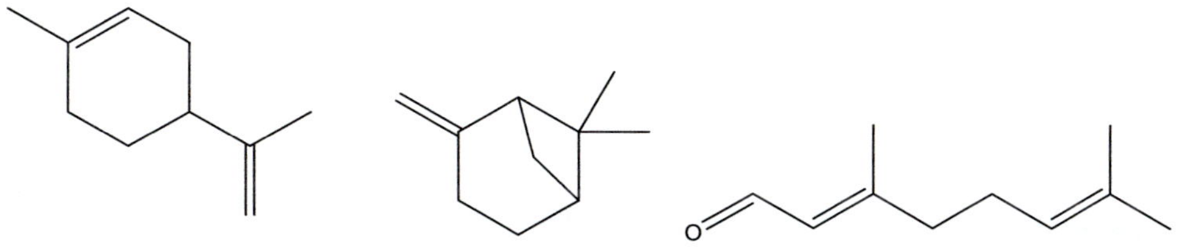

Limonene **ß-pinene** **Citral**

FIGURE 72.1.5 Chemical components of *Citrus aurantiifolia.*

72.1.7 Chemical Composition

High concentrations of coumarins and furocoumarins are reported in *Citrus aurantiifolia* [1108]. Flavonoids, terpenoids, phenolics, limonoids, alkaloids, and essential oils have been identified as secondary metabolites/phytochemicals in *Citrus aurantifolia* [1109]. Essential oil obtained from *Citrus aurantiifolia* contains limonene, β-pinene, γ-terpinene, and citral [1110]. In addition, it also contains germacrene isomers, pineen, linalool dimmer, bornane, anethole, anisole, safrole and demitol [1111].

72.1.8 Uses

The products derived from Omani lime (*Citrus aurantiifolia*) have applications in the beverage, food additive, and cosmetic industries [1112]. Figure 72.1.5 illustrates the various pharmacological effects of *C. aurantium* extracts. The pharmacological effects of *C. aurantium* extracts are illustrated below.

72.1.9 Mechanism of Action

The mechanism of action of coumarins involves competitive antagonism of vitamin K, thereby inhibiting coagulation by impeding prothrombin production [1113].

72.1.10 Toxicity and Clinical Data

The oil obtain from *Citrus aurantifolia* is generally considered as non-toxic to vital organs and cells. The toxicity assessment of the essential oil extract from *Citrus aurantiifolia* involved oral administration through two distinct methodologies: the acute toxicity single-dose study and the repeated dose regimen. The results obtained from this study showed that despite of the consideration related to the safety of lime essential oil, it can have mild hematotoxic, nephrotoxic and hepatotoxic effects [1111].

72.1.11 Most Cited Research

iCite, Database, NIH Open Citation Collection was used on September 1, 2023, to get the details of the most cited research. Filters for research papers were used by using the keywords *Citrus aurantiifolia.* NIH database search revealed natural products for cancer-targeted therapy: citrus flavonoids as potent chemopreventive agents [1114].

72.1.12 Patents

Olivier Buysschaert, Annelieke Nicoline van Ooij, Timothy A. AngleaA. Reza Kamarei—US20100159100A1—Flavonoid-rich citrus extract and compositions thereof, 2008. This invention is related to a flavonoid citrus extract and beverage compositions prepared therefrom that can be added to beverages.

72.1.13 Plant Pathology

Citrus tristeza virus (CTV) serves as the etiological agent responsible for a paramount citrus disease [1115]. As a closterovirus, it encompasses multiple strains characterized by variations in serological and biological attributes, double-stranded RNA (dsRNA) patterns, and genomic sequences [1116].

73 Genus *Clematis*

73.1 *Clematis orientalis* L.

73.1.1 Synonyms

Clematis orientalis var. *vulgaris, Meclatis orientalis* & *Viticella orientalis*

73.1.2 Family

Ranunculaceae

73.1.3 Vernacular Name

Melwaya & mahrqa

73.1.4 Common Name

Bailari cheiti [1117].

73.1.5 Botanical Description

Clematis orientalis is a climbing shrub that possesses tendril-like stems and sheds its leaves annually. This shrub grows densely and reaches a height of 2–8 m. Leaves are divided into three to seven ovate to linear-lanceolate leaflets. Yellow flowers originate from the leaf nodes in small clusters and possess a central cluster of stamens and feathery pistils. The hermaphroditic flower has four yellow-to-yellow-orange sepals (Figures 73.1.1–73.1.4).

73.1.6 Geographical Distribution

The plant's native range lies between Persia and the Himalayas and Iran to Himalayas. This plant is available in Asia and Central Europe. It's also used as an ornamental plant in the United States [1118]. This species is found in the rugged embrace of the Western Hajar Mountains on the cool shade of rocks and boulders, seeks refuge beneath the boughs of Juniper trees.

FIGURE 73.1.1 Leaves of *Clematis orientalis*.

73.1.7 Chemical Components

Plants contain phenolic acid, such as resorcinol, in its chloroform fraction [1119]. This plant also contains triterpene glycosides such as orientaloside F, H, I, J, and K [1120].

C. orientalis has been reported for the presence of gallic acid as a main content in hexane and ethylacetate fractions [1119].

73.1.8 Uses

Traditionally, this plant is used as an analgesic, antimicrobial, anti-inflammatory, antiphlogistic, antibacterial, anticancer, abirritative, and diuretic agent [1119–1122].

Clematis orientalis has been reported for its potential role in the treatment of rheumatoid arthritis by suppressing IL-1β, TNF-α, IL-6, COX-2 and NF-Kβ due to the presence of flavonoids and glycosides [1123].

73.1.9 Mechanism of Action

The inhibition of proinflammatory cytokines and inhibition of cyclooxygenase (COX) enzymes. Specifically, the ethanolic extracts of three *Clematis* species (i.e., *Clematis pickeringii, Clematis macrophilia*, and *Clematis glycosides*) were found, albeit at differing degrees, to inhibit COX-1, COX-2, and 5-lipoxygenase, with the exception that stems and leaf extracts of *C. macrophilia* did not inhibit COX-2. The highest inhibition was noted with *C. pickeringii* [1124,1125].

73.1.10 Toxicity and Clinical Data

Clematis preparation can cause severe mouth pain and ulcers if eaten. Toxic to dogs, cats, and horses. Clinical signs are salivation, vomiting, and diarrhea. In humans, symptoms may include dizziness, confusion, fainting, and convulsions [1126].

73.1.11 Most Cited Research

iCite, Database, NIH Open Citation Collection was used on January 3, 2023, to get the details of the most cited research. Filters for research papers were used by using the keywords *Clematis orientalis.* NIH database search revealed the inhibitory effects of Turkish folk remedies on inflammatory cytokines: interleukin-1alpha, interleukin-1beta, and tumor necrosis factor-alpha [1125].

DOI: 10.1201/9781003295488-73

FIGURE 73.1.2 *Clematis orientalis* at fruiting stage.

FIGURE 73.1.3 Flowers of *Clematis orientalis*.

FIGURE 73.1.4 Feathery seeds of *Clematis orientalis*.

73.1.12 Plant Pathology

It has been observed that plants belonging to the *Clematis* genus are susceptible to infection (leaf spot and stem rot) by *Phoma clematidina*. The plant also showed its association with microfungi, *Haetosphaeronema clematidis* [1127].

74 Genus *Cleome*

74.1 *Cleome amblyocarpa* Barratte & Murb.

74.1.1 Synonyms

Cleome arabica subsp. *amblyocarpa*

74.1.2 Family

Cleomaceae

74.1.3 Vernacular Name

Afainah

74.1.4 Common Name

Clover-leaf Cleome, Spider plants

74.1.5 Botanical Description

Annual or perennial herbs attain a height exceeding 50 cm. Characterized by stiff, upright, and branched stems, it displays alternate trifoliate leaves. The small flowers are clustered amidst foliage, while the fruit manifests as dry, dehiscent siliqua-shaped capsules, elongated and pendant in structure, enclosing seeds adorned with fine hairs (Figures 74.1.1 and 74.1.2) [1128].

74.1.6 Geographical Distribution

Cleome amblyocarpa has a widespread distribution across North Africa, primarily within sandy habitats within the region of Tunisia [1129]. In Oman, it is found in Southern and Northern Oman mainly in sandy areas.

FIGURE 74.1.1 Natural habit of *Cleome amblyocarpa.*

FIGURE 74.1.2 Fruiting shoot of *Cleome amblyocarpa.*

74.1.7 Chemical Composition

Phytochemical screening has revealed that *Cleome* species exhibit a substantial presence of diverse secondary metabolites, mainly terpenoids, flavonoids, phenolic compounds, and alkaloids [1130]. Essential oils of this plant actively contain chemical components such as caryophyllene oxide, hexahydrofarnesyl acetone, alloaromadendrene epoxide, myrtenyl acetate, isoshyobunone, shyobunol, and trans-caryophyllene (Figure 74.1.3) [1131].

74.1.8 Uses

Cleome amblyocarpa traditionally used for the treatment of various diseases such as diabetes and colic, as a stomachic therapy, rheumatic fever, scabies, and inflammation [1131].

DOI: 10.1201/9781003295488-74

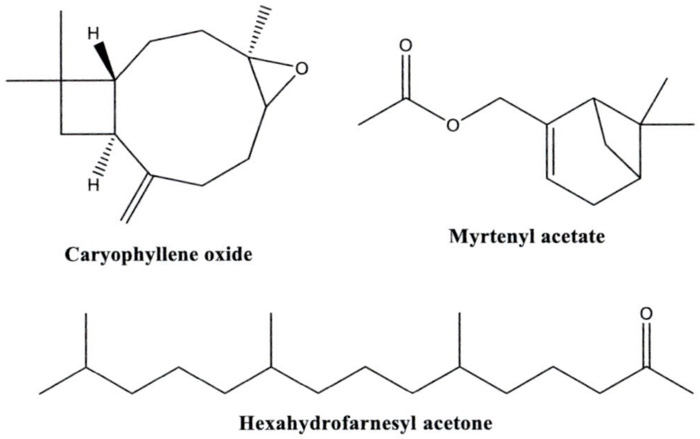

FIGURE 74.1.3 Primary chemical components of *Cleome amblyocarpa.*

Previous investigations have documented the potential utility of various *Cleome* species in the management of conditions encompassing inflammation, pyrexia, pain, ulcerations, depressive disorders, and diarrhea [1132].

74.1.9 Mechanism of Action

One of the significant underlying mechanisms of action of dietary flavonoids and related polyphenols is linked to their capacity to inhibit oxidative stress and its associated downstream responses, including inflammatory diseases [933]. Chemical components present in the plant oil showed significant anti-inflammatory activity via the suppression of lipoxygenase and cyclooxygenases, along with membrane stabilization [1131].

74.1.10 Toxicity and Clinical Data

After drying, this herb serves as a fodder source favored by animals, although the consumption of fresh plants by animals is typically avoided. Its misuse has been associated with potential toxicity, mainly the development of nervous disorders [1133].

74.1.11 Adulterants or Substituents

There is no adulterant or substitute reported for *Cleome amblyocarpa.* The species of genus *Cleome* are similar in chemical composition so they can be substituted with closely resembling other species of genus *Cleome.*

74.1.12 Most Cited Research

iCite, Database, NIH Open Citation Collection was used on October 1, 2023, to get the details of the most cited research. Filters for research papers were used by using the keywords *Cleome amblyocarpa.* NIH database search revealed natural products for cancer-targeted therapy: citrus flavonoids as potent chemopreventive agents [1114]. Antifungal properties of crude extracts of five Egyptian medicinal plants against dermatophytes and emerging fungi were studied [1134].

74.1.13 Patents

René Belle, Francoise Belaubre—US20100119632A1—*Cleome spinosa* extract used in pharmaceutical preparations and cosmetic compositions, 2008. The present invention is related to the pharmaceutical or cosmetic preparations containing Cleome *spinosa* extract to treat hair fall.

74.1.14 Plant Pathology

Leaf spot is a common plant disease caused by fungi. In Egypt, *Cleome herbarum* is infected by leaf spot disease. Similarly, plant species such as *Cleome amblyocarpa, Haplophyllum tuberculatum,* and *Pluchea dioscorides* are also vulnerable to leaf spot disease [1135].

74.2 *Cleome brachycarpa* (Forssk.) Vahl ex DC.

74.2.1 Synonyms

Cleome brevisiliqua & *Cleome diversifolia*

74.2.2 Family

Cleomaceae

74.2.3 Vernacular Name

Kezam

74.2.4 Botanical Description

An erect, aromaticherb, either annual or perennial, reaching a height up to 30 cm, possesses extensive branching and glandular characteristics throughout, characterized by the glands that are subsessile and/or shortly stalked. The plant has 3–5-foliolate leaves, and leaflets are linear-lanceolate to obovate (Figures 74.2.1–74.2.4).

FIGURE 74.2.1 Early growth stage showing young shoots of *Cleome brachycarpa.*

FIGURE 74.2.2 Flowering shoot of *Cleome brachycarpa*.

FIGURE 74.2.3 Apical portion with ripe seeds of *Cleome brachycarpa*.

FIGURE 74.2.4 Phyllotaxy of *Cleome brachycarpa*.

74.2.5 Geographical Distribution

Cleome brachycarpa exhibits a broad geographical distribution, primarily within sandy habitats across regions of Africa and Asia. Apparently, it is available in North Africa, Egypt, Saudi Arabia, Afghanistan, Pakistan, India, and the Hormozgan province of Iran [1136]. It is distributed throughout Oman.

74.2.6 Chemical Composition

Terpenoids are the main chemical component in *Cleome brachycarpa* essential oil [1137]. Essential oils contain triterpenoid dilactone, deacetoxy brachycarpone [1138], triterpenoid cleocarbpone, ursolic acid, and cabralealactone (Figure 74.2.5).

74.2.7 Uses

It exhibits a diverse array of applications, including the alleviation of joint pain, soreness, tenderness, itching, and the treatment of conditions such as leprosy, leucoderma, scabies, intestinal pain, and swelling. Furthermore, it possesses anti-rheumatic, vermicidal, anthelmintic, antioxidant, and antiemetic properties [1139]. *Cleome brachycarpa* has been used for the treatment of various diseases such as diarrhea, malaria, illness, hypertension, and snakebite [1140].

74.2.8 Mechanism of Action

Terpenoids exert inhibitory effects on the production of inflammatory cytokines [1030].

74.2.9 Toxicity and Clinical Data

Cleome brachycarpa, administered at doses as high as 5,000 mg/kg, has exhibited substantial reductions in the levels of ALT (Alanine Aminotransferase), AST (Aspartate Aminotransferase), and ALP (Alkaline Phosphatase), while not eliciting any indications of organ toxicity when compared with acetaminophen [1141].

74.2.10 Adulterants or Substituents

No adulterant or substitute was reported for *Cleome brachycarpa*. The species of genus *Cleome* are similar in chemical composition so they can be substituted with closely resembling other species of genus *Cleome*.

74.2.11 Most Cited Research

iCite, Database, NIH Open Citation Collection was used on October 2, 2023, to get the details of the most cited research. Filters for research papers were used by using the keywords *Cleome brachycarpa*. NIH database search revealed the silico and *in vivo* characterization of cabralealactone, solasodin, and salvadorin in a rat model: potential anti-inflammatory agents [1142].

74.2.12 Patents

René Belle, Francoise Belaubre—US20100119632A1—Cleome spinosa extract used in pharmaceutical preparations and cosmetic compositions, 2008.

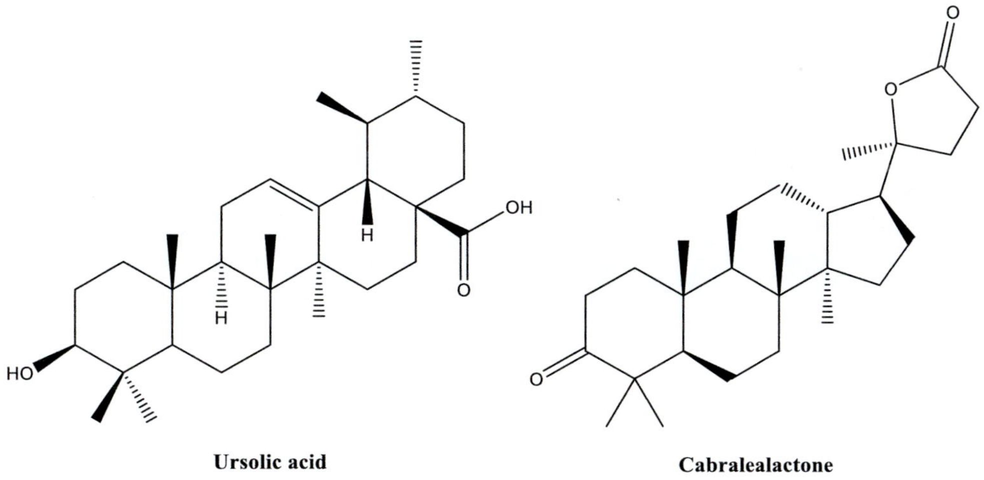

FIGURE 74.2.5 Major chemical components of *Cleome brachycarpa*.

74.3 *Cleome droserifolia* (Forssk.) Delile

74.3.1 Synonyms

Cleome austroarabica subsp. *muscatensis, Rorida droserifolia & Rorida arabica*

74.3.2 Family

Cleomaceae

74.3.3 Vernacular Name

Shagrat al hema

74.3.4 Botanical Description

It is a glandular-viscid and aromatic cushion-like shrub, typically measuring between 25 and 60cm in length. The plant possesses orbicular or broadly ovate leaves with densely glandular-hairy structures on its surface. Flowers: yellow with reddish stripes. Fruit: hairy capsule (Figures 74.3.1–74.3.3).

FIGURE 74.3.1 Flowering shoot of *Cleome droserifolia*.

FIGURE 74.3.2 Phyllotaxy of *Cleome droserifolia*.

74.3.5 Geographical Distribution

It is native to western Asia and northern Africa, which aligns perfectly with the Eastern Sahara to Arabian Peninsula and Syria. In Oman, it is distributed throughout the country.

74.3.6 Chemical Composition

The phytochemical analysis of *C. droserifolia* revealed the existence of flavonoids, terpenoids, and alkaloids [1143]. Plant mainly contain terpenes and sterols such as drosericarpone, buchariol, stigmasterol glucoside, 4(15)-guaiane-6-ol, teucladiol, (6S,9R)-roseoside, daucosterol, guai-7(11) 8-diene, 18-hydroxy-dollabela-8 (17)-ene, and (24 E) stigmasta-5,8-dien-3β-ol [1144] (Figure 74.3.4).

74.3.7 Uses

Cleome species are commonly employed in traditional folk medicine for their stomachic and rubefacient properties, as well as their therapeutic applications in the treatment of conditions such as scabies, rheumatic fever, and inflammation [1145].

FIGURE 74.3.3 Fruiting shoot of *Cleome droserifolia*.

74.3.8 Mechanism of Action

One of the significant underlying mechanisms of action of dietary flavonoids and related polyphenols is linked to their capacity to inhibit oxidative stress and its associated downstream responses, including inflammatory diseases [933].

74.3.9 Toxicity and Clinical Data

Cleome droserifolia is recognized as a toxic plant [1146].

74.3.10 Adulterants or Substituents

No adulterant or substitute was reported for *Cleome droserifolia*. The species of genus *Cleome* are similar in chemical composition so they can be substituted with closely resembling other species of genus *Cleome*.

74.3.11 Most Cited Research

iCite, Database, NIH Open Citation Collection was used on October 2, 2023, to get the details of the most cited research. Filters for research papers were used by using the keywords *Cleome droserifolia*. NIH database search revealed Honey/Chitosan nanofiber wound dressing enriched with *Allium sativum* and *Cleome droserifolia*: enhanced antimicrobial and wound healing activity [1144].

74.3.12 Patents

Jean Pachot—US20220354912A1—*Cleome droserifolia* extract and therapeutic or cosmetic use thereof, 2020. The invention is related to *Cleome droserifolia* extract possessing a therapeutic or cosmetic effect intended for topical application.

74.3.13 Plant Pathology

Apparently, there is no data in the literature that refers to the pathology of *Cleome droserifolia*. However, *A. candida* is a type of fungal disease that usually occurs in the genus *Cleome* [1147].

74.4 *Cleome gynandra* L.

74.4.1 Synonyms:

Pedicellaria gynandra & *Gynandropsis gynandra*

74.4.2 Family

Cleomaceae

74.4.3 Common Name

Wild Spider Flower & Spider wisp

74.4.4 Botanical Description

An erect, annual herb reaching a height of up to 1 m. It exhibits extensive branching, and potential lignification over time. The stem is characterized by glandular trichomes and longitudinal striations. Its leaves are palmately compound, featuring 3–5 leaflets [1148]. The flowers are white, sometimes changing to rose pink as they age (Figure 74.4.1). The fruits of *C. gynandra* are tiny, cylindrical, and siliques with a 1–4 mm long beak and split into two valves, whereas the seeds are brown and circular with rough seed coat and parallel longitudinal line (Figures 74.4.1 and 74.4.2) [1149].

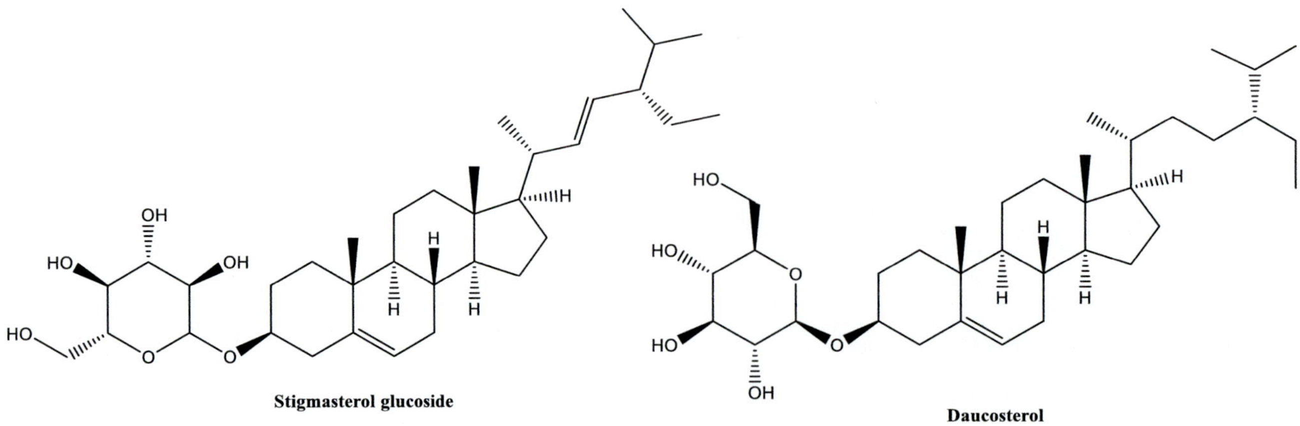

FIGURE 74.3.4 Active chemical components of the plant belonging to the Cleome genus.

FIGURE 74.4.1 Natural habit of *Cleome gynandra*.

FIGURE 74.4.2 Flowering shoot of *Cleome gynandra*.

74.4.5 Geographical Distribution

Cleome gynandra, a prevalent herbaceous plant, is found extensively in southern Africa, covering regions such as Limpopo, North-West, Gauteng, Mpumalanga, KwaZulu-Natal, Free State, the Northern Cape, and Namibia. Its semi-cultivation, as exemplified in the Kentani District of the Eastern Cape, may have contributed to the potential expansion of its distribution in Oman, which is found in some coastal areas of Al Batinah.

74.4.6 Chemical Composition

Preliminary phytochemical analysis of the powdered leaf unveiled the presence of various compounds, including carotenoids, cardiac glycosides, cyanogenic glycosides, flavonoids, saponins, triterpenes, sugars, and tannins, among others [1150].

Cleome gynandra possesses high levels of α-tocopherol, β-tocopherol, and γ-tocopherol, ascorbic acid, α-carotene, β-carotene, lutein, violaxanthin, and β-cryptoxanthin [1151] (Figure 74.4.3).

74.4.7 Uses

This plant is widely consumed wild leafy vegetables. Traditionally, the consumption of *Cleome gynandra* involves boiling its leaves and shoots or putting them in stews. In sub-Saharan Africa, these leaves are frequently dried for preservation purposes and subsequently cooked with milk or butter to mitigate their bitter flavor [1152]. *C. gynandra* is fast emerging as one of the most widely consumed "super vegetables", particularly in sub-Saharan Africa and Asia [1153].

74.4.8 Mechanism of Action

C. gynandra leaves extract is reported for its anti-inflammatory effects as chemical components present in the extract stabilize lysosomal enzyme activities in the plasma and liver [1154].

74.4.9 Toxicity and Clinical Data

C. viscosa and *C. gynandra* were reported as non-toxic [1155].

74.4.10 Adulterants or Substituents

No adulterant or substitute is reported for *Cleome gynandra*. The species of genus *Cleome* are similar in chemical composition so they can be substituted with closely resembling species belonging to same genus *Cleome*.

74.4.11 Most Cited Research

iCite, Database, NIH Open Citation Collection was used on October 2, 2023, to get the details of the most cited research. Filters for research papers were used by using the keywords *Cleome gynandra*. NIH database search revealed an mRNA blueprint for C4 photosynthesis derived from comparative transcriptomics of closely related C3 and C4 species [1156].

74.4.12 Patents

Philippe Msika, Alex Saunois, Caroline Baudouin, Sophie Leclere-Bienfait, Sebastien Debrock—US9370541B2—Extract of the above-ground portions of *Gynandropsis gynandra* or *Cleome gynandra*, and cosmetic, dermatological, or pharmaceutical compositions including the same, 2011. The invention relates to a composition, containing extract of the above-ground parts of *Gynandropsis gynandra,* for cosmetic and pharmaceutical application.

74.4.13 Plant Pathology

C. gynandra is well adapted to diverse habitats (predominantly in warm climates) due to their C4 photosynthetic mechanism [1157].

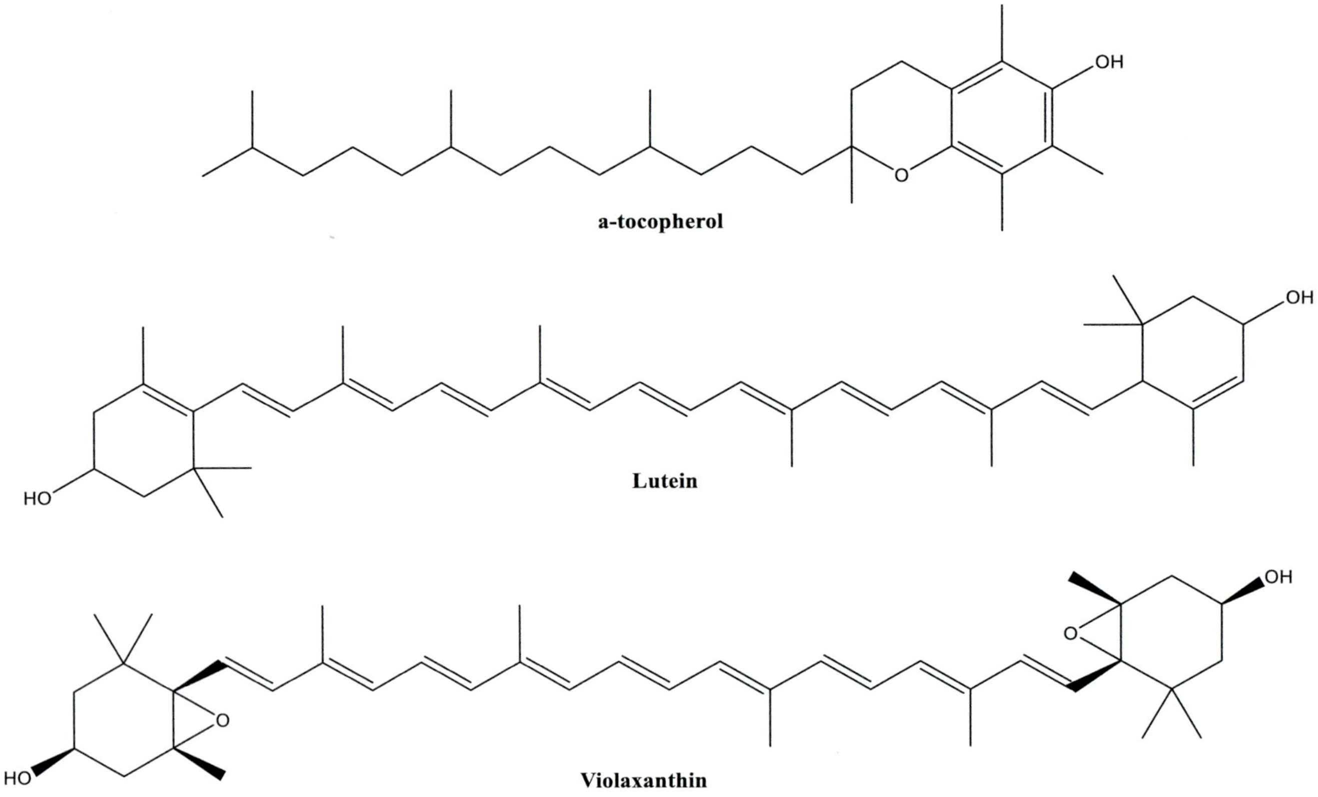

FIGURE 74.4.3 Chemical components of *Cleome gynandra*.

74.5 *Cleome rupicola* Vicary

74.5.1 Synonyms

Cleome oxypetala var. *micrantha* & *Cleome stocksiana*

74.5.2 Family

Cleomaceae

74.5.3 Vernacular Name

Meqablouat shams

74.5.4 Common Name

Rupicola spiderflower, rocky mountain spiderflower

74.5.5 Botanical Description

Cleome rupicola Vicary is a perennial, aromatic herb that can reach a height of up to 70 cm and is characterized by elongated slender stems. Leaves: alternate, oval shape with entire margins. Flowers: dark yellow/orange with brown veins and four petals curving backward (Figures 74.5.1–74.5.3).

FIGURE 74.5.1 Overall growth of *Cleome rupicola*.

74.5.6 Geographical Distribution

Native to Arabian Peninsula, Pakistan, and Iran. In Oman, it thrives throughout the country.

74.5.7 Chemical Composition

Phytochemical screening has revealed that *Cleome* species exhibit a substantial presence of diverse secondary

FIGURE 74.5.2 Inflorescence of *Cleome rupicola*.

FIGURE 74.5.3 Phyllotaxy of *Cleome rupicola*.

metabolites such as terpenoids, flavonoids, phenolic compounds, and alkaloids [1130]. *Cleome rupicola* contains essential oil [1158]. Oil of this plant contains mainly contain 2-methyl isothiocyanate, cubenol, hexanal, selin-11-en-4-alpha-ol, β-himachalene and carotol [1158]. A recent report showed the presence of secodammarane triterpenes: cleomdiolic acid, shoreic acid, foveolins B, and two flavonol glycosides (Figure 74.5.4) [1159].

74.5.8 Uses

The extract of *Cleome rupicola* is employed as eye drops for cataract treatment [192]. Its methanol extracts have demonstrated larvicidal activity against fourth-instar larvae of *Aedes caspius* and *Culex pipiens* mosquitoes [1160]. A recent study showed the cytotoxicity of the cleomdiolic acid [1159].

74.5.9 Mechanism of Action

Terpenoids exert inhibitory effects on the production of inflammatory cytokines [1030]. Moreover, the larvicidal mechanism of action is based on its neurotoxic effects to suppress larval development and midgut damage [1159].

74.5.10 Toxicity and Clinical Data

Apparently, no data is available on the toxicity of *Cleome rupicola*. However, genus *Cleome* flowers are considered toxic [1161].

74.5.11 Adulterants or Substituents

No adulterant or substitute is reported for *Cleome rupicola*. The species of genus *Cleome* are similar in chemical composition so they can be substituted with closely resembling other species of genus *Cleome*.

74.5.12 Most Cited Research

iCite, Database, NIH Open Citation Collection was used on October 18, 2023, to get the details of the most cited research. Filters for research papers were used by using the keywords *Cleome*. NIH database search revealed an mRNA blueprint for C4 photosynthesis derived from comparative transcriptomics of closely related C3 and C4 species [1156].

74.5.13 Patents

René Belle, Francoise Belaubre—US20100119632A1—*Cleome spinosa* extract used in pharmaceutical preparations and cosmetic compositions, 2008.

74.5.14 Plant Pathology

No information in the literature that refers to the pathology of *Cleome rupicola* is available. However, *A. candida* is a type of fungal disease that usually occurs in the genus *Cleome* [1147].

74.6 *Cleome scaposa* DC.

74.6.1 Synonyms

Cleome cordata, *Gilgella scaposa* & *Cleome ehrenbergiana*

74.6.2 Family

Cleomaceae

74.6.3 Vernacular Name

Khezaimet dhabi

74.6.4 Common Name

Heartleaf Tickweed

74.6.5 Botanical Description

Cleome scaposa is an annual herb, typically measuring 10–30 cm in height. Its flowers are actinomorphic, with a

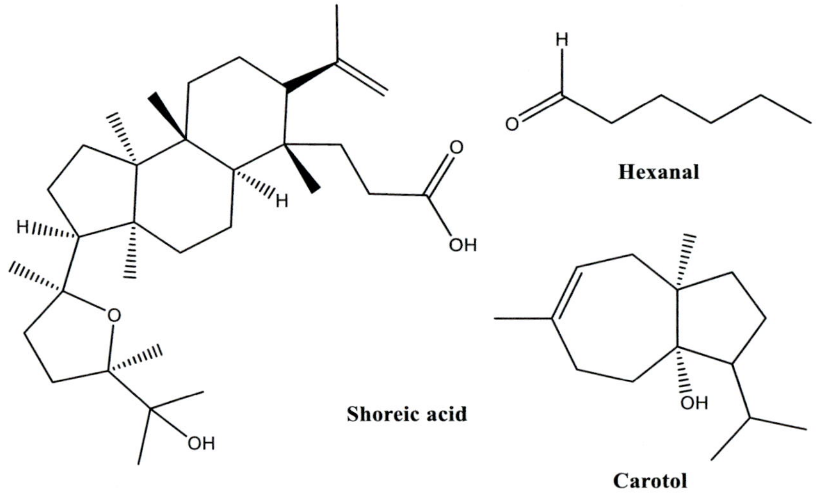

FIGURE 74.5.4 Primary chemical components of *Cleome rupicola*.

diameter ranging from 3 to 4 mm, initially white but transitioning to yellowish or occasionally pinkish [1162]. This is a tiny herb with semi-succulent leaves and small yellow flowers. Its fruit is a slender, cylindrical capsule with numerous minute seeds (Figures 74.6.1–74.6.3).

FIGURE 74.6.1 Fruiting shoot of *Cleome scaposa*.

FIGURE 74.6.2 Inflorescence of *Cleome scaposa*.

FIGURE 74.6.3 Leaf arrangements of *Cleome scaposa*.

74.6.6 Geographical Distribution

Cleome scaposa DC.is widely distributed as a common weed across the plains of Arabia, Egypt, India, Pakistan, and throughout the tropical regions globally. In Oman, it is found all around the country.

74.6.7 Chemical Composition

Preliminary phytochemical analysis of the hydro-methanol extract from *C. scaposa* leaves yielded positive results for the presence of alkaloids, flavonoids, saponins, tannins, and phenols [1163].

74.6.8 Uses

Each component of the *Cleome scaposa* DC plant possesses therapeutic properties capable of alleviating allergic manifestations, diabetes mellitus, and gastric disorders. Additionally, it exhibits diuretic, analgesic, and antipyretic effects [1164]. This plant is a weed however traditionally used as medicinal plant for their properties like analgesic, anti-inflammatory, and anti-pyretic effects [1163].

74.6.9 Mechanism of Action

C. scaposa possesses antipyretic activity comparable to that of standard drug paracetamol This was reported by its comparable efficacy in mitigating fever induced by *Saccharomyces cerevisae* in rabbits. This confirms the traditional application of *C. scaposa* in managing fever [1163].

74.6.10 Toxicity and Clinical Data

Dose fixation investigations indicated the non-toxic nature of *C. scaposa* leaves within the dosage range of 1–5 g/kg body weight when administered to normal rabbits [1163].

74.6.11 Most Cited Research

iCite, Database, NIH Open Citation Collection was used on October 3, 2023, to get the details of the most cited research. Filters for research papers were used by using the keywords *Cleome scaposa*. NIH database search revealed that phytochemical analysis and *in vivo* evaluation of anti-pyretic effects of hydro-methanolic extract of *Cleome scaposa* leaves [1163].

74.6.12 Patents

René Belle, Francoise Belaubre—US20100119632A1—*Cleome spinosa* extract used in pharmaceutical preparations and cosmetic compositions, 2008.

74.7 *Cleome viscosa* L.

74.7.1 Synonyms

Arivela viscosa & *Corynandra viscosa*

74.7.2 Family

Cleomaceae

74.7.3 Common Name

Asian spiderflower and tick weed

74.7.4 Botanical Description

The plant is an annual herb characterized by glandular and simple hairs and strong penetrating odor. It attains a height of approximately 30–90 cm and exhibits branching. The leaves are obovate, obtuse, and 3–5 foliate, with a gradual reduction in size toward the upper part of the plant. Yellow axillary flowers extend into a lax raceme. Fruits are present in the form of compressed and hairy capsules, while the seeds are subglobose and brownish-black hue upon ripening [1165]. This plant bears a linear cylindrical capsule, which opens at maturity only in the upper part (Figures 74.7.1–74.7.4).

FIGURE 74.7.1 Fruiting shoot of *Cleome viscosa*.

FIGURE 74.7.2 Whole plant of *Cleome viscosa*.

FIGURE 74.7.3 Leaves of *Cleome viscosa.*

FIGURE 74.7.4 Fruit of *Cleome viscosa.*

74.7.5 Geographical Distribution

This species is regarded as invasive and widely available in warm and humid habitats in the Americas, Africa, and Asia. Interestingly, in Australia, it is recognized as a native species. Within Oman, it is found both in Western and Eastern Hajar regions.

74.7.6 Chemical Composition

The active compounds identified in this plant include coumarin, cedrene, α-amorphene, ethyl palmitate, coumarinolignoids (cleomiscosin A and B), and eriodictyol-5-rhamnoside [1166] (Figure 74.7.5). A diverse array of compounds have been extracted and identified from various segments of *Cleome viscosa*, encompassing triterpenoids, saponins, tannins, flavonoids, and steroids [1167].

74.7.7 Uses

Cleome viscosa has undergone comprehensive scientific screening for a range of pharmacological activities, resulting in the identification of significant properties, including anthelmintic, antimicrobial, analgesic, anti-inflammatory, immunomodulatory, antipyretic, psychopharmacological, antidiarrheal, and hepatoprotective activities [1165].

74.7.8 Mechanism of Action

The mechanism of action of coumarins involves competitive antagonism of vitamin K, thereby inhibiting coagulation by impeding prothrombin production [1113].

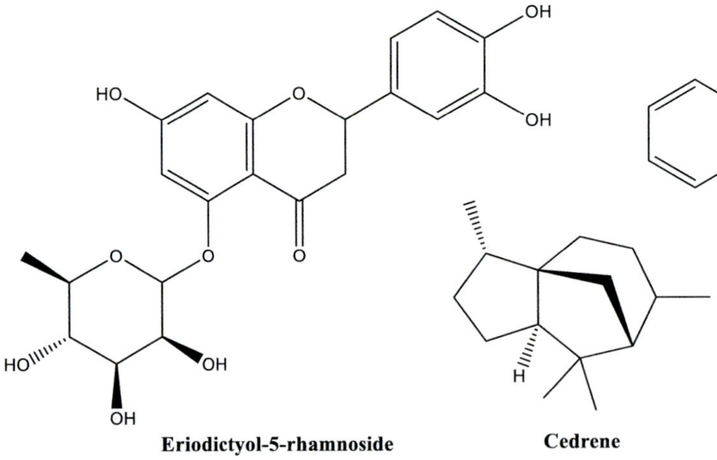

FIGURE 74.7.5 Active chemical components of *Carduus viscosa.*

74.7.9 Toxicity and Clinical Data

None of the components of *Cleome viscosa* are considered toxic or harmful to humans or pets. In fact, a majority of the plant's parts are edible and possess a diverse range of medicinal applications [1168].

74.7.10 Adulterants or Substituents

No adulterant or substitute for *Cleome viscosa* has been reported so far. The species of genus *Cleome* are similar in chemical composition so they can be substituted with closely resembling other species of genus *Cleome*.

74.7.11 Most Cited Research

iCite, Database, NIH Open Citation Collection was used on October 3, 2023, to get the details of the most cited research. Filters for research papers were used by using the keywords *Cleome viscosa*. NIH database search revealed that plants used for the treatment of diabetes in Israel [1169].

74.7.12 Patents

Dongguan Bowden Spices Co Ltd—CN106433974A—*Cleome viscosa* extract and cut tobaccos containing *Cleome viscosa* extract, 2016. This invention is related to plant extract including *Cleome viscosa* and its applied technical field are for their application in tobacco industry.

74.7.13 Plant Pathology

Ralstonia solanacearum is a plant pathogen that is found in *Cleome viscosa* [1170].

75 Genus *Corbichonia*

75.1 *Corbichonia decumbens* (Forssk.) Exell

75.1.1 Synonyms

Orygia decumbens, Portulaca decumbens & *Talinum decumbens*

75.1.2 Family

Lophiocarpaceae

75.1.3 Vernacular Name

Um lahaleeb

75.1.4 Common Name

Corbichonia decumbens is commonly called stone plants or carpet weeds [1171]. In India, it's known as Pater-chatti.

75.1.5 Botanical Description

Corbichonia decumbens is an annual semi-succulent, branched, creeping, and prostrate to procumbent herb [1171,1172] that can be reached up to a height of 50 cm. It is characterized by shiny fruit that are yellow-greenish in appearance [1173]. Flowers; pink, arranged in terminal or pseudo-lateral inflorescences. The staminodes are extended than the real perianth, acting as a great sum of petals (Figures 75.1.1 and 75.1.2).

FIGURE 75.1.1 Leaves of *Corbichonia decumbens.*

FIGURE 75.1.2 Flowers of *Corbichonia decumbens.*

75.1.6 Geographical Distribution

Corbichonia decumbens is widely available in tropical and southern Africa and tropical Asia; it was introduced to tropical American countries [1173]. This species found in widespread areas in Northern and Southern Oman, thrives in sandy stretches, wadi beds, and stony slopes.

75.1.7 Chemical Composition

Phytochemical studies showed the presence of mome-inositol, guanosine, and cis-vaccenic acid in the methanolic extract of root, whereas ethyl-acetate extract of root contain pentadecanoic acid, octadecanoic acid and cis-vaccenic acid. Methanolic and ethyl-acetate extract of stem contain inositol and 1-heptacosanol [1171].

75.1.8 Uses

Plant preparation has been reported for its potential anti-inflammatory and antiulcer effects [1174] whereas leaves are used for healing various diseases [1171,1175].

75.1.9 Most Cited Research

iCite, Database, NIH Open Citation Collection was used on January 3, 2023, to get the details of the most cited research. Filters for research papers were used by using the keywords *Corbichonia decumbens.* NIH database search revealed the gas chromatography-mass spectrometry profiling in methanolic and ethyl-acetate root and stem extract of *Corbichonia decumbens* (Forssk.) Exell from Thar desert of Rajasthan, India [1171].

DOI: 10.1201/9781003295488-75

76 Genus *Commiphora*

76.1 *Commiphora gileadensis* (L.) C. Christ.

76.1.1 Synonyms

Amyris gileadensis, *Balsamea gileadensis* & *Balsamodendrum gileadense*

76.1.2 Family

Burseraceae

76.1.3 Vernacular Name

Sukof

76.1.4 Common Name

Balsam

76.1.5 Botanical Description

Commiphora gileadensis is an aromatic plant that exhibits variable sizes, ranging from small-leaved shrubs to 4 m tall trees. It is typically non-spiny, with peeling bark and aromatic resin [10]. Leaves of this plant are alternate on short, condensed side shoots, pinnate with 3–5 leaflets. Its fruit is dull red, ovoid in shape, and marked with four longitudinal white stripes (Figures 76.1.1–76.1.5).

76.1.6 Geographical Distribution

The native range of *Commiphora gileadensis* encompassing NE Tropical Africa to NE Kenya and the Arabian Peninsula. In Oman, it is restricted to the mountains of Dhofar.

FIGURE 76.1.1 Shoot with leaves of *Commiphora gileadensis.*

76.1.7 Chemical Composition

Phytochemical screening of the aerial parts of *Commiphora gileadensis* identified the presence of various compounds,

FIGURE 76.1.2 Peeling bark of *Commiphora gileadensis.*

FIGURE 76.1.3 Leaf portion of *Commiphora gileadensis.*

FIGURE 76.1.4 Fruits of *Commiphora gileadensis.*

DOI: 10.1201/9781003295488-76

FIGURE 76.1.5 Leafy portion of *Commiphora gileadensis*.

including flavonoids, sterols, triterpenes, saponins, volatile bases, and oils [1176]. Terpenes are the major components found in *Commiphora gileadensis* [1177].

A recent study showed the presence of β-myrcene, nonane, verticiol, β-phellandrene, β-cadinene, and terpinen-4-ol in the essential oil in *Commiphora gileadensis.* Moreover, the oil also contains β-eudesmol, α-pinene, cis-β-copanene, and verticillol [1178].

Another study revealed that *C. gileadensis* oil contains the major components in the form of monoterpenes such as α-pinene, sabinene, β-pinene, p-cymene, limonene, γ-terpinene, and terpinen-4-ol [1177] (Figure 76.1.6).

76.1.8 Uses

Commiphora gileadensis is well known for its valuable perfume, extracted from various parts of the plant, as well as its exceptional medicinal attributes associated with its sap, wood, bark, and seeds. This plant has a long history of recognition as both a perfume and incense source, and it is typically found in specific ecological regions [1179]. This aromatic plant is traditionally used in the Middle East as a common remedy for pain and inflammation. A recent study revealed its anti-proliferative effect on cancer cell lines [1180].

In Dhofar it was used in traditional medicine, the dried and powdered inner wood has been used to treat wounds, skin irritations, insect bites, and for cleansing and purification after childbirth [10].

76.1.9 Mechanism of Action

A recent study revealed that *C. gileadensis* sap extracts contain cytotoxic components that can target proliferating cells in a cell cycle-dependent manner. Thus, it could be used for the treatment of skin disorders characterized by hyperproliferation, such as skin cancer and psoriasis [1180].

76.1.10 Toxicity and Clinical Data

No toxicity was observed when *Artimia salina* was used as a test organism. Additionally, there was no evidence of antitumor activity against Ehrlich ascites carcinoma [1181]. Also, the cytotoxic action of *C. gileadensis* sap extracts on human epidermal cells and skin tissues is reported.

76.1.11 Most Cited Research

iCite, Database, NIH Open Citation Collection was used on October 7, 2023 to get the details of most cited research. Filters for research papers were used by using the keywords *Commiphora gileadensis.* NIH database search revealed that β-caryophyllene, a compound isolated from the biblical balm of gilead (*Commiphora gileadensis*), is a selective apoptosis inducer for tumor cell lines [1182].

76.1.12 Patents

Patrice Andre, Stéphane Lhermite, Françoise Pellicier—US6630177B1—Products extracted from a plant of the genus Commiphora, particularly the *Commiphora mukul* plant, extracts containing same and applications thereof, for example in cosmetics, 2002. This is related to the use in cosmetics of extracts of a plant of the genus Commiphora as an agent with antiwrinkle activity.

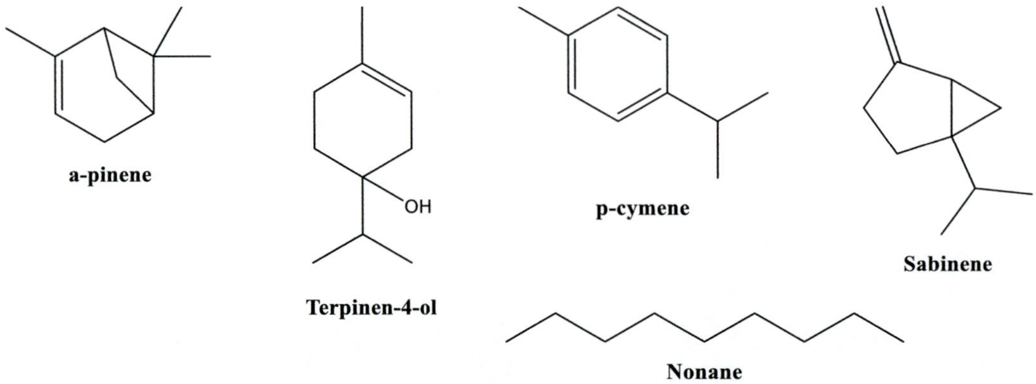

FIGURE 76.1.6 Primary chemical components of *Commiphora gileadensis*.

76.1.13 Plant Pathology

Apparently, there is no pathogenic report against *Commiphora gileadensis*. However, *Alternaria alternata* causes leaf spot disease in genus *Commiphora* [1183].

76.2 *Commiphora foliacea* Sprague

76.2.1 Family

Burseraceae

76.2.2 Vernacular Name

Akebobeh & akebot

76.2.3 Common Name

Leafy Myrrh & Foliacea Myrrh

76.2.4 Botanical Description

Commiphora foliacea is a much-branched, dioecious shrub that can reach a height of 1.5 to 2 meters and may attain a stem diameter of up to 18 cm or more. It possesses 3-foliolate leaves that areglaucous and glabrous with petioles 2–10(–20) mm long. Fruits: ovoide drupe, 1-seeded (Figures 76.2.1–76.2.5) [1184].

76.2.5 Geographical Distribution

The species is available in various regions of Somalia and the southern Arabian Peninsula. It manifests as a shrub or tree predominantly found in arid desert or dry shrubland biomes [1185]. In Oman, it is found in the mountains of Dhofar.

76.2.6 Chemical Composition

The chemical components from the stem of *C. foliacea* Sprague essential oil contain camphene followed by eremophilene [1186] (Figure 76.2.6).

FIGURE 76.2.1 Leaf of *Commiphora foliacea*.

76.2.7 Uses

Resin-based recipes have been employed in indigenous medicinal practices to address a range of health concerns, including gastrointestinal ailments, arthritis, wound healing, obesity management, pain relief, and the treatment of parasitic infections [1187].

FIGURE 76.2.2 Flower of *Commiphora foliacea*.

FIGURE 76.2.3 Fruit of *Commiphora foliacea*.

FIGURE 76.2.4 New shooting tip of *Commiphora foliacea*.

FIGURE 76.2.5 Natural form of *Commiphora foliacea.*

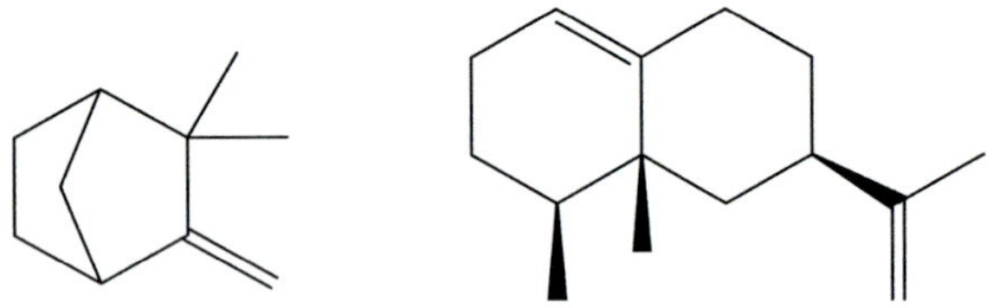

Camphene **Eremophilene**

FIGURE 76.2.6 Active chemical components of *Commiphora foliacea.*

Furthermore, the anti-analgesic and anti-inflammatory potential of the essential oil isolated from the stem of *C. foliacea* has also been reported. Also, essential oil isolated from the stem of this plant was found effective against *B. subtilis*, while maximum antifungal inhibition was observed against *A. parasiticus* [1186].

76.2.8 Mechanism of Action

A recent study showed that considerable antidiabetic and carbonic anhydrase-II potential was observed. The essential oil components showed substantial activity against breast cancer and normal cell lines, and it has been supposed that the essential oil did not cause damage to the normal cell lines [1186].

76.2.9 Toxicity and Clinical Data

No data regarding the toxic effects of *Commiphora foliacea* is available. A recent study showed that no apparent signs of toxicity, genotoxicity, mutagenicity or cytotoxicity were found *in vivo* and *in vitro* experiments on bark and leaves aqueous extract *Commiphora leptophloeos* [1188].

76.2.10 Most Cited Research

iCite, Database, NIH Open Citation Collection was used on October 8, 2023, to get the details of the most cited research. Filters for research papers were used by using the keywords *Commiphora foliacea*. NIH database search revealed the first complete chloroplast genomics and comparative phylogenetic analysis of *Commiphora gileadensis* and *C. foliacea*: Myrrh-producing trees [1187].

76.2.11 Patents

Patrice Andre, Stéphane Lhermite, Françoise Pellicier—US6630177B1—Products extracted from a plant of the genus Commiphora, particularly the *Commiphora mukul* plant, extracts containing same and applications thereof, for example in cosmetics, 2002.

76.3 *Commiphora kua* (R. Br. ex Royle) Vollesen

76.3.1 Synonyms

Balsamodendron habessinicum, Balsamodendrum kua & *Commiphora habessinica* var. *simplicifolia*

76.3.2 Family

Burseraceae

76.3.3 Vernacular Name

Okor & dij

76.3.4 Botanical Description

Commiphora kua displays significant variation in its morphology. In arid regions, it manifests as low, spiny shrubs with small leaves, while in monsoon-affected areas, it grows into medium-sized, non-spiny trees with distinctive trunks and large leaves. The plant is dioecious, characterized by bark that peels or flakes, and when incised, it releases an aromatic resin [1189]. Plants possess asymmetrical obovoid fruits ellipsoid, sometimes somewhat beaked, with a 2-valved pericarp and four lobes pseudaril (Figures 76.3.1–76.3.4) [1189].

FIGURE 76.3.1 Apical fruiting portion of *Commiphora kua.*

FIGURE 76.3.2 Fruit portion of *Commiphora kua*.

FIGURE 76.3.3 Leaf portion of *Commiphora kua*.

FIGURE 76.3.4 Habitat view with *Commiphora kua* in Eastern Haajr.

76.3.5 Geographical Distribution

Commiphora kua is a plant indigenous to northeast Africa and the Arabian Peninsula, with its natural range covering across Djibouti, Eritrea, Ethiopia, Zambia, Malawi, Oman, and Yemen. In Oman, it is found in Southern mountains and Eastern Hajar mountains.

76.3.6 Chemical Composition

Previous phytochemical investigations of the plant resin from *Commiphora kua* have revealed the presence of various compounds, including cycloartane triterpenes, bisabolenes, and furanosesquiterpenoids with a germacrane skeleton, dammarane triterpenes, and octanordammaranes. Some of these compounds exhibit anti-inflammatory and phytofungicidal properties, particularly against *Cladosporium cucumerinum* [1190]. The plant is known for its natural phytosterols, such as lophenol and lathosterol.

76.3.7 Uses

Commiphora kua yields wood suitable for crafting household utensils like cups and pots, as well as furniture such as stools and tools like hammers. During rainy seasons, its trunk is tapped for refreshing sap. In dry periods, the tree exudes a brown resin used for incense [1191]. Phytosterols present in this plant possess hepatoprotective effects [1192].

76.3.8 Mechanism of Action

Treatment of acetaminophen-treated mice with phytosterols isolated from *C. kua* resulted in significant recovery of Liver function tests profile and anti-oxidant enzymes [1192].

76.3.9 Toxicity and Clinical Data

There is no data regarding the toxic effects of *Commiphora kua* available. The genus *Commiphora* does not exert any toxic effects [1193].

76.3.10 Most Cited Research

iCite, Database, NIH Open Citation Collection was used on October 8, 2023, to get the details of the most cited research. Filters for research papers were used by using the keywords *Commiphora kua*. NIH database search revealed bisabolenes and dammarane triterpenes of *Commiphora kua* resin [1190].

76.3.11 Patents

Patrice Andre, Stéphane Lhermite, Françoise Pellicier—US6630177B1—Products extracted from a plant of the genus Commiphora, particularly the *Commiphora mukul* plant, extracts containing same and applications thereof, for example in cosmetics, 2002. The invention is related to the use extracts of a plant of the genus Commiphora in cosmetics as an agent with antiwrinkle activity.

76.4 *Commiphora myrrha* (T.Nees) Engl.

76.4.1 Synonyms

Balsamea myrrha & *Balsamodendrum myrrha*

76.4.2 Family

Burseraceae

76.4.3 Vernacular Name

Mur

76.4.4 Botanical Description

Commiphora myrrha is a sturdy, spiny, and large shrub or small tree, characterized by a well-defined, short trunk that reaches a height of up to 4 m [1194]. The crown is curvy, with the branches rising and then curving down. Fruit; ovoid to oblong drupe, reddish-brown, contains a single hard seed. (Figures 76.4.1–76.4.6).

76.4.5 Geographical Distribution

Commiphora myrrha, located in the southern regions of Arabia, northeastern Africa (primarily Somalia), and the northeastern part of Kenya, is responsible for the production of genuine myrrh. Additionally, other Commiphora species that produce resins can be found in Sudan, southern Arabia, Eritrea, Kenya, Ethiopia, and Somalia [1195]. In Oman, it is restricted to dry, rocky habitats in the southwestern part of the Dhofar region.

76.4.6 Chemical Composition

Triterpenoids are major constituents found in the resins of *Commiphora* species, whereas plant stems of these species commonly contain flavonoids and lignans [1196]. The main chemical components of terpenes in myrrh are furano-sesequiterpenoids with elemane, eudes-mane, guaiane, or runcates, structures. Furanoeudesma 1,3-diene is reported as a key component present in myrrh oil. *C. myrrha* contains

FIGURE 76.4.1 Plant habit of *Commiphora myrrha*.

FIGURE 76.4.2 Fruiting shoot of *Commiphora myrrha*.

FIGURE 76.4.3 Phyllotaxy of *Commiphora myrrha*.

FIGURE 76.4.4 Flower portion of *Commiphora myrrha*.

elemanolide, guaianolide, and cadinanolide as active components (Figure 76.4.7) [1197].

76.4.7 Uses

Commiphora myrrha possesses diverse medicinal properties and historical usage in treating conditions such as amenorrhea, pain, dysmenorrhea, tumors, fever, stomach disorders, gallbladder issues, chest ailments, snake and scorpion bites, as well as skin infections in ancient civilizations like India, China, Rome, Greece, and Babylon [1198]. It's also used as an antiseptic, astringent, anthelmintic, carminative, emmenagogue, and as an expectorant [1197]. Further, it has been used for ages to make incense and as a perfume [1199].

FIGURE 76.4.5 *Commiphora myrrha* renews its leaf growth following rainfall.

76.4.8 Mechanism of Action

Commiphora myrrha increases insulin production from β-cells via activation of atypical protein kinase C and mitogen-activated protein kinase. Thus, it can be used in the management of type 2 diabetes mellitus [1200].

76.4.9 Toxicity and Clinical Data

Commiphora myrrha does not exert toxicity [1201].

76.4.10 Adulterants or Substituents

Commiphora species, which are well-known adulterants of myrrh, such as *C. sphaerocarpa, C. holtziana, C. kataf,* and the other species mentioned by Tucker [1202].

76.4.11 Most Cited Research

iCite, Database, NIH Open Citation Collection was used on October 8, 2023, to get the details of the most cited research. Filters for research papers were used by using the keywords *Commiphora myrrha*. NIH database search revealed a natural product that lowers cholesterol as an antagonist ligand for FXR (Farnesoid X receptor).

FIGURE 76.4.6 Gum of *Commiphora myrrha.*

76.4.12 Patents

Patrice Andre, Stéphane Lhermite, Françoise Pellicier—US6630177B1—Products extracted from a plant of the genus Commiphora, particularly the *Commiphora mukul* plant, extracts containing same and applications thereof, for example in cosmetics, 2002.

76.4.13 Plant Pathology

Apparently, there is no pathogenic report against *Commiphora kua*. However, fungal endophyte (*Nigrospora* sps.) was isolated from this genus, which is responsible for the production of bostrycin and deoxybostrycin known for their antitumor properties [1203].

76.5 *Commiphora wightii* (Arnott) Bhandari

76.5.1 Synonyms

Balsamea mukul

76.5.2 Family

Burseraceae

76.5.3 Vernacular Name

Moqel

76.5.4 Common name

Guggulu & Indian bdellium

76.5.5 Botanical Description

Commiphora wightii is a shrub or small tree with a maximum height of 4 m (13 ft) and characterize by thin papery bark. The branches are armed with thorns. The leaves are either simple or trifoliate, with ovate leaflets measuring 1–5 cm (0.39–1.97 in) in length, 0.5–2.5 cm (0.20–0.98 in) in width, and irregularly toothed (Figures 76.5.1 –76.5.4) [1204].

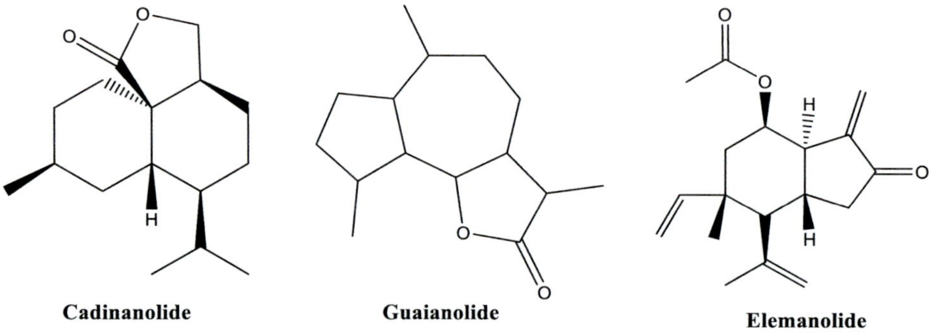

FIGURE 76.4.7 Primary chemical components of *Commiphora myrrha.*

FIGURE 76.5.1 Leaf portion of *Commiphora wightii*.

FIGURE 76.5.2 Phyllotaxy of *Commiphora wightii*.

FIGURE 76.5.3 Fruits of *Commiphora wightii*.

76.5.6 Geographical Distribution

Commiphora wightii exhibits a limited global distribution, predominantly confined to arid regions of Western India and adjacent areas in Pakistan, Yemen, and Oman.

76.5.7 Chemical Composition

The gum is a complex mixture of diverse chemical compounds, including lignans, lipids, diterpenoids, and steroids [1205]. The primary active constituents of this oleogum resin are Z-guggulsterone, E-guggulsterone, Z-guggulsterol, and guggulsterols I–V [1206]. Guggulsterone is a plant steroid and bioactive compound present in gum Guggul of *Commiphora wightii* [1207] (Figure 76.5.5).

FIGURE 76.5.4 Natural habit of *Commiphora wightii*.

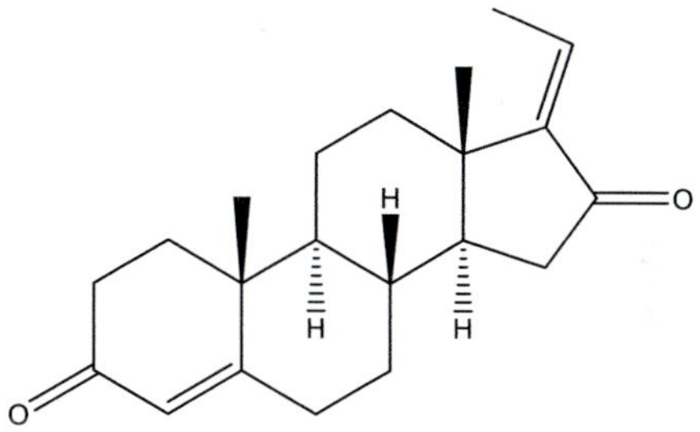

FIGURE 76.5.5 Active chemical components of *Commiphora wightii*.

76.5.8 Uses

Classical Ayurvedic literature asserts the efficacy of guggul in the treatment of rheumatoid arthritis, obesity, neurological disorders, hemorrhoids, urinary disorders, skin ailments, allied conditions, and other therapeutic applications [1208]. Oleogum resin was used traditionally for the treatment of inflammation and hyperlipidemia [1209].

76.5.9 Mechanism of Action

Z-guggulsterone demonstrates antidepressant efficacy by triggering the brain-derived neurotropic factor (BDNF) signaling pathway and promoting the enhancement of hippocampal neurogenesis [1210]. Also, gum resin diminished hyperlipidemia via inhibition of farnesoid X receptor activation as guggulsterone is the only known antagonist of farnesoid X receptor [1209].

76.5.10 Toxicity and Clinical Data

No data is available regarding the toxic effects of *Commiphora wightii*. However, the genus *Commiphora* does not exert any toxic effects [1193].

Administration of guggul preparation in acute and 28 days repeated doses did not display any toxicity or adverse effect at the doses used, and the no-observed-adverse-effect level was found to be at 2,700 mg/kg [1211].

76.5.11 Adulterants or Substituents

Guggal is frequently adulterated with the oleogum resin of *B. serrata* or sometimes with the resin of pinus sp. Recent study showed that adulteration of *Commiphora wightii* with *Mangifera indica* gum [1212].

76.5.12 Most Cited Research

iCite, Database, NIH Open Citation Collection was used on October 19, 2023, to get the details of the most cited research. Filters for research papers were used by using the keywords *Commiphora wightii*. NIH database search revealed that a natural product that lowers cholesterol as an antagonist ligand for FXR [1213].

76.5.13 Patents

Patrice Andre, Stéphane Lhermite, Françoise Pellicier—US6630177B1—Products extracted from a plant of the genus Commiphora, particularly the *Commiphora mukul* plant, extracts containing same and applications thereof, for example in cosmetics, 2002.

76.5.14 Plant Pathology

No plant pathogens are reported on this wild plant. However, plants aged 2–3 years are susceptible to termite infestations, particularly during the drier months when irrigation is limited. Termites target both the roots and stems of these plants, resulting in significant damage [1205].

77 Genus *Convolvulus*

77.1 *Convolvulus arvensis* L.

77.1.1 Synonyms

Convolvulus hastatus, Strophocaulos arvensis & Convolvulus sagittifolius

77.1.2 Family

Convolvulaceae

77.1.3 Vernacular Name

Lewaya & khatmi

77.1.4 Common Name

Field bindweed

77.1.5 Botanical Description

Convolvulus arvensis is a long-lived perennial herb with stems that either twine or sprawl along the ground. Its leaves are typically hairless and range from ovate (egg-shaped with a broader base) to oblong or lanceolate (lance-like), measuring up to 7.5 cm in length and 3 cm in width, featuring a notched base [1214]. Flowers: white to pink (Figures 77.1.1 and 77.1.2).

77.1.6 Geographical Distribution

Field Bindweed, a native perennial plant in Europe and Asia, has become an introduced and troublesome weed in North America. It is prevalent and widely distributed in Britain, except for the northernmost regions [1215]. In Oman, it is found throughout the country among cultivations as an herbal weed.

FIGURE 77.1.1 Inflorescence of *Convolvulus arvensis.*

FIGURE 77.1.2 Flowers of *Convolvulus arvensis.*

77.1.7 Chemical Composition

Phytochemical analyses of *Convolvulus arvensis* L. have identified the presence of alkaloids, phenolic compounds, flavonoids, carbohydrates, sugars, mucilage, sterols, resin, tannins, unsaturated sterols/triterpenes, lactones, and proteins [1216]. Phenolic compounds are the main active chemical compounds found in *Convolvulus arvensis* [1217]. Essential oils of this plant contain O-cymene, thymol, himachalene, longifolene, cuprenene, carvacrol, and cadinene (Figure 77.1.3) [1218].

Recent study showed the presence of rutin, kaempferol-3-O-rutinoside, chlorogenic acid, 3,5-di-O-caffeoylquinic acid, N-trans-p-coumaroyl-tyramine, and N-trans-feruloyl-tyramine in ethyl acetate and methanol extracts [1219].

77.1.8 Uses

Convolvulus arvensis has been historically employed for various medicinal purposes, including its use as a laxative, wound healer, anti-spasmodic, anti-hemorrhagic, anti-angiogenetic agent, and for the treatment of parasites and jaundice [1220]. Additionally, it has been utilized as a diuretic and in the management of skin disorders such as anti-furunculosis, antidandruff, and for treating spider bites [1221]. In some areas in Oman, leaves and roots of this species have been traditionally used to treat various ailments, including fevers, wounds, and constipation. For external application, fresh leaves or roots are crushed with a small amount of water to form a soft paste, which is then applied

DOI: 10.1201/9781003295488-77

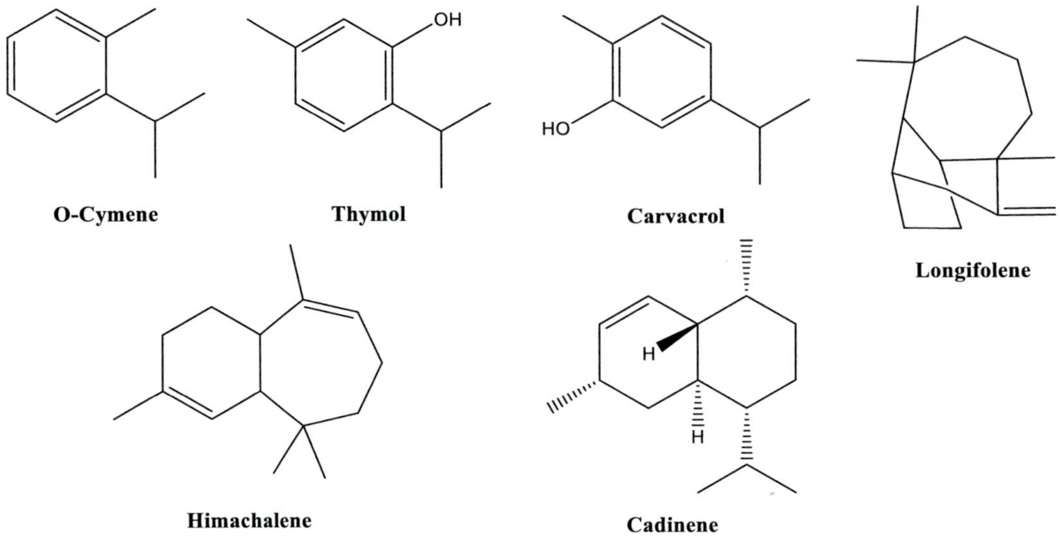

FIGURE 77.1.3 Chemical components of *Convolvulus arvensis*.

directly to wounds to promote healing and stop bleeding. The same paste, when diluted with water, is taken orally once a day as a remedy for constipation and to help reduce fever [911].

77.1.9 Mechanism of Action

Terpenoids exert inhibitory effects on the production of inflammatory cytokines [1030].

77.1.10 Toxicity and Clinical Data

Bindweed contains several alkaloids, including pseudotropine, and lesser amounts of tropine, tropinone, and meso-cuscohygrine. Previous study showed that mice administered with bindweed passed away or were euthanized and had severe hepatic necrosis and gastritis with ulceration or erosions [1222].

Convolvulus arvensis contains alkaloids, which, when consumed in high doses, can be toxic and lead to digestive disturbances in animals such as pigs and horses [1223].

77.1.11 Adulterants or Substituents

Convolvulus arvensis shares similarities with several other species, leading to potential confusion. Similar species include *Calystegia sepium, Calystegia spithamaea, Fagopyrum tataricum, Fallopia convolvulus*, and various species of *Ipomoea* [1224].

77.1.12 Most Cited Research

iCite, Database, NIH Open Citation Collection was used on October 8, 2023, to get the details of the most cited research. Filters for research papers were used by using the keywords *Convolvulus arvensis.* NIH database search revealed the transformation of several species of higher plants by *Agrobacterium rhizogenes*: sexual transmission of the transformed genotype and phenotype [1225].

77.1.13 Patents

Xiaolong Meng, Hugh D. Riordan, Neil H. Riordan—US20020127284A1—High-molecular-weight extracts of *Convolvulus arvensis* (field bindweed) and *Polygonum convolvulus* (wild buckwheat), 2002. This invention is related purified bindweed and buckwheat extracts to inhibit the growth of tumor cells, inhibit the growth of blood vessels, and enhance immune function.

77.1.14 Plant Pathology

Among the many herbaceous plants that serve as food sources for adult *H. obsoletus*, *Urtica dioica* (stinging nettle), and *Convolvulus arvensis* (field bindweed) are the most commonly reported hosts for *Ca. Phytoplasma solani* in western and central Europe [1226]. Recent reports showed that this plant was found to be infected with *Macrophomina phaseolina* and *Alternaria alternate* [1227].

78 Genus *Corchorus*

78.1 *Corchorus depressus* (L.) Pterm.

78.1.1 Synonyms

Antichorus depressus & *Corchorus antichorus*

78.1.2 Family

Malvaceae

78.1.3 Vernacular Name

Selentah & metari

78.1.4 Common Name

Baphuli

78.1.5 Botanical Description

The prostrate, woody perennial, characterized by its radiating branches emanating from a central woody crown, exhibits close depression to the ground, effectively functioning as a proficient sand binder within desert ecosystems [1228]. The plant has alternate and oval leaves and termina, bright yellow flowers (Figures 78.1.1–78.1.3).

78.1.6 Geographical Distribution

The distribution of this species extends across arid and semi-arid regions from India and Pakistan to the north, encompassing tropical Africa [1229]. This species is common all around Oman on the sandy and gravel plains as well as cultivations.

FIGURE 78.1.1 Flower of *Corchorus depressus.*

FIGURE 78.1.2 Flowering shoot of *Corchorus depressus.*

FIGURE 78.1.3 Flowers and fruits of *Corchorus depressus.*

78.1.7 Chemical Composition

The plant *Corchorus depressus*, is abundant in flavonoids, including apigenin, luteolin, depressonol A, depressonol B, quercetin, and kaempferol (Figure 78.1.4) [1230]. Triterpenes and sterols such as depressoside A, depressoside B, and β-sitosterol have been identified in various parts of the plant, including the whole plant, roots, and fruits [1231].

78.1.8 Uses

The plant has been studied for its potential medicinal uses, including general weakness, gonorrhea, diabetes, treachery troubles, and improving sexual vigor, as reported by various researchers [1228]. *Corchorus depressus* Linn. has been used as an aphrodisiac in traditional Indian medicine to treat male sexual dysfunction and impotency [1232].

DOI: 10.1201/9781003295488-78

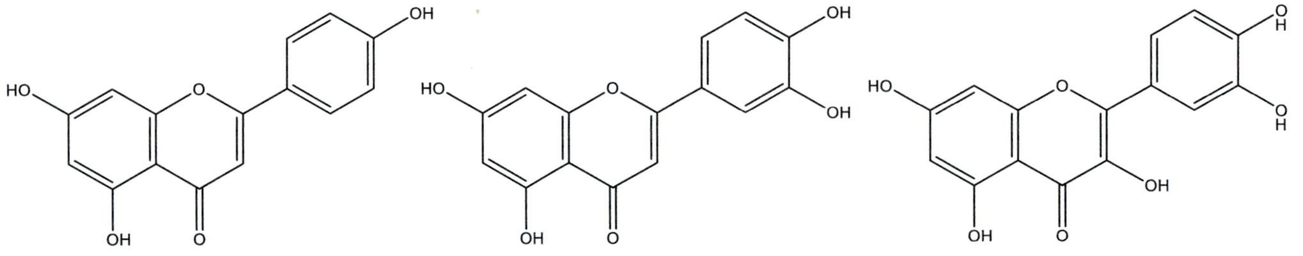

Apigenin Luteolin Quercetin

FIGURE 78.1.4 Chemical components of *Corchorus depressus*.

Traditionally, *Corchorus* is used as a medicine for the ailment of aches, dysentery, enteritis, fever, and tumors [1233]. This plant is also used for the treatment of respiratory infections and acts as a nervine tonic [1234].

78.1.9 Mechanism of Action

Corchorus depressus extract showed a useful therapeutic option as an antioxidant and inhibition of acetylcholinesterase, butyrylcholinesterase, and α-glucosidase activities [1234].

78.1.10 Toxicity and Clinical Data

An acute toxicity study revealed that the chloroform fraction was safe for use. No mortality or toxic symptoms were observed during the study [1232].

78.1.11 Adulterants or Substituents

The conflation of *Corchorus depressus* and *M. chamomilla* under the Unani name "*Babuna*" in various literature sources has led to significant confusion and misuse of the herbal drug [1235].

78.1.12 Most Cited Research

iCite, Database, NIH Open Citation Collection was used on October 8, 2023, to get the details of the most cited research. Filters for research papers were used by using the keywords *Corchorus depressus*. NIH database search revealed antipyretic studies on some indigenous Pakistani medicinal plants: II [1236].

78.1.13 Patents

Mahua Choudhury, Xiaofeng Kang, Allison R. Ficht—US11291750B2—Hydrogel condom, 2017. In this patent researchers demonstrated the ethnopharmacological evidence of *Corchorus depressus* extract (contains quercetin) as a sexual stimulating agent [1237].

78.1.14 Plant Pathology

Apparently, there is no data in the literature that refers to the pathology of *Corchorus depressus*. Also, plants belonging to the same genus were found to be infected by an endophytic Basidiomycete, *Grammothele lineata*, isolated from *Corchorus olitorius*, which produces paclitaxel that shows cytotoxicity [1238].

78.2 *Corchorus aestuans* L.

78.2.1 Family

Malvaceae

78.2.2 Common Name

Jute & east Indian jew's-mallow

78.2.3 Botanical Description

Annual herb exhibits a maximum height range of 50–100 cm, featuring an erect and pubescent stem. Its leaves are ovate, measuring 8 mm in diameter. Fruits; cylindrical capsule reachs up to 4 cm (Figures 78.2.1 and 78.2.2) [1239].

78.2.4 Geographical Distribution

The annual herb is distributed across the warmer regions of the Indian subcontinent, Indochina, Australia, Tropical Africa, the West Indies, and Central America [1240]. It is also distributed in the northern and southern regions of Oman.

FIGURE 78.2.1 Leaves of *Corchorus aestuans*.

FIGURE 78.2.2 Fruits of *Corchorus aestuans*.

78.2.5 Chemical Composition

Several bioactive molecules have been reported in the genus *Corchorus*, including cardiac glycosides, their aglycones, polysaccharides, triterpenoids, phenolics, sterols, and fatty acids [1241]. β-sitosterol and the fatty acids are present in the seed oil. Seeds also contain a glycoside called corchorine and a flavonoid, quercetine [1242] cardenolides, ceryl alcohol, and oligosaccharides [1243] (Figure 78.2.3). Capsules of *C. aestuans* contain β-sitosterol, lupeol, betulin, 2-methyl anthraquinone, scopoletin, and corchoroside-A [1244]. Leaves contain fusidic acid, β-sitosterol, 2-methylanthraquinone, and coumarins. A study revealed the presence of triterpenoid glycosides (corchorusins A, B, C, and D) in the aerial part of *C. acutangulus*. Furthermore, four triterpenoid glycosides (corchorusins C1, D1, D2, and D3) were detected in the aerial parts of *C. acutangulus* [1245,1246].

78.2.6 Uses

The roots and leaves of this plant are known to have medicinal properties for the treatment of gonorrhea and are utilized in the preparation of an injection for urethral discharge [1247]. Additionally, the seeds exhibit stomachic qualities and are employed in the management of pneumonia. This plant is reputed to possess various therapeutic attributes, including anticancer, antipyretic, anticonvulsant, stomachic, and actions resembling those of digitalis glycosides [1243].

78.2.7 Mechanism of Action

Saikosaponin-A found in this plant induces apoptotic mechanism in human breast MDA-MB-231 (independent of the P53/p21 pathway) and MCF-7 (p53/p21-dependent pathway cancer cells) [1248,1249].

78.2.8 Toxicity and Clinical Data

Corchorusin-D present in this plant causes significant cytotoxicity in leukemic cell lines U937 and HL-60. COR-D produced apoptotic cell death via mitochondrial dysfunction [1249].

78.2.9 Adulterants or Substituents

Okra (*Abelmoschus esculentus*) and other *Corchorus* species have been used as adulterants within the genus *Corchorus* [1250].

78.2.10 Most Cited Research

iCite, Database, NIH Open Citation Collection was used on October 9, 2023, to get the details of the most cited research. Filters for research papers were used by using the keywords *Corchorus aestuans*. NIH database search revealed that traditional leafy vegetables in Senegal: diversity and medicinal uses [1251].

78.2.11 Patents

Maqsudul Alam, Mohammed Shahidul Islam, Borhan Ahmed, Mohammed Samiul Haque, Mohammed Monjurul Alam—US10301639B2—Nucleotide sequence encoding homeobox-leucine zipper protein HAT22 (HD-ZIP protein 22) from *Corchorus olitorius* and *Corchorus capsularis* and methods of use, 2014. The present invention is related to the method for enhancing the fiber length, height and/or biomass of plants by modulating expression of a nucleic acid encoding homeobox-leucine zipper HAT22 polypeptide.

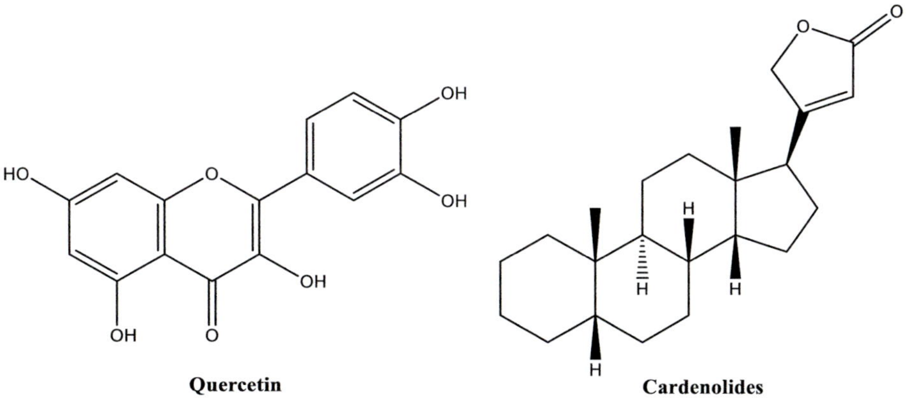

Quercetin **Cardenolides**

FIGURE 78.2.3 Active chemical components of *Corchorus aestuans*.

78.3 *Corchorus olitorius* L.

78.3.1 Synonyms

Corchorus catharticus, Corchorus decemangularis & *Corchorus lanceolatus*

78.3.2 Family

Malvaceae

78.3.3 Common Name

Jew's mallow & Nalta jute

78.3.4 Botanical Description

Corchorus olitorius is an annual herbaceous reaching a height of 80cm. Its leaves are arranged alternately, possessing a simple, lanceolate shape with finely serrated or lobed margins [1252]. It has spindle-shaped dehiscent fruits that are separated into transversal segments via five valves (Figures 78.3.1 and 78.3.2).

FIGURE 78.3.1 Fruit of *Corchorus olitorius*.

FIGURE 78.3.2 Leaves of *Corchorus olitorius*.

78.3.5 Geographical Distribution

This species is distributed across various regions, including Egypt, Southern Asia, Japan, India, China, Lebanon, Palestine, Syria, Jordan, Tunisia, Nigeria, Côte d'Ivoire, Benin, Cameroon, Sudan, Kenya, Uganda, and Zimbabwe. Additionally, it is cultivated as a leafy vegetable in the Caribbean, Brazil, India, Bangladesh, China, and the Middle East and also grown for its fiber in Asia (India, Bangladesh, and China) [1253]. In Oman, it is common in the southern regions.

78.3.6 Chemical Composition

Several bioactive molecules have been reported in *Corchorus olitorius*, including cardiac glycosides, their aglycones, polysaccharides, triterpenoids, phenolics, sterols, and fatty acids [1241].

Digitoxigenin glycosides, coroloside, and glucoevatromonoside, as well as strophanthidin glycosides (erysimoside, olitoriside, biosides, trioside, corchoroside-A, and helveticoside), were isolated from the methanolic extract of seeds, which were the major cardiac glycosides. β-amyrin represents the major compound in (23.18%) of the stems (Figure 78.3.3). Cis-3-Hexen-1-ol, cis-4-hexen-1-ol, terpinolene, sabinene, and phytol were the major compounds reported in fresh leaves. Apigenin, astragalin, caffeic acid, naringin, botulin, chlorogenic acid, cichoriin, quinic acid, capsin, ursolic acid, stigmasterol, scopolin, scopoletin, luteolin, rosamarinic acid, galactuonic acid, raffinose, cirsiliol, and cirsilineol are the common components present in *C. capsularis* and *C. olitorius* [1254].

78.3.7 Uses

Leaves are traditionally used as soup. Plant possess antioxidant, cardiovascular, antitumor, anticancer, antiulcer activity, antidiabetic, bioadsorbent, antiobesity, antifertility,

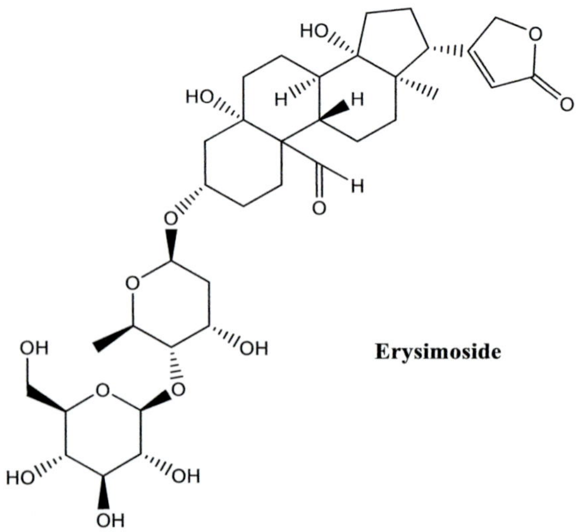

FIGURE 78.3.3 Chemical component of *Corchorus olitorius*.

antinociceptive, anti-inflammatory, antipyretic, anti-inflammatory, antipyretic, hepatobiliary, renal, hematological, antibacterial phytoalexin, and neuromodulatory activities [1254].

Wild okra is utilized in traditional medicine for the management of conditions such as gonorrhea, chronic cystitis, pain, fever, and tumors [1255]. This plant is recognized for its elevated concentrations of iron and folate, which confer benefits in the prevention of anemia [1256].

78.3.8 Mechanism of Action

Plants possess a gastroprotective effect with significant increases in the production of gastric mucosa in rats via a cytoprotective mechanism similar to that of endogenous prostaglandins [1257].

78.3.9 Toxicity and Clinical Data

The present study did not observe any acute toxicity effects from *C. olitorius* extracts [1258], consistent with prior research conducted in Bangladesh and Nigeria [1259]. Additionally, in a related study, *C. olitorius* exhibited no toxicity when assessing enzyme markers in alloxan-induced diabetic rats [1260].

A previous study showed a high safety index (no toxicity or fatality) after oral administration of 5,000 mg/kg body weight of the ethanol extract of *Corchorus olitorius* [1261].

However, the leaf of *C. capsularis* contains HCN, which may be responsible for cytotoxicity and corchoroside, a bioactive component, is known for its potential cytotoxicity [1262].

78.3.10 Adulterants or Substituents

Several jute substitutes, including Bimli (derived from *Hibiscus cannabinus*) and China jute (obtained from *Abutilon theophrasti*), have been identified [1263].

78.3.11 Most Cited Research

iCite, Database, NIH Open Citation Collection was used on October 9, 2023, to get the details of the most cited research. Filters for research papers were used by using the keywords *Corchorus olitorius*. NIH database search revealed medicinal foodstuffs. V. Moroheiya. (1): Absolute stereostructures of corchoionosides A, B, and C, histamine release inhibitors from the leaves of Vietnamese *Corchorus olitorius* L. (Tiliaceae) [1264].

78.3.12 Patents

Maqsudul Alam, Mohammed Shahidul Islam, Borhan Ahmed, Mohammed Samiul Haque, Mohammed Monjurul Alam—US10301639B2—Nucleotide sequence encoding homeobox-leucine zipper protein HAT22 (HD-ZIP protein 22) from *Corchorus olitorius* and *Corchorus capsularis* and methods of use, 2014.

78.3.13 Plant Pathology

Seven endophytic fungal strains belonging to three different including *Penicillium, Fusarium*, and *Aspergillus* genera, were isolated from *Corchorus olitorius* that produced different metabolites such as astellatol, aspergillipeptide A, and emericellamide C [1265].

78.4 *Corchorus trilocularis* L.

78.4.1 Synonyms

Corchorus aestuans & *Corchorus asplenifolius*

78.4.2 Family

Malvaceae

78.4.3 Common Name

Wild Jute

78.4.4 Botanical Description

These annual herbs, ranging from 30 to 150 cm in height, are characterized by erect or suberect growth and extensive branching. Their leaves are alternate, narrowly oblong-lanceolate to broadly oblong-elliptic, with crenate-serrate margins and basal serrations extending into filiform processes [1266]. The fruit is a fusiform capsule, opening in 3–4 valves comprising many angular seeds (Figures 78.4.1 and 78.4.2).

78.4.5 Geographical Distribution

Its distribution covers all the way from Cape Verde and tropical and southern Africa, across the Arabian Peninsula to Iraq, and all the way east to India and even a corner of China (northwestern Yunnan province). In Oman, it is distributed in the northern and southern regions.

FIGURE 78.4.1 Fruiting shoot of *Corchorus trilocularis*.

FIGURE 78.4.2 Leaf of *Corchorus trilocularis*.

78.4.6 Chemical Composition

n-Hexadecanoic acid followed by 10-Methyl-E-11-tridecen-1-ol propionate is mainly found in the leaves of *Corchorus trilocularis* L [1267] (Figure 78.4.3). However, several bioactive molecules have been reported in the genus *Corchorus*, including cardiac glycosides, their aglycones, polysaccharides, triterpenoids, phenolics, sterols, and fatty acids [1241].

78.4.7 Uses

Corchorus trilocularis exhibits anti-inflammatory [1268] and demulcent properties [1269]. In traditional Indian medicine, it is employed in the treatment of syphilis.

78.4.8 Mechanism of Action

The gastroprotective effects of *Corchorus trilocularis* Linn are reported against diclofenac-induced gastric ulcers in Sprague Dawley rats via offering protection to the gastric mucosal barrier, increase in gastric pH and inhibition of gastric acid secretion through the histamine H_2 and acetylcholine M3 extracellular pathways on the parietal cell [1270].

78.4.9 Toxicity and Clinical Data

The chloroform extract of *Corchorus trilocularis* (CECT) demonstrated non-toxicity at doses of up to 2,000 mg/kg body weight of Albino Wistar rats [1271].

78.4.10 Adulterants or Substituents

Corchorus trilocularis can be substituted with *Corchorus olitorius* or okra, particularly in arid regions, where *Corchorus tridens L.* is the predominant replacement [1272].

78.4.11 Most Cited Research

iCite, Database, NIH Open Citation Collection was used on October 9, 2023, to get the details of the most cited research. Filters for research papers were used by using the keywords *Corchorus trilocularis*. NIH database search revealed xyloglucan endotransglycosylase/hydrolase genes from a susceptible and resistant jute species show opposite expression patterns following *Macrophomina phaseolina* infection [1273].

78.4.12 Patents

Maqsudul Alam, Mohammed Shahidul Islam, Borhan Ahmed, Mohammed Samiul Haque, Mohammed Monjurul Alam—US10301639B2—Nucleotide sequence encoding homeobox-leucine zipper protein HAT22 (HD-ZIP protein 22) from *Corchorus olitorius* and *Corchorus capsularis* and methods of use, 2014.

78.4.13 Plant Pathology

Previous reports show the upregulation of an XTH gene in a resistant jute (*Corchorus trilocularis*) variety during infection with *Macrophomina phaseolina* may reinforce the cell wall by cross-linking [1273].

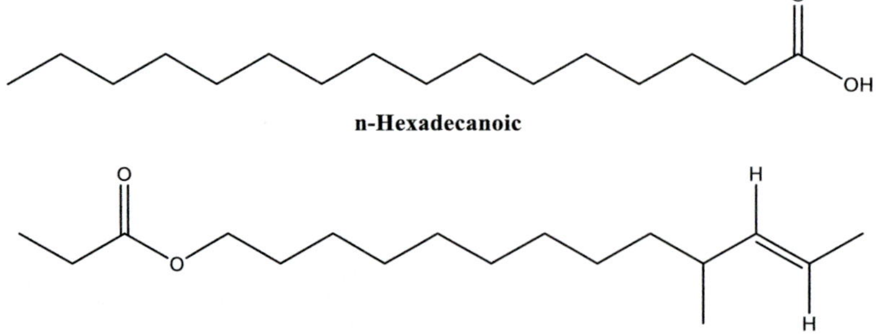

n-Hexadecanoic

10-Methyl-E-11-tridecen-1-ol propionate

FIGURE 78.4.3 Chemical components of *Corchorus trilocularis*.

79 Genus *Cordia*

79.1 *Cordia myxa* L.

79.1.1 Synonyms

Gerascanthus myxus & *Lithocardium myxa*

79.1.2 Family

Boraginaceae

79.1.3 Vernacular Name

Qaw & humbo

79.1.4 Common Name

Assyrian plum, Bumber, Makhet, Dabag, Benber sebstan, Megsas

79.1.5 Botanical Description

This dioecious shrub or small tree grows up to a height of 12 m and exhibits a tortuous or straight bole, gray cracked bark, and spreading branches forming a dense crown. Its alternate, simple leaves are broadly ovate to orbicular, sometimes obovate, with glabrous upper surfaces and velvety hairy undersides. The inflorescence is a lax terminal or short lateral panicle, 3–8.5 cm long, without bracts [1274]. This tree produces edible globular-ovoid fruits that is light pale to brown or even pink in color. Fruit is generally sticky when squashed, the pericarp having the consistency of glue (Figures 79.1.1–79.1.3).

79.1.6 Geographical Distribution

Cordia myxa, a plant native to temperate and tropical regions of Asia, originated in the Eastern Mediterranean to Eastern India. It was historically introduced to tropical Africa, Asia, Australia, and more recently, the Americas, and is extensively available in paleo tropics [1274]. It is found throughout the country, primarily cultivated in gardens and on farms.

FIGURE 79.1.1 Fruits of *Cordia myxa*.

FIGURE 79.1.2 Developing fruits of *Cordia myxa*.

FIGURE 79.1.3 Natural habit of *Cordia myxa*.

79.1.7 Chemical Composition

Flavonoids are the main active chemical compounds. The preliminary phytochemical screening carried out on *Cordia myxa* fruit extract revealed the presence of oil, glycosides, flavonoids, sterols, saponins, terpenoids, alkaloids, phenolic acids, coumarins, tannins, resins, gums, and mucilage [1275].

Recent study showed the presence of components such as 1-naphthaleneacetic-5-carboxy-1,2,3,4,4a,7,8,8a-octahydro-1,2,4a-trimethyl-[1S-(1α,2β, 4a,8aα)]-acid, hexacosanoate-1-glyceryl, 3β-urs-12,20(30)-diene-27,28-dioic acid (3), 3β-D-glucopyranosylurs-12,20(30)-diene-27,28-dioic acid, stigmasterol, stigmasterol-3-O-β-D-glucopyranoside, oleanolic acid (7), 3-O-acetyl-oleanolic acid, betulin

DOI: 10.1201/9781003295488-79

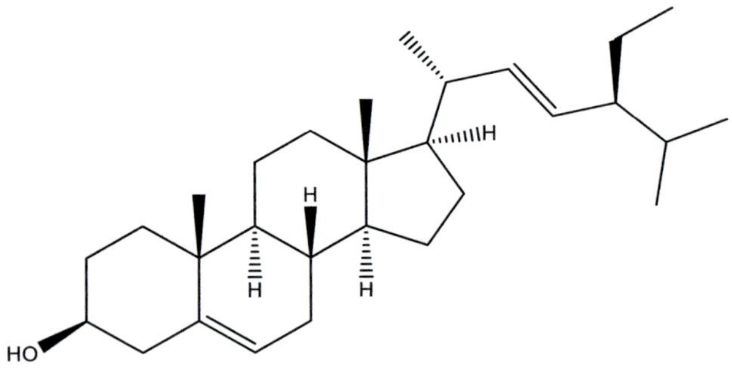

FIGURE 79.1.4 Chemical component of *Cordia myxa*.

and spinasterol-3β-O-D-glucopyranoside (Figure 79.1.4) in *Cordia myxa* L [1276].

79.1.8 Uses

It was consumed as a remedy to alleviate cough, treat respiratory infections, and soothe sore throats due to its demulcent qualities. Additionally, the pulp was employed topically as an emollient for mature abscesses, to alleviate rheumatic pain, and as an anthelminthic agent [1274]. *Cordia myxa* fruit extract possesses low cytotoxicity, antibacterial activity, and antioxidant properties in vitro [1277].

79.1.9 Toxicity and Clinical Data

The safety studies on *C. myxa* L. plant revealed no mortality up to a dose level of 2,500 mg/Kg [1278].

79.1.10 Adulterants or Substituents

Cordia myxa L. and *C. sebestena* L. are considered substituents of each other [1279].

79.1.11 Most Cited Research

iCite, Database, NIH Open Citation Collection was used on October 9, 2023, to get the details of the most cited research. Filters for research papers were used by using the keywords *Cordia myxa*. NIH database search revealed extraction, chemical composition, rheological behavior, antioxidant activity, and functional properties of *Cordia myxa* mucilage [1280].

79.1.12 Patents

Philippe Ragot, Esther Pons, Bernard Mompon, Cedric Rousseau—US20200390696A1—Edible Product Comprising Reconstituted Plant Material, 2020.

79.1.13 Plant Pathology

This plant is reported for leaf spot and blight disease caused by *Alternaria tenuissima* and *A. alternata* in Egypt [1281].

79.2 *Cordia monoica* Roxb.

79.2.1 Synonyms

Gerascanthus monoicus,

79.2.2 Family

Boraginaceae

79.2.3 Vernacular Name

Shebxid & shoxid

79.2.4 Common Name

C. monoica is also known as sandpaper saucer berry or snot berry.

79.2.5 Botanical Description

Small tree or large shrub grows up to 5 m with flaking bark. The leaves are arranged alternately on the branches, sometimes clustered on short shoots. Their shape varies from broadly oval to obovate. The small flowers produce a pleasant fragrance, either unisexual or bisexual, and appear in clusters at the ends of branches (terminal) or along the stems (axillary). The petals come in shades of greenish-white to pale yellow. Fruit is a delightful reward, boasting a bright yellow or orange color and a fleshy texture.

79.2.6 Geographical Distribution

Its native range includes the warmth of Tropical & Southern Africa, Southwest Madagascar, the Arabian Peninsula's southwest corner, India stretching to Myanmar, and finally, the southern reaches of Malaysia. In Oman, it is restricted to the southern escarpment mountains.

79.2.7 Chemical Composition

Plants have been reported for the presence of carotene, lycopene, benzofuran, nonacosane, hexadecanoic acid, and neophytadiene (Figure 79.2.1) [1282].

79.2.8 Uses

Leaves are used in eye diseases, whereas the plant is known for its antimicrobial, antioxidant, and anti-ulcer effects [1283].

It has been reported that many species of *Cordia* have been used in traditional medicine for centuries to treat a range of ailments, including *C. monoica*, which showed significant anti-ulcer activity. Additionally, *C. monoica* leaves are used as a vapor bath for leprosy, its roots for vomiting, and its stem bark for chest pains [1283–1287].

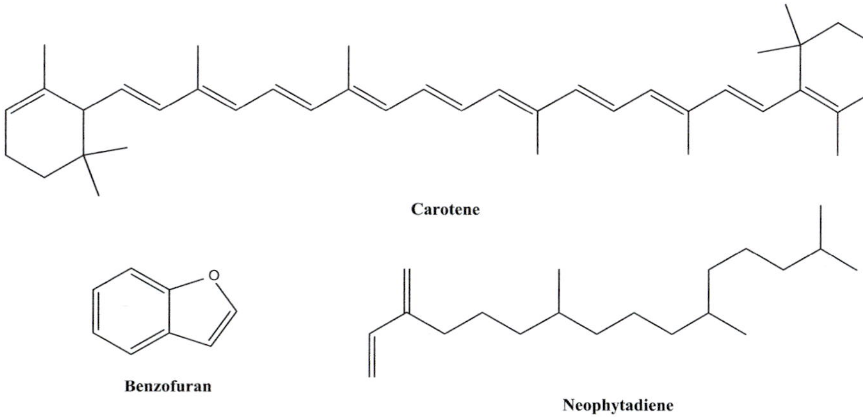

FIGURE 79.2.1 Primary chemical components of *Cordia monoica*.

79.2.9 Toxicity and Clinical Data

Chloroform, ethyl acetate and methanol bark extract of *C. monoica* presented no signs of toxicity after administration of 2,000 mg/kg dose orally in rats [1283].

79.2.10 Most Cited Research

iCite, Database, NIH Open Citation Collection was used on January 3, 2023, to get the details of the most cited research. Filters for research papers were used by using the keywords *Cordia monoica*. NIH database search revealed the traditional use of wild edible plants in pastoral and agro-pastoral communities of Mieso District, eastern Ethiopia [1288].

79.3 *Cordia sinensis* Lam.

79.3.1 Synonyms

Gerascanthus sinensis & *Quarena sinensis*

79.3.2 Family

Boraginaceae

79.3.3 Vernacular Name

Shabhkait

79.3.4 Common Name

In English, is called gray-leaved saucer berry.In India, it's known as "Goondi" [1289].

79.3.5 Botanical Description

Cordia sinensis is a small tree that reaches up to a height of 5–7 m high. *Cordia sinensis* possesses elliptic-oblong to inverted-lance-shaped leaves with slightly wavy margins. This tree has small, drupe, red, or orange (when ripe) fruits present in clusters, and each fruit possesses a visible long tip and 1–4 hard and rough seeds (Figures 79.3.1 and 79.3.2) [1289,1290].

FIGURE 79.3.1 Leaves of *Cordia sinensis*.

FIGURE 79.3.2 Fruits of *Cordia sinensis*.

79.3.6 Geographical Distribution

This plant is widely distributed in the drier regions of Saudi Arabia, Africa, and India [1291]. In Oman, it is found in southern and central regions.

79.3.7 Chemical Composition

This plant has been known for its several chemical components, such as protocatechuic acid [1292], trans-caffeic acid [1293], methyl rosmarinate [1294], rosmarinic acid [1294], kaempferide-3-O-β-D-glucopyranoside [1295], kaempferol-3-O-β-D-glucopyranoside [1296], quercetin-3-O-β-D-glucopyranoside [1291], and kaempferide derivatives [1296].

79.3.8 Uses

Plants, especially roots, have been used for stomach-related problems. Decoction of this plant is used in the treatment of malaria. Plant has been also used to treat ocular problems such as conjunctivitis in animals [1297]. This plant has been also known for antimicrobial effects [1093,1298].

This plant has also been reported to have antioxidant, anti-glycation, and anti-inflammatory effects [1299].

79.3.9 Mechanism of Action

Chemical components present in the essential oil of *Cordia sinensis* suppress fungal growth. Essential oils of this plant inhibit fungal growth, possibly by disrupting the cell wall and cell membrane [1300].

79.3.10 Toxicity and Clinical Data

Its methanolic toxicity in the brine shrimp lethality test has been reported. A methanolic extract and ethyl acetate soluble sub-fractions showed strong toxicity in the brine shrimp lethality test [1299].

79.3.11 Most Cited Research

iCite, Database, NIH Open Citation Collection was used on January 3, 2023, to get the details of the most cited research. Filters for research papers were used by using the keywords *Cordia monoica*. NIH database search revealed the ethnomedicines used in Trinidad and Tobago for reproductive problems [1301].

80 Genus *Cornulaca*

80.1 *Cornulaca monacantha* Delil

80.1.1 Synonyms

Cornulaca arabica & *Cornulaca amblyacantha* & *Kochia monacantha*

80.1.2 Family

Amaranthaceae

80.1.3 Vernacular Name

Selej

80.1.4 Common Name

Had, Djouri

80.1.5 Botanical Description

Cornulaca monacantha is a straggling, branched, woody shrub growing to a height of 60 cm. The stubby, bluish-green leaves are scale-like and clasp the grayish, wiry stems. They each have a single stiff spine at the tip, hence the specific name monacantha. The leaves turn yellow or white when the plant is dead (Figures 80.1.1–80.1.3) [1302].

80.1.6 Geographical Distribution

Cornulaca monacantha is native to North Africa, Arabia, the Middle East, Iran, and Baluchistan. It is a desert plant found in the Middle East and the Sahara, and the southern end of its range is considered to delineate the edge of the desert [1302]. Widespread in Oman's coastal regions, it thrives on the stabilized dunes, gravelly and sandy wadis.

FIGURE 80.1.1 Seeds of a cottony appearance, covered with dense, woolly hairs of *Cornulaca monacantha*.

FIGURE 80.1.2 Stems of *Cornulaca monacantha*.

FIGURE 80.1.3 Fruits of *Cornulaca monacantha*.

80.1.7 Collection and Cultivation

Cornulaca monacantha is a desert plant that grows in arid conditions on sandy and stony ground. The porous activated carbon retains high adsorption capacity, and adsorption also plays a crucial role in the photodegradation of organic contaminants [1303].

80.1.8 Chemical Composition

The aerial parts of the Egyptian herb *Cornulaca monacantha* have yielded two new galloyltannin analogs (named monacanthin A and monacanthin B) characterized by protoctechuoyl moiety at C-6 and a new flavonol glycoside quercetin-40-O-b-d-galactoside. The structures of the new compounds were established by conventional analytical methods and confirmed by spectral analyses.

DOI: 10.1201/9781003295488-80

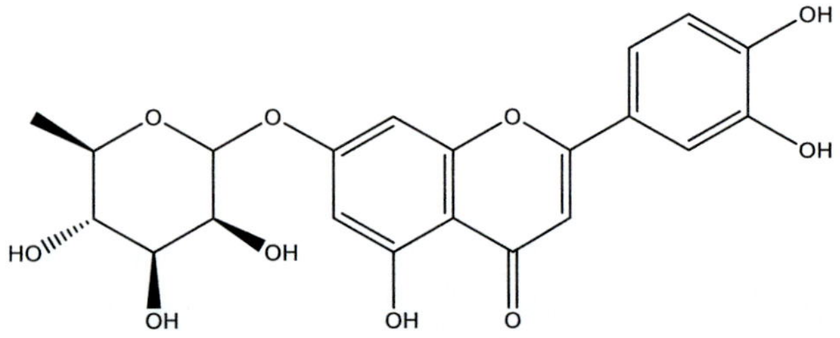

Luteolin-7-O-rhamnoside

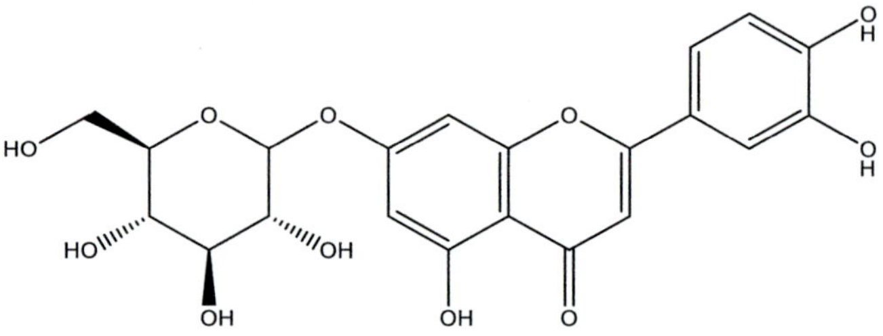

Luteolin-7-O-glucoside

FIGURE 80.1.4 Chemical components of *Cornulaca monacantha*.

Two known flavonoids, luteolin-7-O-rhamnoside, luteolin-7-O-glucoside (Figure 80.1.4), and two known gallotannins, penta-O-galloyl-b-dglucose and 1,2,3,6-tetra-O-galloyl-b-d-glucose, were also isolated [1304].

In the course of phytochemical and chemotaxonomical investigations of *Cornulaca monacantha*(Amaranthaceae), two new isoflavones, 3-(2-hydroxyphenyl)-5,7-dimethoxy-6-(methoxymethyl)-4H-1-benzopyran-4-one and 7-hydroxy-3-(4-hydroxyphenyl)-5-methoxy-6-(methoxymethyl)-4H-1-benzopyran-4-one, were isolated from the fresh aerial parts of *C. monacantha* among with three known compounds named vanillic acid, N-cis-feruloyltyramine, and N-trans-feruloyltyramine [1305].

80.1.9 Uses

A decoction of the leaves of *Cornulaca monacantha* is used as a traditional medicine for jaundice and liver problems and as a purgative. Externally, it is used to treat scabies. Despite its spiny leaves, it provides good grazing for camels and is said to increase the supply of milk produced by lactating females [1302].

80.1.10 Most Cited Research

iCite, Database, NIH Open Citation Collection was used on January 3, 2023, to get the details of the most cited research. Filters for research papers were used by using the keywords *Cornulaca monacantha*. NIH database search revealed the polyphenols from *Cornulaca monacantha* [1304].

80.1.11 Plant Pathology

The endophytic fungus *Paecilomyces* sp. (AUMC 15510) was isolated from healthy stem samples of the Egyptian medicinal plant *Cornulaca monacantha* [1306].

81 Genus *Corallocarpus*

81.1 *Corallocarpus epigaeus* (Rottler) Hook.f.

81.1.1 Synonyms

Aechmandra Epigaea, Bryonia sinuate, & *Corallocarpus bedquartii*

81.1.2 Family

Cucurbitaceae

81.1.3 Common Name

Redfruit Creeper

81.1.4 Botanical Description

A trailing or climbing perennial herb, featuring sinuately 3-lobed leaves, lanceolate calyx lobes, a dull or greenish-yellow corolla, and ovoid or ellipsoid fruit with a red hue, except for the greenish base and a 2.5–5-mm long beak (Figures 81.1.1–81.1.3) [1307].

81.1.5 Geographical Distribution

It is prevalent in tropical regions such as India, Ceylon, Deccan, and South Maratha [1308]. It is distributed throughout Oman, where it climbs over larger plants and is sometimes seen hanging from rocks.

FIGURE 81.1.1 Fruits of *Corallocarpus epigaeus.*

FIGURE 81.1.2 Leaf of *Corallocarpus epigaeus.*

FIGURE 81.1.3 Leaf and fruits of *Corallocarpus epigaeus.*

81.1.6 Chemical Composition

In the roots of *Corallocarpus epigaeus*, a pyridine carboxylic ester, an aliphatic C32 keto diol, a sesterpene lactone, and a p-hydroxybenzoyl ester have been identified. Additionally, a glycoside known as bryonin has been isolated from the same source [1309].

Cucurbitacin B was isolated from the rhizome of *Corallocarpus epigaeus* possesses anti-melanoma potential [1309]. N-methyl asparagine was also isolated from *C. epigaeus* seeds, whereas components ishwarane and ishwarone were isolated from the roots [1310] (Figure 81.1.4).

DOI: 10.1201/9781003295488-81

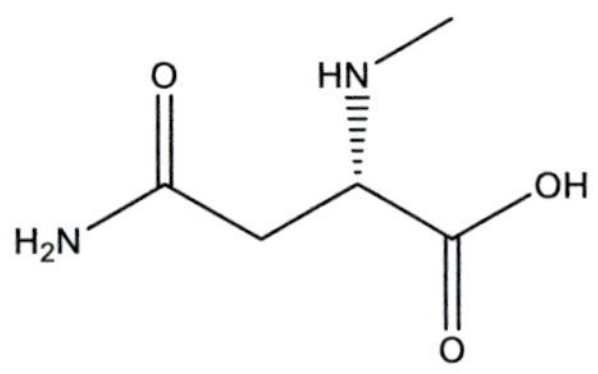

N-methyl asparagine

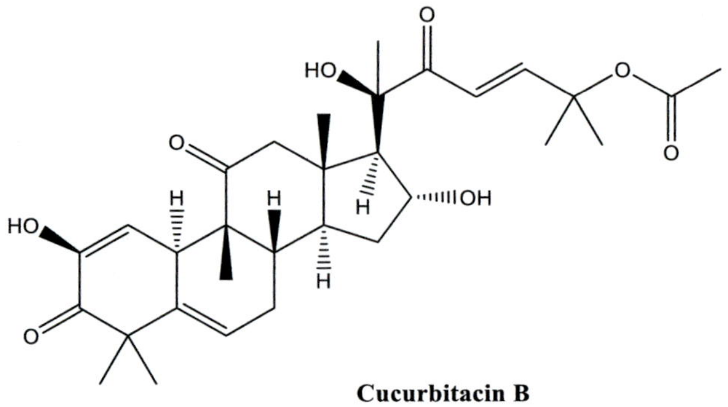

Cucurbitacin B

FIGURE 81.1.4 Primary chemical components of *Corallocarpus epigaeus.*

81.1.7 Uses

Corallocarpus epigaeus demonstrates varying biological activities against different diseases, including antifungal, antibacterial, anticancer, antidiabetic, and antioxidant properties [1311].

It was traditionally used in the treatment of acute dysentery, venereal diseases, skin diseases, and snake bite [1309].

81.1.8 Mechanism of Action

The anti-melanoma potential of cucurbitacin B by targeting the mitogen-activated protein kinase (MAPK) pathway is reported [1312].

81.1.9 Toxicity and Clinical Data

No mortality was observed in the animals tested up to 2,000 mg/kg test doses of *Corallocarpus epigaeus* rhizomes ethanol extract [1313].

81.1.10 Most Cited Research

iCite, Database, NIH Open Citation Collection was used on October 10, 2023, to get the details of the most cited research. Filters for research papers were used by using the keywords *Corallocarpus epigaeus.* NIH database search revealed cucurbitacin B, purified and characterized from the rhizome of *Corallocarpus epigaeus* exhibits anti-melanoma potential [1309].

81.1.11 Patents

Madhavan P.N. Nair, Zainulabedin M. Saiyed, Nimisha H. Gandhi— WO2011143218A2—Plant-derived formulations for treatment of HIV, 2015.

82 Genus *Ceropegia*

82.1 *Ceropegia bulbosa* Roxb.

82.1.1 Synonyms

Ceropegia acuminata, Ceropegia brosima & Ceropegia edulis

82.1.2 Family

Apocynaceae

82.1.3 Botanical Description

Ceropegia bulbosa Roxb a twining herb. Leaves are almost stalkless to stalked, round to ovate leaves. The flowers are pale gray, candles-shaped, becoming grayish purple toward the mouth. It bears paired, slender follicles fruits (Figures 82.1.1–82.1.3).

82.1.4 Geographical Distribution

The genus of climbers, herbs, and occasionally subshrubs is found in tropical and subtropical regions of Asia, Africa, Australia, and Malaysia, as well as in the Canary and Pacific islands [1314]. In Oman, it is distributed in Southern mountains within the monsoon-influenced range.

82.1.5 Chemical Composition

The alkaloid cerpegin, found in the root tubers of this plant, has been identified as a pyridone-type alkaloid, a relatively uncommon compound in nature [1315] (Figure 82.1.4).

82.1.6 Uses

The tubers of *Ceropegia bulbosa* are employed in culinary preparations and traditional fertility-enhancing practices by tribal women. These tubers possess medicinal attributes for treating diarrhea, dysentery, urinary bladder stones, and kidney diseases, making them a valuable source of therapeutic compounds [1316].

FIGURE 82.1.1 Leaf portion of *Ceropegia bulbosa.*

FIGURE 82.1.2 Inflorescence of *Ceropegia bulbosa.*

FIGURE 82.1.3 Paired fruits of *Ceropegia bulbosa.*

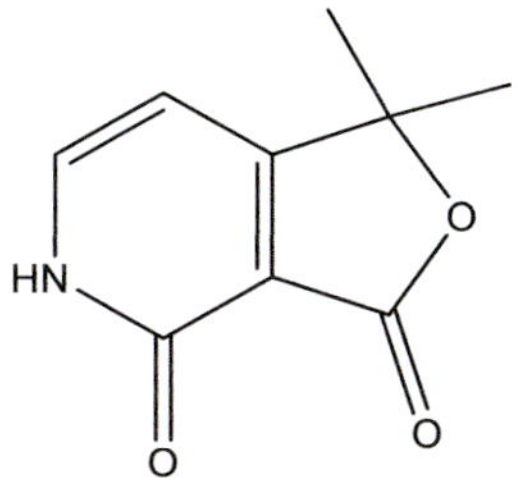

Cerpegin

FIGURE 82.1.4 Chemical component of *Ceropegia bulbosa.*

DOI: 10.1201/9781003295488-82

82.1.7 Mechanism of Action

Alkaloids mitigate the progression of neurodegenerative diseases (NDDs) by various mechanisms, including inhibiting acetyl-cholinesterase (AChE) activity, elevating gamma-aminobutyric acid (GABA) levels, and acting as NMDA receptor antagonists [1317].

82.1.8 Toxicity and Clinical Data

According to the chemical profile of *Ceropegia bulbosa*, this plant can be considered toxic.

82.1.9 Plant Pathology

No data has been reported in the literature regarding the pathology of *Ceropegia bulbosa*. However, the genus *Ceropegia* is susceptible to root rot, with common pests that are likely to infest and attack, including aphids, mealybugs, and scale insects [1318].

83 Genus *Cressa*

83.1 *Cressa cretica* L.

83.1.1 Synonyms

Cressa monosperma

83.1.2 Family

Convolvulaceae

83.1.3 Vernacular Name

Shewailah

83.1.4 Common Name

Molleih, Nadewa & Rosin weed

83.1.5 Botanical Description

Cressa cretica L. is a small perennial herb reaching a maximum height of 38 cm. Its densely branched subshrub possesses gray-greenish leaves with lanceolate to ovate shape that cordate to cuneate at the base. Flower: tiny white to whitish-pink, clustered together at the branch tips. Each cluster is guarded by small green or petal-like structures (Figures 83.1.1 and 83.1.2).

83.1.6 Geographical Distribution

This plant exhibits a distribution range spanning India, Timor, and various regions in Australia, including Western Australia, Northern Territory, Southern Australia, Queensland, New South Wales, and Victoria [1319]. It is distributed throughout Oman.

FIGURE 83.1.1 Whole plant of *Cressa cretica*.

FIGURE 83.1.2 Flower part of *Cressa cretica*.

83.1.7 Chemical Composition

Five flavonoids, namely quercetin, quercetin-3-O-glucoside, kaempferol-3-O-rhamnoglucoside, and rutin, were isolated from the aerial parts of *Cressa cretica* [1320] (Figure 83.1.3). A phytochemical screening of plants detected the presence of alkaloids, coumarins, and sterols [1321].

83.1.8 Uses

Cressa cretica exhibits antibilious, antitubercular, and expectorant properties [1321]. It serves as an anthelmintic, stomachic, tonic, and aphrodisiac, promoting blood enrichment and addressing conditions such as constipation, leprosy, asthma, urinary discharges, diabetes, and general debility [1322]. In Bahrain, it is traditionally employed as an expectorant and antibilious agent [1321], while in Sudan, dry *C. cretica* leaves mixed with sugar function as an emetic [1323].

83.1.9 Mechanism of Action

Ethylacetate fraction of *Cressa cretica* possess bronchodilator activity and mast cell stabilizing activity via relaxing histamine and K^+ and calcium-induced contraction [1324].

83.1.10 Toxicity and Clinical Data

The acute toxicity study based on methanolic extraction of *Cressa cretica* in rats produced no death or signs of toxicity, even at a high dose (5 g/kg) of the extract [1325].

83.1.11 Most Cited Research

iCite, Database, NIH Open Citation Collection was used on October 10, 2023, to get the details of the most cited research.

DOI: 10.1201/9781003295488-83

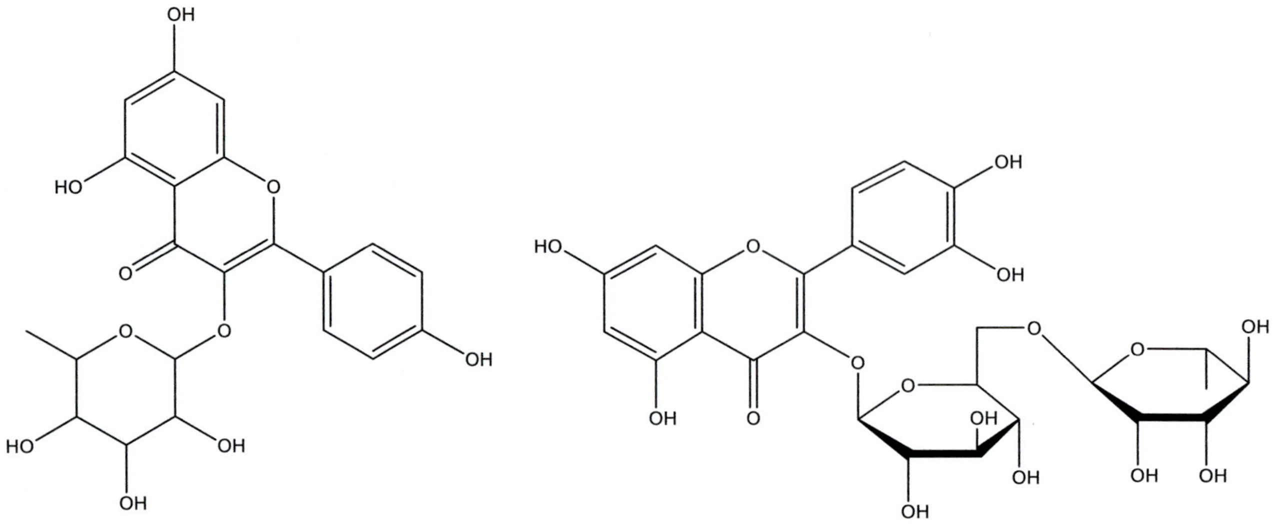

Kaempferol-3-O-rhamnoside **Rutin**

FIGURE 83.1.3 Chemical components of *Cressa cretica.*

Filters for research papers were used by using the keywords *Cressa cretica* L. NIH database search revealed prospecting for *Cressa cretica* to treat COVID-19 via in silico molecular docking models of the SARS-CoV-2 [1326].

83.1.12 PATENTS

Sabrina Arpini, Giacomo Mombelli, Massimo Ronchi. WO2025027106A1. Halophytic plant extract stabilised with a lipophilising agent, and process for the manufacture thereof 2024. This invention is related to a composition with low hygroscopicity comprising a halophytic plant extract and a lipophilising agent.

83.1.13 PLANT PATHOLOGY

Previous studies have reported the isolation and identification of endophytic fungi and bacteria from medicinal plants such as *Cressa cretica* with the most common fungal infection caused by *Penicillium chrysogenum* [1327].

84 Genus *Crotalaria*

84.1 *Crotalaria retusa* L.

84.1.1 Synonyms

Crotalaria retusifolia & Dolichos cuneifolius, Crotalaria hostmannii & Crotalaria cuneifolia

84.1.2 Family

Fabaceae

84.1.3 Common Name

Rattlebox

84.1.4 Botanical Description

Annual herb, reaching heights of 60–150 cm, features erect, angular, green branches. Its alternate, narrowly oblong or wedge-shaped leaves are dark green on the upper surface and lighter underneath [1328]. Brush-shaped yellow colour characterized by a rostrate keel, highly dimorphic anthers, stylar trichomes and elaborate callosities on the standard petal (Figures 84.1.1–84.1.4) [1329].

84.1.5 Geographical Distribution

A botanical remedy extensively distributed in tropical and subtropical regions [1330]. In Oman, it is distributed in northern and southern regions.

FIGURE 84.1.1 Flower of *Crotalaria retusa*.

FIGURE 84.1.2 Phyllotaxy of *Crotalaria retusa*.

FIGURE 84.1.3 Habit of *Crotalaria retusa*.

FIGURE 84.1.4 Fruits portion of *Crotalaria retusa*.

DOI: 10.1201/9781003295488-84

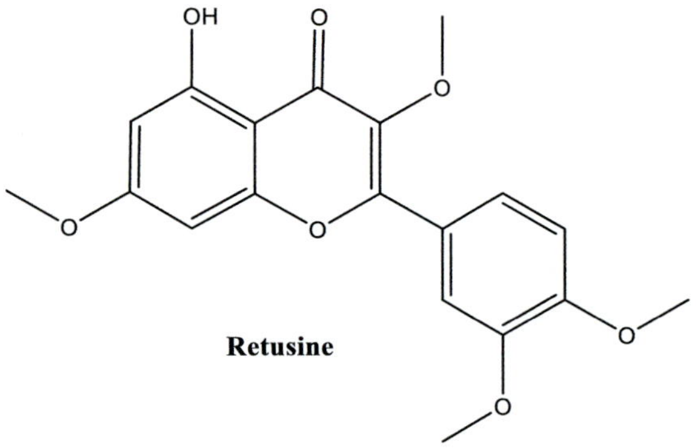

FIGURE 84.1.5 Chemical components of *Crotalaria retusa*.

84.1.6 Chemical Composition

Alkaloids are the active chemical compound in *Crotalaria retusa* L [1331]. *Crotalaria retusa* is rich in alkaloids, with pyrrolizidine alkaloids as the predominant compound (Figure 84.1.5) [1332].

84.1.7 Uses

C. retusa has historically been employed in traditional medicine. It treats prenatal syphilis, malaria, yellow fever, convulsions, ear inflammation, aches, and constipation [1333]. Leaves and flowers treat fever, colds, and lung diseases. Powdered seeds boiled in milk serve as remedies for leprosy, flatulence, fever, and analgesics for scorpion stings [1330].

84.1.8 Mechanism of Action

Seeds are used in popular medicine because of their pharmacological properties. Anti-inflammatory and antinociceptive effects possessed by the seed of this plant are probably due to the inhibitory effect on neutrophil migration of lectin present in albumins of the seeds [1330].

84.1.9 Toxicity and Clinical Data

Crotalaria retusa L. was initially identified as a poisonous plant in the United States of America [1334]. Due to the presence of pyrrolizidine-derived alkaloids in the seeds of *Crotalaria cleomifolia*, raw consumption is not safe [1335].

84.1.10 Adulterants or Substituents

No data is reported in the literature regarding the adulteration or substitution of *Crotalaria retusa*. However, the genus *Crotalaria* can be substituted by other closely resembling species within its own genus [1336].

84.1.11 Most Cited Research

iCite, Database, NIH Open Citation Collection was used on October 11, 2023, to get the details of the most cited research. Filters for research papers were used by using the keywords *Crotalaria retusa L.* NIH database search revealed an assessment of neurotoxicity of monocrotaline, an alkaloid extracted from *Crotalaria retusa* in astrocyte/neuron co-culture system [1337].

84.1.12 Patents

Huachun Wang Larue, Li Lin— US12256685B2 — Cross-pollination through liquid-mediated delivery of pollen to enclosed stigmas of flowers from recipient plants, 2019. This invention is related to the field of methods of improving cross-pollination efficiency via liquid-mediated delivery of donor plant pollen grains.

84.1.13 Plant Pathology

No report in the literature is available on the pathology of *Crotalaria retusa*. However, the genus *Crotalaria* is susceptible to mosaic virus [1338].

85 Genus *Croton*

85.1 *Croton confertus* Baker

85.1.1 Family

Euphorbiaceae

85.1.2 Vernacular Name

Hor

85.1.3 Botanical Description

It is a shrub or small tree standing tall up to 5 m; it boasts smooth, gray bark, a testament to its maturity. But a closer look reveals a youthful charm in its young branches, adorned with a soft, white fuzz that resembles tiny stars (stellate-pubescent). Leaves: yellow-green hue, adding a dash of sunshine to any space. Their shape varies from ovate, like a plump egg, to broadly ovate. Flower: small flowers gather at the branch tips. Female flowers take the stage first, positioned at the base, while the male flowers bloom above them in a clever display of monoecious reproduction. Fruits: covered in a soft, star-shaped felt (stellate-tomentose). The monoecious flowers exhibit a composition of 20 stamens, and six sepals, which contribute to their distinctive floral structure. The pistils have five sepals, linear divisions, and a scaly-covered ovary (Figures 85.1.1 and 85.1.2).

85.1.4 Geographical Distribution

Genus *Croton* is a large genus of Euphorbiaceae, comprising around 1,300 species of trees, shrubs and herbs distributed in tropical and subtropical regions of both hemispheres [1339]. This species is distributed in Oman, Yemen, and Ethiopia. Within Oman, this species is restricted to southern region.

FIGURE 85.1.1 Leafy branches of *Croton confertus*.

FIGURE 85.1.2 Leaf portion of *Croton confertus*.

85.1.5 Chemical Composition

No data in the literature is available that refers to the chemical composition of *Croton confertus*. The genus *Croton* is abundant in biologically active compounds, predominantly diterpenoids such as phorbol esters, clerodane, labdane, kaurane, trachylobane, and pimarane [1340] (Figure 85.1.3). Additionally, *Croton* exhibits a substantial presence of active alkaloids [1341]. Diterpenoids are characteristic constituents of *Croton* species [1342].

85.1.6 Uses

Traditional medicine holds the leaves and new shoots in high regard for their ability to soothe coughs, ease pain and inflammation, purify the blood, and even act as a general tonic. An infusion made from its fruits is said to offer relief from constipation and stomachache, making it a potentially valuable addition to any herbal remedy cabinet.

Popular uses include treatment of cancer, constipation, diabetes, digestive problems, dysentery, external wounds, fever, hypercholesterolemia, hypertension, inflammation, intestinal worms, malaria, pain, ulcers, and weight loss [1339].

85.1.7 Mechanism of Action

No data is reported in literature. However, in literature focused on *Croton* species, 17 distinct varieties exhibited pharmacological effects potentially beneficial in managing diabetes. These

DOI: 10.1201/9781003295488-85

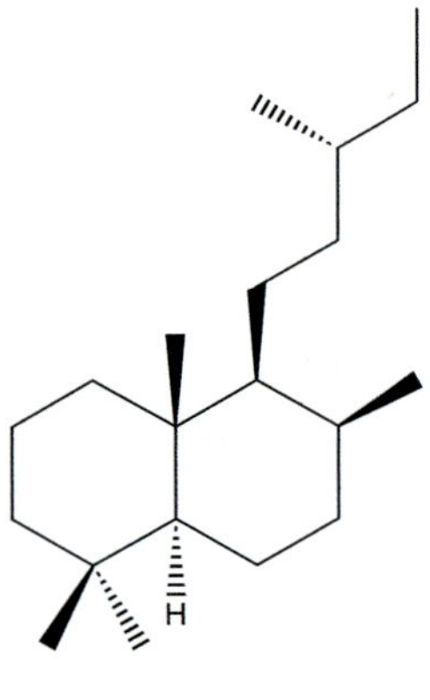

Labdane

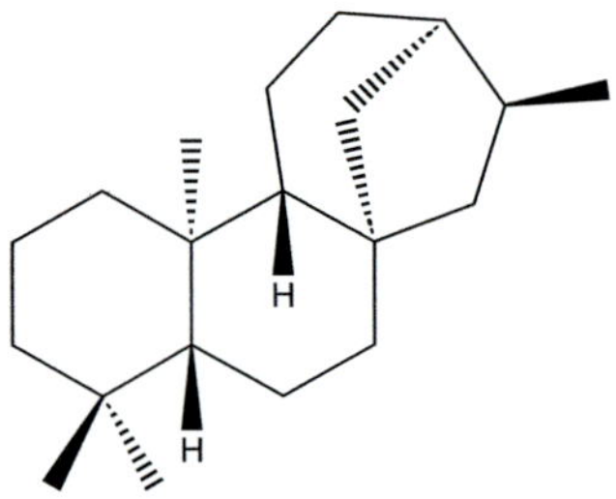

Kaurane

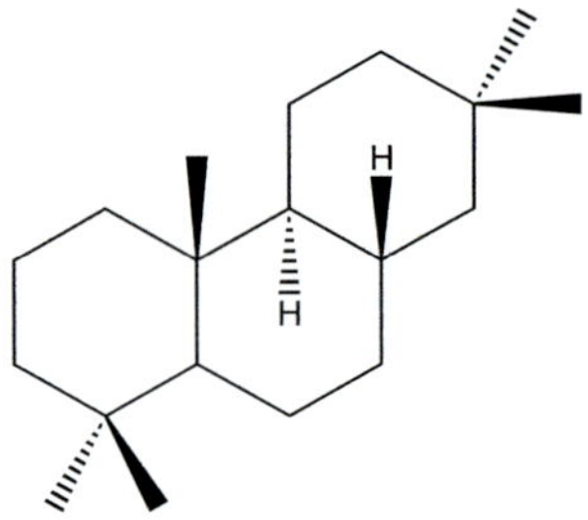

Pimarane

FIGURE 85.1.3 Chemical components of plants belonging to *Croton* genus.

include hypoglycemic, antihyperglycemic, antilipidemic, antihypertensive, antioxidant, and anti-inflammatory properties. These therapeutic effects mediated by several mechanisms via such as the inhibition of α-glucosidases, stimulation of insulin secretion, and enhancement of glucose uptake [1343].

85.1.8 Toxicity and Clinical Data

No data in the literature is present that refers to the toxicity of *Croton confertus*. However, genus *Croton* is considered toxic [1344].

85.1.9 Patents

No information related to the patent is available on this plant. Gary Michael PEKOE, Jazmyne Kristyne MINK, Steven Aaron PENTELNIK, Neal G. Koller—US20220105147A1—*Croton lechleri* compositions for use in the topical treatment of acute bacterial skin or skin structure infection, 2020. In this invention, plant related to the same genus, *Croton lechleri* was used in pharmaceutical compositions to treat acute bacterial skin or skin structure infections.

85.1.10 Plant Pathology

The discovery of a rhabdovirus in the genus *Croton* constitutes a novel occurrence of viral infection within this ornamental plant species [1345]. A recent study showed the presence of *Fusarium* sp. in the roots, stem, and leaves of *Croton sparsiflorus* [1346].

.... (*For additional images refer to the supplementary figures section; page no.* 301)

86 Genus *Cucumis*

86.1 *Cucumis canoxyi* Thulin & Al-Gifri

86.1.1 Family

Cucurbitaceae

86.1.2 Botanical Description

Cucumis canoxyi is an annual herbaceous plant with a maximum height of 0.6 m. It exhibits small yellow flowers with pentamerous symmetry, featuring five petals and five sepals. Its seeds are diminutive, round, and black in color. Seedlings of *C. canoxyi* are characterized by their small size and typically possess a solitary pair of leaves at an early growth stage (Figures 86.1.1 and 86.1.2) [1347].

86.1.3 Geographical Distribution

It is distributed in Yemen and Oman. In Oman, it is distributed in the central desert and Dhofar.

86.1.4 Chemical Composition

Plants within the *Cucumis* genus encompass a diverse array of chemical constituents, including cucurbitacin, phenolic compounds, vitamins, minerals, essential oils, and fatty acids [1348]. Various bioactive compounds have been extracted from cucumbers, include cucurbitacins, cucumegastigmanes I and II, cucumerin A and B, vitexin, orientin, isoscoparin 2″-O-(6‴-(E)-p-coumaroyl) glucoside, and apigenin 7-O-(6″-O-p-coumaroylglucoside) [1349].

Seeds of *Cucumis sativus* contain 24-ethylcholesta-7, 22, 25-trienol, 24-ethylcholesta-7, 25-dienol, avenasterol, spinasterol (Figure 86.1.3), karounidiol, and isokarounidiol identified from the unsaponifiable matter, whereas linoleic acid and arachidic acid were identified in the fatty acids fraction [1350].

86.1.5 Uses

It exhibits potential antidiabetic, lipid-lowering, and antioxidant properties. It also possess cleansing action within the

FIGURE 86.1.1 Fruits of *Cucumis canoxyi*.

FIGURE 86.1.2 Leaves of *Cucumis canoxyi*.

DOI: 10.1201/9781003295488-86

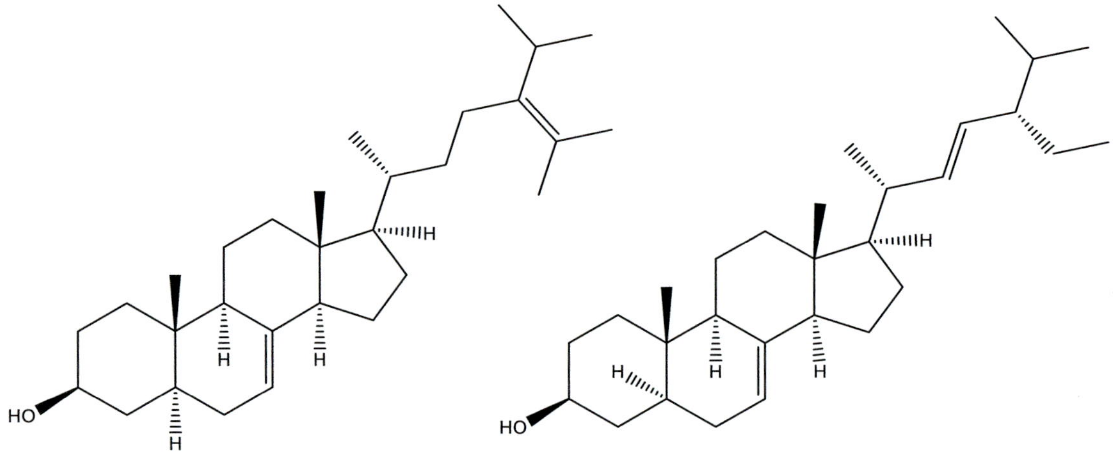

FIGURE 86.1.3 Chemical components of plant belonging to *Cucumis* genus.

body, eliminating accumulated waste materials and chemical toxins. Fresh cucumber fruit juice nourishes the skin, providing a soothing effect on skin irritations and reducing inflammation [1349].

86.1.6 Toxicity and Clinical Data

The seedlings exhibit pronounced toxicity and are unsuitable for consumption [1351]. Cucurbitacins present in this plant are tetracyclic terpenes with steroidal structures recognized for their toxic properties [1099].

86.1.7 Adulterants or Substituents

The Eastern Asian *Citrus hystrix* is considered as an adulterant of *Cucumis sativus* (cucumber) [1352].

86.1.8 Most Cited Research

iCite, Database, NIH Open Citation Collection was used on October 20, 2023, to get the details of the most cited research. Filters for research papers were used by using the keywords *Cucumis sativus.* NIH database search revealed the genome of the cucumber, *Cucumis sativus* L. [1353].

86.1.9 Patents

Jaap Mazereeuw, Brigit Van Kampen, Ronald Wilterdink, Nanne Faber— EP2453730A1—Cucumber vein yellowing virus (CVYV)–resistant cucumber plants (*Cucumis sativus* L.), 2009. This invention is related to methods for providing Cucumber Vein Yellowing Virus (CVYV) resistant cucumber plants (*Cucumis sativus* L.), molecular markers for use in the method.

86.1.10 Plant Pathology

Less information is available on *Cucumis canoxyi. Podosphaera xanthii* is a plant pathogen that causes powdery mildew in *Cucumis sativus* [1354].

86.2 *Cucumis prophetarum* L.

86.2.1 Synonyms

Cucumis foetidus

86.2.2 Family

Cucurbitaceae

86.2.3 Vernacular Name

Lomyet ghanaem, sheri & sheru

86.2.4 Common Name

wild cucumber

86.2.5 Botanical Description

It is a climber annual with pubescent stems and leaves with ovate to round, cordate-based, 2–4 cm long leaves, characterized by dentate or lobulated margins [1355]. The yellow color fruit is somewhat ovoid and covered in spike-like pustules (Figures 86.2.1 and 86.2.2).

86.2.6 Geographical Distribution

Its native distribution includes a vast region stretching from Mauritania on the African continent all the way to the

FIGURE 86.2.1 Fruit of *Cucumis prophetarum.*

FIGURE 86.2.2 Developing fruits of *Cucumis prophetarum.*

Arabian Peninsula, encompassing western Asia and reaching India. In Oman, it is found throughout the country.

86.2.7 Chemical Composition

Terpenoids are the main constituents of *Cucumis prophetarum* [1356]. Plants within the *Cucumis* genus encompass a diverse array of chemical constituents, including cucurbitacin, phenolic compounds, vitamins, minerals, essential oils, and fatty acids [1348].

Previous reports showed the presence of cucurbitacin compounds [E, B, D, F-25 acetate and Hexanorcucurbitacin D] in ethyl acetate fraction of *Cucumis prophetarum* [1357]. Moreover, leaves stems and roots of *Cucumis prophetarum* are also reported for the presence of hentriacontane, neophytadiene and octacosane (Figure 86.2.3) [1358]. Recent report showed the presence of spinasterol, cucurbitacin B, and 2-O-β-D-glucopyranosylcucurbitacin E [1359].

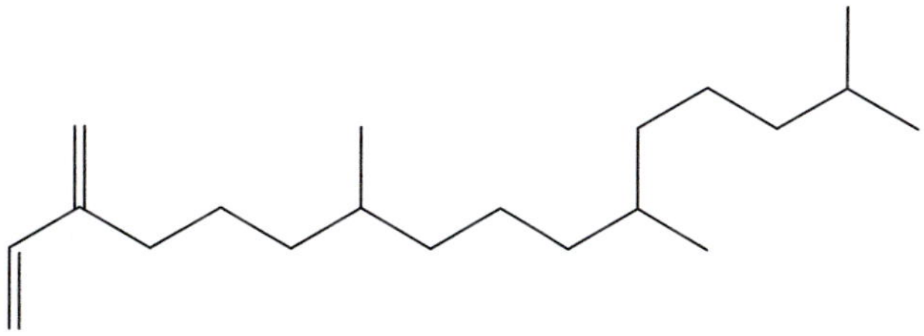

Neophytadiene

FIGURE 86.2.3 Chemical component of plant belonging to genus *Cucumis.*

86.2.8 Uses

The fruit is utilized in Saudi Arabian folk medicine for liver disorder treatment, and its extract demonstrates cytotoxicity against six cancer cell lines [1357].

86.2.9 Mechanism of Action

The anti-inflammatory activity of cucurbitacin B could be due to the partly inhibition of neutrophil and macrophage infiltration into prostatic tissues [1359].

86.2.10 Toxicity and Clinical Data

The isolated compound cucurbitacin B (a terpenoid in *Cucumis prophetarum*) exhibits a moderate level of toxicity, as indicated by its LD_{50} value [1360]. Cytotoxic activity of cucurbitacins is reported against human breast cancer cell line MDA-MB-231 [1357].

86.2.11 Most Cited Research

iCite, Database, NIH Open Citation Collection was used on October 20, 2023, to get the details of the most cited research. Filters for research papers were used by using the keyword *Cucumis prophetarum.* NIH database search revealed biosynthesis of silver nanoparticles using *Cucumis prophetarum* aqueous leaf extract and their antibacterial and antiproliferative activities against cancer cell lines [1361].

86.2.12 Patents

Patent related information is not available on *Cucumis prophetarum.* Cornelis Maria Petrus Van Dun, Magdalena Barbara LASTDRAGER, Lena Johanna HUIJBREGTS-DOORDUIN—US10941411B2—Modified gene resulting in parthenocarpic fruit set, 2017. The present invention is related to a modified gene which is capable of inducing parthenocarpy in *Cucumis sativus.*

86.2.13 Plant Pathology

Apparently, there is no data in the literature that refers to the pathology of *Cucumis prophetarum*. However, the most common pathogens reported in cucurbit plants *Rhizoctonia solani*, *Sclerotinia sclerotiorum*, and *Fusarium oxysporum*. Also, pathogens belonging to *Fusarium*, *Chaetomium*, *Colletotrichum*, and *Acrocalymma* are commonly found in cucurbit plants [1362].

86.3 *Cucumis melo* L.

86.3.1 Synonyms

Cucumis deliciosus

86.3.2 Family

Cucurbitaceae

86.3.3 Vernacular Name

Fekawz

86.3.4 Common Name

Melon, Muskmelon

86.3.5 Botanical Description

Cucumis melo is often exhibiting climbing or trailing growth patterns. Commonly referred to as muskmelon or sweet melon, its nomenclature reflects its inherent characteristics and affiliations within the plant kingdom [1363]. This plant possesses bright yellow color flowers that are composed of five egg-shaped to oval, wrinkled petals and possesses ovate to oblong yellowish-white seeds. The plant possesses bean-shaped leaves with prominent veins. Stems and leaves are covered with long, white hairs (Figures 86.3.1–86.3.2).

FIGURE 86.3.1 Flowering shoot of *Cucumis melo*.

FIGURE 86.3.2 Fruit of *Cucumis melo*.

86.3.6 Geographical Distribution

It is distributed from the lush highlands of Ethiopia all the way to the sunny shores of South Africa! Its travels extend further, encompassing the southwestern corner of Syria, the vast Arabian Peninsula, the diverse Indian Subcontinent, and even reaching the island paradise of New Guinea before settling down in the vibrant landscapes of Northern and Central Australia. In Oman, it is restricted to the woodlands of Dhofar.

86.3.7 Chemical Composition

Chemical analysis revealed a high content of moisture, carbohydrates, dietary fibers, and minerals, along with notable concentrations of carotenoids and phenolic compounds in the examined pulp. Chromatographic profiling identified amentoflavone and gallic acid as the predominant phenolic compounds [1364].

Amentoflavone and gallic acid were the most abundant phenolic compounds found in *Cucumis melo* L. fruit (Figure 86.3.3) [1364]. 3-Hydroxybenzoic acid is found to be major phenolic compound in the melon peels, followed by apigenin-7-glycoside [1365]. Furthermore, the oil obtained from the seeds contains linoleic and oleic acids, vitamin E, and γ-tocopherol [1366].

86.3.8 Uses

It has historically been employed for the management of various medical conditions, including kidney stones, flatulence, leprosy, fever, jaundice, diabetes, obesity, cough, bronchitis, ascites, anemia, constipation, and other abdominal disorders [1367].

86.3.9 Mechanism of Action

Terpenoids exert inhibitory effects on the production of inflammatory cytokines [1030]. Bronchodilator and antidiarrheal effects of *C. melo* are due to the presence of rutin,

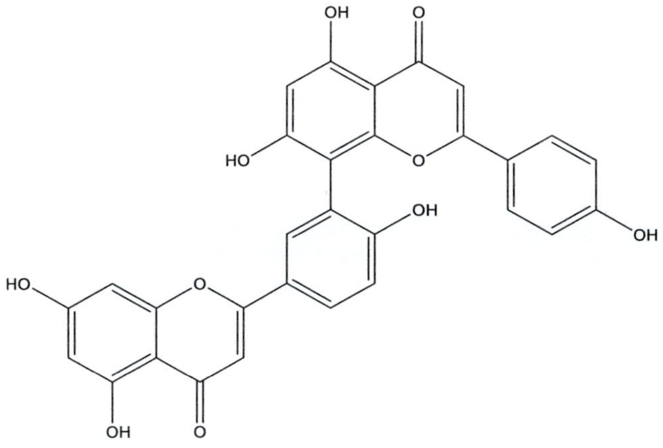

FIGURE 86.3.3 Chemical component of *Cucumis melo.*

quercitrin, and kaempferol mediated by altering the regulatory genes of calcium-mediated smooth muscle contraction [1368].

86.3.10 Toxicity and Clinical Data

Cucumis melo fruit is edible; thus, no toxicity is reported in the literature. So, it can be safe to assume that it is non-toxic. Extracts derived from *Cucumis melo* did not reveal any mortality up to 1,000 mg/kg [1369].

86.3.11 Most Cited Research

iCite, Database, NIH Open Citation Collection was used on October 20, 2023, to get the details of the most cited research. Filters for research papers were used by using the keywords *Cucumis melo.* NIH database search revealed the genome of the cucumber, *Cucumis sativus* L [1353].

86.3.12 Patents

Paul Johan Diergaarde, Leonora Johanna Gertruda Van Enckevort, Karin Ingeborg Posthuma, Marinus Willem Prins—US10093944B2—Powdery mildew resistance providing genes in *Cucumis melo*, 2017. The present invention is related to a modified gene which is capable of inducing parthenocarpy in *Cucumis sativus.*

86.3.13 Plant Pathology

A large, compiled plant pathogens are well documented on different melons that include viral, bacterial, fungal, and nematodes. *Trialeurodes vaporariorum* (Hemiptera: Aleyrodidae), a prevalent pest, commonly recognized as the greenhouse whitefly, poses a significant threat to Oriental melon (*Cucumis melo* var L.) [1370].

The endophyte *F. clavum* EeR24 is reported as an effective salt stress reducer and *C. melo* growth promoter and thus can be considered an eco-friendly bio-fertilizer to confer salt stress tolerance in melon [1371].

87 Genus *Cuscuta*

87.1 *Cuscuta planiflora* Ten.

87.1.1 Synonyms

Cuscuta planiflora var. *holstii* & *Cuscuta abyssinica*

87.1.2 Family

Convolvulaceae

87.1.3 Vernacular Name

Shebakah, alhamool & melwiah

87.1.4 Common Name

Dodder

87.1.5 Botanical Description

It is a parasitic annual. Its slender, thread-like stems, devoid of leaves and boasting a yellowish-green hue, serve as the primary identifier of this parasitic nature. These wispy tendrils function not for nutrient creation but rather for host acquisition. It has small, clustered flowers along its stems and even develops capsule-like fruits containing seeds (Figures 87.1.1 and 87.1.2).

87.1.6 Geographical Distribution

It has a vast native range spanning from Macaronesia and the Mediterranean to Northwest India, and further south to Eritrea, South Africa, and even Madagascar. In Oman, it found throughout the country.

FIGURE 87.1.1 Flower portion of *Cuscuta planiflora.*

FIGURE 87.1.2 Natural habit growth of *Cuscuta planiflora.*

87.1.7 Chemical Composition

Flavonoids are the active chemical compounds found in genus *Cuscuta.* This species has been documented to possess various phytoconstituents, such as flavonoids, glycosides, alkaloids, tannins, lignins, and organic substances [1372].

87.1.8 Uses

Genus *Cuscuta* has gathered significant attention from researchers for its diverse medicinal properties, including anticancer, antibacterial, anti-inflammatory, antimicrobial, antifungal, antiproliferative, antioxidant, hepatoprotective, and antiulcer activities [1372].

87.1.9 Mechanism of Action

One of the significant underlying mechanisms of action of dietary flavonoids and related polyphenols is linked to their capacity to inhibit oxidative stress and its associated downstream responses, including inflammatory diseases [933].

87.1.10 Toxicity and Clinical Data

Based on an analysis of reported toxicity data, it can be concluded that species within the genus *Cuscuta* are deemed non-toxic [1373].

87.1.11 Adulterants or Substituents

Plants of the *Cuscuta* genus closely resemble each other, rendering the adulteration of *Cuscuta planiflora* can be done with a congenerous plant of comparable appearance [1374].

DOI: 10.1201/9781003295488-87

87.1.12 Most Cited Research

iCite, Database, NIH Open Citation Collection was used on October 12, 2023, to get the details of the most cited research. Filters for research papers were used by using the keywords *Cuscuta planiflora*. NIH database search revealed the effectiveness of *Cuscuta planiflora* Ten. and *Nepeta menthoides* Boiss. Buhse in major depression: a triple-blind randomized controlled trial study [1375].

87.1.13 Patents

Beth Irwin Johnson, Mark Christopher Mescher, Consuelo Moreira De Moraes—US20160021830A1—Manipulation of light spectral quality to reduce parasitism by *Cuscuta* and other plant parasites, 2014. This invention is related to preventing infestation of plants by parasitic plants, by light manipulation to disrupt host location or host attachment by the parasite.

88 Genus *Cymbopogon*

88.1 *Cymbopogon commutatus* (Steud.) Stapf

88.1.1 Synonyms

Andropogon commutatus, Sorghum commutatum & *Gymnanthelia commutata*

88.1.2 Family

Poaceae

88.1.3 Vernacular Name

Sakhbar

88.1.4 Common Name

Lemongrass

88.1.5 Botanical Description

It is a perennial tuft that boasts a delightful aromatic fragrance and reaches up to 1 m high. Leaves: Flat and linear, they gracefully curl outwards from the base, adding a touch of movement. Basal leaves are adorned with soft hairs, giving them a unique texture. Inflorescence, branched cluster of paired woolly spikelets. It is characterized by a twist and comes with a bent bristle-like appendage called an awn, attached to each floret pair (Figures 88.1.1 and 88.1.2).

88.1.6 Geographical Distribution

It has a wide range includes sun-drenched tropics of Africa, stretching eastward to encompass Iraq and India, the Arabian Peninsula specifically, found in Oman, Bahrain, Qatar, Saudi Arabia, and Yemen. In Oman, it found in Northern Oman, primarily at higher elevations.

FIGURE 88.1.1 Inflorescences of *Cymbopogon commutatus*.

FIGURE 88.1.2 Plant habit of *Cymbopogon commutatus*.

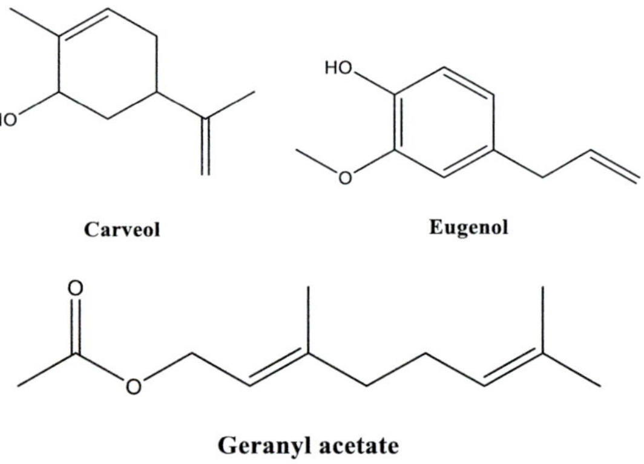

FIGURE 88.1.3 Chemical components of plants belonging to *Cymbopogon* genus.

88.1.7 Chemical Composition

Citronellal is a terpenoid is the active chemical compound found in the genus *Cymbopogon* [1376]. Phytochemical analysis of the genus *Cymbopogon* revealed the presence of phenols, flavonoids, tannins, saponins, anthraquinones, and alkaloids in the sample. Additionally, the essential oil is composed of aldehydes, alcohols, esters, and various terpenoids [1377]. It contains limonene and carveol [1378]. α-Citral, β-citral, geranyl acetate, ylangene, δ-elemene, and eugenol were found to be major constituents of *Cymbopogon citratus* [1379] (Figure 88.1.3).

DOI: 10.1201/9781003295488-88

88.1.8 Uses

The fresh herb of genus *Cymbopogon* is employed in salads and traditional culinary preparations and serves as a substitute for other tea beverages owing to its agreeable taste [1380]. The oil is used for Perfumery and cosmetics applications [1381].

88.1.9 Toxicity and Clinical Data

Usually, *Cymbopogon* is safe when consumed orally, topically, or used as aromatherapy.

The previous research suggested that the essential oil derived from Ethiopian *Cymbopogon commutatus* is relatively safe and non-toxic [1382]. However, plants from a similar genus are reported for toxicological biomarkers, and genotoxicity of *C. citratus* essential oil has a low toxicity level but is suitable and safe in long-term treatment at doses up to 100 mg/kg [1383]

Also causes skin allergic reactions when used directly on the skin, toxic alveolitis when inhaled, and also cause elevation in amylase, and bilirubin [1384]

88.1.10 Adulterants or Substituents

Apparently, there is no data in the literature that refers to the adulteration or substitution of *Cymbopogon commutatus*. However, Lemon zest can be used as a substitute for *Cymbopogon commutatus* [1385]. Several adulterants of *Cymbopogon distans* such as *C. citratus, C. flexuosus,* and *Imperata cylindrical* were identified [1386].

88.1.11 Most Cited Research

iCite, Database, NIH Open Citation Collection was used on October 12, 2023, to get the details of the most cited research. Filters for research papers were used by using the keywords *Cymbopogon commutatus*. NIH database search revealed chitosan nanoparticles encapsulating lemongrass (*Cymbopogon commutatus*) essential oil: physicochemical, structural, antimicrobial, and in vitro release properties [1387].

88.1.12 Patents

Ashok Shahi, Satya Sharma, Swadesh Pal, Suresh Chandra, Prabhu Dutt, Surendra Balyan, Maharaj Bhan, Subhash Taneja, Ravinder Raina, Vijeshwar Verma, Shazia Zahoor, Maharaj Kaul, Ghulam Qazi—US20030154532P1—Drought-tolerant plant strain of *Cymbopogon commutatus,* 2002.

88.1.13 Plant Pathology

Collar rot and wilt disease of the plants from the genus *Cymbopogon*, caused by *Fusarium moniliforme*, poses a significant threat to the health and yield of these plants [1388]. Three fungal endophytes such as *Trichoderma atrobrunneum*, *Trichoderma afroharzianum*, and *Fusarium pernambucanum* were isolated from *Cymbopogon nardus* [1389].

88.2 *Cymbopogon jwarancusa* (Jones ex Roxb.) Schult.

88.2.1 Family

Poaceae

88.2.2 Vernacular Name

Sakhbar

88.2.3 Common Name

Izkhir & Jwarankush Grass

88.2.4 Botanical Description

The perennial plant exhibits a caespitose growth habit with glabrous butt sheaths that persist and envelop the base of the culm, featuring either compacted or curly dead sheaths. The erect or geniculately ascending culms reach heights of 30–150 cm. The ligule is an eciliate membrane measuring 0.5–4 mm, while the aromatic leaf blades are 15–30 cm long and 3–5 mm wide, with a filiform, attenuate apex [1390].

88.2.5 Geographical Distribution

Its native range covers regions with very different climates and habitats from western Asia to South Central China; it is also found in Yemen, Socotra, and Oman. In Oman, it is found in the northern mountains.

88.2.6 Chemical Composition

Cymbopogon species mainly contain geraniol, geranial, citral, myrcene, neral, nerol, citronellol, and limonene, etc. [1391]. Generally, citronellal, a terpenoid, is an active chemical compound found in the genus *Cymbopogon* [1376]. Additionally, the essential oil is composed of aldehydes, alcohols, esters, and various terpenoids [1377]. Piperitone was the major component of essential oil extracted from *Cymbopogon jwarancusa* [1392,1393] (Figure 88.2.1).

88.2.7 Uses

In recent years, numerous researchers have documented the utilization of this plant in the treatment of various ailments, including vomiting, abdominal tumors, unconsciousness, blood impurities, and skin disorders [1388].

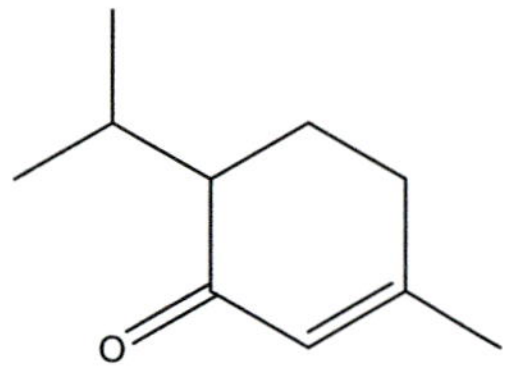

Piperitone

FIGURE 88.2.1 Primary chemical component of *Cymbopogon jwarancusa.*

Diethyl ether fraction isolated from plants demonstrated highest inhibitory activity against *Streptococcus pneumoniae* compared with gentamicin and penicillin [1392]. The diuretic effect is reported [1394].

88.2.8 Mechanism of Action

The hepatoprotective effect of the hydroalcoholic extract of *Cymbopogon jwarancusa* is reported, which could be associated with the synergistic effect of phenolics, flavonoids, glycoside, tannins, steroids, fatty acids, sterols, esters, saponins, terpenes/terpenoids, and essential oil [1395].

88.2.9 Toxicity and Clinical Data

Apparently, there is no data in the literature that refers to the toxicity of *Cymbopogon jwarancusa. Cymbopogon citratus* at doses generally higher than 1,500 mg/kg body weight showed significant functional damage to the stomach and liver of rat [1396].

88.2.10 Adulterants or Substituents

No data in the literature that refers to the adulteration or substitution of *Cymbopogon jwarancusa.* The most common adulteration practices occur through the incorporation of other low-value oils (of either mineral or vegetable origin) as well as synthetic solvents in order to increase profit. For instance, the addition of poly (ethylene glycol) to sandalwood oil (*Santalum* spp.) has been reported. A recent study showed the presence of dipropylene glycol as an adulterant in commercial essential oils of Cymbopogon ssp., and due to this reason, it was not possible to quantify citronellal using 1 H-NMR [1397].

88.2.11 Most Cited Research

iCite, Database, NIH Open Citation Collection was used on October 12, 2023, to get the details of the most cited research. Filters for research papers were used by using the keywords *Cymbopogon jwarancusa.* NIH database search revealed microscopic implications and evaluation of herbaceous melliferous plants of southern Khyber Pakhtunkhwa-Pakistan using light and scanning electron microscope [1398].

88.2.12 Patents

Ashok Shahi, Satya Sharma, Swadesh Pal, Suresh Chandra, Prabhu Dutt, Surendra Balyan, Maharaj Bhan, Subhash Taneja, Ravinder Raina, Vijeshwar Verma, Shazia Zahoor, Maharaj Kaul, Ghulam Qazi—US20030154532P1—Drought-tolerant plant strain of *Cymbopogon commutatus,* 2002. This present invention is related to development of drought tolerant strain of Cymbopogon rich in geraniol and geranyl acetate designated as Cymbopogon RLJCC1.

88.2.13 Plant Pathology

P. cymbopogonis has been reported for causing rust on lemongrass in Brazil. Researchers reported the various disease symptoms and pathogens of lemongrass. Lemon grass is also reported for the infection caused by *Curvularia* spp. Pathogens such as *C. cymbopogonis, C. verruciformis, C. andropogonis, C. asianensis, C. lunata,* and *C. microspora* caused severe infection to lemongrass resulting in foliage blight and change in oil yield [1399].

88.3 *Cymbopogon schoenanthus* (L.) Spreng.

88.3.1 Synonyms

Andropogon schoenanthus & *Cymbopogon citriodorus*

88.3.2 Family

Poaceae

88.3.3 Vernacular Name

Sakhbar

88.3.4 Common Name

El bekhirai & Lemongrass

88.3.5 Botanical Description

Cymbopogon schoenanthus features erect, tufted stems of 90 cm in length, characterized by slender, glabrous culms with 3–4 nodes. Its leaves are simple, alternate, and linear, measuring 5–7 cm in length and 1 cm in width, with sheathed bases and spiny, entire apices. Inflorescence: arranged on branching stalks that spread outward. Each spike has a pair of spikes; each spike has only one spikelet (Figures 88.3.1 and 88.3.2).

88.3.6 Geographical Distribution

This species is distributed across a broad swath of arid and semi-arid regions, ranging from the Sahara and Sahel zones of North and West Africa, through the Horn of Africa, and extending eastward to Kenya in East Africa and Iran in the Middle East. Within Oman it found throughout the country.

FIGURE 88.3.1 Inflorescence of *Cymbopogon schoenanthus*.

FIGURE 88.3.2 Natural habit of *Cymbopogon schoenanthus*.

88.3.7 Chemical Composition

Cymbopogon schoenanthus essential oil contains piperitone, δ-2-carene, elemol, and limonene as major components [1400]. Another study showed the presence of cis- and trans-p-menth-2-en-1-ols (22.6%±3.6% and 14.3%±1.7%, respectively) [1401]. It was also found that essential oil contains α-eudesmol, transdauca-4(11),7-diene, α-muurolol, hinesol, allo-ocimene, and δ-2-carene as the major components [1402] (Figure 88.3.3).

88.3.8 Uses

In folk medicine, this species has different applications

Culinary: Lemongrass is a popular herb in Southeast Asian cuisine, known for its citrusy and lemony flavor. The stalks are typically used in soups, curries, stir-fries, and stews. Lemongrass oil, also known as citronella oil, is used as a flavoring agent in some foods and beverages.

Insect Repellent: Citronella oil, extracted from lemongrass, is a natural insect repellent. It is commonly used in candles, torches, and insect sprays.

Medicinal: Lemongrass has various traditional medicinal uses. People have used it to treat digestive issues, fever, coughs, and skin problems. While more research is needed to confirm its effectiveness for all these conditions, lemongrass tea is a popular home remedy for some ailments.

Cymbopogon schoenanthus is employed in the management of colds, epilepsy, abdominal cramps, and pains. Additionally, it finds application in culinary and perfume products [1403]. Antimicrobial activity was measured against *Staphylococcus aureus* strain, while the antioxidant potential was low [1401].

Essential *Cymbopogon schoenanthus* is used as a diuretic, anti-rheumatic, asthmatic, antipyretic agent, anti-infectious, anti-intestinal, antispasmodic, anti-malarial, and anti-helminthic and also used in the treatment of colds, epilepsy, and abdominal cramps and pains [1404].

C. citratus and *C. schoenanthus* essential oils have been reported for the inhibition aflatoxins [1405].

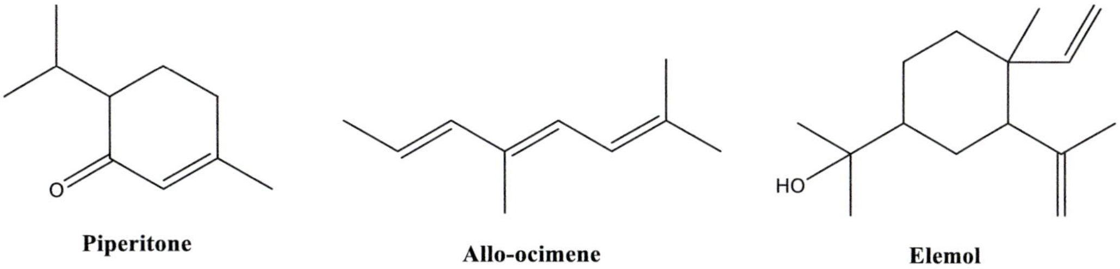

FIGURE 88.3.3 Chemical components of *Cymbopogon schoenanthus*.

Locally, the leaves and floral parts of this species have been used in traditional medicine for both internal and external treatments, particularly for digestive disorders and skin conditions. A tea prepared from fresh plant parts is traditionally consumed three times daily for three days to relieve fever and abdominal pain. To address chesty coughs and throat irritation, leaves are often added to the tea. For external use, the leaves are mashed into a paste and applied directly to the skin to treat sunburn and skin redness [911].

88.3.9 Mechanism of Action

Essential oil reduced the proliferation of LNCaP and HeLa cells and thus prevented LNCaP cell migration resulting in the arrest of their cell cycle in the G2/M phase [1400].

88.3.10 Toxicity and Clinical Data

Cymbopogon schoenanthus showed low toxic effects in the liver [1406].

88.3.11 Most Cited Research

iCite, Database, NIH Open Citation Collection was used on October 12, 2023, to get the details of the most cited research. Filters for research papers were used by using the keywords *Cymbopogon schoenanthus.* NIH database search revealed the in vitro efficacy of plant extracts and synthesized substances on Rhipicephalus (Boophilus) Microplus (Acari: Ixodidae) [1407].

88.3.12 Patents

Joseph R. Wegner, Cheryl A. Littau—US8383686B2—Foaming alcohol compositions with selected dimethicone surfactants, 2018.

89 Genus *Cyphostemma*

89.1 *Cyphostemma ternatum* (Forssk.) Desc.

89.1.1 Synonyms

Cissus gmelinii, Saelanthus ternatus & *Vitis apodophylla*

89.1.2 Family

Vitaceae

89.1.3 Vernacular Name

Kum

89.1.4 Common Name

Wild Grape or Bush Grape

89.1.5 Botanical Description

Cyphostemma ternatum is a succulent climbing vine, reaching 2 m in height. Its leaves are alternate, simple, and trifoliolate, measuring up to 25 cm x 30 cm, with serrated edges and a petiole up to 5 cm in length. Leaves are 3–5-foliolate, glabrous, sessile, or lower leaves with petiol with toothed margins. Fruits: fleshy round berry (Figures 89.1.1 and 89.1.2).

89.1.6 Geographical Distribution

It is distributed from Northeast Tropical Africa to Eastern Kenya and the southern Arabian Peninsula, which encompasses parts of Ethiopia, Somalia, Kenya, Yemen, Oman, and Saudi Arabia. In Oman, it thrives in the southern mountains.

FIGURE 89.1.1 Phyllotaxy of *Cyphostemma ternatum*.

FIGURE 89.1.2 Fruits of *Cyphostemma ternatum*.

89.1.7 Chemical Composition

The composition of genus *Cyphostemma* was predominantly characterized by terpenoid compounds [1408]. The phytochemical constituents of the genus *Cyphostemma encompass* flavonoids, alkaloids, coumarins, saponins, stilbenoids, and tannins [1409]. *C. juttae* essential oil contains phytol (30%) as the main chemical component [1408].

C. adenocaule commonly known as Swahili, reported to contain ceanothane-type triterpenoids [1410]. Previous research showed the presence of betulin, betulinic acid, cyphostemmic acid A, cyphostemmic acid B, cyphostemmic acid C, cyphostemmic acid D, epigouanic acid A, lupeol, zizyberanal acid, β-sitosterol, and its glucoside, and 3β, 28-dihydroxy-30-norlupan-20-one (10) in the methanol extract of bark and wood of *Cyphostemma adenocaule* (Figure 89.1.3) [1409].

89.1.8 Uses

In traditional herbal medicine, the plant was utilized in Arabia to remedy foot rot. A poultice composed of the cooked leaves' juice, combined with salt, was topically applied to the afflicted area [10]. Leaves have a peppery and acrid taste. The fruits and roots of the *C. ternatum* are edible. Antiplasmodial, anti-inflammatory, and antimicrobial effects of *Cissus adenocaulis* are reported [1409].

89.1.9 Mechanism of Action

C. juttae essential oil interferes with the NF-κB pathway and thus can be used in triple-negative breast cancers [1408].

DOI: 10.1201/9781003295488-89

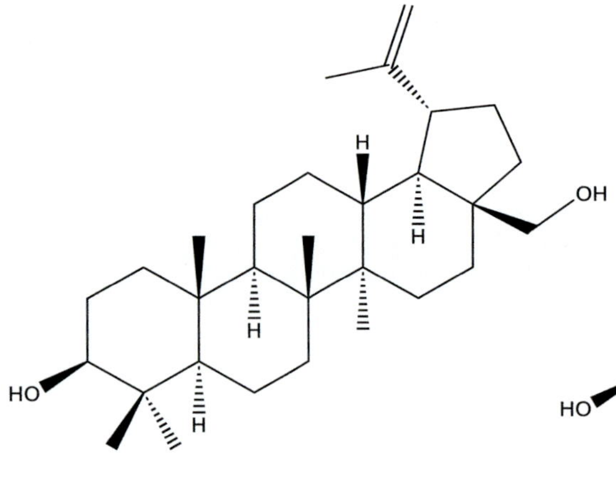

Betulin

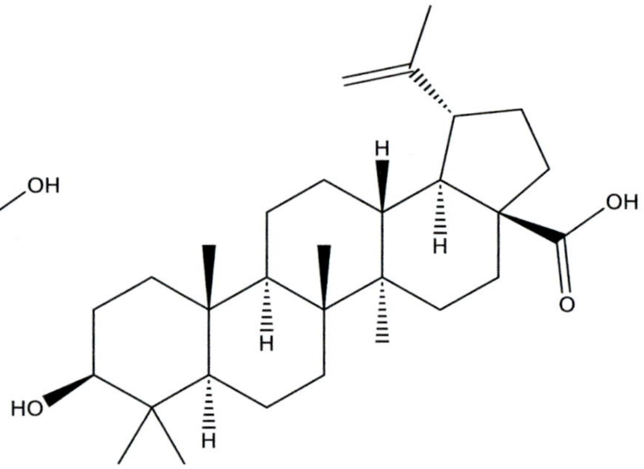

Betulinic acid

FIGURE 89.1.3 Chemical components of plants belonging to *Cyphostemma* genus.

89.1.10 Toxicity and Clinical Data

Cyphostemma juttae leaves and fruit possess elevated concentrations of toxic tannins. Consumption of any plant part is unequivocally contraindicated for both human and animal ingestion [1411].

89.1.11 Most Cited Research

iCite, Database, NIH Open Citation Collection was used on October 15, 2023, to get the details of the most cited research. Filters for research papers were used by using the keywords *Cyphostemma ternatum*. NIH database search revealed an ethnobotanical study of medicinal plants used by people in Zegie Peninsula, Northwestern Ethiopia [1412].

89.1.12 Patents

Reda HADJ-SLIMANE, Yves Lepelletier, Tewfik Hadj-Slimane—US20130295199A1—Composition Comprising OPC and Omega-3 for Preventing and/or Inhibiting the Development of Diabetic Retinopathy, 2012. The present invention is related to a composition for improving eye health, and particularly to reduce the risk of eye damage associated with diabetes.

89.1.13 Plant Pathology

No plant diseases are reported on this wild plant. However, the genus *Cyphostemma* is susceptible to infestations of mealy bugs, necessitating vigilant monitoring for the presence of fuzzy white specks within the caudex folds [1411].

90 Genus *Cyperus*

90.1 *Cyperus aucheri* Jaub. & Spach

90.1.1 Synonyms

Cyperus conglomeratus var. aucheri

90.1.2 Family

Cyperaceae

90.1.3 Vernacular Name

Theda

90.1.4 Common Name

Sedges

90.1.5 Botanical Description

It is a tufted perennial with short or long-creeping rhizomes. Stems range from 2 to 90 cm in length and are terete or subtrigonous. Basal leaves vary in length compared to the stem and may be flat to nearly unifacial, with basal sheaths that can be either soft or stiff and hard (Figures 90.1.1 and 90.1.2) [1413].

90.1.6 Geographical Distribution

Arid regions in Southern Iraq, Egypt, the Arabian Peninsula, Iran, Pakistan, and Afghanistan. In Oman, it is distributed throughout the countery, primarily in sand dunes.

FIGURE 90.1.1 Inflorescence of *Cyperus aucheri.*

FIGURE 90.1.2 Whole plant of *Cyperus aucheri.*

90.1.7 Chemical Composition

Cyperus is chemically characterized by the presence of flavonoids, stilbenes, coumarins, sesquiterpenes, triterpenes, sterols, and quinines [1414]. Cyperene was found to be major component of plants belonging to genera such as *Cyperus conglomeratus* [1415].

In the essential oil of *Cyperus conglomeratus*, eugenol, α-cyperone, and cyperotundone were the major compounds [1416]. *Cyperus conglomeratus* also contains α-amyrin and β-sitosterol that show anti-Candidal activity [1417].

90.1.8 Uses

The genus *Cyperus* demonstrates efficacy in treating asthma, stomach ache, arthritis, dysentery, malaria, and nervous diseases, and possesses antitumor, breast cancer risk reduction, antidiarrheal, and antifungal properties [1414].

Anti-inflammatory, anthelmintic, antioxidant, antimicrobial, and anticancer properties of the *Cyperus aucheri* plant are reported. The plant is reported for its antioxidant, anticancer, antigenotoxic, antimicrobial, anti-inflammatory, anthelmintic, nematicide, antihistaminic, antieczemic, and insectifuge properties and is also used to treat asthma, stomach ache, arthritis, dysentery, malaria, nervous disease, tumor, diarrhea, and fungal diseases [1416].

90.1.9 Mechanism of Action

One of the significant underlying mechanisms of action of dietary flavonoids and related polyphenols is linked to their capacity to inhibit oxidative stress and its associated downstream responses, including inflammatory diseases [933].

90.1.10 Toxicity and Clinical Data

The genus *Cyperus* comprises perennial, grass-like plants with edible properties. So the species of this genus is considered non-toxic [1418]. Alcohol extract of *Cyperus conglomeratus* has not shown any sub-chronic toxicity to rats at a dose of 1,000 mg/kg [1417].

90.1.11 Most Cited Research

iCite, Database, NIH Open Citation Collection was used on October 15, 2023, to get the details of the most cited research. Filters for research papers were used by using the keywords *Cyperus*. NIH database search revealed the anti-candida activity of Brazilian medicinal plants [1419].

90.1.12 Patents

Patent related information is not available on this plant. Kyu Hyuck CHUNG, Jong Hwan Kwak— US20150110901A1—Sesquiterpenoid-based compounds, extracts of *Cyperus rotundus* comprising the same, and use thereof, 2014. The present invention is related to use of a sesquiterpenoid-based compounds containing extracts of *Cyperus rotundus* for the prevention and treatment of menopausal.

90.1.13 Plant Pathology

Tiger nut tuber rot, caused by *Rhizoctonia necatrix*, was initially documented in 1998 in Valencia province. Subsequently, it emerged as a prominent and detrimental disease afflicting the genus *Cyperus* [1420].

90.2 *Cyperus conglomeratus* Rottb.

90.2.1 Family

Cyperaceae

90.2.2 Vernacular Name

Qasqas

90.2.3 Botanical Description

Tufted perennial grows up to 50 cm. Papery leaf sheaths and stiff leaves point toward a plant. Spherical clusters of flowers or seeds at the end of long floral stalks. Fruits: brown to black (Figures 90.2.1–90.2.3).

90.2.4 Geographical Distribution

The native range of *Cyperus conglomeratus* encompassing West Tropical Africa to Northwest India and the Western Indian Ocean. In Oman, it found throughout the country.

FIGURE 90.2.1 Plant habit of *Cyperus conglomeratus*.

FIGURE 90.2.2 Inflorescence of *Cyperus conglomeratus*.

FIGURE 90.2.3 Whole plant of *Cyperus conglomeratus*.

90.2.5 Chemical Composition

Multiple metabolites, including flavonoids, triterpenes, steroids, and essential oils, were extracted and identified from *Cyperus conglomeratus* [1421]. Essential oil of *C. conglomerates* contains cyperene as the major constituent, followed by cyperotundone, cyperol, α-cyperone, isorotundene, and α-cubebene [1415] (Figure 90.2.4).

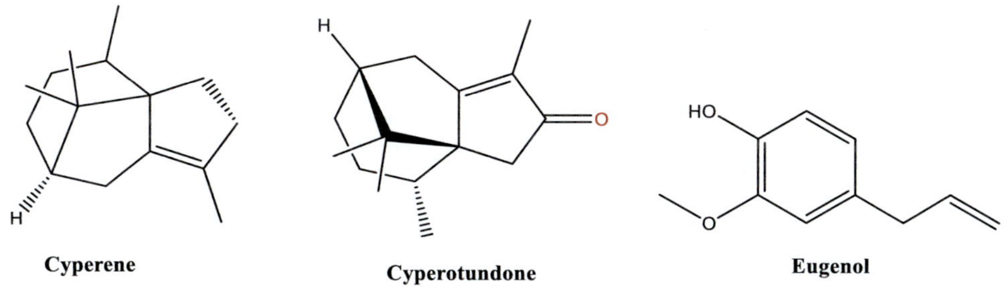

FIGURE 90.2.4 Chemical components of *Cyperus conglomeratus*.

90.2.6 Uses

Cyperus conglomeratus serves as a vital medicinal plant in traditional medicine, employed for its analgesic, diuretic, stimulant, pectoral, emollient, and anthelmintic properties [1422].

90.2.7 Mechanism of Action

One of the significant underlying mechanisms of action of dietary flavonoids and related polyphenols is linked to their capacity to inhibit oxidative stress and its associated downstream responses, including inflammatory diseases [933].

C. conglomeratus extract provides significant gastroprotective activity in ethanol-induced gastric ulcers via reducing galactin-3 and TNF-α [1421].

90.2.8 Toxicity and Clinical Data

The genus *Cyperus* comprises perennial, grass-like plants with edible properties. So, the species of this genus is considered non-toxic [1418]. Alcoholic extracts of tubers of *Cyperus rotundus* and essential oil showed less cytotoxicity [1423].

90.2.9 Most Cited Research

iCite, Database, NIH Open Citation Collection was used on October 15, 2023, to get the details of the most cited research. Filters for research papers were used by using the keywords *Cyperus conglomeratus*. NIH database search revealed UPLC-qTOF-MS phytochemical profile and antiulcer potential of *Cyperus conglomeratus* Rottb. alcoholic extract [1421].

90.2.11 Plant Pathology

Klebsiella aerogenes M2017452 infect *Cyperus rotundus* and can degrade malachite green to non-toxic substances, e.g., N, N-dimethylaniline and 2-(4-dimethylamino-phenyl)-phenyl-methanone [1424].

90.3 *Cyperus longus* L.

90.3.1 Family

Cyperaceae

90.3.2 Vernacular Name

Saed

90.3.3 Common Name

Sweet Cyperus & Galingale

90.3.4 Botanical Description

Cyperus longus is a perennial that reaches a height of 60 cm tall. The leaves are 2–10 mm wide, typically shorter or equivalent in length to the stems [1425] Inflorescence exhibits a simple or compound umbel structure composed of many spikelets.

90.3.5 Geographical Distribution

Cyperus longus exhibits a broad distribution native across the entire landmass of Africa, Europe, India, and the Middle East. In Oman, it is distributed in Southern mountains.

90.3.6 Chemical Composition

Numerous flavonoids, terpenoids, and stilbenes have been successfully extracted and identified from *Cyperus longus* [1426]. α-Humulene, γ-himachalene, and β-himachalene were found to be the main constituents of *Cyperus longus* [1426].

Essential oils of *Cyperus longus* contain mainly β-himachalene, α-caryophyllene oxide, irisone, β-caryophyllene oxide, humulene oxide, viridiflorol, aristolone, and longiverbenone (Figure 90.3.1) [1426]. Another report showed that *C. longus* EO contains β-himachalene, α-humulene, and γ-himachalene [1427]. Previous reports revealed the presence of sesquiterpenes, such as cyperusols A1 (1), A2 (2), B1 (3), B2 (4), C (5), and D (6), in *Cyperus longus* [1428].

90.3.7 Uses

The genus *Cyperus* demonstrates efficacy in treating asthma, stomach ache, arthritis, dysentery, malaria, and nervous diseases and possesses antitumor, breast cancer risk reduction, antidiarrheal, and antifungal properties [1414]. Viridiflorol and longiverbenone can be used as chemotherapeutic agents in cancer treatment [1426]

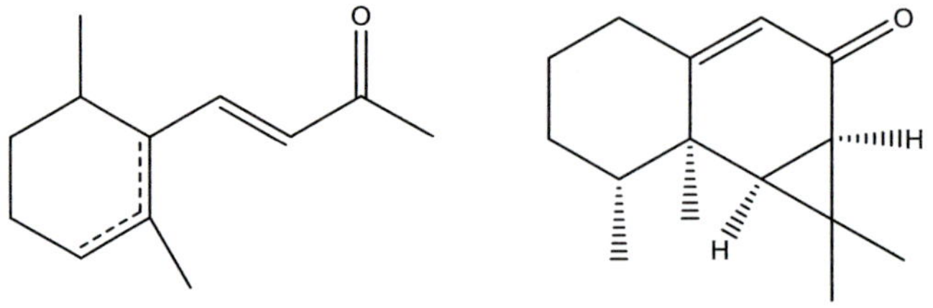

FIGURE 90.3.1 Chemical components of *Cyperus longus*.

Cyperus longus is another source of stilbenoids particularly scirpusins A. Research has shown that the methanol extract of *C. longus* exhibits an antiallergic effect, specifically reducing ear passive cutaneous anaphylaxis reactions in mice [1429].

90.3.8 Mechanism of Action

One of the significant underlying mechanisms of action of dietary flavonoids and related polyphenols is linked to their capacity to inhibit oxidative stress and its associated downstream responses, including inflammatory diseases [933].

90.3.9 Toxicity and Clinical Data

This species is considered as no toxic [1425].

90.3.10 Most Cited Research

iCite, Database, NIH Open Citation Collection was used on October 15, 2023, to get the details of the most cited research. Filters for research papers were used by using the keywords *Cyperus longus*. NIH database search revealed structures of new sesquiterpenes and hepatoprotective constituents from the Egyptian herbal medicine *Cyperus longus* [1428].

90.4 *Cyperus rotundus* L.

90.4.1 Family

Cyperaceae

90.4.2 Vernacular Name

Sa'ed

90.4.3 Common Name

Nutgrass

90.4.4 Botanical Description

Cyperus rotundus, a perennial sedge, grows up to 50 cm; it is alternate, simple, glossy, with leave up to 25 cm long. Inflorescence has 3–8 redish to brown spikes (Figures 90.4.1 and 90.4.2).

FIGURE 90.4.1 Inflorescence of *Cyperus rotundus*.

FIGURE 90.4.2 Whole plant of *Cyperus rotundus*.

90.4.5 Geographical Distribution

The distribution range of the mentioned entity spans across multiple continents, including Africa, Western Asia, the Caucasus, Middle Asia, Eastern Asia, Europe, and North America [1430]. In Oman, it is distributed all over the country as a weed plant.

90.4.6 Chemical Composition

Flavonoids are the active chemical compounds in *Cyperus rotundus* [1431]. This plant mainly contains sesquiterpenoids such as patchoulenone, isopatchoulenone, sugeonyl acetate, sugetriol triacetate, and sugebiol, as well

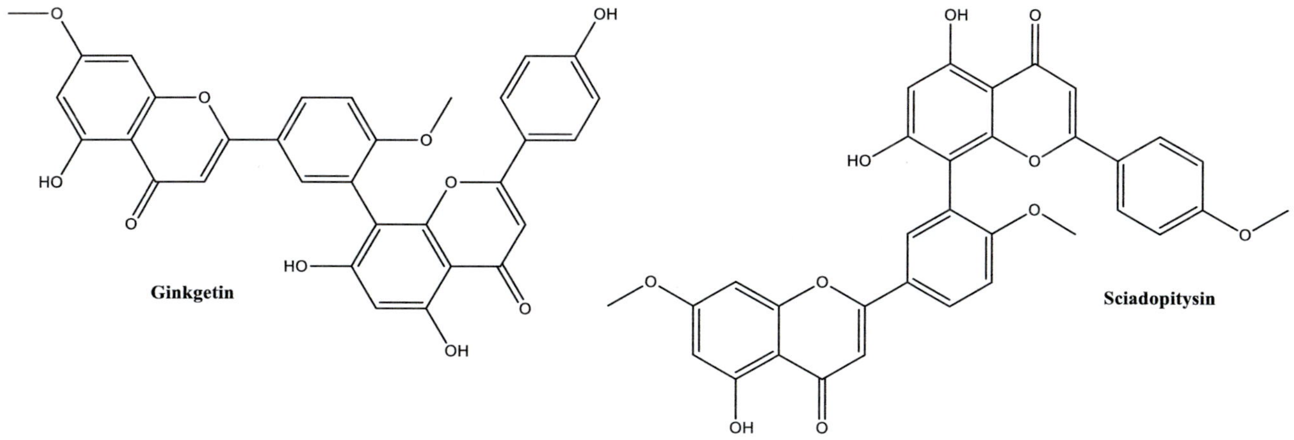

FIGURE 90.4.3 Active chemical components of *Cyperus rotundus*.

as flavonoids such as kaempferol, luteolin, and quercetin [1432]. From *Cyperus rotundus* L., four biflavone constituents have been isolated and identified. These constituents are amentoflavone, ginkgetin, isoginkgetin, and sciadopitysin [1432] (Figure 90.4.3).

90.4.7 Chemical Tests

A method utilizing HPLC-DAD was optimized and validated to generate both qualitative and quantitative profiles of cupressuflavone and amentoflavone.

90.4.8 Uses

Cyperus rotundus has been traditionally employed for its medicinal properties in addressing various gastrointestinal ailments, including spasms, stomach disorders, nausea, vomiting, intestinal parasites, food poisoning, indigestion, and bowel irritation [1430].

Previous studies revealed the potential of *C. rotundus* as an antiandrogenic, antibacterial, anticancerous, anticonvulsant, antidiabetic, antidiarrheal, antigenotoxic, anti-inflammatory, antilipidemic, antimalarial, antimutagenic, antiobesity, antioxidant, anti-uropathogenic, hepatoprotective, cardioprotective, neuroprotective, and nootropic agent [1433].

90.4.9 Mechanism of Action

Amentoflavone appears to induce its effects by upregulating the expression of the pro-apoptotic protein Bax, concurrently suppressing the expression of the anti-apoptotic protein Bcl-2. This action potentially leads to the formation of Bax/Bax homodimers, facilitating the promotion of apoptosis in uterine fibroid cells. Additionally, amentoflavone may contribute to apoptosis by decreasing the levels of plasma estradiol and progesterone within the body [1432].

90.4.10 Toxicity and Clinical Data

Histopathological analysis reported the safety and absence of toxicity associated with the administered drug [1434]. *Cyperus rotundus* steam distillate containing components such as p-cymene, nerol, linalool, o-cymene, (S)-(-)-citronellal, (1S)-(-)-camphor, terpinolene, and m-cymenec can be used as potential insecticide, especially for *Blattella germanica* [1435].

In the acute toxicity study of mice, oral ethanol extract of *Cyperi rhizoma* 5,000 mg/kg b.w. did not produce any toxic symptoms, visceral gross appearance changes, behavioral changes, or signs of death. In subacute poisoning, repeated oral administration of 1,000 mg/kg b.w. ethanol extract of *Cyperi rhizoma* did not cause changes in general behavior, mortality, weight gain, hematology, and clinical hematological parameters [1436].

90.4.11 Adulterants or Substituents

Cyperus amabilis Vahl, *C. esculentus* Linn., *C. platystylis* R. Br., and *C. scariosus* R. Br. have been identified as suitable substitutes for *Cyperus rotundus* [1437].

90.4.12 Most Cited Research

iCite, Database, NIH Open Citation Collection was used on October 15, 2023, to get the details of the most cited research. Filters for research papers were used by using the keywords *Cyperus rotundus*. NIH database search revealed anti-Candidal activity of Brazilian medicinal plants.

90.4.13 Patents

Kyu Hyuck CHUNG, Jong Hwan Kwak—US9795649B2—Sesquiterpenoid-based compounds, extracts of *Cyperus rotundus* comprising the same, and use thereof, 2014.

The present invention is related to use of a sesquiterpenoid-based compounds containing extracts of *Cyperus rotundus* for the prevention and treatment of menopausal.

90.4.14 Plant Pathology

Rainfall facilitated the proliferation of monocotyledonous species, specifically *Cyperus rotundus*, serving as an optimal host for the pathogenic fungus *Puccinia graminis* [1438]. The endophytic fungi *A. tereus* and *F. oxyporum* were found in *Cyperus rotundus* [1439]. Fifteen endophytic bacteria were isolated from nutgrass (*Cyperus rotundus*). Eleven isolates were Gram-negative, and the rest were Gram-positive [1440].

91 Genus *Dalbergia*

91.1 *Dalbergia sissoo* Roxb. Ex DC.

91.1.1 Synonyms

Amerimnon sissoo & *Pterocarpus sissoo*

91.1.2 Family

Fabaceae

91.1.3 Vernacular Name

Mais

91.1.4 Common Name

Sasam, Sarsoa, & Indian rosewood

91.1.5 Botanical Description

Dalbergia sissoo, a deciduous tree, exhibits a substantial stature, reaching approximately 25 m in height, characterized by a gray-yellow trunk with a diameter ranging from 2 to 3 m. The leathery leaves are pinnately compound, featuring alternate, broad, ovate leaflets with an acuminate apex. These leaflets are glabrescent and petiolate, and possess fine-pointed tips (Figures 91.1.1–91.1.4) [1441].

91.1.6 Geographical Distribution

The native range of *Dalbergia sissoo* encompasses a diverse array of regions, including southern Arabian Peninsula Eastwood to Himalayas. It has also been introduced to many parts of the world. In Oman, it found in scattered area in northern Oman near springs.

FIGURE 91.1.1 Shoot apex displaying newly developing leaves of *Dalbergia sissoo*.

FIGURE 91.1.2 Inflorescence of *Dalbergia sissoo*.

FIGURE 91.1.3 Branches of *Dalbergia sissoo*.

91.1.7 Chemical Composition

Dalbergin (a flavonoid) is an active chemical component of *Dalbergia sissoo* found in stem [1442]. Phytochemical analysis of the ethanolic extract of *Dalbergia sissoo* revealed the presence of carbohydrates, proteins, amino acids, phenolic compounds, flavonoids, alkaloids, saponins, phytosterols, steroids, and tannins [1443].

Another report showed the presence of that isoflavone glucoside, caviunin 7-O-[β-d-apiofuranosyl-(1→6)-β-d-glucopyranoside] and a new itaconic derivative, (E)-4-methoxy-2-(3,4-dihydroxybenzylidene)-4-oxobutanoic acid, isoflavones and flavonols, glucosides and a lignan glucoside in the ethanolic extract of *Dalbergia sissoo* leaves [1444]. Rhamnose, galactose, and glucuronic acid have been isolated from the leave, whereas stem-bark contains dalberjenone, dalbergin, and methyl dalbergin, and a new 4-phenyl chromene, dalbergichromene [1445] (Figure 91.1.5).

DOI: 10.1201/9781003295488-91

91.1.8 Uses

Dalbergia sissoo wood exhibits potential applications in the treatment of various medical conditions, including blood disorders, scabies, eye and nose disorders, burning sensations, scalding urine, stomach problems, syphilis, boils, leprosy, and nausea [1446].

91.1.9 Mechanism of Action

Prunetin, tectorigenin, and prunetin 4′-O-galactoside show efficient binding to the EGFR binding domain, suggesting its potential role as antioxidant and anticancer agents [1447].

91.1.10 Toxicity and Clinical Data

Dalbergia sissoo does not exhibit any toxic effects [1442]. The acute toxicity study reported earlier did not reveal any toxic symptoms or death in any of the animals up to the dose level of 3,000 mg/kg body weight [1448].

FIGURE 91.1.4 Natural habit of *Dalbergia sissoo.*

91.1.11 Most Cited Research

iCite, Database, NIH Open Citation Collection was used on October 16, 2023, to get the details of the most cited research. Filters for research papers were used by using the keywords *Dalbergia sissoo.* NIH database search revealed relationships between chromium biomagnification ratio, accumulation factor, and mycorrhizae in plants growing on tannery effluent-polluted soil [1449].

91.1.12 Patents

Rakesh Maurya, Preety Dixit, Ritu Trivedi, Vikram Khedgikar, Jyoti Gautam, Avinash Kumar, Divya Singh, Sheelendra Pratap Singh, (none) Wahajuddin, Girish Kumar Jain, Naibedya Chattopadhyay—US20140079834A1—Bioactive fractions and compounds from *Dalbergia sissoo* for the prevention or treatment of osteo-health related disorders, 2012. This invention is related to bioactive fractions and compounds from *Dalbergia sissoo* for the prevention or treatment of osteo-health related disorders.

91.1.13 Plant Pathology

Dieback disease, prevalent in South Asia, cause substantial damage upon *Dalbergia sissoo* trees of varying age groups, with a high susceptibility observed across all developmental stages [1450]. Wilt of this plant is reported as Wilt of Shisham (*Dalbergia sissoo* Roxb.) due to *Fusarium solani* [1451].

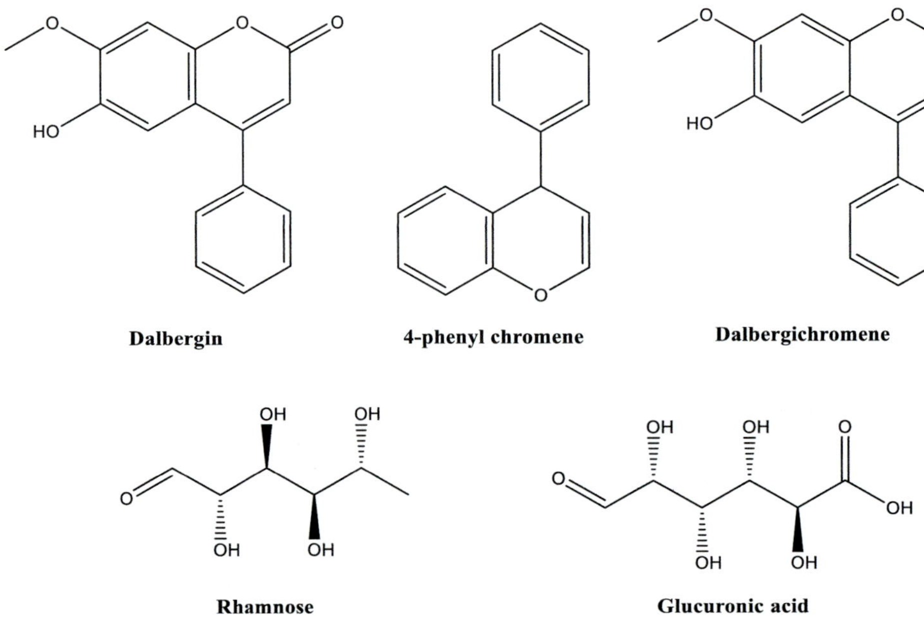

FIGURE 91.1.5 Chemical components of *Dalbergia sissoo.*

92 Genus *Dalechampia*

92.1 *Dalechampia scandens* L.

92.1.1 Synonyms

Dalechampia scandens var. genuina

92.1.2 Family

Euphorbiaceae

92.1.3 Common Name

Gujarat Spurge-Creeper

92.1.4 Botanical Description

Dalechampia scandens is a climbing perennial with stems that reach up to a height of 3 m. It is velvet-hairy and is usually covered with hairs. Leaves are 3-lobed and deeply heart-shaped at the base. Flowers: small, unisexual, both staminate and pistillate flowers enclosed by a pair of pale green 3-lobed leaf-like bracts (Figures 92.1.1 and 92.1.2) [1452].

92.1.5 Geographical Distribution

It is a neotropical vine distributed from Mexico to Argentina. It's native to Asia-Tropical, Southern America, Asia-Temperate, Africa, and Northern America [1453]. It found throughout Northern and Southern Oman; this adaptable plant thrives in diverse habitats. In the foothills of the Eastern Hajar mountains, it seeks shelter among the rugged slopes, often nestled near the shade of other trees.

92.1.6 Chemical Composition

This plant has been reported for the presence of phenolic content, a high flavonoid content, and antioxidant properties [1454].

92.1.7 Uses

It is used as animal food, a poison, and a medicine [1455].

92.1.8 Most Cited Research

iCite, Database, NIH Open Citation Collection was used on January 3, 2023, to get the details of the most cited research. Filters for research papers were used by using the keywords *Dalechampia scandens.* NIH database search revealed the evolvability and genetic constraints in Dalechampia blossoms: components of variance and measures of evolvability [1456].

FIGURE 92.1.1 Fruit of *Dalechampia scandens.*

FIGURE 92.1.2 Shoot with flowers and fruits of *Dalechampia scandens.*

DOI: 10.1201/9781003295488-92

93 Genus *Daphne*

93.1 *Daphne mucronate* Royle

93.1.1 Synonyms

Daphane angustifolia, Daphane argentea & Daphane cachemireana

93.1.2 Family

Thymelaeaceae

93.1.3 Vernacular Name

Zekam & Menten, Bakhteet

93.1.4 Botanical Description

The evergreen shrub, characterized by slender pale green branches, attains a height of 2.5 m. The younger branches frequently exhibit tomentose characteristics, with leaves arranged in an alternate or scattered fashion, measuring 3–5.8 cm in length and 0.4-1 cm in width. These leaves are elliptic-oblong to lanceolate, often mucronate, occasionally obtuse, and possess a coriaceous texture, being sessile in nature [1457]. Flowers; terminal form in clusteres; white to green. Fruits: orange berry (Figures 93.1.1–93.1.6).

93.1.5 Geographical Distribution

Daphne mucronata is distributed in Arabian Penesila and from SE. Türkiye eastward to Himalaya in Oman it found in Western and Eastern Hajar Mountains.

FIGURE 93.1.1 Apical portion showing fruits of *Daphne mucronata*.

FIGURE 93.1.2 Inflorescence of *Daphne mucronata*.

FIGURE 93.1.3 Apical part with closed flowers of *Daphne mucronata*.

93.1.6 Chemical Composition

Coumarins and diterpenes have been identified as predominant constituents of the plant [1458]. *Daphne mucronata* is a rich source of biologically active natural products, comprising a diverse array of compounds such as coumarins, flavonoids, triterpenoids, lignin, and coumarinolignans. Significant constituents of this plant include glucosides, daphnecin, coumarin, aquillochin, daphnine, and umbelliferone, contributing to its scientific significance in natural product research [1459]. Oil obtained from leaves contains pentadecane, 2-methyl hexadecane, 7,9-dimethyl hexadecane, tetradecane, 5-Propyl decane, 2,3,5,8 tetramethyl hexadecane, 2-methyl-6-propyl dodecane, and 5-methyl tetradecane [1460].

DOI: 10.1201/9781003295488-93

Plants also contain phenolic components such as malic acid, gallic acid, chlorogenic acid, epigallocatechin gallate, quercetin, morin, ellagic acid, and rutin (Figure 93.1.7) [1461].

FIGURE 93.1.4 Apical structure of *Daphne mucronata*.

Another report showed the presence of hexanoic acid (32.05%) in *D. mucronata* [1462].

93.1.7 Uses

In Oman, its leaves were used to treat and alleviate abdominal colic, a condition characterized by severe abdominal pain, as well as to treat coughs, potentially due to their demulcent properties that soothe irritated tissues. *Daphne mucronata*, used as a remedy for skeletomuscular ailments, offers relief from rheumatism when utilized as a dressing created from its fruit and leaves [1463]. Moreover, its ointment demonstrates efficacy in the treatment of infectious wounds. Anti-angiogenic effect due to the potential components such as syringin and daphnodorin from the roots of *Daphne mucronata* is reported [1464].

Furthermore, chemical components such as syringin, pinoresinol 4-O-β-D-glucopyranoside, daphnodorin B, lariciresinol 4,4′-bis-O-β-D-glucopyranoside, and genkwanol A revealed a strong antioxidant effect [1464].

FIGURE 93.1.5 Shows the outer skin bark of *Daphne mucronata*.

FIGURE 93.1.6 Whole plant of *Daphne mucronata*.

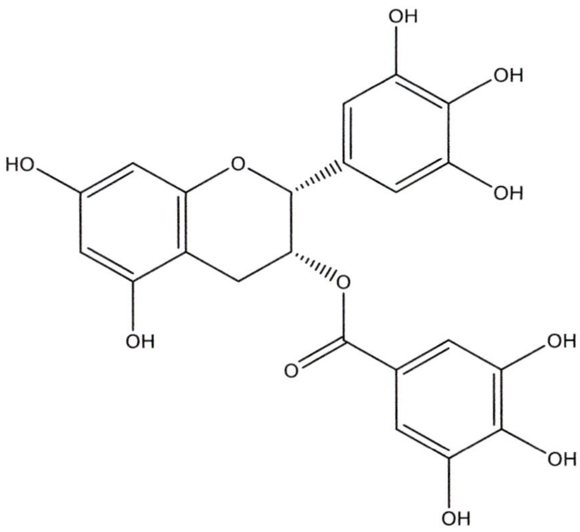

Epigallocatechin gallate

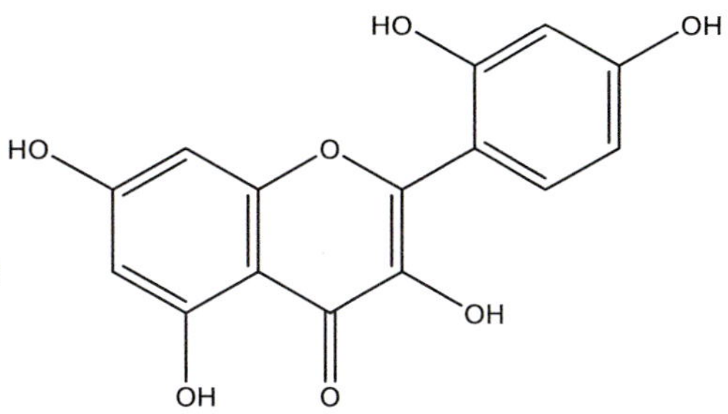

Morin

FIGURE 93.1.7 Chemical components of *Daphne mucronata*.

93.1.8 Mechanism of Action

Terpenoids exert inhibitory effects on the production of inflammatory cytokines [1030]. Syringin and daphnodorin B presented the best anti-angiogenic response on the chorioallantoic membrane, causing a considerable decrease of the microvasculature and hemoglobin content, resulting in an anti-angiogenic effect [1464].

93.1.9 Toxicity and Clinical Data

In a prior study, cytotoxic effects of the hydroalcoholic extract of *D. mucronata* were observed across various cell lines. Remarkably, the extract exhibited its most pronounced cytotoxicity, specifically on breast cancer cell lines [1465]. Leaves of this plant are poisonous and applied as insect-repulsive abscesses for sore and glue is used for muscular and nerve troubles [1460]. Also, Daphne is highly irritant due to the presence of tigliane diterpenoids [1464].

93.1.10 Most Cited Research

iCite, Database, NIH Open Citation Collection was used on October 16, 2023, to get the details of the most cited research. Filters for research papers were used by using the keywords *Daphne mucronata*. NIH database search revealed an ethnobotanical survey of herbal remedies traditionally used in Kohghiluyeh va Boyer Ahmad province of Iran.

93.1.11 Patents

Jae Wha Kim, Kyung Seop Ahn, Ho Bum Kang, Sei Ryang Oh, Jae Jong Go, Joo Heon Kim, Jang Mi Sun, Jae Sung Song, Hyun Woo Oh. Da Jung Ji—US9428520B2—*Daphne genkwa* extracts, and pharmaceutical composition containing fractions of the extracts or compounds separated from the extracts as active ingredients for preventing or treating atopic dermatitis, 2012.

93.1.12 Plant Pathology

The plant pathogenic fungal species *Coniothyrium ferrarisianum* has been isolated from leaves of *Daphne mucronata* [1466].

94 Genus *Datura*

94.1 *Datura innoxia* Mill.

94.1.1 Synonyms

Datura guayaquilensis, Datura meteloides & Datura velutinosa

94.1.2 Family

Solanaceae

94.1.3 Vernacular Name

Marnha

94.1.4 Common Name

Thorn-Apple, Downy Thorn-Apple

94.1.5 Botanical Description

It is a low-growing, spreading perennial plant characterized by hairy leaves measuring 2–5 inches in length. It produces white flowers and bears a spiny fruit (Figures 94.1.1 and 94.1.2) [1467].

94.1.6 Geographical Distribution

The species under consideration exhibits a native distribution encompassing the Southwestern United States, Central and South America, with introduced populations in Africa, Asia, Australia, and Europe [1468]. It is distributed throughout Oman.

FIGURE 94.1.1 Fruit portion of *Datura innoxia*.

FIGURE 94.1.2 Inflorescence of *Datura innoxia*.

94.1.7 Chemical Composition

Datura innoxia Mill. is known for its synthesis of tropane alkaloids, specifically hyoscyamine and scopolamine (Figure 94.1.3) [1469]. A preliminary phytochemical analysis was conducted on the methanolic and hydroalcoholic extracts of dried *Datura* seeds, detecting the existence of alkaloids, tannins, cardiac glycosides, flavonoids, carbohydrates, amino acids, and phenolic compounds [1470].

94.1.8 Uses

Genus *Datura* dried leaves, flowers, and roots of the plant were historically employed for their narcotic, antispasmodic, antitussive, bronchodilator, anti-asthmatic, and hallucinogenic properties [1470]. Insecticidal effects of *D. innoxia* acetone extracts against *Tribolium castaneum, Trogoderma granarium,* and *Sitophilus granarius* are also reported [1471,1472].

94.1.9 Mechanism of Action

Tropane alkaloids exert their mechanism of action through competitive antagonism at muscarinic acetylcholine receptors, thereby impeding the binding of acetylcholine [1473]. The extract derived from *D. innoxia* has been assessed for its anti-inflammatory activity to prepare an active herbal pain-relieving drug. The mechanism of action behind this anti-inflammatory and analgesic action is based on cyclooxygenase (COX1 and COX2) inhibition, followed by suppression of prostaglandin-synthesis or probably the narcotic effects of *Datura* species [1474,1475].

DOI: 10.1201/9781003295488-94

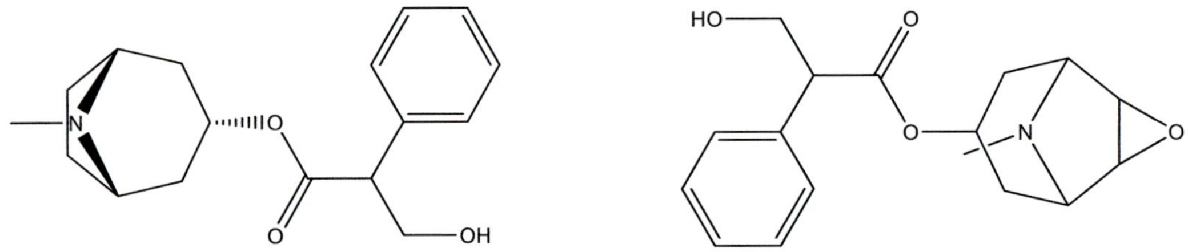

Hyoscyamine **Scopolamine**

FIGURE 94.1.3 Major chemical components of *Datura innoxia*.

94.1.10 Toxicity and Clinical Data

The indiscriminate use of the extract may result in acute poisoning [1476]. The ethanolic extracts of *D. innoxia* possess cytotoxic effects on different cancer cells, particularly against human colon cancer cells, inducing cell-cycle arrest at the sub-G1 phase and apoptosis [1477].

94.1.11 Most Cited Research

iCite, Database, NIH Open Citation Collection was used on October 16, 2023, to get the details of the most cited research. Filters for research papers were used by using the keywords *Datura innoxia* Mill. NIH database search revealed extraction optimization of medicinally important metabolites from *Datura innoxia* Mill.: an in vitro biological and phytochemical investigation [1476].

94.1.12 Patents

Andrew H. Segal, Elihu Young US9951113B2. Lectin compositions and methods for modulating an immune response to an antigen. This invention is related to Cannabis oils and its composition including with vitamin E, and methods of preparation thereof.

94.1.13 Plant Pathology

Potato tuber necrotic ringspot disease was observed in *Daturan innoxia* [1478].

94.2 *Datura metel* L.

94.2.1 Synonyms

Datura fastuosa & *Datura alba*

94.2.2 Family

Solanaceae

94.2.3 Vernacular Name

Marnha & benj

FIGURE 94.2.1 Flower of *Datura metel*.

94.2.4 Common Name

Tatura, Joz mashel, Joz mathel

94.2.5 Botanical Description

The annual herbaceous plants exhibit a height range of 0.5–1.5 m, characterized by a lack of trichomes. The stems often display a dark violet hue. The petioles measure 2–6 cm, while the leaf blades are ovate or broadly ovate, with dimensions of 5–20×4–15 cm. They possess a membranous texture, and the leaf base is either truncate or cuneate, displaying asymmetry (Figures 94.2.1 and 94.2.2) [1470].

94.2.6 Geographical Distribution

Datura metel, a plant species originating from the Americas, has been disseminated throughout tropical and subtropical regions [1479]. It is distributed throughout Oman.

94.2.7 Chemical Composition

Tropane alkaloids are active chemical compounds found in *Datura metel* [1480]. A preliminary phytochemical analysis was conducted on the methanolic and hydroalcoholic extracts of dried Datura seeds, detecting the existence of alkaloids, tannins, cardiac glycosides, flavonoids, carbohydrates, amino acids, and phenolic compounds [1470].

FIGURE 94.2.2 Plant habit of *Datura metel*.

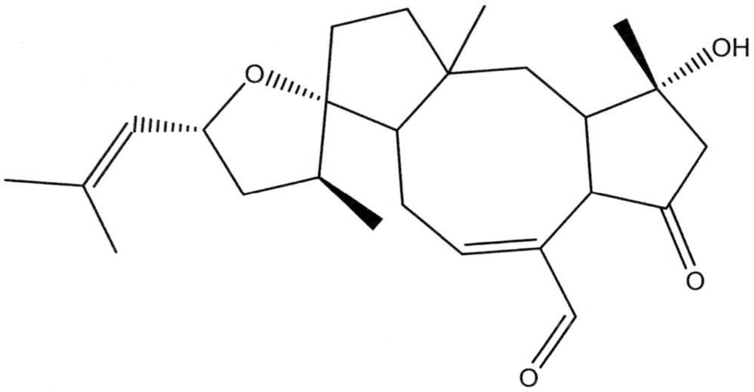

Ophiobolin A

FIGURE 94.2.3 Chemical component of *Datura metel*.

This plant contains chemical compounds like flavonoids, tropane alkaloids, tannins, and saponins. Plants contain withanolides, daturaolone, datumetine, daturglycosides, ophiobolin A, and baimantuoluoline A as an active component (Figure 94.2.3) [1481,1482].

Another study showed the presence of citroside A, coniferin, paeoniflorin, cannabisin D, cannabisin D, cannabisin E, melongenamide B, and paprazine in the roots of *Datura metel* [1483]. Flowers of this plant contain yangjinhualine A and megastigmane sesquiterpenes [1484].

Seeds of this plant have been reported for the presence of isofraxidin, scopatone, daturadiol,1,4-benzenediol, arenarine D, vanillin, n-trans-feruloyl-tyramine, scopoletin, g-sitosterol, and hyoscyamilactol [1485]. Essential components of this plant include ketone, palmitic acid, phytol, palmitic acid, ethyl linoleate, and ethyl palmitate [1486].

94.2.8 Chemical Tests

Vitali-Morin's Reaction is used to detect tropane alkaloids.

94.2.9 Uses

In traditional medicine, dried leaves and flowers have been used to treat asthma and ease breathing difficulties. Some studies suggest potential bronchodilatory effects in the leaves, which could help relax airways as well as an anesthetic.

The plant found application in treating diarrhea, skin disorders, epilepsy, hysteria, rheumatic pain, hemorrhoids, dysmenorrhea, skin ulcers, wounds, and burns [1470]. Plant is known for various therapeutic effects such as antioxidant, anti-inflammatory, anti-microbial, insecticidal, anti-cancer, anti-diabetic, analgesic, antipyretic, neurological, contraceptive, and wound-healing capacity [1481].

94.2.10 Mechanism of Action

Tropane alkaloids exert their mechanism of action through competitive antagonism at muscarinic acetylcholine receptors, thereby impeding the binding of acetylcholine [1473].

94.2.11 Toxicity and Clinical Data

The entire plant, including its foliage and seeds, exhibits toxic properties [1470]. This plant contains toxic alkaloids that cause gastrointestinal and central nervous system symptoms when consumed [1487].

The whole plant, especially the foliage and seeds, is toxic due to the presence of tropane alkaloids. The maximum amount of toxins is found in the seeds around 0.1 mg of atropine per seed or 3–6 mg/50–100 seeds. Phytochemicals such as atropine, L-hyoscyamine, and L-scopolamine cause anticholinergic syndrome, leading to inhibition of central and peripheral muscarinic neurotransmission. It's important to note that toxicity caused by this plant typically ensues within 60 min after consumption, and the clinical symptoms may persist up to 24–48 h [1488].

94.2.12 Adulterants or Substituents

Datura metel can be substituted with *Datura stramonium* in certain contexts. Both species belong to the Datura genus and exhibit similar chemical profiles [1489].

94.2.13 Most Cited Research

iCite, Database, NIH Open Citation Collection was used on October 16, 2023, to get the details of the most cited research. Filters for research papers were used by using the keywords *Datura metel*. NIH database search revealed the ethnobotanical survey of medicinal plants commonly used by Kani tribals in Tirunelveli hills of Western Ghats, India [1490].

94.2.14 Patents

Constance Finley, Haley Poole Bestwick. US20240285549A1. Cannabis oil compositions and methods for preparation thereof. This invention is related to Cannabis oils and its composition including with vitamin E, and methods of preparation thereof.

94.2.15 Plant Pathology

Datura metel serves as a host plant for Carlavirus [1491].

95 Genus *Delonix*

95.1 *Delonix elata* (L.) Gamble

95.1.1 Synonyms

Caesalpinia elata & Poinciana elata

95.1.2 Family

Fabaceae

95.1.3 Vernacular Name

Erir & eyrir

95.1.4 Common Name

White gold mohur

95.1.5 Botanical Description

It is characterized by its erect stature, reaching a height range of 6–9 m. The trunk exhibits a smooth texture and displays an ash-colored appearance. The leaves are compound, featuring a slender main rachis measuring 10–20 cm in length and a bipinnate structure. [1492]. Flowers: large white flowers turn yellow when pollinated. Fruits: large, flattened bodies with brown seeds (Figures 95.1.1 and 95.1.2).

95.1.6 Geographical Distribution

The deciduous tree is native to parts of Eastern Africa, Arabian Peninsula, and the Indian subcontinent. In Oman it thrives in Southern mountains among wet escarpment woodlands.

FIGURE 95.1.1 Inflorescence of *Delonix elata.*

FIGURE 95.1.2 *Delonix elata* in its natural habitat.

95.1.7 Chemical Composition

Delonix elata, a plant species, exhibits a physiological system marked by high levels of phenolic compounds within its leaves [1493]. *Delonix elata* is rich in flavonoids, phenolic compounds, tannins, cardiac glycosides, terpenoids, quinones, and coumarins [1494]. Plant stem bark ethanolic extract contains gallic acid, coumaric acid, ellagic acid, flavonoid compounds, rutin, quercetin, and myricetin [1495].

n-Butanol fraction of *D. elata* contain quercetin 3-o-rutinoside-7-O-glucoside, quercetin 3,7-diglucoside, quercetin 3-O-rutinoside and kaempferol 3-o-rutinoside, and rutin (Figure 95.1.3) [1496].

95.1.8 Uses

Delonix elata is traditionally employed in folklore to alleviate joint pain and address flatulence. In Indo-China, the bark of this plant serves as a febrifuge and antiperiodic agent. Local populations employ a paste derived from its leaves and bark to mitigate inflammation and pain [1492].

The *in vitro* antioxidant and hepatoprotective effects of *Delonix elata* stem bark extracts and flavonoid-isolated quercetin against CCl4-induced liver damage in rats are also reported [1495].

95.1.9 Mechanism of Action

Stem bark extract and its isolated constituent quercetin-3-rhamnopyranosyl-(1–6) glucopyranoside enhance wound healing by increasing collagen synthesis through up-regulation of Col 1 α (I) [1497].

DOI: 10.1201/9781003295488-95

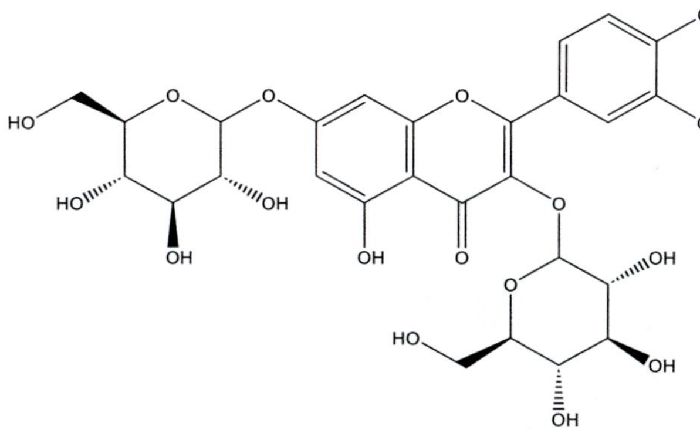

Quercetin 3,7-diglucoside

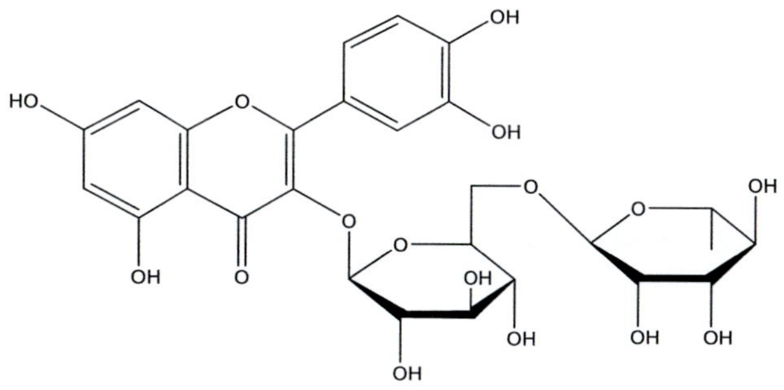

Quercetin 3-O-rutinoside

FIGURE 95.1.3 Chemical components of *Delonix elata*.

95.1.10 Toxicity and Clinical Data

The plant extract exhibited negligible toxicity even at elevated concentrations [1494]. The plant extract was non-toxic, even at higher concentrations to copepod [1498].

95.1.11 Most Cited Research

iCite, Database, NIH Open Citation Collection was used on October 16, 2023, to get the details of the most cited research. Filters for research papers were used by using the keywords *Delonix elata*. NIH database search revealed the trypsin inhibitor from the seeds of *Acacia confuse* [1499].

95.1.12 Patents

LeneFeldskovNielsen,MonicaRamosGallego—US2021017 7641A1—An Adhesive Wafer with a Neutralizer Matrix, 2018. This invention is related to an adhesive wafer (with neutralizer from plants such as *Delonix*) for an ostomy appliance.

95.1.13 Plant Pathology

Ganoderma colossum is a pathogen that attacks the genus *Delonix* [1500].

96 Genus *Dionysia*

96.1 *Dionysia mira* (Jaub. & Spach.) Wendelbo

96.1.1 Synonyms

Primula aucheri

96.1.2 Family

Primulaceae

96.1.3 Vernacular Name

Shejrat al wal (Ibex plant)

96.1.4 Botanical Description

A small shrub forming dense cushion-forming reaches high up to 30 cm. Oval-shaped leaves display a distinctive combination of colors. Their green upper surface is adorned with a distinct white coating, while the underside reveals a contrasting white hue. Flowers: clusters of cheerful yellow flowers grace the tips of the stalks, adding a vibrant touch. Fruit brown capsules hold numerous tiny black seed flowers native to the Caucasus and Iran, thriving in rocky, mountainous habitats (Figures 96.1.1–96.1.3) [1501].

96.1.5 Geographical Distribution

This species is endemic to Oman and restricted to the Western and Eastern Hajar mountains.

96.1.6 Chemical Composition

The principal components of the genus *Dionysia* include 2′-hydroxy-5′-methoxyacetophenone, myrtenol, globulol, eudesm-6-en-4-ol, 2′-hydroxy-4′-methoxyacetophenone, δ-cadinene, δ-cadinol (3.4%), and benzaldehyde dimethyl acetal (Figure 96.1.4) [1502].

FIGURE 96.1.1 Phyllotaxy of *Dionysia mira*.

FIGURE 96.1.2 Inflorescence of *Dionysia mira*.

FIGURE 96.1.3 Plant habit of *Dionysia mira*.

96.1.7 Mechanism of Action

Terpenoids exert inhibitory effects on the production of inflammatory cytokines [1030].

96.1.8 Most Cited Research

iCite, Database, NIH Open Citation Collection was used on October 21, 2023, to get the details of the most cited research. Filters for research papers were used by using the keywords *Dionysia mira*. NIH database search revealed activity-dependent structural and functional plasticity of astrocyte-neuron interactions [1503].

DOI: 10.1201/9781003295488-96

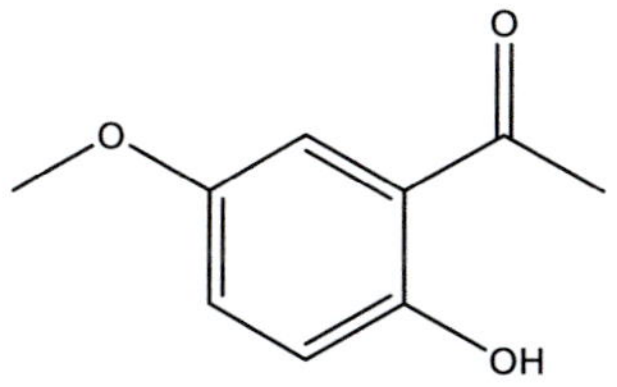

2'-hydroxy-5'-methoxyacetophenone

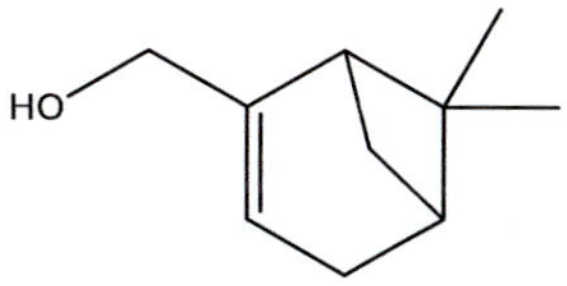

Myrtenol

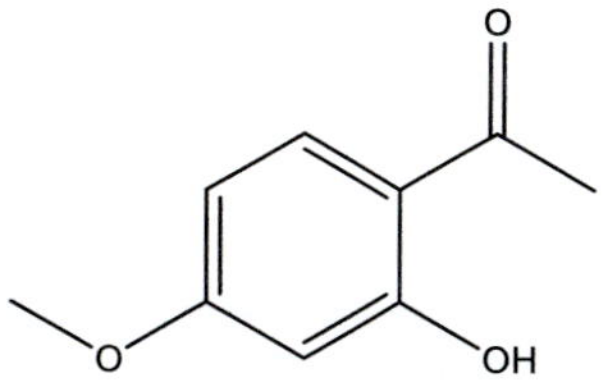

2'-hydroxy-4'-methoxyacetophenone

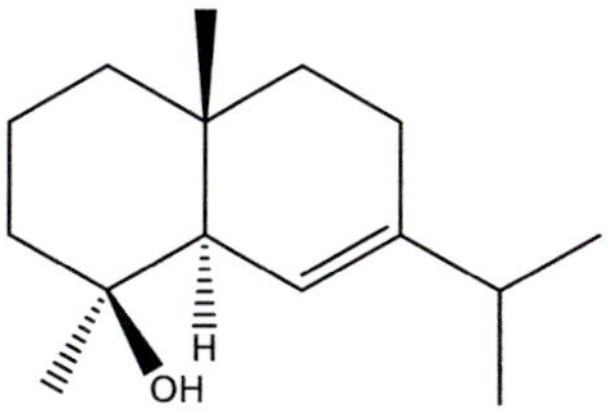

Eudesm-6-en-4-ol

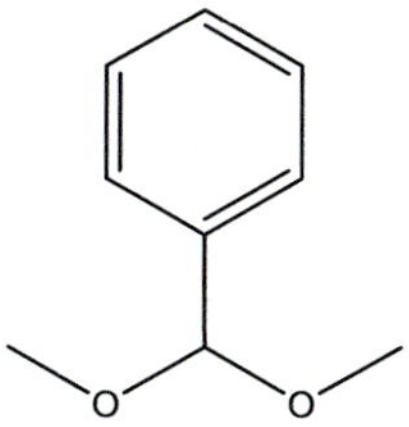

Benzaldehyde dimethyl acetal

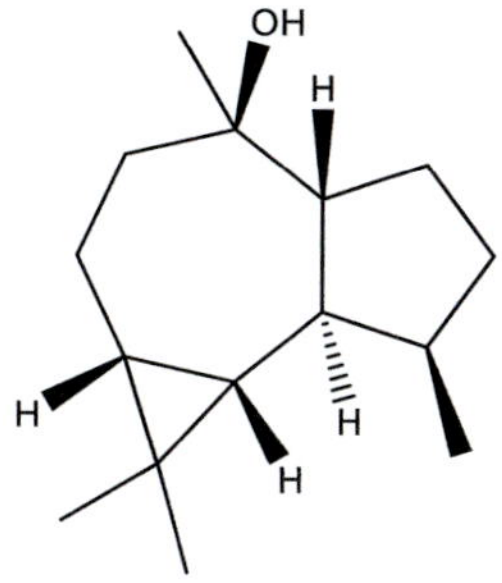

Globulol

FIGURE 96.1.4 Chemical components of plant belonging to genus *Dionysia*.

97 Genus *Dipcadi*

97.1 *Dipcadi erythraeum* Webb & Berthel.

97.1.1 Synonyms

Ornithogalum erythraeum & *Uropetalon erythraeum*

97.1.2 Family

Asparagaceae

97.1.3 Vernacular Name

Hansalait

97.1.4 Botanical Description

A bulbous, annual plant reaches up to 25 cm with linear and succulent leaves. It has white to dark flowers arranged along a long floral stem (18 cm). Fruits: large capsule divided into three parts containing black seeds.

97.1.5 Geographical Distribution

Dipcadi erythraeum is distributed from Egypt eastward through the Arabian Peninsula to Western India. In Oman, it is found in the Northern Oman Mountains, including Musandam.

97.1.6 Chemical Composition

Phytochemical analysis of various Dipcadi species has unveiled the existence of alkaloids, tannins, saponins, and flavonoids [1504]. The bulb extract of *D. erythraeum* yielded the identification of 22 phenolic compounds, representing the initial characterization of these constituents within this species. Apparently, 14 of these compounds were recognized as C-glycosyl flavonoids [1505].

Defatted aqueous methanol extract of *Dipcadi erythraeum* is also reported for the presence of kaempferol, quercetin, quercetin 3-O-(6″-α-rhamnopyranosyl)-β-glucopyranoside-7-O-α-rhamnopyranoside, vitexin, isovitexin, orientin and isoorientin (Figure 97.1.1) [1504]. A plant belonging to a similar genus, such as *Drypetes littoralis* have been reported for the presence of new tricyclic diterpenes, drypetenones A, B, and C, and one new xanthone [1506].

97.1.7 Uses

The bulbs and capsules are consumed in their raw form as a food source during periods of famine. The leaves exhibit laxative properties and are employed topically as an ointment for wound care. The entire plant is utilized in traditional medicine for its efficacy in treating coughs, biliousness, diabetes, urinary issues, and discharges [1505].

97.1.8 Adulterants or Substituents

The bulbs of *Dipcadi erythraeum* Webb & Berthel are occasionally employed as a substitute and adulterant in the formulations of *D. indica* [1507].

97.1.9 Most Cited Research

iCite, Database, NIH Open Citation Collection was used on October 21, 2023, to get the details of the most cited research. Filters for research papers were used by using the keywords *Dipcadi erythraeum*. NIH database

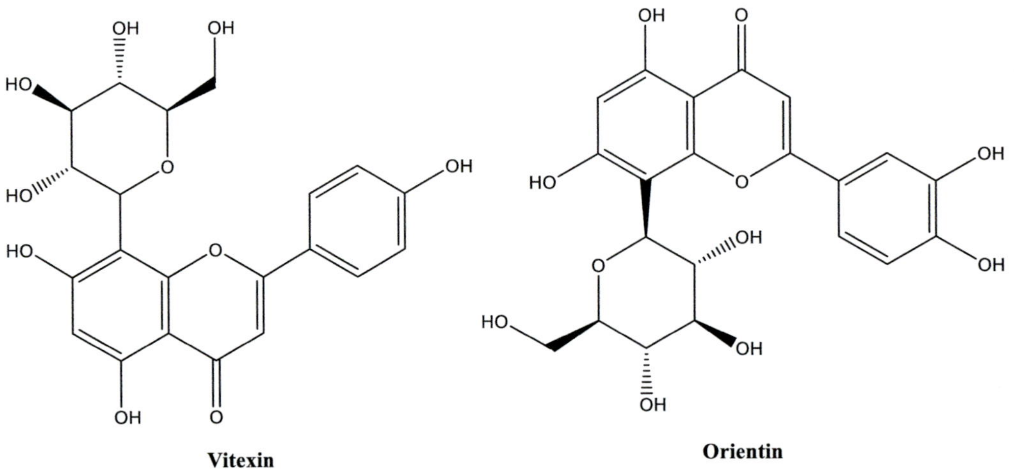

FIGURE 97.1.1 Major chemical components of *Dipcadi erythraeum*.

DOI: 10.1201/9781003295488-97

search revealed comparative cytogenetics and fluorescent chromosome banding in five Indian species of *Dipcadi* Medik [1508].

97.1.10 Patents

Rachid Ennamany—US20170319467A1—Topical composition and method of preparation, 2017. The invention is related to a method for the preparation of in vitro plant cells containing a mixture of molecules for treatment of problems associated with the skin.

97.1.11 Plant Pathology

Teliospore samples from four *Hordeum* taxa infected *Ornithogalum eigii, Ornithogalum trichophyllum*, and *Ornithogalum brachystachys*, as well as *Dipcadi erythraeum* and *Leopoldia eburnean* were reported [1509].

97.2 *Dipcadi viride* (L.) Moench.

97.2.1 Synonyms

Hyacinthus viridis, Lachenalia viridis & Scilla viridis

97.2.2 Family

Asparagaceae

97.2.3 Common Name

Dainty green bells

97.2.4 Botanical Description

The plants exhibit morphological variability, with heights ranging from 15 to 95 cm, arising from bulbs that may occasionally possess scales. Each shoot bears 1–5 leaves, which may emerge after the flowers. The leaves are linear to linear-lanceolate in shape, measuring 6–60 cm in length and 3–15 mm in width. They typically clasp at the base and display indistinct veining patterns [1510].

97.2.5 Geographical Distribution

Dipcadi viride is a widely distributed species, spanning from the Arabian Peninsula to tropical Africa and extending southwards within Africa. In Oman, it is distributed in Dhofar.

97.2.6 Chemical Composition

Various secondary metabolites, including flavonoids, alkaloids, and steroids, have been identified from distinct *Dipcadi* species through analytical techniques such as HPLC, GC-MS, and NMR spectroscopy [1511]. The phenolic and flavonoid compositions, along with the antioxidant activity, were assessed in the ethanolic extract obtained from genus *Dipcadi* bulb specimens [1512].

97.2.7 Uses

Dipcadi viride has been ethno-medically documented for its efficacy in treating jaundice, ulcers, diuretic, and purgative properties. Furthermore, plant extracts from *Dipcadi viride* have shown potential for cardiac disease management and exhibit insecticidal properties [1511].

97.2.8 Toxicity and Clinical Data

No data in the literature that refers to the toxicity of *Dipcadi viride* is reported. However, genus *Dipcadi* toxicity depends upon growth stage and geographical locality [1513].

97.2.9 Most Cited Research

iCite, Database, NIH Open Citation Collection was used on October 21, 2023, to get the details of the most cited research. Filters for research papers were used by using the keyword *Dipcadi*. NIH database search revealed leaf rust of cultivated barley: pathology and control.

97.2.10 Patents

Rachid Ennamany—US20170319467A1—Topical composition and method of preparation, 2017.

97.2.11 Plant Pathology

Leaf rust of barley is caused by the macrocyclic, heteroecious rust pathogen *Puccinia hordei*, in *Dipcadi* species [1514].

.... (*For images refer to the supplementary figures section; page no.* 301)

98 Genus *Diplotaxis*

98.1 *Diplotaxis harra* (Forssk.) Boiss.

98.1.1 Synonyms

Sinapis harra

98.1.2 Family

Brassicaceae

98.1.3 Vernacular Name

Khoshian

98.1.4 Botanical Description

Diplotaxis harra is an annual herb with erect stems that reach high up to 60 cm. Leaves grow in a circular cluster at the base of the flowering stem, elongated and egg-shaped with toothed edges. Flowers are typical daisy-like structures with bright yellow ray florets and a central disc of yellow florets. Fruits: long and narrow with beaked ends (Figures 98.1.1 and 98.1.2).

98.1.5 Geographical Distribution

Diplotaxis harra has a wide range of distribution encompassing southern and southeastern Spain, Sicily, North Africa, and stretches eastward to northern Somalia and southwestern Pakistan. It is also distributed throughout the Oman.

98.1.6 Chemical Composition

Phytochemical analysis of Diplotaxis species, particularly *Diplotaxis harra*, has revealed the presence of flavonoids,

FIGURE 98.1.1 Plant habit of *Diplotaxis harra*.

FIGURE 98.1.2 Inflorescence and fruits of *Diplotaxis harra*.

isothiocyanate-releasing glucosinolates, phenolic acids, and phytosterols. However, flavonoids are the most abundant chemical compound found in *Diplotaxis harra* [1515].

The plant contains isopropyl isothiocyanate, 3-butenyl isothiocyanate, 2-hydroxy-3-butenyl isothiocyanate, and arachidonic and palmitic acids, nonadecane, cholesterol, stigmasterol and B-sitosterol [1516]. 3-O-β-glucopyranoside-4′-O-β-xylopyranoside and kaempferol 3-O-β-glucopyranoside -4′-O-β-xylopyranoside were identified in defatted aqueous methanol extract of the whole plant *Diplotaxis harra* along with flavonols [1517].

Diplotaxis species, including *Diplotaxis acris, Diplotaxis erucoides, Diplotaxis harra*, and *Diplotaxis muralis*, also contain luteolin, kaempferol, kaempferol 3-O-β-glucopyranoside-7-O-α-rhamnopyranoside, quercetin 3-O-β-glucopyranoside, quercetin 7-O-β-glucopyranoside, isorhamnetin, and isorhamnetin 3-O-β-glucopyranoside-7-O-α-rhamnopyranoside (Figure 98.1.3) [1517].

98.1.7 Uses

Diplotaxis seeds are administered internally for endogenic thermogenesis and used topically as a rubefacient. Decoctions of *Diplotaxis harra* and *Diplotaxis pitardiana* seeds or leaves serve as remedies for scabs in animals. Glucosinolates and their derivatives exhibit bactericidal, fungicidal, and nematocidal properties [1518].

Diplotaxis harra (Forssk.) Boiss. (Brassicaceae) is traditionally used as an antidiabetic, anti-inflammatory, or anticancer agent [1519].

98.1.8 Mechanism of Action

Isorhamnetin-3-O-β-D-glucoside present in the plant showed anti-inflammatory effects mainly by suppressing

DOI: 10.1201/9781003295488-98

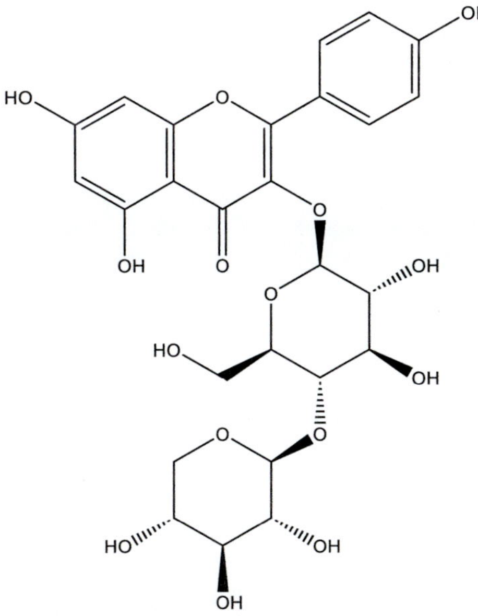

Kaempferol 3-O-β-glucopyranoside -4'-O-β-xylopyranoside

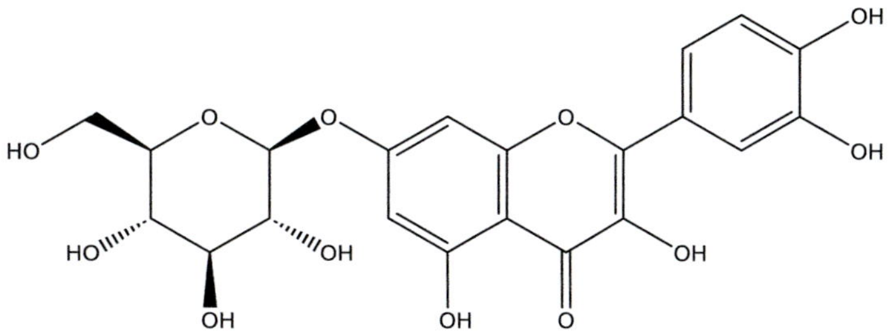

Quercetin 7-O-β-glucopyranoside

FIGURE 98.1.3 Chemical components of *Diplotaxis harra.*

the upregulation of iNOS expression and displayed a strong neuroprotectivity [1515].

Flavonoids such as isorhamnetin-3,7-di-O-glucoside present in this plant inhibit glycogen synthase kinase 3 β, overactivated in various pathologies [1519].

98.1.9 Toxicity and Clinical Data

The plant has been reported to possess analgesic properties while exhibiting a moderate level of toxicity [1520]. The alcoholic extracts and certain isolated flavonoids from the flowers and aerial parts possesses cytotoxic effects against intestinal cancer [1519].

98.1.10 Most Cited Research

iCite, Database, NIH Open Citation Collection was used on October 22, 2023, to get the details of the most cited research. Filters for research papers were used by using the keywords *Diplotaxis harra.* NIH database search revealed antimicrobial components of some Cruciferae plants (*Diplotaxis harra* Forsk. and *Erucaria microcarpa* Boiss.) [1516].

98.1.11 Patents

Danièle Hosemans, Rémi Levieil—US20150368664A1—Plants of the genus *Diplotaxis* having cytoplasmic male sterility, 2015.

98.1.12 Plant Pathology

Rhizoctonia solani, a prevalent pathogen affecting Brassicaceae, was initially documented on *Diplotaxis tenuifolia* in Italy in 2004 [1521].

99 Genus *Dipterygium* (Cleome)

99.1 *Cleome pallida* Kotschy

99.1.1 Synonyms

Dipterygium glaucum

99.1.2 Family

Cleomaceae

99.1.3 Vernacular Name

Alqa

99.1.4 Common Name

Safrawi

99.1.5 Botanical Description

The plant, reaching a maximum height of 0.6m, exhibits slender stems with limited branching and features sessile minute glands. Its oblong or elliptic leaf-blades, measuring up to 15 mm×3.5mm, bear sessile or subsessile glands on both surfaces, with a raised midrib underneath. The petiole is typically 1–1.5mm in length, although it may be longer in extremely young specimens (Figures 99.1.1 and 99.1.2) [1522].

99.1.6 Geographical Distribution

Dipterygium glaucum is distributed in regions encompassing Saudi Arabia, Egypt, Sudan, Pakistan, and India [1523]. In Oman, it is distributed throughout the country.

FIGURE 99.1.1 Seed of *Cleome pallida*.

FIGURE 99.1.2 Plant habit of *Cleome pallida*.

99.1.7 Chemical Composition

Dipterygium glaucum contains volatile alkaloids, flavonoids, coumarins, and cyanides [1524]. *D. glaucum exhibits a chemical composition characterized by the presence of steroids, terpenes, terpenoids, fatty acids,* and their esters, hydrocarbons, as well as phenolic and alcoholic compounds [1525]. Previous reports showed the presence of pigenin 6, 8-di-C-glucopyranoside (vicenin-2), quercetin-3′-O-methyl-3-O-glucopyranoside, quercetin-3′-O-methyl-3-O-galactopyranoside, quercetin-3-O-β-D-glucopyranoside, and quercetin-3-O-β-D-galactopyranoside were isolated from methanolic extract and butanol fraction of *D. glaucum* [1523] (Figure 99.1.3).

99.1.8 Uses

Traditional applications of a decoction and infusion prepared from a certain botanical source include treatment for jaundice, as a blood purifier, for managing psoriasis and ringworm infestations, and as an antiasthma remedy [1523]. *D. glaucum* possesses significant antioxidant, cytotoxic, and antimicrobial activities, which could be ascribed to its flavonoid content [1523].

99.1.9 Mechanism of Action

Terpenoids exert inhibitory effects on the production of inflammatory cytokines [1030]. The plant extract showed antidiabetic activity by inhibiting α-glucosidase and anti-Alzheimer potential by inhibiting acetylcholinesterase [1526].

DOI: 10.1201/9781003295488-99

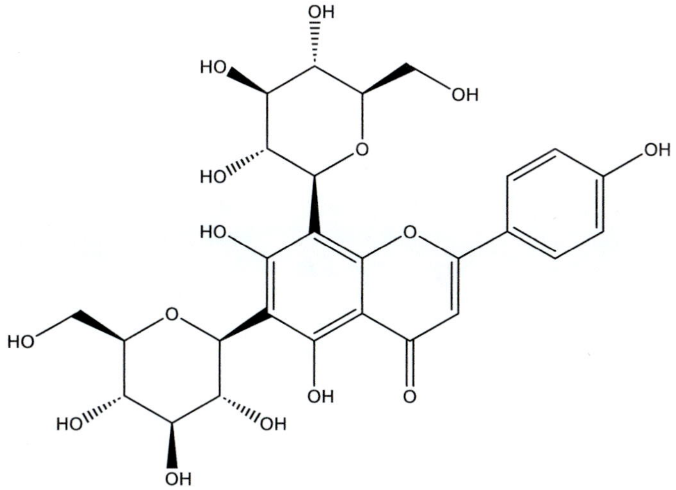

Vicenin-2

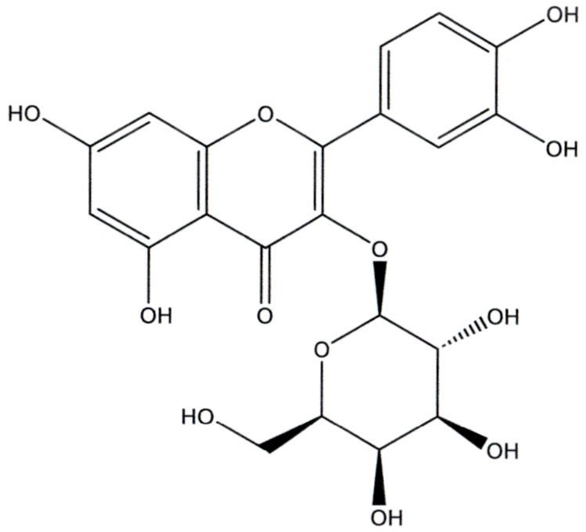

Quercetin-3-O-β-D-galactopyranoside

FIGURE 99.1.3 Chemical components of *Dipterygium glaucum*.

99.1.10 Toxicity and Clinical Data

The plant can be considered a safe and non-toxic alternative for therapeutic applications, showcasing robust antioxidant properties and significant pharmacological potential [1526]. *D. glaucum* extracts showed 45%–100% phytotoxicity as confirmed by the inhibition of Lemna minor plant growth [1527]. Genotoxicity of the plant is also reported where cytogenetic effects of *D. glaucum* on mitotic in roots of *Vicia faba* showed reduction in mitotic activity [1524].

99.1.11 Most Cited Research

iCite, Database, NIH Open Citation Collection was used on October 22, 2023, to get the details of the most cited research. Filters for research papers were used by using the keywords *Dipterygium glaucum*. NIH database search revealed the palyno-morphological investigations of subtropical endangered flora of Capparidaceae through light and scanning electron microscopy.

.... (*For additional images refer to the supplementary figures section; page no.* 302)

100 Genus *Dodonaea*

100.1 *Dodonaea viscosa* Jacq.

100.1.1 Synonyms

Dodonaea angustifolia & *Dodonaea viscosa* var. *vulgaris*

100.1.2 Family

Sapindaceae

100.1.3 Vernacular Name

Shehas

100.1.4 Common Name

Hopbush

100.1.5 Botanical Description

A shrub or small evergreen tree, attaining heights of up to 5 m, features simple, alternate, spiral leaves clustered at the twig ends. The leaves exhibit a petiole approximately 0.2 cm long, characterized by a stout, swollen base and a planoconvex cross-sectional shape. The leaves possess a raised midrib on the upper surface, accompanied by around 16 pairs of nearly parallel secondary nerves and closely reticulate tertiary nerves (Figures 100.1.1–100.1.5) [1528].

100.1.6 Geographical Distribution

Dodonaea viscosa is believed to originate in Australia, yet it exhibits a widespread distribution across the tropics and subtropics, encompassing regions in Africa, Asia, Australasia, North America, and South America [1529]. In Oman, it is found in southern and northern mountains.

FIGURE 100.1.2 Inflorescence of *Dodonaea viscosa*.

FIGURE 100.1.3 Flowering shoot of *Dodonaea viscosa*.

FIGURE 100.1.1 Flowers of *Dodonaea viscosa*.

FIGURE 100.1.4 *Dodonaea viscosa* in fruits.

DOI: 10.1201/9781003295488-100

FIGURE 100.1.5 Apical shoot with developing flowers and leaves of *Dodonaea viscosa*.

100.1.7 Chemical Composition

The plant was found to possess alkaloids, flavonoids, fixed oils and fats, steroids, phenolics, saponins, tannins, gums, mucillages, carbohydrates, reducing sugars, glycosides, and trace elements [1529]. The chemical composition of the plant is characterized by a high content of antioxidant compounds, including flavonoids and phenols. Additionally, it serves as a rich source of fatty acids, with oleic acid being the predominant constituent [1530]. The previous report showed the presence of 3, 5, 7, 3′,4′-pentahydroxyflavone and 6-hydroxy-5, 8, 9-trimethyl-18-carboxylclerodane in the leaves of *Dodonaea viscosa* [1531].

Phytochemical analysis of *Dodonaea viscosa* flowers revealed that the plant contains gallic acid, epicatechin, cumeric acid, flavonoids, oleic acid, stearic acid, ricinoleic acid, and cedrol (Figure 100.1.6) [1530]. Clerodane diterpenoids, phenolics, and three crystals were isolated from *D. viscosa* [1532]. *D. viscosa* has also been reported for the presence of isorhamnetin, kaempferol, quercetin, methyl dodovisate B, hardwickiic acid, viscosol, and dodonic acid [1533].

100.1.8 Uses

Dodonaea viscosa has been employed traditionally in the treatment of various conditions, including rheumatism, skin infections, diarrhea, abdominal pain of hepatic or splenic origin, uterine colic, and disorders related to smooth muscles. Additionally, it serves as an antipruritic agent for skin rashes, dermatitis, hemorrhoids, and sore throat. The leaf infusion is utilized to address rheumatism, gout, hemorrhoids, fractures, and snakebites [1534].

100.1.9 Mechanism of Action

Dodonaea viscosa showed cytotoxic and antiproliferative effects in colorectal cancer cells via regulation of caspase three and p53 [1533]. Saponin A isolated from Seeds of *Dodonaea viscosa* demonstrated the antifeedant mechanism [1535].

100.1.10 Toxicity and Clinical Data

The ethanol+ethyl acetate (1:1) leaf extract presented notable toxicity against brine shrimps, while a notable antiproliferative activity against THP-1 and Hep G2 cell lines was shown by ethanol+ethyl acetate extracts (1:1) of stem and root [1536].

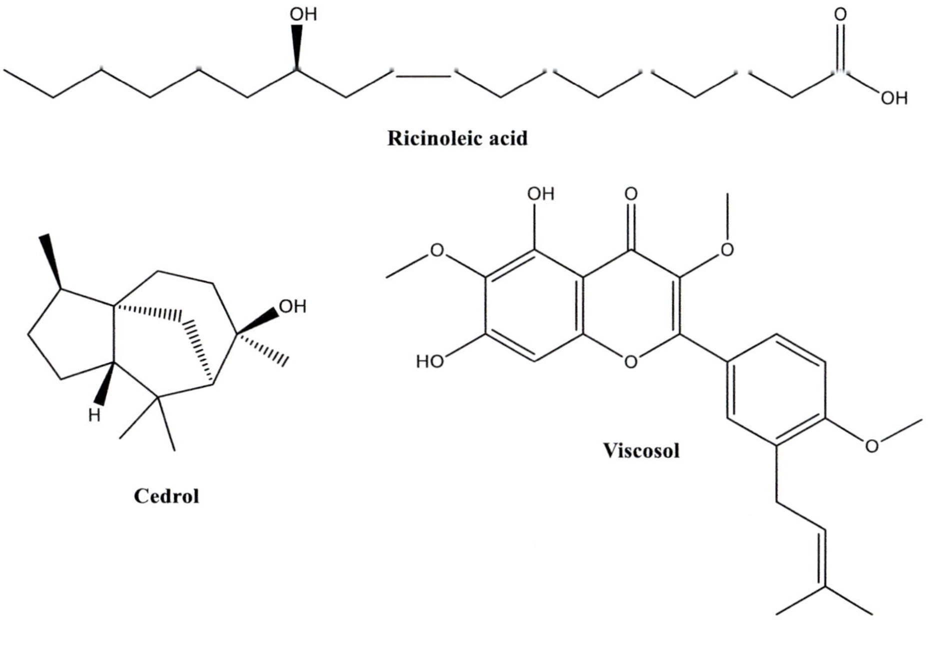

FIGURE 100.1.6 Chemical components of *Dodonaea viscosa*.

100.1.11 Most Cited Research

iCite, Database, NIH Open Citation Collection was used on October 23, 2023, to get the details of the most cited research. Filters for research papers were used by using the keywords *Dodonaea viscosa*. NIH database search revealed an evaluation of the anti-microbial and anti-inflammatory activities of the medicinal plants *Dodonaea viscosa*, *Rumex nervosus*, and *Rumex abyssinicus*.

100.1.12 Patents

Nicholas Benoit Shipley, Russ Allen Buhrow—USPP33514P3—*Prosopis* plant named *Dodonaea viscosa* "Emerald Ice," 2016. The present invention is related to a new and distinct cultivar of *Dodonaea*.

100.1.13 Plant Pathology

Dodonaea yellow disease, a phytoplasma-like affliction akin to sandalwood's "spike" disease (*Santalum album* L.), impacts *Dodonaea viscosa*. Infected plants manifest a witch's broom-like morphology with reduced leaf size and cessation of flowering and fruit production. Furthermore, *Dodonaea viscosa* serves as a host for the root-knot nematode *Meloidogyne incognita*, which poses a threat to numerous cultivated crops [1537].

101 Genus *Dorstenia*

101.1 *Dorstenia foetida* subsp. *Foetida*

101.1.1 Synonyms

Dorstenia crispa, Dorstenia radiata, & Dorstenia arabica

101.1.2 Family

Moraceae

101.1.3 Vernacular Name

Kertib

101.1.4 Common Name

Grendelion

101.1.5 Botanical Description

The plant exhibits perennial characteristics and is classified as a succulent subshrub characterized by an evergreen or caducous nature. Its stem, succulent in composition, may or may not be branched. The stem itself manifests as a robust, conical trunk with dimensions of up to 15 cm in diameter and 30–40 cm in height. Apparently, older segments of the stem often display peeling bark and possess a dark green to mahogany coloration [1282]. Leaves appeared glossy at the top of the stems. Flowerheads consist of a disc-like receptacle in the center, surrounded by ray florets.

101.1.6 Geographical Distribution

The succulent subshrub, a species found in Yemen, Saudi Arabia, Oman (in Dhofar), Somalia, Ethiopia, and Tanzania [1538].

101.1.7 Chemical Composition

D. foetida's latex is primarily attributed to the occurrence of bioactive secondary metabolites, as determined through phytochemical screening, which includes alkaloids, tannins, phenolic compounds, and coumarins [1539]. *Dorstenia foetida* is distinguished by its significant coumarin content, representing a predominant class of secondary metabolites within the plant. Plant leaves are reported to contain furanocoumarins, psoralen, bergapten, isopimpinellin, phellopterin, 5-methoxychalepensin, and turbinatocoumarin (Figure 101.1.1) [1538].

101.1.8 Uses

The juice extracted from the aerial parts of the plant is utilized in Yemen's local traditional medicine for the treatment of skin ailments, with a specific focus on leishmaniasis. In the northwest region of Ethiopia, the root is administered orally to address leprosy, liver disorders, and intestinal worm infestations among the Amhara social and cultural community. In Oman, the cooked tuberous roots serve as a dietary source [1538]. The extracts of *D. foetida* display antifungal, antibacterial, and cytotoxic activity; however,

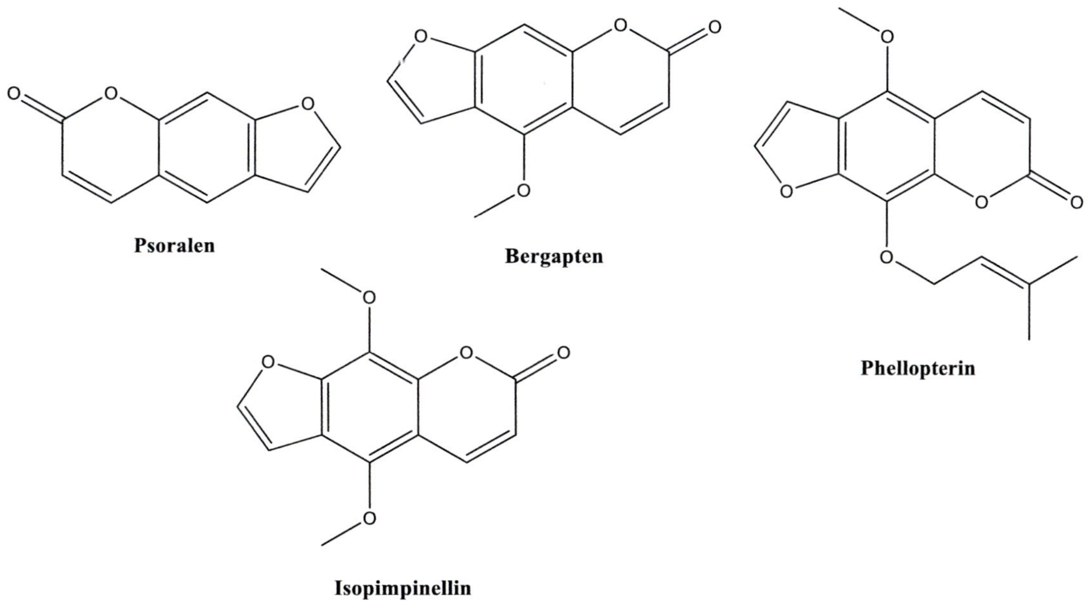

FIGURE 101.1.1 Chemical components of *Dorstenia foetida*.

DOI: 10.1201/9781003295488-101

no anthelminthic activity [1538]. Leaves of this plant have been used traditionally for skin diseases, leishmaniasis, leprosy, liver disease, intestinal worms [1538].

101.1.9 Mechanism of Action

Coumarins exert their mechanism of action through competitive antagonism of vitamin K, thereby the production of prothrombin and consequently inhibiting coagulation within the body. Due to this, coumarins act as a significant group of drugs with potent anticoagulant properties [1113].

101.1.10 Toxicity and Clinical Data

All components of this plant exhibit extreme toxicity [1540].

101.1.11 Most Cited Research

iCite, Database, NIH Open Citation Collection was used on October 23, 2023, to get the details of the most cited research. Filters for research papers were used by using the keywords *Dorstenia foetida*. NIH database search revealed the furanocoumarins from *Dorstenia foetida* [1538].

101.1.12 Patents

Philippe Ragot, Esther Pons, Bernard Mompon, Cedric Rousseau—US20200390696A1—Edible Product Comprising Reconstituted Plant Material, 2020. The present invention relates to an edible product, which comprises a fibrous plant product and a plant extract applied thereto. Further, the invention relates to a corresponding method for producing said edible product and its use in at least one of food, food supplement, medicinal, cosmetic, well-being, nutraceutical or phytotherapeutical applications. The plants used may be all plants comprising one or more substances of interest for an edible product.

101.1.13 Plant Pathology

The *Pythium* blight, affecting *Dorstenia foetida*, is a plant disease caused by a complex of fungal pathogens, including *Helminthosporium*, *Pythium*, and *Fusarium* species [1541].

.... (*For images refer to the supplementary figures section; page no.* 302)

102 Genus *Dracaena*

102.1 *Dracaena serrulata* Baker

102.1.1 Synonyms

Dracaena neoserrulata

102.1.2 Family

Asparagaceae

102.1.3 Vernacular Name

Ariyeb

102.1.4 Common Name

Arabian Dragon Tree

102.1.5 Botanical Description

Dracaena serrulata is a tree with a single trunk reaching up to 5 m. Leaves exhibit rigidity, possess an elongated and slender morphology, attach directly to the stems without stalks, and demonstrate a dense clustering pattern at the branch extremities (Figures 102.1.1 and 102.1.2).

102.1.6 Geographical Distribution

Dracaena serrulata exhibits a disjunct distribution, primarily observed along the southwestern periphery of the Arabian Peninsula. It is present in the hilly regions of southern Medina and the El Asir mountains in Saudi Arabia. Additionally, this species occurs in the foothills of Yemen and on the dry, rocky escarpments and seaward-facing cliffs particularly on north-facing slopes of Dhofar in southern Oman [1542].

FIGURE 102.1.1 Adult trees of *Dracaena serrulata*.

FIGURE 102.1.2 General growth form of *Dracaena serrulata*.

102.1.7 Chemical Composition

Dracaena species have been found to contain a diverse array of specialized metabolites, encompassing flavonoids, triterpenes, steroids, steroidal saponins, lignans, and phenolics. Apparently, the main chemical constituents of genus *Dracaena* dragon's blood are flavonoids [1543]. *Dracaena serrulata* Baker (Arabian Dragon Tree) is closely related to *D. cinnabari* [1544].

Eight components were isolated from plant belonging to the same genus plant, *Dracaena cambodiana* such as syringaresinol, pinoresinol, balanophonin, 2-(4-hydroxyphenyl)-6-(3-methoxy-4-hydroxyphenyl)-3, 7-dioxabicyclo[3.3.0]octane, 5, 7-dihydroxy-4′-methoxy-8-methylflavane, 7, 4′-dihydroxy-8-methoxyhomoisoflavane, (2R)-7, 4′-dihydroxy-8-methylflavane, and (2S)-7,3′-dihydroxy-4′-methoxyflavane [1545].

Another study showed the presence of biflavonoid cinnabarone, triflavonoids, metacyclophanes, chalcones, chalcanes, dihydrochalcones (Figure 102.1.3), sterols, and terpenoids in *D. cinnabari* (Figure 102.1.3) [1546]. SPME-GC×GC-MS analysis of resin showed the presence of δ-2-carene, p-cymene, and limonene were specific to *D. serrulata*, whereas α-pinene, camphene, β-pinene, and δ-3- carene (M6) were typical of *D. cinnabari* [1543].

102.1.8 Uses

It is commonly consumed as a vegetable, and its taste is bitter. *Dracaena serrulata* fibers were employed in historical applications, serving as essential components in camel tackle, baggage ropes, and rope pulleys for lowering heavy sacks of frankincense. These sturdy leaves also provided a significant source of fibrous material. Furthermore, the

DOI: 10.1201/9781003295488-102

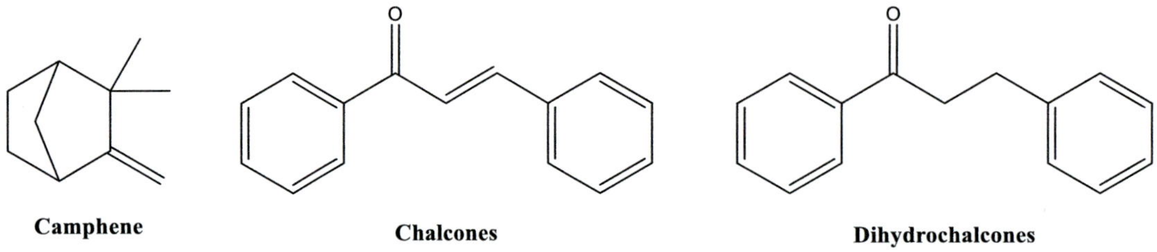

FIGURE 102.1.3 Chemical components of plants belonging to genus *Dracaena*.

wood of *Dracaena serrulata* was sporadically utilized in the construction of drums [1547].

102.1.9 Mechanism of Action

Dolichandrone serrulata flower extract ameliorates male reproductive damage in type 2 diabetic rats. It was found that treatment with *Dolichandrone serrulata* flower extract reduced high, fasted blood glucose and testicular malondialdehyde and increased testosterone levels of T2DM-treated rats [1548].

102.1.10 Toxicity and Clinical Data

Apparently, there was no data in the literature that refers to the toxicity of *Dracaena serrulata*. In the context of acute oral toxicity of *Dracaena* species, no treatment-related mortality or toxicity signs were detected [1549].

A recent study showed that *D. serrulata* flower extract had no sub-chronic toxicity on vital and reproductive structures nevertheless marginally changed some liver and kidney functions [1550].

102.1.11 Adulterants or Substituents

Durán et al. reported that based on barcoding genes (rbcL and matK) and intergenic spacers (trnQ-rps16 and rpl32-trnL), *D. cinnabari* (Socoteran) is more closely related to *D. serrrulata* (Omani species) than *D. draco* (Macaronesian) [1551,1552].

102.1.12 Most Cited Research

iCite, Database, NIH Open Citation Collection was used on October 23, 2023, to get the details of the most cited research. Filters for research papers were used by using the keywords *Dorstenia foetida*. NIH database search revealed that terpenoid profiles of resin in the genus *Dracaena* are species-specific.

102.1.13 Patents

Benoit Cyr—US9364508B2—Inhibitors of extracellular proteases, 2014. This invention pertains to the field of protease inhibitors, specifically inhibitors of extracellular proteases.

102.1.14 Plant Pathology

Panonychus citri infestations induce damage to the *Dracaena* species, affecting their growth, photosynthetic efficiency, and overall health [1553]. Endophytic fungi (*Fusarium* spp.) was isolated from *Dracaena cambodiana* [1554].

103 Genus *Ducrosia*

103.1 *Ducrosia anethifolia* (DC.) Boiss.

103.1.1 Synonyms

Zosima anethifolia & *Ducrosia olivieri*

103.1.2 Family

Apiaceae

103.1.3 Vernacular Name

Bisbas

103.1.4 Botanical Description

The herbaceous biennial plant exhibits a stature ranging from 10 to 30cm. Its stems are devoid of trichomes, presenting branching and ramified structures. The leaves are characterized by an ovate-oblong morphology. The plant bears white florets in an umbel inflorescence arrangement (Figures 103.1.1 and 103.1.2) [1555].

103.1.5 Geographical Distribution

Ducrosia centifolia is distributed from the Eastern Mediterranean and Arabian Peninsula eastward to Western Pakistan. It is found in the northern Hajar mountains, including Musandam, in Oman.

103.1.6 Chemical Composition

Phytochemical analysis of *Ducrosia anethifolia* indicates that its essential oil primarily comprises aliphatic

FIGURE 103.1.1 Whole plant of *Ducrosia anethifolia*.

FIGURE 103.1.2 Fruits of *Ducrosia anethifolia*.

aldehydes and other monoterpene hydrocarbons, while its aerial parts are characterized by the presence of coumarins, including pangolin [1556]. Furanocoumarins, prominent natural products, are abundant in diverse genera within the Apiaceae family [1557].

The phytochemical analysis of *D. anethifolia* essential oil unveiled the prevalence of α-pinene (59.2%), myrcene (11.6%), and limonene (7.5%) as its principal constituents [1558] (Figure 103.1.3).

103.1.7 Chemical Tests

The primary analytical method employed for the identification of furanocoumarins is liquid chromatography [1559].

103.1.8 Uses

The aerial parts of this herb have been utilized in Iranian traditional medicine as an analgesic, addressing conditions such as headache, backache, colic, and colds. It is purportedly efficacious in mitigating anxiety and insomnia in specific regions of Iran. Moreover, it is incorporated into numerous Persian culinary preparations for flavor enhancement [1556].

103.1.9 Mechanism of Action

Upon intercalation into the DNA double helix and subsequent molecular complexing, light-activated furanocoumarins engage in specific reactions with pyrimidine bases, apparently thymine [1308].

D. anethifolia essential oil and its major constituent α-pinene demonstrated anticonvulsant and antioxidant

DOI: 10.1201/9781003295488-103

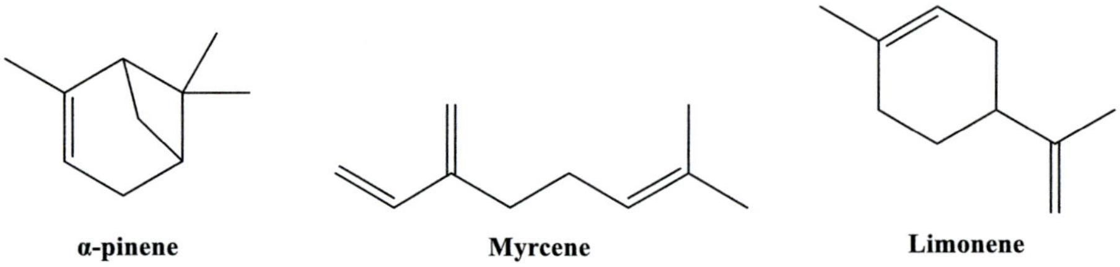

FIGURE 103.1.3 Chemical components of *Ducrosia anethifolia*.

effects in rat model of pentylenetetrazole-induced convulsions by suppressing enzymatic changes caused by pentylenetetrazole [1560].

103.1.10 Toxicity and Clinical Data

The essential oil of *Ducrosia anethifolia* has demonstrated negligible toxicity in both cuttings and layering propagation methods [1561].

103.1.11 Most Cited Research

iCite, Database, NIH Open Citation Collection was used on October 24, 2023, to get the details of the most cited research. Filters for research papers were used by using the keywords *Ducrosia anethifolia*. NIH database search revealed the preliminary in vitro and in vivo evaluation of the antidiabetic activity of *Ducrosia anethifolia* Boiss. and its linear furanocoumarins.

103.1.12 Patents

German, French—EP3217966A1—Compositions having biocidal properties and containing compounds extracted from tropical plants, 2015. This invention is related to the use of a biocidal agent with as composition containing a combination of active agents such as 2-dodecanal, nonanal, dodecanal, or decanal.

104 Genus *Plumbago* (formally know as Dyerophytum)

104.1 (*Plumbago arabica* (Boiss.) *Christenh. & Byng*)

104.1.1 Synonyms

Dyerophytum indicum

104.1.2 Family

Plumbaginaceae

104.1.3 Vernacular Name

Melehlah, gheli, melhan, & emloh Mellah

104.1.4 Botanical Description

This erect or climbing undershrub, reaching a maximum height of 1.5 m, is a glaucous and perennial plant with spherical branches. Its leaves, measuring up to 6×4 cm, are characterized by an ovate to nearly round shape, with blunt or flat apices and wavy margins. They are leathery and possess a rough scale covering [1562]. While the inflorescence consists of panicle-like spikes, approximately 2.8 cm in length, bearing yellow flowers with bracts. The calyx is tubular, measuring 8–9 mm long, prominently ribbed, and adorned with five lance-shaped teeth, each reaching half the length of the tube. The flower tube is cylindrical and extends to 1.5 cm, with petals being ovate-oblong, 8 mm in length, and either notched or terminating in a minute point. The stamens protrude prominently, and the style is characterized by its thread-like structure and a hairy base (Figures 104.1.1–104.1.3) [1562].

FIGURE 104.1.1 Inflorescence of *Plumbago arabica*.

FIGURE 104.1.2 Branches of *Plumbago arabica*.

FIGURE 104.1.3 New growing leaves of *Plumbago arabica*.

104.1.5 Geographical Distribution

It exhibits a wide distribution across the Arabian Peninsula and Western India. In the Sultanate of Oman, this plant is frequently encountered within wadis, which are desert and mountain erosion valleys characterized by intermittent stream flows and can thrive at altitudes as high as 1,800 m [1563].

DOI: 10.1201/9781003295488-104

104.1.6 Chemical Composition

Phytochemicals, including alkaloids, glycosides, tannins, carbohydrates, terpenoids, flavonoids, phenolic acids, and saponins, have been identified in various plant species [1564].

104.1.7 Chemical Tests

The receptor RferOr6 functions as an antenna-biased receptor (antenna-biased olfactory receptor) in Red Palm Weevil for the detection of the volatile compound α-pinene present in non-palm plant species [1565].

104.1.8 Uses

Dyerophytum indicum root extracts have been found to have applications in traditional medicine for treating asthma and chest infections. Pastes derived from *Dyerophytum indicum* roots serve as remedies for skin infections, scabies, and moles, while forehead application is employed to alleviate headaches. Early shoots and stems of *Dyerophytum indicum* are primarily utilized for therapeutic purposes, often in the form of dried powders, aiding in various respiratory conditions [1566].

The aromatic aerial parts of *Dyerophytum indicum* have been used in the traditional folk medicine in Saudi Arabia to treat skin infections and as a repellent agent for insects and reptilians [1567]. Usually, plants belonging to this genus are used as pain killers and in common cold treatments in traditional medicine.

104.1.9 Mechanism of Action

α-Pinene can stimulate apoptosis through several mechanisms. These include the initial disruption of mitochondrial function, the generation of reactive oxygen species (ROS), enhancement of caspase-3 activity, aggregation of heterochromatin, disintegration of DNA, and the exposure of phosphatidylserine on the cell surface [1568].

104.1.10 Toxicity and Clinical Data

Apparently, there is no data in the literature that refers to the toxicity of *Ducrosia anethifolia*. However, keeping in view the chemical profile of species, it might be considered toxic.

104.1.11 Adulterants or Substituents

Dyerophytum indicum can be substituted with *Plumbago zeylanica* [1566]

104.1.12 Most Cited Research

iCite, Database, NIH Open Citation Collection was used on October 24, 2023, to get the details of the most cited research. Filters for research papers were used by using the keywords *Dyerophytum indicum*. An NIH database search revealed the characterization of a mineral coating of the *Dyerophytum indicum* plant.

Appendix Miscellaneous medicinal plants

1. *Anabasis setifera* Moq.

Latin Name: *Anabasis setifera* Moq.

Synonyms: *Salsola setifera, Salsola lutea & Seidlitzia lanigera*
Family: Amaranthacea
Vernacular name: Shenan & shearan

FIGURE 1 Flowering shoot of *Anabasis setifera*.

FIGURE 2 Flowering shoot of *Anabasis setifera*.

FIGURE 3 Plant habit of *Anabasis setifera*.

BOTANICAL DESCRIPTION

A perennial herb, branched from the base, with fleshy stems growing up to 70 cm tall. Leaves are opposite and succulent. Flowers are greenish white, formed in clusters in the leaf axils. Fruits are pink or reddish-brown, each with five wings (Figures 1–3).

GEOGRAPHICAL DISTRIBUTION

The native range of this species extends from Egypt eastwards through the Arabian Peninsula to northwestern India. Within Oman, it is found throughout the country, primarily in sandy and gravelly habitats, particularly in saline environments such as sabkhas.

USES

Anabasis setifera was reported from a few localities in Oman, where, in the past, its aerial parts were used as a cleaning agent and for traditional medicinal purposes. Several related species within the genus *Anabasis* have documented traditional medicinal uses in parts of the Middle East and North Africa.

2. *Cynomorium coccineum* L.

Latin Name: *Cynomorium coccineum* L.

Synonyms: *Cynomorium coccineum* subsp. *songaricum* and *Cynomorium coccineum* subsp. *coccineum*

Family: Cynomoriaceae

Vernacular name: Tarthuth

Common Name: Desert Thumb, Maltese mushroom, or Maltese fungus

BOTANICAL DESCRIPTION

A non-photosynthetic parasitic plant with reddish-brown stems, growing up to approximately 30 cm tall. Stems and flowers are dark red to purplish in color. It has fleshy underground rhizomes that function as storage organs and facilitate nutrient absorption from host plants (Figures 4–6).

GEOGRAPHICAL DISTRIBUTION

It is naturally distributed from the Canary Islands and the Mediterranean region to Mauritania and Mongolia. In Oman, it occurs throughout the country, particularly in sandy and coastal habitats.

FIGURE 4 *Cynomorium coccineum* individuals in flowering season.

USES

Cynomorium coccineum is recorded from some parts of Oman, where it has been traditionally used for both medicinal purposes and as a source of dyes. Its underground parts (rhizome/root) were historically harvested for these uses.

FIGURE 5 Natural habit of *Cynomorium coccineum* in flowering time.

FIGURE 6 *Cynomorium coccineum* individuals emerging in groups following seasonal rainfall.

Supplementary figures

CHAPTER 4

Figure (4a) Plant habit of Acalypha indica

FIGURE 4A Plant habit of *Acalypha indica*

FIGURE 30B Shows a dense community of *Artemisia sieberi* growing on an abandoned terrace in Musandam

CHAPTER 30

Figure (30a-c)

FIGURE 30A Fruiting Shoot of *Artemisia sieberi*

FIGURE 30C *Artemisia sieberi* in fruits

CHAPTER 32

Figure (32a-c) Fruit, Flower, and plant habit of *Asperugo procumbens*

FIGURE 32A Fruit of *Asperugo procumbens*

FIGURE 32B Plant habit of *Asperugo procumbens*

FIGURE 32C Flowers of *Asperugo procumbens*

CHAPTER 34

Figure (34a-c) Capsules, flowering branches and flowers of *Asphodelus fistulosus*

FIGURE 34A Ripe capsules of *Asphodelus fistulosus*

FIGURE 34B Flowering shoots of *Asphodelus fistulosus*

FIGURE 34C Flowering stalk of *Asphodelus fistulosus*

CHAPTER 50

Figure (50a, b) Habit and flowering shoot of *Cadaba linearifolia*

FIGURE 50A Habit of *Cadaba linearifolia*

FIGURE 50B Flowering shoot of *Cadaba linearifolia*

CHAPTER 85

Figure (85 a, b) Fruit and flower of *Croton confertus*

FIGURE 85A Fruit of *Croton confertus*

FIGURE 85B Flowers of *Croton confertus*

CHAPTER 97

Figure (97 a, b and c) Inflorescence, Fruit and bulb of *Dipcadi erythraeum*

FIGURE 97A Inflorescence of *Dipcadi erythraeum*

FIGURE 97B Fruit of *Dipcadi erythraeum*

FIGURE 97C Bulb of *Dipcadi erythraeum*

CHAPTER 99

Figure (99 a) Flower of *Cleome pallida*

FIGURE 99A Flower of *Cleome pallida*

CHAPTER 101

Figure (101 a, b) Disc-like receptacle and plant habit of *Cleome pallida*

FIGURE 101A Disc-like receptacle of *Dorstenia foetida*

FIGURE 101B Habit of *Dorstenia foetida*

References

1. Vavilov, N.I., *Studies on the Origin of Cultivated Plants*. 1926: Institut de Botanique Appliquée et d'Amélioration des Plantes.
2. Lamont, W.J., Okra-A versatile vegetable crop. *Horttechnology-Alexandria Va-*, 1999. **9**: pp. 179–184.
3. Tindall, H.D., *Vegetables in the Tropics*. 1983: Macmillan Press Ltd.
4. Saran, S., et al., *Indian Bioresource Information Network (IBIN)*. Remote Sensing of Northwest Himalayan Ecosystems, 2019: pp. 251–264.
5. Durazzo, A., et al., Abelmoschus esculentus (L.): bioactive components' beneficial properties – Focused on antidiabetic role – For sustainable health applications. *Molecules*, 2018. **24**(1): p. 38.
6. Savello, P.A., F.W. Martin, and J.M. Hill, Nutritional composition of okra seed meal. *Journal of Agricultural and Food Chemistry*, 1980. **28**(6): pp. 1163–1166.
7. Çalışır, S., et al., A study on some physico-chemical properties of Turkey okra (Hibiscus esculenta L.) seeds. *Journal of Food Engineering*, 2005. **68**(1): pp. 73–78.
8. Adelakun, O., et al., Influence of pre-treatment on yield chemical and antioxidant properties of a Nigerian okra seed (Abelmoschus esculentus moench) flour. *Food and Chemical Toxicology*, 2009. **47**(3): pp. 657–661.
9. Islam, M.T., Phytochemical information and pharmacological activities of Okra (Abelmoschus esculentus): a literature-based review. *Phytotherapy Research*, 2019. **33**(1): pp. 72–80.
10. Miller, A.G. and M. Morris, *Plants of Dhofar, the Southern Region of Oman: Traditional, Economic, and Medicinal Uses*. Prepared and published by the Office of the Adviser for Conservation of the Environment, Diwan of Royal Court, Sultanate of Oman, 1998.
11. Liao, H., et al., Analysis and comparison of the active components and antioxidant activities of extracts from Abelmoschus esculentus L. *Pharmacognosy Magazine*, 2012. **8**(30): p. 156.
12. Khomsug, P., et al., Antioxidative activities and phenolic content of extracts from okra (Abelmoschus esculentus L.). *Research Journal of Biological Sciences*, 2010. **5**(4): pp. 310–313.
13. *Uses of Abelmoschus esculentus*. Available from: https://www.researchgate.net/publication/313572725_Okrra_Abelmoschus_esculentus_L_Moench
14. Kumar, A., P. Kumar, and R. Nadendla, A review on: Abelmoschus esculentus (Okra). *International Research Journal of Pharmaceutical and Applied Sciences*, 2013. **3**(4): pp. 129–132.
15. Archana, G., et al., Evaluation of fat uptake of polysaccharide coatings on deep-fat fried potato chips by confocal laser scanning microscopy. *International Journal of Food Properties*, 2016. **19**(7): pp. 1583–1592.
16. Costantino, A. and J. Romanchik-Cerpovicz, Physical and sensory measures indicate moderate fat replacement in frozen dairy dessert is feasible using okra gum as a milk-fat ingredient substitute. *Journal of the American Dietetic Association*, 2004(104): p. 44.
17. Yuennan, P., T. Sajjaanantakul, and H.D. Goff, Effect of okra cell wall and polysaccharide on physical properties and stability of ice cream. *Journal of Food Science*, 2014. **79**(8): pp. E1522–E1527.
18. Elkhalifa, A.E.O., et al., Okra (Abelmoschus esculentus) as a potential dietary medicine with nutraceutical importance for sustainable health applications. *Molecules*, 2021. **26**(3): p. 696.
19. Abbas, A., et al., Possible antidiabetic mechanism of action of ex-maradi okra fruit variety (Abelmoscus esculentus) on alloxan induced diabetic rats. *Nigerian Journal of Basic and Applied Sciences*, 2017. **25**(2): pp. 101–113.
20. Majd, N.E., et al., Okra (Abelmoscus esculentus) improved islets structure, and down-regulated PPARs gene expression in pancreas of high-fat diet and streptozotocin-induced diabetic rats. *Cell Journal (Yakhteh)*, 2018. **20**(1): p. 31.
21. Dantas, T.L., F.C. Alonso Buriti, and E.R. Florentino, Okra (Abelmoschus esculentus L.) as a potential functional food source of mucilage and bioactive compounds with technological applications and health benefits. *Plants*, 2021. **10**(8): p. 1683.
22. Panighel, G., et al., Investigating the in vitro mode of action of okra (Abelmoschus esculentus) as hypocholesterolemic, anti-inflammatory, and antioxidant food. *Food Chemistry: Molecular Sciences*, 2022. **5**: p. 100126.
23. Chandra, S., R. Saha, and P. Pal, Arsenic uptake and accumulation in okra (Abelmoschus esculentus) as affected by different arsenical speciation. *Bulletin of Environmental Contamination and Toxicology*, 2016. **96**: pp. 395–400.
24. Uebelhack, R., et al., Double-blind, randomized, three-armed, placebo-controlled, clinical investigation to evaluate the benefit and tolerability of two dosages of IQP-AE-103 in reducing body weight in overweight and moderately obese subjects. *Journal of Obesity*, 2019. (1): p. 3412952.
25. Saatchi, A., et al., Anti-hyperglycemic effect of Abelmoschus culentesus (Okra) on patients with diabetes type 2: a randomized clinical trial. *Phytotherapy Research*, 2022. **36**(4): pp. 1644–1651
26. Moradi, A., et al., The effect of okra (Abelmoschus esculentus) on lipid profiles and glycemic indices in Type 2 diabetic adults: randomized double blinded trials. *Phytotherapy Research*, 2020. **34**(12): pp. 3325–3332.
27. Akoroda, M., Allogamy, varietal adulteration and the breeding of okra (Abelmoschus esculentus (L.) Moench.) in Nigeria. *The Journal of Agricultural Science*, 1986. **106**(2): pp. 313–321.
28. Sabitha, V., et al., Antidiabetic and antihyperlipidemic potential of Abelmoschus esculentus (L.) Moench. in streptozotocin-induced diabetic rats. *Journal of Pharmacy and Bioallied Sciences*, 2011. **3**(3): pp. 397.
29. Jose, J. and R. Usha, Bhendi yellow vein mosaic disease in India is caused by association of a DNA β satellite with a begomovirus. *Virology*, 2003. **305**(2): pp. 310–317.
30. *Botanical Description of Abutilon fruticosum*. Available from: https://plants.jstor.org/stable/10.5555/al.ap.flora.flos000282
31. Bano, I. and G. Deora, Preliminary phytochemical screening and GC-MS analysis of methanolic leaf extract of

Abutilon pannosum (Forst. F.) Schlect. from Indian Thar desert. *Journal of Pharmacognosy and Phytochemistry*, 2019. **8**(1): pp. 894–899.
32. *Uses of Abutilon fruticosum*. Available from: https://www.wildflower.org/plants/result.php?id_plant=abfr3
33. Khadabadi, S. and N. Bhajipale, A review on some important medicinal plants of Abutilon spp. *Research Journal of Pharmaceutical, Biological and Chemical Sciences*, 2010. **1**(4): pp. 718–729.
34. Baquar, S.R., *Medicinal and Poisonous Plants of Pakistan* 1989. Printas.
35. Pingale, S. and P. Virkar, Evaluation of acute toxicity for Abutilon indicum. *Der Pharmacia Lettre*, 2011. **3**(3): pp. 37–42.
36. Alzahrani, D.A., Complete chloroplast genome of Abutilon fruticosum: genome structure, comparative and phylogenetic analysis. *Plants*, 2021. **10**(2): p. 270.
37. Vrandecic, K., et al., Isolation of Diaporthe phaseolorum var. caulivora from Abutilon theophrasti in Croatia. *Plant Pathology*, 2005. **54**(4): pp. 576–576.
38. *Botanical Description of Abutilon bidentatum*. Available from: https://plants.jstor.org/collection/FLOS
39. Tripathi, R., Phytochemical and pharmacological studies of abutilon bidentatum hochst. *International Journal of Current Pharmaceutical Research*, 2017. **9**(3): pp. 69–70.
40. Gomaa, A.A.-R., et al., Phytochemistry and pharmacological activities of genus Abutilon: a review (1972-2015). *Journal of Advanced Biomedical and Pharmaceutical Sciences*, 2018. **1**(2): pp. 56–74.
41. Migahid, A.M., @ *Flora of Saudi Arabia: Dicotyledons*. University, 1978.
42. Jain, R., S. Jain, and R. Arora, A new cholestane derivative of Abutilon bidentatum Hochst. and its bioactivity. *Pharmazie*, 1996. **51**(4): pp. 253–253.
43. Al-Hamoud, G.A., et al., Abubidentin A, New Oleananetype Triterpene Ester from Abutilon bidentatum and its antioxidant, cholinesterase and antimicrobial activities. *PeerJ*, 2022. **10**: p. e13040.
44. Gomaa, A., et al., Pharmacognostical studies of leaf, stem, root and flower of Abutilon hirtum (Lam.) Sweet. *International Journal of Pharmacognosy and Phytochemical Research*, 2016. **8**(1): pp. 199–216.
45. Ali, B., et al., Pakistamide C, a new sphingolipid from Abutilon pakistanicum. *Revista Brasileira de Farmacognosia*, 2014. **24**: pp. 277–281.
46. Agrawal, T., Abutilon a vulnerable genus: a review. *Research and Reviews: Journal of Pharmacognosy and Phytochemistry*, 2017. **5**(1): pp. 44–46.
47. Al-Shanwani, M., *Plants used in Saudi folk medicine*, vol. 146. Riyadh: King Abdul Aziz City for Science and Technology, 1996.
48. Yasmin, S., M. Akram Kashmiri, and K. Anwar, Screening of Aerial Parts of Abutilon bidentatum for Hepatoprotective Activity in Rabbits. *Journal of Medicinal Plants Research*, 2011. 5(3): pp. 349–353.
49. Survase, S., M. Jamdhade, and S. Chavan, Antibacterial activity of Abutilon bidentatum (Hochst.) leaves. *Science Research Reporter*, 2012. **2**(1): pp. 38–40.
50. Shahwar, D., et al., Antioxidant activities of the selected plants from the family Euphorbiaceae, Lauraceae, Malvaceae and Balsaminaceae. *African Journal of Biotechnology*, 2010. **9**(7): pp. 1086–1096.
51. Akapa, T.C., et al., Antipyretic activity of Abutilon mauritianum (Jacq.) roots in Wistar rats. *International Journal of Pharmaceutical Sciences and Research*, 2014. **5**(2): pp. 42–45.
52. Singh, D. and R.S. Gupta, Modulatory influence of Abutilon indicum leaves on hepatic antioxidant status and lipid peroxidation against alcohol-induced liver damage in rats. *Pharmacology Online*, 2008. **1**: pp. 253–262.
53. Mei, B.C. and J.W. Lyga, *Abutilon Indicum Extracts and Methods of Use*. 2017.
54. Liu, Y., et al., Role of endophytic bacteria in improving plant stress resistance. *Frontiers in Plant Science*, 2022. **13**: p. 1106701.
55. Kassem, H., Study of polyphenolic components and macro- and micromorphological characters of Abutilon hirtum (Lam.) Sweet. *Bulletin of Faculty of Pharmacy*, 2007. **45**(3): pp. 173–83.
56. Gomaa, A.-R., et al., Chemical constituents of the leaves of abutilon hirtum and their antimicrobial activity. *Chemistry of Natural Compounds*, 2022. **58**(2): pp. 368–370.
57. Vivekraj, P., A. Vijayan, and V. Anandgideon, Phytochemical analysis of leaf extract of Abutilon hirtum (Lam.) sweet by GC-MS method. *International Journal of Pharmacological Research*, 2015. **5**(8): pp. 167–171.
58. Vivekraj, P., A. Vijayan, and V. Anandgideon, Phytochemical analysis of Leaf extract of Abutilon hirtum (Lam.) sweet by GC-MS method. *International Journal of Pharmacological Research*, 2015. **5**(8): pp. 167–171.
59. Mohammed, N.M., et al., Comparative lipoidal matter investigation of three abutilon species aerial parts. *Journal of Advanced Pharmacy Research*, 2021. **5**(4): pp. 362–370.
60. Wesley, P.S., et al., In vitro phytochemical screening, free radical scavenging activity and anticancer activity of Abutilon hirtum (Lam.) sweet (Malvaceae). *International Journal of PharmTech Research*, 2013. **5**(1): pp. 155–161.
61. Brink, M. and E. Achigan-Dako, *Fibres*, vol. 16. PROTA, 2012.
62. Reddy, S., Hepatoprotective potential of Abutilon hirtum Sweet leaves in carbon tetrachloride induced hepatotoxicity. *Asian Journal of Biomedical and Pharmaceutical Sciences*, 2011. **1**(3).
63. Kassem, H., Investigation of lipids, mucilage and cytotoxic activity of Abutilon hirtum (Lam.) Sweet grown in Egypt. *Bulletin of Faculty of Pharmacy*, 2001. **39**: pp. 156–159.
64. Gomaa, A.A.-R., et al., Anti-inflammatory, analgesic, antipyretic and antidiabetic activities of Abutilon hirtum (Lam.) Sweet. *Clinical phytoscience*, 2018. **4**(1): pp. 1–7.
65. HA, K., Investigation of lipids, mucilage and cytotoxic activity of the leaves of Abutilon hirtum lam. *Growing in Egypt*, 2001. 39(1): pp.165–69.
66. Maryami, M., et al., Preparation of the Ag/RGO nanocomposite by use of Abutilon hirtum leaf extract: a recoverable catalyst for the reduction of organic dyes in aqueous medium at room temperature. *International Journal of Hydrogen Energy*, 2016. **41**(46): pp. 21236–21245.
67. Gomaa, A., et al., Antimicrobial, antimalarial and antileishmanial activities of Abutilon hirtum. *Journal of advanced Biomedical and Pharmaceutical Sciences*, 2021. **4**(3): pp. 177–181.
68. Hossain, S., et al., Pesticidal action of Abutilon hirtum (Lam.) sweet against Tribolium castaneum (Hbst.) and Lymnaea acuminata Lamarck through dose-mortality assay. *Journal of Pharmacognosy and Phytochemistry*, 2018. **7**(2): pp. 1453–1455.
69. Vélez-Gavilán, J., *Abutilon hirtum (Indian mallow)*, Vol. CABI Compendium. CABI, 2022.
70. Paul, K., et al., Structural activity comparison of Abutilon indicum and synthetic NSAIDs by using spectroscopy and

chromatography techniques. *Materials Today: Proceedings*, 2022. **51**: pp. 651–656.
71. Musthafa, S.A., et al., Syringaldehyde isolated from Abutilon indicum Linn. leaves exhibits broad spectrum anti-microbial activity. *Materials Today: Proceedings*, 2022. **50**: pp. 335–339.
72. Kailasam, K.V., Abutilon indicum L (Malvaceae)-medicinal potential review. *Pharmacognosy Journal*, 2015. **7**(6).
73. Wu, X. and S. Dhanasekaran, Protective effect of leaf extract of Abutilon indicum on DNA damage and peripheral blood lymphocytes in combating the oxidative stress. *Saudi Pharmaceutical Journal*, 2020. **28**(8): pp. 943–950.
74. Khan, S.A., et al., Phytomolecules-coated NiO nanoparticles synthesis using Abutilon indicum leaf extract: antioxidant, antibacterial, and anticancer activities. *International Journal of Nanomedicine*, 2021: pp. 1757–1773.
75. Umavathi, S., et al., Facile synthesis and characterization of ZnO nanoparticles using Abutilon indicum leaf extract: an eco-friendly nano-drug on human microbial pathogens. *Journal of Drug Delivery Science and Technology*, 2021. **66**: p. 102917.
76. Krisanapun, C., et al., Aqueous extract of abutilon indicum sweet inhibits glucose absorption and stimulates insulin secretion in rodents. *Nutrition Research*, 2009. **29**(8): pp. 579–587.
77. Dashputre, N. and N. Naikwade, Immunomodulatory activity of Abutilon indicum linn on albino mice. *International Journal of Pharmaceutical Sciences and Research*, 2010. **1**(3): pp. 178–184.
78. Kaushik, P., D. Kaushik, and S.L. Khokra, In vivo antioxidant activity of plant Abutilon indicum. *Journal of Pharmaceutical Education and Research*, 2011. **2**(1): p. 50.
79. Shama, I.A., A.A. Abobakr, and E.E. Kamal, Toxicity of Abutilon glacum seeds' extracts (water and methanol) on Wistar Rats. *British Journal of Pharmacology and Toxicology*, 2012. **3**(6): pp. 273–277.
80. Ijaz, F., et al., Green synthesis of copper oxide nanoparticles using Abutilon indicum leaf extract: antimicrobial, antioxidant and photocatalytic dye degradation activitie. *Tropical Journal of Pharmaceutical Research*, 2017. **16**(4): pp. 743–753.
81. Abdul Rahuman, A., et al., Isolation and identification of mosquito larvicidal compound from Abutilon indicum (Linn.) Sweet. *Parasitology Research*, 2008. **102**: pp. 981–988.
82. Porchezhian, E. and S. Ansari, Hepatoprotective activity of Abutilon indicum on experimental liver damage in rats. *Phytomedicine*, 2005. **12**(1–2): pp. 62–64.
83. Banso, A. and S. Adeyemo, Phytochemical screening and antimicrobial assessment of Abutilon mauritianum, Bacopa monnifera and *Datura stramonium*. *Biokemistri*, 2006. **18**(1): pp. 39–44.
84. Uroko, R.I., et al., Combined ethanol extract of Funtumia africana and Abutilon mauritianium leaves improves the lipid profile and kidney function indices of benign prostatic hyperplasia in rats. *Acta Scientiarum Polonorum Technologia Alimentaria*, 2020. **19**(4): pp. 395–404.
85. Uroko, R.I., et al., Effects of combined ethanolicic extracts of Funtumia Africana and Abutilon mauritianum leaves on sex Hormones and prostate indices of benign prostatic Hyperplasia induced rats. *Animal Research International*, 2020. **17**(3): pp. 3892–3902.
86. Suroowan, S., K. Pynee, and M. Mahomoodally, A comprehensive review of ethnopharmacologically important medicinal plant species from Mauritius. *South African Journal of Botany*, 2019. **122**: pp. 189–213.
87. Khalil, I., et al., Evaluation of biological activities and in vivo amelioration of CCl4 induced toxicity in lung and kidney with Abutilon pannosum (G. Forst.) Schltdl. in rat. *Journal of Ethnopharmacology*, 2020. **249**: p. 112395.
88. Aadesariya, M., V. Ram, and P. Dave, Investigation of phytochemicals in methanolic leaves extracts of Abutilon pannosum and Grewia tenax by Q-TOF LC/MS. *Progress in Chemical and Biochemical Research*, 2019. **2**(1): pp. 13–19.
89. Kittur, M., et al., Characteristics and composition ofAbutilon pannosum andHibiscus panduriformis seeds and oils. *Journal of the American Oil Chemists' Society*, 1982. **59**(3): pp. 123–124.
90. Survase, S., B. Sarwade, and D. Chavan, Antibacterial activity of various extracts of Abutilon pannosum (Forst. f.) Schlecht. leaves. *African Journal of Plant Science*, 2013. **7**(4): pp. 128–30.
91. Khalil, I., et al., Abutilon pannosum stem bark enhances the aphrodisiac activities and spermatogenesis in rat. *Andrologia*, 2019. **51**(10): p. e13404.
92. *Description of Acacia senegal*. Available from: https://apps.worldagroforestry.org/treedb/AFTPDFS/Acacia_senegal.PDF
93. Phillips, G.O. and P.A. Williams, *Handbook of Hydrocolloids*. Elsevier, 2009.
94. Babiker, R., et al., Effects of gum Arabic ingestion on body mass index and body fat percentage in healthy adult females: two-arm randomized, placebo controlled, double-blind trial. *Nutrition journal*, 2012. **11**(1): pp. 1–7.
95. Jarrar, A.H., et al., The effect of gum arabic (Acacia senegal) on cardiovascular risk factors and gastrointestinal symptoms in adults at risk of metabolic syndrome: a randomized clinical trial. *Nutrients*, 2021. **13**(1): p. 194.
96. Verbeken, D., S. Dierckx, and K. Dewettinck, Exudate gums: occurrence, production, and applications. *Applied Microbiology and Biotechnology*, 2003. **63**: pp. 10–21.
97. Patel, S. and A. Goyal, Applications of natural polymer gum arabic: a review. *International Journal of Food Properties*, 2015. **18**(5): pp. 986–998.
98. El-Garawani, I., et al., The ameliorative role of Acacia senegal gum against the oxidative stress and genotoxicity induced by the radiographic contrast medium (ioxitalamate) in albino rats. *Antioxidants*, 2021. **10**(2): p. 221.
99. Ghazanfar, S. A. (2003). *Flora of the Sultanate of Oman* (Vol. 4). National Botanic Garden (Belgium).
100. Ali, B.H., A. Ziada, and G. Blunden, Biological effects of gum arabic: a review of some recent research. *Food and Chemical Toxicology*, 2009. **47**(1): pp. 1–8.
101. Gado, A.M. and B.A. Aldahmash, Antioxidant effect of Arabic gum against mercuric chloride-induced nephrotoxicity. *Drug Design, Development and Therapy*, 2013: pp. 1245–1252.
102. Kaddam, L., et al., Gum Arabic as novel anti-oxidant agent in sickle cell anemia, phase II trial. *BMC Hematology*, 2017. **17**: pp. 1–6.
103. Tiss, A., F. Carrière, and R. Verger, Effects of gum arabic on lipase interfacial binding and activity. *Analytical Biochemistry*, 2001. **294**(1): pp. 36–43.
104. Glover, D.A., et al., Acacia (sen) SUPERGUM™(Gum Arabic): an evaluation of potential health benefits in human subjects. *Food Hydrocolloids*, 2009. **23**(8): pp. 2410–2415.
105. Ali, B.H., et al., Effect of Acacia gum on blood pressure in rats with adenine-induced chronic renal failure. *Phytomedicine*, 2011. **18**(13): pp. 1176–1180.
106. Matsumoto, N., et al., Butyrate modulates TGF-β1 generation and function: potential renal benefit for Acacia (sen)

SUPERGUM™(gum arabic)? *Kidney International*, 2006. **69**(2): pp. 257–265.

107. Calame, W., et al., Evaluation of satiety enhancement, including compensation, by blends of gum arabic. A methodological approach. *Appetite*, 2011. **57**(2): pp. 358–364.
108. Nasir, O., Renal and extrarenal effects of gum arabic (Acacia senegal)-what can be learned from animal experiments? *Kidney and Blood Pressure Research*, 2013. **37**(4-5): pp. 269–279.
109. Additives, E.P.o.F., et al., Re-evaluation of acacia gum (E 414) as a food additive. *EFSA Journal*, 2017. **15**(4): p. e04741.
110. Abu-Serie, M.M., A.F. Hamouda, and N.H. Habashy, Acacia senegal gum attenuates systemic toxicity in CCl4-intoxicated rats via regulation of the ROS/NF-κB signaling pathway. *Scientific Reports*, 2021. **11**(1): p. 20316.
111. Haukka, K., K. Lindström, and J.P.W. Young, Three phylogenetic groups of nodA and nifH genes in Sinorhizobium and Mesorhizobium isolates from leguminous trees growing in Africa and Latin America. *Applied and Environmental Microbiology*, 1998. **64**(2): pp. 419–426.
112. Morris, M., Biology of the Acacia gall rust, Uromycladium tepperianum. *Plant Pathology*, 1987. **36**(1): pp. 100–106.
113. Bakhoum, N., et al., Senegalia senegal (synonym: Acacia senegal), its importance to sub-Saharan Africa, and its relationship with a wide range of symbiotic soil microorganisms. *South African Journal of Botany*, 2018. **119**: pp. 362–368.
114. *Description of Acacia tortilis.* Available from: https://apps.worldagroforestry.org/treedb/AFTPDFS/Acacia_tortilis.PDF
115. Andersen, G.L. and K. Krzywinski, Longevity and growth of Acacia tortilis; insights from 14C content and anatomy of wood. *BMC ecology*, 2007. **7**: pp. 1–14.
116. Kumar Bhateja, P. and R. Singh, Antidiabetic activity of Acacia tortilis (Forsk.) Hayne ssp. raddiana polysaccharide on streptozotocin-nicotinamide induced diabetic rats. *BioMed Research International*, 2014. **2014:** PP. 9. http://dx.doi.org/10.1155/2014/572013
117. Ogunwande, I.A., et al., Constituents of the essential oil from the leaves of Acacia tortilis (Forsk.) Hayne. *Journal of Essential Oil Research*, 2008. **20**(2): pp. 116–119.
118. Ziani, B.E., et al., Phenolic profiling, biological activities and in silico studies of Acacia tortilis (Forssk.) Hayne ssp. raddiana extracts. *Food Bioscience*, 2020. **36**: p. 100616.
119. Seigler, D.S., Phytochemistry of Acacia – sensu lato. *Biochemical Systematics and Ecology*, 2003. **31**(8): pp. 845–873.
120. Moustafa, A. and S.R. Mansour, Improving soil physical properties and its effect on Acacia tortilis seedlings growth under field conditions. *Asian Journal of Plant Sciences*, 2003.
121. Embaby, H.E. and A.M. Rayan, Chemical composition and nutritional evaluation of the seeds of Acacia tortilis (Forssk.) Hayne ssp. raddiana. *Food Chemistry*, 2016. **200**: pp. 62–68.
122. Hidosa, D. and S. Kibiret, Research article effects of acacia tortilis pod feeding in improving performance of goats fed grass hay. *Asian Journal of Animal and Veterinary Advances*, 2020. **15**: pp. 50–59.
123. Agrawal, N.K. and U. Gupta, Evaluation of hypoglycemic and antihyperglycemic effects of Acacia tortilis seed extract in normal and diabetic rats. *International Journal of PharmTech Research*, 2013. **5**(2): pp. 330–336.
124. Gabr, S., et al., Characterization and optimization of phenolics extracts from Acacia species in relevance to their anti-inflammatory activity. *Biochemical Systematics and Ecology*, 2018. **78**: pp. 21–30.
125. de Lajudie, P., et al., Characterization of tropical tree rhizobia and description of Mesorhizobium plurifarium sp. nov. *International Journal of Systematic and Evolutionary Microbiology*, 1998. **48**(2): pp. 369–382.
126. Alqarni, A.S., A.M. Awad, and A.A. Owayss, Evaluation of Acacia gerrardii Benth.(Fabaceae: Mimosoideae) as a honey plant under extremely hot-dry conditions: flowering phenology, nectar yield and honey potentiality. *Journal of Animal and Plant Sciences*, 2015. **25**(6): pp. 1667–1674.
127. Aldhahrani, A., Suadian Acacia Gerrardii: antidiabetic effect in rats suffering from diabetic nephropathy and DNA fingerprinting using ISSR. *Pakistan Journal of Biological Sciences: PJBS*, 2020. **23**(9): pp. 1162–1175.
128. Sanchez, C., et al., Acacia gum: history of the future. *Food Hydrocolloids*, 2018. **78**: pp. 140–160.
129. Batiha, G.E.-S., et al., Bioactive compounds, pharmacological actions, and pharmacokinetics of genus Acacia. *Molecules*, 2022. **27**(21): pp. 7340.
130. Owayss, A.A., et al., In vitro antimicrobial activities of Saudi honeys originating from Ziziphus spina-christi L. and Acacia gerrardii Benth. trees. *Food Science & Nutrition*, 2020. **8**(1): pp. 390–401.
131. Al-Khalifa, A. and I. Al-Arify, Physicochemical characteristics and pollen spectrum of some Saudi honeys. *Food Chemistry*, 1999. **67**(1): pp. 21–25.
132. Aref, I., K. Elkhalifa, and L. El-Juhany, A dendrological key for identification of Acacia species growing in Saudi Arabia and Northern Sudan. *Journal of King Abdulaziz University for Meteorology, Environment and Arid Land*, 2003. **14**: pp. 87–94.
133. Dharani, N., *Field Guide to Acacias of East Africa*. 2006.
134. Awad, A.M., A.A. Owayss, and A.S. Alqarni, Performance of two honey bee subspecies during harsh weather and Acacia gerrardii nectar-rich flow. *Scientia Agricola*, 2017. **74**: pp. 474–480.
135. *Toxicity of Acacia gerrardii.* Available from: https://apps.cals.arizona.edu/arboretum/taxon.aspx?id=723
136. Hashem, A., et al., The interaction between arbuscular mycorrhizal fungi and endophytic bacteria enhances plant growth of Acacia gerrardii under salt stress. *Frontiers in Microbiology*, 2016. **7**: pp. 1089.
137. Lingaraju, D., M. Sudarshana, and N. Rajashekar, Ethnopharmacological survey of traditional medicinal plants in tribal areas of Kodagu district, Karnataka, India. *Journal of Pharmacy Research*, 2013. **6**(2): pp. 284–297.
138. Zahidin, N.S., et al., A review of Acalypha indica L.(Euphorbiaceae) as traditional medicinal plant and its therapeutic potential. *Journal of Ethnopharmacology*, 2017. **207**: pp. 146–173.
139. Jayaprakasam, R. and T. Ravi, Evaluation of anti arthritic activity of the root extract of Acalypha indica Linn. using in vitro techniques. *International Journal of Phytopharmacy*, 2012. **2**(6): pp. 169–173.
140. Takle, V., et al., Pharmacognostic and Phytochemical investigations of aerial parts of Acalypha indica Linn. *Pharmacognosy Journal*, 2011. **3**(21): pp. 33–35.
141. Gurib-Fakim, A., *Medicinal Plants of Mauritius and of the World*. A. Gurib Fakim, 2007.
142. Duke, J. and M.J. Bogenschutz, Dr. *Duke's Phytochemical and Ethnobotanical Databases*. Washington, DC: USDA, Agricultural Research Service, 1994.
143. Francisco, I.A. and M.H.P. Pinotti, Cyanogenic glycosides in plants. *Brazilian Archives of Biology and Technology*, 2000. **43**: pp. 487–492.
144. Ma, Y.T., et al., Phenolics from Acalypha indica. *Journal of the Chinese Chemical Society*, 1997. **44**(5): pp. 499–502.

145. Beena, J. and M. Molly, Anti-oxidant studies and chemical investigation of ethanol extract of Acalypha indica Linn. *Drug plants I*, 2010: pp. 261–279.
146. Sanseera, D., et al., *Antioxidant and Anticancer Activities from Aerial Parts of Acalypha indica Linn.* University of Wollongong. Faculty of Science, Medicine and Health - Papers: part A. 88. Chiang Mai University Journal of Natural Sciences 2012. **11**(2): pp. 157–168.
147. Sundararaman, P. and C. Djerassi, A convenient synthesis of progesterone from stigmasterol. *The Journal of Organic Chemistry*, 1977. **42**(22): pp. 3633–3634.
148. Nahrstedt, A., J.-D. Kant, and V. Wray, Acalyphin, a cyanogenic glucoside from Acalypha indica. *Phytochemistry*, 1982. **21**(1): pp. 101–105.
149. Stefanus, G., *Efek neuroterapi ekstrak air akar Acalypha indica Linn. pada dosis 5 mg dan 10 mg secara eks vivo pada persambungan saraf-otot gastroknemius katak Bufo melanostictus Schneider= Neuro-therapy effect of water extract from the roots of Acalypha indica Linn. in dose of 5 mg and 10 mg in neuromuscular junction of gastrocnemius muscle of frog Bufo melanostictus Schneider ex vivo.* Faculty of Medicine, University of Indonesia, Depok, 2009.
150. Taylor, J.C., L. Rapport, and G.B. Lockwood, Octacosanol in human health. *Nutrition*, 2003. **19**(2): p. 192.
151. Marwah, R.G., et al., Antioxidant capacity of some edible and wound healing plants in Oman. *Food Chemistry*, 2007. **101**(2): pp. 465–470.
152. Basha, S.K. and G. Sudarshanam, Multiple herbal therapy-Antimicrobial activity of wound healing paste (Pasuru) used by Sugali tribes of Yerramalais of Kurnool district., Andhra Pradesh, India. *International Journal of PharmTech Research*, 2011. **3**(3): pp. 1238–1241.
153. Ghazanfar, S.A., *Handbook of Arabian Medicinal Plants.* CRC Press, 1994.
154. Alam, M.I. and A. Gomes, Viper venom-induced inflammation and inhibition of free radical formation by pure compound (2-hydroxy-4-methoxy benzoic acid) isolated and purified from anantamul (Hemidesmus indicus R. BR) root extract. *Toxicon*, 1998. **36**(1): pp. 207–215.
155. Steyn, D.G., The presence of hydrocyanic acid in stock feeds and other plants. *Journal of the South African Veterinary Association*, 1938. **9**(2): pp. 60–64.
156. Watt, J.M. and M.G. Breyer-Brandwijk, *The Medicinal and Poisonous Plants of Southern and Eastern Africa being an Account of their Medicinal and Other Uses, Chemical Composition, Pharmacological Effects and Toxicology in Man and Animal*, 2nd edn. The Medicinal and Poisonous Plants of Southern and Eastern Africa being an Account of their Medicinal and other Uses, Chemical Composition, Pharmacological Effects and Toxicology in Man and Animal., 1962.
157. Hungeling, M., et al., Cyanogenic and non-cyanogenic pyridone glucosides from Acalypha indica (Euphorbiaceae). *Phytochemistry*, 2009. **70**(2): pp. 270–277.
158. Yoshida, N. and T. Tamada, Host range and molecular analysis of Beet leaf yellowing virus, Beet western yellows virus-JP and Brassica yellows virus in Japan. *Plant Pathology*, 2019. **68**(6): pp. 1045–1058.
159. Srivastav, S., et al., Achyranthes aspera-An important medicinal plant: a review. *ournal of Natural Product and Plant Resources*, 2011. **1**(1): pp. 1–14.
160. Anonymous. *The Wealth of India – Raw Materials*, Council of Scientific & Industrial Research, New Delhi, 2005, pp. 55–57.
161. Zafar, R.. *Medicinal Plants of India.* CBS Publishers & Distributors, 2009, pp. 1–15.
162. Gupta, R.K.. *Medicinal & Aromatic Plants.* CBS Publishers & Distributors, 2010, p. 190.
163. Hariharan, V. and S. Rangaswami, Structure of saponins A and B from the seeds of Achyranthes aspera. *Phytochemistry*, 1970. **9**(2): pp. 409–414.
164. Ali, M. Chemical investigation of Achyranthes Aspera Linn. *Oriental Journal of Chemistry* 1993. **9**(1): p. 84.
165. Rastogi, R. and B. Mehrotra, *Compendium of Medicinal Plants Central Drug Research Institute (CDRI),* vol. **406**. New Delhi: Lucknow and National institute of Communication and Information Resources, 2004.
166. Bhosale, U.A., et al., Effect of aqueous extracts of Achyranthes aspera Linn. on experimental animal model for inflammation. *Ancient Science of Life*, 2012. **31**(4): pp. 202.
167. Tuyet Lan, H.T., et al., Four new flavonoid C-Glycosides isolated from Achyranthes aspera and their nitric oxide production inhibitory activities. *Chemistry & Biodiversity*, 2023. **20**(8): p. e202300853.
168. Pai, S.R., et al., Determination of betulinic acid, oleanolic acid and ursolic acid from Achyranthes aspera L. using RP-UFLC-DAD analysis and evaluation of various parameters for their optimum yield. *Indian Journal of Experimental Biology*, 2016. **54**(3): pp. 196–202.
169. Nadkarni, A., *KM Nadkarni's Indian Materia Medica. Bombay Popular Prakashana*, vol. 2. Popular Prakashan Pvt.Bombay India, 1976.
170. Khare, C., *Indian Medicinal Plants.* Springer-Verlag Berlin, Heidelberg, 2007.
171. Arora, S. and S. Tandon, Achyranthes aspera root extracts induce human colon cancer cell (COLO-205) death by triggering the mitochondrial apoptosis pathway and S phase cell cycle arrest. *The Scientific World Journal*, 2014. **2014**.
172. Paul, D., et al., Comparative study on the spermicidal activity of organic solvent fractions from hydroethanolic extracts of Achyranthes aspera and Stephania hernandifolia in human and rat sperm. *Contraception*, 2010. **81**(4): pp. 355–361.
173. Akhtar, M.S. and J. Iqbal, Evaluation of the hypoglycaemic effect of Achyranthes aspera in normal and alloxan-diabetic rabbits. *Journal of Ethnopharmacology*, 1991. **31**(1): pp. 49–57.
174. Chakraborty, A., et al., Cancer chemopreventive activity of Achyranthes aspera leaves on Epstein–Barr virus activation and two-stage mouse skin carcinogenesis. *Cancer Letters*, 2002. **177**(1): pp. 1–5.
175. Teshome, D., et al., Developmental toxicity of ethanolic extracts of leaves of Achyranthes aspera, Amaranthaceae in rat embryos and fetuses. *Journal of Experimental Pharmacology*, 2021: pp. 555–563.
176. Han, S.-T. and C.-C. Un, Cardiac toxicity caused by Achyranthes aspera. *Veterinary and Human Toxicology*, 2003. **45**(4): pp. 212–213.
177. Mandefro, B., et al., Analysis of Achyranthes aspera leaf extract and acute toxicity study on fingerlings of Nile tilapia, Oreochromis niloticus. *Biochemistry and Biophysics Reports*, 2024. **37**: p. 101624.
178. Rao, Y.V., et al., Effect of Achyranthes aspera on the immunity and survival of Labeo rohita infected with Aeromonas hydrophila. *Fish & Shellfish Immunology*, 2006. **20**(3): pp. 263–273.
179. Raj, S., et al., Association of 'Candidatus Phytoplasma asteris' (16SrI group) with yellows of Achyranthes aspera in India. *Plant Pathology*, 2009. **58**(2).
180. Singh, N., P. Sharma, and O. Verma, First report of Achyrenthes aspera leaf spot disease caused by Curvularia prasadii in India. *New Disease Reports*, 2011. **23**(1): pp. 22–22.

181. Kisksi, T., C. Guenaoui, and N. Fawzi, Early growth stages of the rare Acridocarpus orientalis in the UAE-A first step towards conservation. *Natural Resources*, 2012. **3**: pp. 1–5.
182. Jongbloed, M., et al., *The Comprehensive Guide to the Wild Flowers of the United Arab Emirates*. Environmental research and wildlife development agency, 2003.
183. Malebo, H.M., et al., Anti-protozoal and structure-activity relationships of chemical constituents of Acridocarpus chloropterus Oliver (Malpighiaceae) from Tanzania. *International Journal of Natural Products Research*, 2013. **3**(4): pp. 74–81.
184. Cao, S., et al., Cytotoxic triterpenoids from Acridocarpus vivy from the Madagascar rain forest. *Journal of Natural Products*, 2004. **67**(6): pp. 986–989.
185. Hussain, J., et al., Isolation and bioactivities of the flavonoids morin and morin-3-O-β-D-glucopyranoside from Acridocarpus orientalis – A wild Arabian medicinal plant. *Molecules*, 2014. **19**(11): pp. 17763–17772.
186. Ali, L., et al., Total polyphenols quantification in Acridocarpus orientalis and Moringa peregrina by using NIR spectroscopy coupled with PLS regression. *Chemical Data Collections*, 2018. **13**: pp. 104–112.
187. Rehman, N.U., et al., Chemical constituents and carbonic anhydrase II activity of essential oil of Acridocarpus orientalis A. Juss. in comparison with stem and leaves. *Journal of Essential Oil Bearing Plants*, 2021. **24**(1): pp. 68–74.
188. Rehman, N.U., et al., Chemical constituents of Acridocarpus orientalis and their chemotaxonomic significance. *Chemistry of Natural Compounds*, 2019. **55**: pp. 586–588.
189. Ijaz, M.U., et al., Ameliorative effects of morin on cisplatin-induced toxicity in renal mitochondria isolated from rats. *Journal of King Saud University-Science*, 2021. **33**(8): p. 101655.
190. Wan, J., et al., The effects of morin on bone regeneration to accelerate healing in bone defects in mice. *International Journal of Immunopathology and Pharmacology*, 2020. **34**: p. 2058738420962909.
191. Rehman, N.U., et al., Evaluation of biological potential and physicochemical properties of Acridocarpus orientalis (Malpighiaceae). *Pakistan Journal of Botany*, 2019. **51**(3): pp. 1099–106.
192. Ghazanfar, S.A. and A.M. Al-Al-Sabahi, Medicinal plants of northern and central Oman (Arabia). *Economic Botany*, 1993. **47**: pp. 89–98.
193. Hammiche, V. and K. Maiza, Traditional medicine in Central Sahara: pharmacopoeia of Tassili N'ajjer. *Journal of Ethnopharmacology*, 2006. **105**(3): pp. 358–367.
194. Hellyer, P. and S. Aspinall, *The Emirates: A Natural History*. Trident Press Ltd., 2005.
195. Balhamar, S.O.M.S., et al., Differential cytotoxic potential of Acridocarpus orientalis leaf and stem extracts with the ability to induce multiple cell death pathways. *Molecules*, 2019. **24**(21): p. 3976.
196. Jamshidi-Adegani, F., et al., Secondary metabolites from Acridocarpus orientalis inhibits 4T1 cells and promotes mesenchymal stem cells (MSCs) proliferation. *Molecular Biology Reports*, 2020. **47**: pp. 5421–5430.
197. Ksiksi, T. and A.A. Hamza, Antioxidant, lipoxygenase and histone deacetylase inhibitory activities of Acridocarbus orientalis from al ain and oman. *Molecules*, 2012. **17**(11): pp. 12521–12532.
198. Lotfy, M., et al., Hepatoprotective potentials of Acridocarpus orientalis in mice. *Clinical Phytoscience*, 2020. **6**(1): pp. 1–9.
199. Lotfy, M., et al., Anti-diabetic Effect of. *The Open Medicinal Chemistry Journal*, 2020. **14**(1).
200. Kamatou, G., I. Vermaak, and A. Viljoen, An updated review of Adansonia digitata: a commercially important African tree. *South African Journal of Botany*, 2011. **77**(4): pp. 908–919.
201. De Caluwé, E., K. Halamová, and P. Van Damme, *Baobab (Adansonia digitata L.): A Review of Traditional Uses, Phytochemistry and Pharmacology*, American Chemical Society, 2009.
202. Abiona, D., Z. Adedapo, and M. Suleiman, Proximate analysis, phytochemical screening and antimicrobial activity of baobab (Adansonia digitata) leaves. *IOSR Journal of Applied Chemistry*, 2015. **8**(5): pp. 60–65.
203. Namratha, V. and P. Sahithi, Baobab: a review about "the tree of life". *International Journal of Advance Herbal Technology*, 2015. **1**: pp. 20–26.
204. Talari, S., et al., In vitro free radical scavenging activity of different extracts of Adansonia digitata L. *International Journal of Environment, Agriculture and Biotechnology*, 2017. **2**(3): pp. 1169–1172.
205. Muthai, K.U., et al., Nutritional variation in baobab (Adansonia digitata L.) fruit pulp and seeds based on Africa geographical regions. *Food Science & Nutrition*, 2017. **5**(6): pp. 1116–1129.
206. Hyacinthe, T., et al., Variability of vitamins B1, B2 and minerals content in baobab (Adansonia digitata) leaves in East and West Africa. *Food Science & Nutrition*, 2015. **3**(1): pp. 17–24.
207. Busson, F., P. Deniel, and J. Toury, Amino acid composition of leaves and pulp of fruits of baobab (Adansonia digitata. L.). *Bulletin de la Societe de Chimie Biologique*, 1958. **40**: pp. 711–714.
208. Li, X.-N., et al., Profiling hydroxycinnamic acid glycosides, iridoid glycosides, and phenylethanoid glycosides in baobab fruit pulp (Adansonia digitata). *Food Research International*, 2017. **99**: pp. 755–761.
209. Irondi, E.A., et al., Blanching influences the phenolics composition, antioxidant activity, and inhibitory effect of Adansonia digitata leaves extract on α-amylase, α-glucosidase, and aldose reductase. *Food Science & Nutrition*, 2017. **5**(2): pp. 233–242.
210. Ismail, B.B., et al., Characterizing the phenolic constituents of baobab (Adansonia digitata) fruit shell by LC-MS/QTOF and their in vitro biological activities. *Science of the Total Environment*, 2019. **694**: p. 133387.
211. Yazzie, D., et al., The amino acid and mineral content of baobab (Adansonia digitata L.) leaves. *Journal of Food Composition and Analysis*, 1994. **7**(3): pp. 189–193.
212. Braca, A., et al., Phytochemical profile, antioxidant and antidiabetic activities of Adansonia digitata L.(Baobab) from Mali, as a source of health-promoting compounds. *Molecules*, 2018. **23**(12): p. 3104.
213. Al-Juhaimi, F., et al., Assessment of oxidative stability and physicochemical, microbiological, and sensory properties of beef patties formulated with baobab seed (Adansonia digitata) extract. *Meat Science*, 2020. **162**: p. 108044.
214. Trela-Makowej, A., et al., Acylserotonins–a new class of plant lipids with antioxidant activity and potential pharmacological applications. *Biochimica et Biophysica Acta (BBA)-Molecular and Cell Biology of Lipids*, 2021. **1866**(12): p. 159044.
215. Assogbadjo, A.E., et al., Variation in biochemical composition of baobab (Adansonia digitata) pulp, leaves and seeds in relation to soil types and tree provenances. *Agriculture, Ecosystems & Environment*, 2012. **157**: pp. 94–99.

216. Darr, D., et al., Processed Baobab (Adansonia digitata L.) food products in Malawi: from poor men's to premium-priced specialty food? *Forests*, 2020. **11**(6): p. 698.
217. Babatunde, D.J., et al., Aqueous extract of Adansonia digitata reversed cotton seed extract-induced testicular damage in Wistar rats. *JBRA Assisted Reproduction*, 2021. **25**(2): p. 257.
218. Ebaid, H., et al., Efficacy of a methanolic extract of Adansonia digitata leaf in alleviating hyperglycemia, hyperlipidemia, and oxidative stress of diabetic rats. *BioMed Research International*, 2019. **2019**.
219. Barakat, H., Nutritional and rheological characteristics of composite flour substituted with baobab (Adansonia digitata l.) pulp flour for cake manufacturing and organoleptic properties of their prepared cakes. *Foods*, 2021. **10**(4): p. 716.
220. Oyewole, A.L., et al., Adansonia digitata and its use in neuropathic pain: prostaglandins and beyond, in Rajkumar Rajendram, Vinood B. Patel, Victor R. Preedy, Colin R. Martin, (eds.), *Treatments, Mechanisms, and Adverse Reactions of Anesthetics and Analgesics*. 2022, Elsevier. pp. 329–350.
221. Owoyele, B.V. and A.O. Bakare, Analgesic properties of aqueous bark extract of Adansonia digitata in Wistar rats. *Biomedicine & Pharmacotherapy*, 2018. **97**: pp. 209–212.
222. Nwokocha, L.M. and P.A. Williams, Rheological properties of a polysaccharide isolated from Adansonia digitata leaves. *Food Hydrocolloids*, 2016. **58**: pp. 29–34.
223. Mohammed, M.A., et al., Comprehensive metabolomic, lipidomic and pathological profiles of baobab (Adansonia digitata) fruit pulp extracts in diabetic rats. *Journal of Pharmaceutical and Biomedical Analysis*, 2021. **201**: p. 114139.
224. Besco, E., et al., The use of photochemiluminescence for the measurement of the integral antioxidant capacity of baobab products. *Food Chemistry*, 2007. **102**(4): pp. 1352–1356.
225. Mulaudzi, R., et al., Anti-inflammatory and mutagenic evaluation of medicinal plants used by Venda people against venereal and related diseases. *Journal of Ethnopharmacology*, 2013. **146**(1): pp. 173–179.
226. Dzoyem, J.P., et al., The 15-lipoxygenase inhibitory, antioxidant, antimycobacterial activity and cytotoxicity of fourteen ethnomedicinally used African spices and culinary herbs. *Journal of Ethnopharmacology*, 2014. **156**: pp. 1–8.
227. Musila, M., et al., In vivo antimalarial activity, toxicity and phytochemical screening of selected antimalarial plants. *Journal of Ethnopharmacology*, 2013. **146**(2): pp. 557–561.
228. Ryan, L., A. Fraser, and P. Thondre, BAOBAB (Adansonia digitata l.) Reduces the Glycaemic and Insulin Response in Healthy Volunteers, *Journal of Nutrition & Intermediary* 2014.
229. Coe, S.A., et al., The polyphenol-rich baobab fruit (Adansonia digitata L.) reduces starch digestion and glycemic response in humans. *Nutrition Research*, 2013. **33**(11): pp. 888–896.
230. Obizoba, I. and J. Anyika, Nutritive value of baobab milk (gubdi) and mixtures of baobab (Adansonia digitata L.) and hungry rice, acha (Digitaria exilis) flours. *Plant Foods for Human Nutrition*, 1994. **46**: pp. 157–165.
231. Komane, B.M., et al., Beauty in Baobab: a pilot study of the safety and efficacy of Adansonia digitata seed oil. *Revista Brasileira de Farmacognosia*, 2017. **27**: pp. 1–8.
232. Jour, I., Muhammad, G. M., Mu'azu, Abubakar, Acute toxicity and antidiabetic studies of adansonia digitata stem bark extracts on streptozotocin induced diabetic rats. *Researchgate* 2021. **89**: p. 5.
233. Cicolari, S., et al., Hydromethanolic extracts from Adansonia digitata L. Edible parts positively modulate pathophysiological mechanisms related to the metabolic syndrome. *Molecules*, 2020. **25**(12): p. 2858.
234. Shehu, A., et al., Methanol stem bark extract of Adansonia digitata ameliorates chronic unpredictable mild stress-induced depression-like behavior: involvement of the HPA axis, BDNF, and stress biomarkers pathways. *Journal of Basic and Clinical Physiology and Pharmacology*, 2019. **30**(3): p. 20180153.
235. Quartey, A.K., et al., In-vivo anti-inflammatory activity of different parts of Adansonia digitata. *Journal of Medicinal Plants Research*, 2021. **15**(6): pp. 252–259.
236. Ujor, A., E. Oche, and J. Oloche, Performance and toxicological effects of cooked baobab (Adansonia digitata L.) seed meal on West African dwarf goats. *Nigerian Journal of Animal Science*, 2020. **22**(3): pp. 296–306.
237. Chepngeno, J., et al., Baobab pulp authenticity and quality control by multi-imaging high-performance thin-layer chromatography. *Food Chemistry*, 2022. **390**: p. 133108.
238. Halfeld-Vieira, B., L. Ferreira, and K. Nechet, Bombacopsis quinata: a new host for Oidiopsis haplophylli in Brazil. *Plant Pathology*, 2007. **56**(6): p. 1040.
239. Cherotich, S., et al., Fungal microflora biodiversity of healthy and diseased Adansonia digitata and Sclerocarya birrea trees in Kenya. *Topola*, 2020. **205**: p. 5–13.
240. Attafuah, A. and T. Tinsley, Virus diseases of Adansonia digitata L.(Bombacaceae) and their relation to cacao in Ghana. *Annals of Applied Biology*, 1958. **46**(1): pp. 20–22.
241. Al Hatmi, S, Ethnobotany of Central Desert. (2017). [Unpublished report].
242. YAMAUCHI, T. and F. ABE, Cardiac glycosides and pregnanes from Adenium obesum (studies on the constituents of Adenium. I). *Chemical and Pharmaceutical Bulletin*, 1990. **38**(3): pp. 669–672.
243. Tijjani, A., I. Ndukwe, and R. Ayo, Studies on antibacterial activity of Adenium obesum (Apocynaceae) stem-bark. *Continental Journal of Microbiology*, 2011. **5**(1): pp. 12–17.
244. Ebrahim, N., R. Kershi, and L. Rastrelli, Free radical scavenging activity and anthocyanin in flower of Adenium obesum collected from Yemen. *Journal of Pharmacology and Phytotherapy*, 2013. **1**: pp. 5–7.
245. Akhtar, M.S., M.A. Hossain, and S.A. Said, Isolation and characterization of antimicrobial compound from the stem-bark of the traditionally used medicinal plant Adenium obesum. *Journal of Traditional and Complementary Medicine*, 2017. **7**(3): pp. 296–300.
246. Tatsuzawa, F., et al., Flavonoids from the Flowers of Adenium obesum (Forssk.) Roem. & Schult., Mandevilla sanderi (Hemsl.) Woodson, and Nerium oleander L.(Apocynaceae). *Biochemical Systematics and Ecology*, 2021. **99**: p. 104347.
247. Hossain, M.A., et al., Two new flavonoids from Adenium obesum grown in Oman. *Journal of King Saud University-Science*, 2017. **29**(1): pp. 62–69.
248. Hoffmann, J.J. and J.R. Cole, Phytochemical investigation of Adenium obesum Forskal (Apocynaceae): isolation and identification of cytotoxic agents. *Journal of Pharmaceutical Sciences*, 1977. **66**(9): pp. 1336–1338.
249. Ahmed, S.K., et al., Cytotoxic cardiac glycosides from the fruit (pods) of Adenium obesum (Forssk.) Roem. & Schult. *Natural Product Research*, 2017. **31**(10): pp. 1205–1208.
250. Ali, A.Q., et al., Cytogenotoxic effects of Adenium obesum seeds extracts on breast cancer cells. *Saudi Journal of Biological Sciences*, 2019. **26**(3): pp. 547–553.

251. Kiyohara, H., et al., In vitro anti-influenza virus activity of a cardiotonic glycoside from Adenium obesum (Forssk.). *Phytomedicine*, 2012. **19**(2): pp. 111–114.
252. Al-Sarar, A., et al., Molluscicidal activity of methomyl and cardenolide extracts from Calotropis procera and Adenium arabicum against the land snail Monacha cantiana. *Molecules*, 2012. **17**(5): pp. 5310–5318.
253. Alshehri, A., et al., In vitro evaluation of antioxidant, anticancer, and anti-inflammatory activities of ethanolic leaf extract of Adenium obesum. *Frontiers in Pharmacology*, 2022. **13**: p. 847534.
254. Al Rashdi, R.S.Y., M.A. Hossain, and S.S.J. Al Touby, Antioxidant and antibacterial activities of leaves crude extracts of Adenium obesum grown in Oman National Botanical Garden. *Advances in Biomarker Sciences and Technology*, 2021. **3**: pp. 8–14.
255. Hossain, M.A., et al., A comparison of the antimicrobial effectiveness of different polarities crude extracts from the leaves of Adenium obesum used in Omani traditional medicine for the treatment of microbial infections. *Asian Pacific Journal of Tropical Disease*, 2014. **4**: pp. S934–S937.
256. Farah, M.A., et al., Silver nanoparticles synthesized from Adenium obesum leaf extract induced DNA damage, apoptosis and autophagy via generation of reactive oxygen species. *Colloids and Surfaces B: Biointerfaces*, 2016. **141**: pp. 158–169.
257. Gurung, A.B., et al., Molecular docking elucidates the plausible mechanisms underlying the anticancer properties of acetyldigitoxigenin from Adenium obesum. *Saudi Journal of Biological Sciences*, 2020. **27**(7): pp. 1907–1911.
258. Abalaka, S.E., et al., Gills and skin histopathological evaluation in African sharptooth catfish, Clarias gariepinus exposed to ethanol extract of Adenium obesum stem bark. *Egyptian Journal of Aquatic Research*, 2015. **41**(1): pp. 119–127.
259. Chaturvedi, Y., et al., First report of association of 'Candidatus Phytoplasma asteris' (16SrI group) with little leaf disease of rose (Rosa alba) in India. *Plant Pathology*, 2009. 788, **58**(4).
260. Cadamuro, R.D., et al., Bioactive compounds from mangrove endophytic fungus and their uses for microorganism control. *Journal of Fungi*, 2021. **7**(6): p. 455.
261. Shahin, S.M., A. Jaleel, and M.A.M. Alyafei, Yield and in vitro antioxidant potential of essential oil from Aerva javanica (burm. F.) juss. Ex schul. flower with special emphasis on seasonal changes. *Plants*, 2021. **10**(12): p. 2618.
262. Hemmati, C., M. Nikooei, and A. Bertaccini, Identification and transmission of phytoplasmas and their impact on essential oil composition in Aerva javanica. *3 Biotech*, 2019. **9**(8): p. 310.
263. Saleem, H., et al., New insights into the phytochemical composition, enzyme inhibition and antioxidant properties of desert cotton (Aerva javanica (Bum. f) Shult.-Amaranthaceae). *Natural Product Research*, 2021. **35**(4): pp. 664–668.
264. Arbab, A.H., et al., Therapeutic efficacy of ethanolic extract of Aerva javanica aerial parts in the amelioration of CCl4-induced hepatotoxicity and oxidative damage in rats. *Food & Nutrition Research*, 2016. **60**(1): p. 30864.
265. Samejo, M.Q., et al., Comparison of chemical composition of Aerva javanica seed essential oils obtained by different extraction methods. *Pakistan Journal of Pharmaceutical Sciences*, 2013. **26**(4): pp. 757–760.
266. Saleh, N.A., R.M. Mansour, and K.R. Markham, An acylated isorhamnetin glycoside from Aerva javanica. *Phytochemistry*, 1990. **29**(4): pp. 1344–1345.
267. Mussadiq, S., et al., New acylated flavonoid glycosides from flowers of Aerva javanica. *Journal of Asian Natural Products Research*, 2013. **15**(7): pp. 708–716.
268. Saleem, M., et al., Ecdysteroids from the flowers of Aerva javanica. *Steroids*, 2013. **78**(11): pp. 1098–1102.
269. Khan, A.W., et al., Phytochemical analysis and enzyme inhibition assay of Aerva javanica for ulcer. *Chemistry Central Journal*, 2012. **6**(1): pp. 1–6.
270. Shad, A.A., et al., Screening of Aerva javanica and Linum ustitatissimum for their antidiabetic and anti-oxidant activity. *Pakistan Journal of Pharmaceutical Sciences*, 2017. **30**(1).
271. Mufti, F.U.D., et al., Antimicrobial activities of Aerva javanica and Paeonia emodi plants. *Pakistan Journal of Pharmaceutical Sciences*, 2012. **25**(3): pp. 565–569.
272. Hussain, L., et al., Analgesic and anti-inflammatory activity of aqueous-methanolic extract of Aerva javanica. *Pakistan Journal of Pharmaceutical Sciences*, 2017. **30**(1): pp. 213–215.
273. Dedar, R.K., et al., Leaf extract of Aerva javanica suppresses excessive growth of granulation tissue in horses. *Journal of Equine Veterinary Science*, 2020. **93**: p. 103193.
274. Amin, F., et al., Green synthesis of copper oxide nanoparticles using Aerva javanica leaf extract and their characterization and investigation of in vitro antimicrobial potential and cytotoxic activities. *Evidence-Based Complementary and Alternative Medicine*, 2021. **2021**: pp. 12.
275. Khalid, M., et al., Antimicrobial, antioxidant, cytotoxicity and LC-MS analyses of Aerva javanica: an ethnomedicinally important plant. *J. Biol. Regul. Homeost. Agents*, 2017. **31**(4): pp. 963–969.
276. Kamalanathan, D. and D. Natarajan, Anticancer potential of leaf and leaf-derived callus extracts of Aerva javanica against MCF-7 breast cancer cell line. *Journal of Cancer Research and Therapeutics*, 2018. **14**(2): pp. 321–327.
277. Mookherjee, A., et al., Characterization of endo-metabolome exhibiting antimicrobial and antioxidant activities from endophytic fungus Cercospora sp. PM018. *South African Journal of Botany*, 2020. **134**: pp. 264–272.
278. Al-Laith, A.A., J. Alkhuzai, and A. Freije, Assessment of antioxidant activities of three wild medicinal plants from Bahrain. *Arabian Journal of Chemistry*, 2019. **12**(8): pp. 2365–2371.
279. Phoboo, S., K. Shetty, and T. ElObeid, In vitro assays of anti-diabetic and anti-hypertensive potential of some traditional edible plants of Qatar. *Journal of Medicinally Active Plants*, 2015. **4**(3): pp. 22–29.
280. Freije, A., J. Alkhuzai, and A.A. Al-Laith, Fatty acid composition of three medicinal plants from Bahrain: new potential sources of γ-linolenic acid and dihomo-γ-linolenic. *Industrial Crops and Products*, 2013. **43**: pp. 218–224.
281. Bakr, R.O., et al., New adenosine derivatives from Aizoon canariense L.: in vitro anticholinesterase, antimicrobial, and cytotoxic evaluation of its extracts. *Molecules*, 2021. **26**(5): p. 1198.
282. Abuzaid, H., et al., Liquid chromatography high-resolution mass spectrometry analysis, phytochemical and biological study of two aizoaceae plants: a new kaempferol derivative from Trianthema portulacastrum L. *Pharmacognosy Research*, 2020. **12**(3).
283. Adam, M., et al., In vitro antioxidant and cytotoxic activities of 18 plants from the Erkowit region, Eastern Sudan. *Natural Products and Bioprospecting*, 2018. **8**: pp. 97–105.
284. AlQathama, A., et al., In vitro cytotoxic activities of selected Saudi medicinal plants against human malignant melanoma cells (A375) and the isolation of their active

principles. *European Journal of Integrative Medicine*, 2022. **49**: p. 102083.
285. Kingsbury, J.M., *Poisonous Plants of the US and Canada* Prentice Hall, 1964.
286. Dai, J., et al., Metabolites from the fungus Phoma sp. 7210, associated with Aizoon canariense. *Natural Product Communications*, 2010. **5**(8): p. 1934578X1000500804.
287. Patil, S.P., Proton NMR and HR-LC/MS based phytochemical analysis of methanolic fraction of Alectra parasitica A. Rich. rhizomes. *Heliyon*, 2020. **6**(1): 1–7.
288. Kakpure, M., S. Rothe, and S. Shivaji, Phytochemical screening of Alectra parasitica A. Rich–a rare medicinal parasitic plant. *Advance Research in Pharmacy & Biological Science*, 2012. **2**(1): pp. 103–111.
289. Ranjana, et al., Stimulation of glucose uptake by glycosides from Alectra parasitica subsp. chitrakutensis (MA Rau) KK Khanna & An. Kumar: an in-vitro study of TNF-α induced insulin resistance in L6 myoblasts. *Natural Product Research*, 2023. **37**(12): p. 2024–2030.
290. Agrawal, P., K. Laddha, and A. Tiwari, Isolation and HPLC method development of azafrin from Alectra parasitica var. chitrakutensis. *Natural Product Research*, 2014. **28**(13): pp. 940–944.
291. Rajagopalan, T. and T. Seshadri, Chemical components of Alectra parasitica. *Current Science*, 1964. **33**(6): pp. 174–175.
292. Ghosh, S. and B. Chakraborty, Alectra parasítica-a Rich, var. chitraku-tensis in the Treatment of Leprosy. *Bulletin of the Calcutta School of Tropical Medicine*, 1964. **12**(4): pp. 179–80.
293. Prasad, B., Alectra parasitica A. Rich (Variety Chitrakutensis). An indigenous drug in the treatment of leprosy in Bihar-a preliminary observation. *Leprosy Review*, 1962. **33**(3): pp. 207–209.
294. Bedi, R., Alectra parasitica var. chitrakutensis. *Economic Botany*, 1967. **21**(3): pp. 277–283.
295. Quattrocchi, U., *CRC World Dictionary of Medicinal and Poisonous Plants: Common Names, Scientific Names, Eponyms, Synonyms, and Etymology*, 5 volume Set. CRC press, 2012.
296. Riches, C. R. (1988). *The biology and control of Alectra vogelii Benth. (Scrophylariaceae) in Botswana* (Doctoral dissertation, University of Reading).
297. Maela, M.P., H. van der Walt, and M.H. Serepa-Dlamini, The antibacterial, antitumor activities, and bioactive constituents' identification of ALECTRA sessiliflora bacterial endophytes. *Frontiers in Microbiology*, 2022. **13**: p. 870821.
298. Al-Khulaidi, A.W.A., Preliminary evaluation of the status of aloe Dhufarensis Lavranos (asphodelaceae Juss.) in the Arabian Peninsula. *Electronic Journal of University of Aden for Basic and Applied Sciences*, 2021. **2**(1): pp. 14–21.
299. Al-Rashdi, F.K.H., et al., Endophytic fungi from the medicinal plant Aloe dhufarensis Lavranos exhibit antagonistic potential against phytopathogenic fungi. *South African Journal of Botany*, 2022. **147**: pp. 1078–1085.
300. Dhanalekshmi, U., T. Alam, and S.A. Khan, The folkloric uses and economic importance of some selected edible medicinal plants native to Oman: a brief overview, in *Edible Plants in Health and Diseases: Volume 1: Cultural, Practical and Economic Value* Springer. 2022: pp. 1–29.
301. Singh, R.J., *Genetic Resources, Chromosome Engineering, and Crop Improvement: Vegetable Crops*, vol. 3. CRC Press, 2006.
302. Sherwani, N., Aloe dhufarensis – anitoxidant and anti-inflammatory properties. *Journal of Pharmacognosy & Natural Products*, 2021. https://www.hilarispublisher.com/open-access/aloe-dhufarensis--anitoxidant-and-antiinflammatory-properties.pdf
303. Cock, I., The genus Aloe: phytochemistry and therapeutic uses including treatments for gastrointestinal conditions and chronic inflammation. Novel natural products: therapeutic effects in pain, arthritis and gastro-intestinal diseases. 2015: pp. 179–235.
304. Sánchez, M., et al., Pharmacological update properties of Aloe vera and its major active constituents. *Molecules*, 2020. **25**(6): p. 1324.
305. Hamman, J.H., Composition and applications of Aloe vera leaf gel. *Molecules*, 2008. **13**(8): pp. 1599–1616.
306. Boudreau, M.D. and F.A. Beland, An evaluation of the biological and toxicological properties of Aloe barbadensis (miller), Aloe vera. *Journal of Environmental Science and Health Part C*, 2006. **24**(1): pp. 103–154.
307. Carvalho, N.C., et al., Analysis of Aloe vera cytotoxicity and genotoxicity associated with endodontic medication and laser photobiomodulation. *Journal of Photochemistry and Photobiology B: Biology*, 2018. **178**: pp. 348–354.
308. Guo, X. and N. Mei, Aloe vera: a review of toxicity and adverse clinical effects. *Journal of Environmental Science and Health, Part C*, 2016. **34**(2): pp. 77–96.
309. Mandal, K. and S. Maiti, Bacterial soft rot of aloe caused by Pectobacterium chrysanthemi: a new report from India. *Plant Pathology*, 2005. **54**(4).
310. Añibarro-Ortega, M., et al., Extraction of aloesin from Aloe vera rind using alternative green solvents: process optimization and biological activity assessment. *Biology*, 2021. **10**(10): p. 951.
311. Foster, M., D. Hunter, and S. Samman, Evaluation of the Nutritional and Metabolic Effects of Aloe vera. in Benzie IFF, Wachtel-Galor S,(eds.), *Herbal Medicine: Biomolecular and Clinical Aspects*. 2nd edn, 2011. Boca Raton (FL): CRC Press/Taylor & Francis.
312. Esua, M.F. and J.-W. Rauwald, Novel bioactive maloyl glucans from Aloe vera gel: isolation, structure elucidation and in vitro bioassays. *Carbohydrate Research*, 2006. **341**(3): pp. 355–364.
313. Ulbricht, C., et al., An evidence-based systematic review of Aloe vera by the natural standard research collaboration. *Journal of Herbal Pharmacotherapy*, 2008. **7**(3–4): pp. 279–323.
314. *Toxicity of Aloe vera*. Available from: https://www.niehs.nih.gov/health/topics/agents/aloe/index.cfm
315. Wu, J., et al., Safety evaluation of Aloe vera soft capsule in acute, subacute toxicity and genotoxicity study. *PloS One*, 2021. **16**(3): p. e0249356.
316. Shah, A., et al., Toxicity studies on six plants used in the traditional Arab system of medicine. *Phytotherapy Research*, 1989. **3**(1): pp. 25–29.
317. Panel, C.I.R.E., Final report on the safety assessment of aloeandongensis extract, aloe andongensis leaf juice, aloe arborescens leaf extract, aloe arborescens leaf juice, aloe arborescens leaf protoplasts, aloe barbadensis flower extract, aloe barbadensis leaf, aloe barbadensis leaf extract, aloe barbadensis leaf juice, aloe barbadensis leaf polysaccharides, aloe barbadensis leaf water, aloe ferox leaf extract, aloe ferox leaf juice, and aloe ferox leaf juice extract. *International Journal of Toxicology*, 2007. **26**: pp. 1–50.
318. Brusick, D. and U. Mengs, Assessment of the genotoxic risk from laxative senna products. *Environmental and Molecular Mutagenesis*, 1997. **29**(1): pp. 1–9.

319. Pelley, R., et al., Multiparameter analysis of commercial "Aloe vera" materials and comparison to Aloe barbadensis Miller extracts. Subtrop. *Plant Science*, 1998. **50**: pp. 1–14.
320. Lachenmeier, K., et al., Quality control of Aloe vera beverages. *Electronic Journal of Environmental, Agricultural and Food Chemistry*, 2005. **4**(4): pp. 1033–1042.
321. Chandran, S.P., et al., Synthesis of gold nanotriangles and silver nanoparticles using Aloevera plant extract. *Biotechnology Progress*, 2006. **22**(2): pp. 577–583.
322. Kee, Y.J., L. Zakaria, and M.H. Mohd, Morphology, phylogeny and pathogenicity of Fusarium species from Sansevieria trifasciata in Malaysia. *Plant Pathology*, 2020. **69**(3): pp. 442–454.
323. *Flora of Oman*. Available from: https://www.pemberleybooks.com/product/flora-of-the-sultanate-of-oman-volume-3-loganiaceaeasteraceae/38234/
324. Hegnauer, R., Distribution of hydrocyanic acid in cormophytes. 4. New research on the distribution of cyanogenesis. *Pharmaceutisch Weekblad*, 1961. **96**: pp. 577–596.
325. Kumar, M., et al., Anti-osteoporotic constituents from Indian medicinal plants. *Phytomedicine*, 2010. **17**(13): pp. 993–999.
326. David, J.P., I.D. dos Santos, and J.M. David, A new sesquiterpene from the fruits of Allophylus laevigatus. *Fitoterapia*, 2004. **75**(7-8): pp. 795–798.
327. Johnson, T., *CRC Ethnobotany Desk Reference*. CRC Press, 2019.
328. Agnew, A., A. G. Miller & M. Morris 1988. Plants of Dhofar (The southern region of Oman; traditional, economic and medicinal uses). *Journal of Tropical Ecology*, 1990. **6**(1): pp. 102–102.
329. Ferreres, F., et al., Leaves and stem bark from Allophylus africanus P. Beauv.: an approach to anti-inflammatory properties and characterization of their flavonoid profile. *Food and Chemical Toxicology*, 2018. **118**: pp. 430–438.
330. Adebayo, A.H., T.A. Ishola, and O.F. Yakubu, Acute toxicity and antimalarial studies of extract of Allophylus spicatus in animals. *Toxicological Research*, 2021. **37**: pp. 345–354.
331. Kariuki, H.G., et al., Screening extracts of fungal endophytes isolated from Allophylus abyssinicus (Hochst. Radlk.) for control of bean anthracnose. *African Journal of Biotechnology*, 2023. **22**(10): pp. 235–246.
332. Pickering, H. and A. Patzelt, *Field Guide to the Wild Plants of Oman*. Royal Botanic Gardens, 2008.
333. Geering, A. and J. Thomas, Characterisation of a virus from Australia that is closely related to papaya mosaic potexvirus. *Archives of Virology*, 1999. **144**: pp. 577–592.
334. Rajagopalan, A., M.T. Devi, and A. Raghavendra, Patterns of phospho enol pyruvate carboxylase activity and cytosolic pH during light activation and dark deactivation in C 3 and C 4 plants. *Photosynthesis Research*, 1993. **38**: pp. 51–60.
335. Jakhar, S. and P. Dahiya, Antimicrobial, antioxidant and phytochemical potential of Alternanthera pungens HB&K. *Journal of Pharmaceutical Sciences and Research*, 2017. **9**(8): p. 1305.
336. Kazmi, Z., N. Safdar, and A. Yasmin, Biological screening of three selected folklore medicinal plants from Pakistan. *Pakistan Journal of Pharmaceutical Sciences*, 2019. **32**(4).
337. Singla, R.K., et al., The genus alternanthera: phytochemical and ethnopharmacological perspectives. *Frontiers in Pharmacology*, 2022. **13**: p. 769111.
338. Kone, W.M., et al., Evaluation of 17 medicinal plants from Northern Côte d'Ivoire for their in vitro activity against Streptococcus pneumoniae. *African Journal of Traditional, Complementary and Alternative Medicines*, 2007. **4**(1): pp. 17–22.
339. Djohan, Y., et al. Interest of antioxidants in the care of the patients infected by the HIV: the experience of long term administration of Alternanthera pungens herb tea. *Annales de Biologie Clinique*. 2009. **67**(5): pp. 563–568. https://doi.org/10.1684/abc.2009.0362
340. Campos-Navarro, R. and G.F. Scarpa, The cultural-bound disease "empacho" in Argentina. A comprehensive botanico-historical and ethnopharmacological review. *Journal of Ethnopharmacology*, 2013. **148**(2): pp. 349–360.
341. Gayathri, J., et al., Immunological characteristics of PEP carboxylase from leaves of C 3, C 4-and C 3 C 4 intermediate species of Alternanthera-Comparison with selected C 3 and C 4-plants. *Indian Journal of Experimental Biology*, 2001. **39**(7): pp. 643–649.
342. Al-Dosari, M.S. and M.K. Parvez, Novel plant inducers of PXR-dependent cytochrome P450 3A4 expression in HepG2 cells. *Saudi Pharmaceutical Journal*, 2018. **26**(8): pp. 1069–1072.
343. Ahmad, M. and M. Zafar, Conversion of waste seed oil of Citrus aurantium into methyl ester via green and recyclable nanoparticles of zirconium oxide in the context of circular bioeconomy approach. *Waste Management*, 2021. **136**: pp. 310–320.
344. Ishtiaq, M., et al., Medicinal plant diversity in the flora of Leepa Valley, Muzaffarabad (AJK), Pakistan. *African Journal of Biotechnology*, 2012. **11**(13): pp. 3087–3098.
345. Ishtiaq, M., et al., Ethnobotanical inventory and medicinal perspectives of herbal flora of Shiwalik mountainous range of District Bhimber, Azad Jammu and Kashmir, Pakistan. *Plos One*, 2022. **17**(3): p. e0265028.
346. Semenya, S., M. Potgieter, and L. Erasmus, Exotic and indigenous problem plants species used, by the Bapedi, to treat sexually transmitted infections in Limpopo Province, South Africa. *African Health Sciences*, 2013. **13**(2): pp. 320–326.
347. Kasthuri, O. and B. Ramesh, Toxicity studies on leaf extracts of Alternanthera brasiliana (L.) Kuntze and Alternanthera bettzickiana (Regel) Voss. *Journal of Applied Pharmaceutical Science*, 2018. **8**(10): pp. 82–89.
348. Thanabhorn, S., et al., Acute and Subacute Toxicities of the Ethanol Extract from Alternanthera Philoxeroides Griseb. *Pharmaceutical Sciences Asia*, 2005. **31**(1–2): pp. 7–14.
349. Ciuffo, M. and M. Turina, A potexvirus related to Papaya mosaic virus isolated from moss rose (Portulaca grandiflora) in Italy. *Plant Pathology*, 2004. **53**(4).
350. Xu, X.-Y., et al., The complete chloroplast genome sequence of spleen amaranth (Amaranthus dubius Mart. ex Thell., Amaranthaceae). *Mitochondrial DNA Part B*, 2021. **6**(11): pp. 3267–3268.
351. Wang, Q., et al., Amaranthus dubius Mart. ex Thell.-a newly naturalized plant of mainland of China. *Journal of Tropical and Subtropical Botany*, 2015. **23**(3): pp. 284–288.
352. Viljoen, E., et al., Application of chloroplast phylogenomics to resolve species relationships within the plant genus Amaranthus. *Journal of Molecular Evolution*, 2018. **86**(3–4): pp. 216–239.
353. Mailloux, J., et al., Influence of ground cover management on diversity and density of phytoseiid mites (Acari: Phytoseiidae) in Guadeloupean citrus orchards. *Experimental and Applied Acarology*, 2010. **52**: pp. 275–290.
354. Singh, A., U. Pathre, and P. Sane, Purification and characterization of leucine aminotransferase from green leaves of Amaranthus dubius. *Biochemie und Physiologie der Pflanzen*, 1991. **187**(5): pp. 337–345.
355. Akter, S., et al., Elemental analysis of Basella alba, Spinacia oleracea, Abelmoschus esculentus (L.), Ipomoea aquatica,

Colocasia esculenta, Amaranthus dubius, and Raphanus sativus vegetables using the PIXE technique in a saline region of Bangladesh, Rampal Area. *Biological Trace Element Research*, 2022. **200**(6): pp. 2999–3008.
356. Aguilar, E.G., et al., Multielemental analysis and classification of amaranth seeds according to their botanical origin. *Journal of Agricultural and Food Chemistry*, 2011. **59**(17): pp. 9059–9064.
357. Nabulo, G., H. Oryem-Origa, and M. Diamond, Assessment of lead, cadmium, and zinc contamination of roadside soils, surface films, and vegetables in Kampala City, Uganda. *Environmental Research*, 2006. **101**(1): pp. 42–52.
358. Nyonje, W.A., et al., Exploring the status of preference, utilization practices, and challenges to consumption of Amaranth in Kenya and Tanzania. *Journal of Nutrition and Metabolism*, 2022. **2022**. https://doi.org/10.1155/2022/2240724.
359. Kinyi, H.W., et al., Effect of cooking method on vitamin C loses and antioxidant activity of indigenous green leafy vegetables consumed in Western Uganda. *International Journal of Food Science*, 2022. **2022**. https://doi.org/10.1155/2022/2088034.
360. Moore, A.N., et al., Red spinach extract increases ventilatory threshold during graded exercise testing. *Sports*, 2017. **5**(4): p. 80.
361. Chandrashekhar, U. and M. Newman, Antidiabetic activity of Amaranthus dubious ethanolic leaf extract on alloxan induced diabetic mice. *International Journal of Current Microbiology and Applied Sciences*, 2014. **3**(2): pp. 584–590.
362. Koua, G.A., et al., Studies on nutritive value and antioxidant properties of three neglected leafy vegetables of Cote d'Ivoire. *Pakistan Journal of Nutrition*, 2015. **14**(12): p. 877.
363. Rauter, A.P., et al., Antihyperglycaemic and protective effects of flavonoids on streptozotocin–induced diabetic rats. *Phytotherapy Research*, 2010. **24**(S2): pp. S133–S138.
364. Nderitu, K.W., et al., Antiobesity activities of methanolic extracts of Amaranthus dubius, Cucurbita pepo, and Vigna unguiculata in progesterone-induced obese Mice. *Evidence-Based Complementary and Alternative Medicine*, 2017. **2017**.
365. Yong, Y.Y., et al., Biofilm inhibiting activity of betacyanins from red pitahaya (Hylocereus polyrhizus) and red spinach (Amaranthus dubius) against Staphylococcus aureus and Pseudomonas aeruginosa biofilms. *Journal of Applied Microbiology*, 2019. **126**(1): pp. 68–78.
366. Odhav, B., et al., Preliminary assessment of nutritional value of traditional leafy vegetables in KwaZulu-Natal, South Africa. *Journal of Food Composition and Analysis*, 2007. **20**(5): pp. 430–435.
367. Akula, U.S. and B. Odhav, In vitro 5-Lipoxygenase inhibition of polyphenolic antioxidants from undomesticated plants of South Africa. *Journal of Medicinal Plants Research*, 2008. **2**(9): pp. 207–212.
368. Tufts, H., et al., Antioxidant and anti-inflammatory activities of Kenyan leafy green vegetables, wild fruits, and medicinal plants with potential relevance for kwashiorkor. *Evidence-Based Complementary and Alternative Medicine*, 2015. **2015**. https://doi.org/10.1155/2015/807158.
369. Ramesar, S., et al., Angiotensin I-converting enzyme inhibitor activity of nutritive plants in KwaZulu-Natal. *Journal of Medicinal Food*, 2008. **11**(2): pp. 331–336.
370. Gordon, A., Case study: FSQS in solving market access prohibition for a vegetable product – callaloo (Amaranthus sp.), in André Gordon (ed.), Food Safety and Quality Systems in Developing Countries. 2017, Elsevier. pp. 217 244.
371. Kumaresan, V., et al., Fungal endophytes of some green leafy vegetables. *World Journal of Agricultural Sciences*, 2013. **9**(6): pp. 415–420.
372. Zhou, Y., et al., AtSWEET1 negatively regulates plant susceptibility to root-knot nematode disease. *Frontiers in Plant Science*, 2023. **14**: p. 1010348.
373. Ishtiaq, S., et al., Phytochemical and in vitro antioxidant evaluation of different fractions of Amaranthus graecizans subsp. silvestris (Vill.) Brenan. *Asian Pacific Journal of Tropical Medicine*, 2014. **7**: pp. S342–S347.
374. Ishtiaq, S., et al., Phytochemical and biological evaluations of methanolic extract of Amaranthus graecizans subsp. silvestris (Vill.) Brenan. *British Journal of Pharmaceutical Research*, 2017. **15**(3): pp. 1–11.
375. Grubben, G. and O.A. Denton, Plant resources of tropical Africa 2. Vegetables. Plant resources of tropical Africa 2. Vegetables. *Economic Botany*, **59**(4): pp. 401–402.
376. Ishtiaq, S., M.S.K. Afridi, and N. Masood, Amelioration of isoniazid and rifampicin-induced liver toxicity by Amaranthus graecizans subsp. silvestris in rat. *Bangladesh Journal of Pharmacology*, 2017. **12**(3): pp. 354–358.
377. Strzelecka, M., et al., Anti-inflammatory effects of extracts from some traditional Mediterranean diet plants. *Journal of Physiology and Pharmacology*, 2005. **56**(1): pp. 139–156.
378. Fuhrmann, M., et al., Uptake and release of cesium-137 by five plant species as influenced by soil amendments in field experiments. *Journal of Environmental Quality*, 2003. **32**(6): pp. 2272–2279.
379. Teri, J. and D. Mlasani, Choanephora blight and Alternaria leaf spot of Amaranth in Tanzania. *Plant Pathology*, 1994. **43**(1): pp. 228–229.
380. Kumari, S., R. Elancheran, and R. Devi, Phytochemical screening, antioxidant, antityrosinase, and antigenotoxic potential of Amaranthus viridis extract. *Indian Journal of Pharmacology*, 2018. **50**(3): p. 130.
381. Kaur, N., et al., A novel antiproliferative and antifungal lectin from Amaranthus viridis Linn seeds. *Protein and Peptide Letters*, 2006. **13**(9): pp. 897–905.
382. Ragasa, C.Y., et al., Chemical constituents of Amaranthus viridis. *Chemistry of Natural Compounds*, 2015. **51**: pp. 146–147.
383. Sarker, U. and S. Oba, Nutraceuticals, antioxidant pigments, and phytochemicals in the leaves of Amaranthus spinosus and Amaranthus viridis weedy species. *Scientific Reports*, 2019. **9**(1): p. 20413.
384. Silva, A.D., et al., In vitro bioaccessibility of proteins, phenolics, flavonoids and antioxidant activity of Amaranthus viridis. *Plant Foods for Human Nutrition*, 2021. **76**: pp. 478–486.
385. Oluwagunwa, O.A., A.M. Alashi, and R.E. Aluko, Inhibition of the in vitro activities of α-Amylase and Pancreatic Lipase by aqueous extracts of Amaranthus viridis, Solanum macrocarpon and Telfairia occidentalis Leaves. *Frontiers in Nutrition*, 2021. **8**: p. 772903.
386. Kwon, S.-Y., et al., A Ribosome–Inactivating Protein from Amaranthus viridis. *Bioscience, Biotechnology, and Biochemistry*, 1997. **61**(9): pp. 1613–1614.
387. Adetutu, A., et al., Inhibition of in vivo growth of Plasmodium berghei by Launaea taraxacifolia and Amaranthus viridis in mice. *Malaria Research and Treatment*, 2016. pp. 1–8.
388. Ayoka, O.A., et al., Neuro-endocrine effects of aqueous extract of Amaranthus viridis (Linn.) leaf in male Wistar rat model of cyclophosphamide-induced reproductive toxicity. *Toxicology Reports*, 2016. **3**: pp. 608–619.
389. Salvamani, S., et al., Anti-HMG-CoA reductase, antioxidant, and anti inflammatory activities of Amaranthus

viridis leaf extract as a potential treatment for hypercholesterolemia. *Evidence-Based Complementary and Alternative Medicine*, 2016. pp. 1–10.

390. Saravanan, G. and P. Ponmurugan, Amaranthus viridis Linn., a common spinach, modulates C-reactive protein, protein profile, ceruloplasmin and glycoprotein in experimental induced myocardial infarcted rats. *Journal of the Science of Food and Agriculture*, 2012. **92**(12): pp. 2459–2464.
391. Saravanan, G., et al., Cardioprotective activity of Amaranthus viridis Linn: effect on serum marker enzymes, cardiac troponin and antioxidant system in experimental myocardial infarcted rats. *International Journal of Cardiology*, 2013. **165**(3): pp. 494–498.
392. Ngweme, G.N., et al., Occurrence of organic micropollutants and human health risk assessment based on consumption of Amaranthus viridis, Kinshasa in the Democratic Republic of the Congo. *Science of the Total Environment*, 2021. **754**: p. 142175.
393. Yap, C.K., et al., Potentially toxic metals in the high-biomass non-hyperaccumulating plant Amaranthus viridis: human health risks and phytoremediation potentials. *Biology*, 2022. **11**(3): p. 389.
394. Azi, F., et al., Heavy metal and microbial safety assessment of raw and cooked pumpkin and Amaranthus viridis leaves grown in Abakaliki, Nigeria. *Food Science & Nutrition*, 2018. **6**(6): pp. 1537–1544.
395. Hasnain, S.M., K. Fatima, and A. Al-Frayh, Prevalence of airborne allergenic Amaranthus viridis pollen in seven different regions of Saudi Arabia. *Annals of Saudi Medicine*, 2007. **27**(4): pp. 259–263.
396. Singh, H.P., et al., α-Pinene inhibits growth and induces oxidative stress in roots. *Annals of Botany*, 2006. **98**(6): pp. 1261–1269.
397. Nascimento, D.M., et al., Survival of Curtobacterium flaccumfaciens pv. flaccumfaciens in weeds. *Plant Pathology*, 2020. **69**(7): pp. 1357–1367.
398. Rusinque, L., et al., Root-knot nematode assessment: species identification, distribution, and new host records in Portugal. *Frontiers in Plant Science*, 2023. **14**: pp. 1230968.
399. *Description of Anagallis arvensis*. Available from: https://en.wikipedia.org/wiki/Anagallis_arvensis
400. Salam, I., A. Moinuddin, and T.-A. Syed, Allelopathic effect of Scarlet pimpernel (Anagallis arvensis) on seed germination and radical elongation of mung bean and pearl millet. *Pakistan Journal of Botany*, 2011. **43**(1): pp. 351–355.
401. Sharifi-Rad, J., et al., Exploration of phytochemical and antibacterial potentiality of Anagallis arvensis L. extract against methicillin-resistant Staphylococcus aureus (MRSA). *British Biotechnology Journal*, 2016. **10**(2): pp. 1–8.
402. Yasmeen, Z., A. Basit, and S. Tahir, Traditional uses and pharmacological effects of Anagallis arvensis: a review: Anagallis arvensis: a review. *The International Journal of Frontier Sciences*, 2020. **4**(2): pp. 97–100.
403. Morozowska, M. and M. Wesołowska, In vitro clonal propagation of Primula veris L. and preliminary phytochemical analysis. *Acta Biologica Cracoviensia Series Botanica*, 2004. **46**: pp. 169–175.
404. Asolkar, L. and R. Chopra, *Second Supplement to Glossary of Indian medicinal Plants with Active Principles*. Publications & Information Directorate, 1992.
405. Yamada, Y., et al., Cucurbitacins from Anagallis arvensis. *Phytochemistry*, 1978. **17**(10): p. 1798.
406. Viegi, L., et al., A review of plants used in folk veterinary medicine in Italy as basis for a databank. *Journal of Ethnopharmacology*, 2003. **89**(2–3): pp. 221–244.
407. Akerreta, S., et al., Analyzing factors that influence the folk use and phytonomy of 18 medicinal plants in Navarra. *Journal of Ethnobiology and Ethnomedicine*, 2007. **3**: pp. 1–18.
408. López, V., R.Y. Cavero, and M.I. Calvo, Cytotoxic effects of Anagallis arvensis and Anagallis foemina in neuronal and colonic adenocarcinoma cell lines. *Pharmacognosy Journal*, 2013. **5**(1): pp. 2–5.
409. Bhatia, H., et al., Ethnomedicinal plants used by the villagers of district Udhampur, J&K, India. *Journal of Ethnopharmacology*, 2014. **151**(2): pp. 1005–1018.
410. López, V., et al., Pharmacological properties of Anagallis arvensis L.("scarlet pimpernel") and Anagallis foemina Mill. ("blue pimpernel") traditionally used as wound healing remedies in Navarra (Spain). *Journal of Ethnopharmacology*, 2011. **134**(3): pp. 1014–1017.
411. Shabbir, U., et al., Uroprotective and hepatoprotective potential of Anagallis arvensis against the experimental animal model. *Journal of Tropical Medicine*, 2022. **2022**.
412. Saleem, H., et al., Therapeutic propensities, phytochemical composition, and toxicological evaluation of Anagallis arvensis (L.): a wild edible medicinal food plant. *Food Research International*, 2020. **137**: p. 109651.
413. Agrawal, S.K., et al., Anagallis arvensis induces apoptosis in HL-60 cells through ROS-mediated mitochondrial pathway. *Nutrition and Cancer*, 2021. **73**(11–12): pp. 2720–2731.
414. Ahsan,H.,etal.,HepatoprotectivepotentialofAnagallisarvensis (L.) extract against carbon tetrachloride (CCl 4)-induced hepatic injury and oxidative stress in rabbits. *Pakistan Journal of Pharmaceutical Sciences*, 2023. **36**(4).
415. Clarke, E.G.C. and M.L. Clarke, *Veterinary Toxicology*. Bailliere Tindall, 1975.
416. Pttllar, E., Studies on five suspected poisonous plants. *Australian Veterinary Journal*, 1930. **15**: pp. 19–23.
417. Al-Sultan, S., Y. Hussein, and A. Hegazy, Toxicity of Anagallis arvensis plant. *Pakistan Journal of Nutrition*, 2003. **2**(3): pp. 116–122.
418. Saleem, H., et al., A comprehensive review on the botany, traditional uses, phytochemistry, pharmacology and toxicity of Anagallis arvensis (L).: a wild edible medicinal food plant. *Food Bioscience*, 2023. **52**: p. 102328.
419. Apak, R., et al., The cupric ion reducing antioxidant capacity and polyphenolic content of some herbal teas. *International Journal of Food Sciences and Nutrition*, 2006. **57**(5–6): pp. 292–304.
420. Stevens, M., H. Smith, and P. Hallsworth, The host range of beet yellowing viruses among common arable weed species. *Plant Pathology*, 1994. **43**(3): pp. 579–588.
421. Sharf, W., et al., Induction of resistance in chili against Sclerotium rolfsii by plant-growth-promoting rhizobacteria and Anagallis arvensis. *Egyptian Journal of Biological Pest Control*, 2021. **31**: pp. 1–11.
422. Khare, C., Indian Medicinal Plants-An Illustrated Dictionary. 1st Indian Reprint. Springer (India) Pvt. Ltd., New Delhi, India, 2007. p. 28.
423. Usmani, Q.I., et al., Aatrilal (Ammi majus L.), an important drug of Unani system of medicine: a review. *Journal of Ethnopharmacology*, 2021. **276**: p. 114144.
424. Selim, Y.A. and N.H. Ouf, Anti-inflammatory new coumarin from the Ammi majus L. *Organic and Medicinal Chemistry Letters*, 2012. **2**(1): pp. 1–4.
425. *Toxicity of Ammi majus*. Available from: https://plants.ces.ncsu.edu/plants/ammi-majus/
426. Hamerski, D. and U. Matern, Elicitor-induced biosynthesis of psoralens in Ammi majus L. suspension cultures:

microsomal conversion of demethylsuberosin into (+) marmesin and psoralen. *European Journal of Biochemistry*, 1988. **171**(1-2): pp. 369–375.
427. Curri, S.B. and E. Bombardelli, Pharmaceutical Compositions Having Activity on the Cutaneous Microcirculation. 1993.
428. Tsror, L., et al., Characterization of weeds and rotational crops as alternative hosts of Spongospora subterranea, the causal agent of powdery scab in Israel. *Plant Pathology*, 2020. **69**(2): pp. 294–301.
429. Tovar-Pedraza, J., et al., First report of powdery mildew caused by Erysiphe heraclei on Ammi majus in Mexico. *Plant Disease*, 2016. **100**(3): pp. 647–647.
430. Baker, C., et al., Bidens mottle virus and Apium virus Y identified in Ammi majus in Florida. *Plant Disease*, 2008. **92**(6): pp. 975–975.
431. Yoshikawa, M., et al., Anastatins A and B, new skeletal flavonoids with hepatoprotective activities from the desert plant Anastatica hierochuntica. *Bioorganic & Medicinal Chemistry Letters*, 2003. **13**(6): pp. 1045–1049.
432. Md Zin, S.R., et al., Mutagenicity evaluation of Anastatica hierochuntica L. aqueous extract in vitro and in vivo. *Experimental Biology and Medicine*, 2018. **243**(4): pp. 375–385.
433. Rameshbabu, S., et al., Anastatica hierochuntica (L.) methanolic and aqueous extracts exert antiproliferative effects through the induction of apoptosis in MCF-7 breast cancer cells. *Saudi Pharmaceutical Journal*, 2020. **28**(8): pp. 985–993.
434. Zin, S.R.M., et al., Potential toxicity effects of Anastatica hierochuntica aqueous extract on prenatal development of Sprague-Dawley rats. *Journal of Ethnopharmacology*, 2019. **245**: p. 112180.
435. Nakashima, S., et al., Melanogenesis inhibitors from the desert plant Anastatica hierochuntica in B16 melanoma cells. *Bioorganic & Medicinal Chemistry*, 2010. **18**(6): pp. 2337–2345.
436. El-Bondkly, A.M. and M.M. El-Gendy, Keratinolytic activity from new recombinant fusant AYA2000, derived from endophytic Micromonospora strains. *Canadian Journal of Microbiology*, 2010. **56**(9): pp. 748–760.
437. Ahmed, M., et al., Treatment of common ailments by plant-based remedies among the people of district Attock (Punjab) of Northern Pakistan. *African Journal of Traditional, Complementary and Alternative Medicines*, 2007. **4**(1): pp. 112–120.
438. Mill, S. and C. Hootelé, Alkaloids of Andrachne aspera. *Journal of Natural Products*, 2000. **63**(6): pp. 762–764.
439. Abebe, D. and A. Ayehu, Medicinal Plants and Enigmatic Health Practices of Northern Ethiopia. 1993. Berhanena Selam Printing Press.
440. Arshad, M., et al., An ethnobiological study in Kala Chitta hills of Pothwar region, Pakistan: multinomial logit specification. *Journal of Ethnobiology and Ethnomedicine*, 2014. **10**: pp. 1–17.
441. Devi, K.A., et al., The endophytic symbiont – Pseudomonas aeruginosa stimulates the antioxidant activity and growth of Achyranthes aspera L. *Frontiers in Microbiology*, 2017. **8**: p. 1897.
442. Brahmachari, G., D. Gorai, and R. Roy, Argemone mexicana: chemical and pharmacological aspects. *Revista Brasileira de Farmacognosia*, 2013. **23**: pp. 559–567.
443. Simoes-Pires, C., et al., Reverse pharmacology for developing an anti-malarial phytomedicine. The example of Argemone mexicana. *International Journal for Parasitology: Drugs and Drug Resistance*, 2014. **4**(3): pp. 338–346.
444. Nancy, A. and A. Praveena, Argemone mexicana: a boon to medicinal and pharmacological approaches in current scenario. *Cardiovascular & Hematological Agents in Medicinal Chemistry (Formerly Current Medicinal Chemistry-Cardiovascular & Hematological Agents)*, 2017. **15**(2): pp. 78–90.
445. Singh, S., et al., Cytotoxicity of alkaloids isolated from Argemone mexicana on SW480 human colon cancer cell line. *Pharmaceutical Biology*, 2016. **54**(4): pp. 740–745.
446. Sharma, B., et al., Epidemic dropsy: observations on pathophysiology and clinical features during the Delhi epidemic of 1998. *Tropical Doctor*, 2002. **32**(2): pp. 70–75.
447. Verma, S., et al., Argemone mexicana poisoning: autopsy findings of two cases. *Forensic Science International*, 2001. **115**(1–2): pp. 135–141.
448. Das, M. and S.K. Khanna, Clinicoepidemiological, toxicological, and safety evaluation studies on argemone oil. *Critical Reviews in Toxicology*, 1997. **27**(3): pp. 273–297.
449. Pahwa, R. and V. Chatterjee, The toxicity of Mexican poppy (Argemone mexicana L) seeds to rats. *Veterinary and Human Toxicology*, 1989. **31**(6): pp. 555–558.
450. Steyn, D.G., Poisoning with the seeds of Argemone mexicana (Mexican poppy) in human beings: Indian epidemic dropsy in South Africa. *South African Medical Journal*, 1950. **24**(18): pp. 333–339.
451. Singh, R., J. Kaur, and M. Thakar, Detection of argemone oil in the adulterated mustard oil by using colour tests and thin layer chromatography. *Anil Aggrawal's Internet Journal of Medicine and Toxicology*, 2004. **5**(2): p. 76.
452. Cicero, A.F. and A. Baggioni, Berberine and its role in chronic disease. *Anti-inflammatory Nutraceuticals and Chronic Diseases*, 2016. *928*: pp. 27–45.
453. Singh, P., et al., Antimicrobial, antioxidant, GC-MS analysis and molecular docking analysis of bioactive compounds of endophyte aspergillus flavus from argemone Mexicana. *Journal of Microbiology, Biotechnology and Food Sciences*, 2023. *13*: p. e9970.
454. Shokoohinia, Y., et al., Sesquiterpenes and Flavonoids of Anthemis odontostephana var. odontostephana. *Chemistry of Natural Compounds*, 2015. **51**: pp. 491–494.
455. Sajjadi, S.E., et al., Essential oil composition of flowers of Anthemis odontostephana Boiss. var. odontostephana. *Journal of Essential Oil Bearing Plants*, 2013. **16**(2): pp. 247–251.
456. Zebarjad, Z. and M.H. Farjam, Evaluation of antimicrobial activity of essential oils from different parts of Anthemis odontostephana Boiss. var. odontostephana. *International Journal of Pharmacognosy and Phytochemical Research*, 2015. **7**(3): pp. 579–84.
457. Zangeneh, M., Hematoprotective and nephroprotective properties of ethanolic extract of Anthemis odontostephana Boiss in Streptozotocin-induced diabetic mice. *Journal of Medicinal Plants*, 2018. **17**(65): : pp. 47–.
458. Zangeneh, A., et al., A stereological and biochemical examination: hepatoprotective activity of anthemis odontostephana Boiss. Ethanol extract against CCl4-induced hepatotoxicity in mice. *Research Journal of Pharmacognosy*, 2019. **6**(2): pp. 67–75.
459. Gough, H., The tortrix Cnephasia longana on fruit trees in Essex. *Plant Pathology*, 1952. **1**(1): pp. 31–31.
460. Al- Noumani, A.J., et al., Antimicrobial, antioxidant and cytotoxic activities of anogeissusdhofarica from oman. *International Journal of Recent Advances in Pharmaceutical Research*, 2013. **3**: pp. 35–38.
461. Singh, D., et al., The genus Anogeissus: a review on ethnopharmacology, phytochemistry and pharmacology. *Journal of Ethnopharmacology*, 2016. **194**: pp. 30–56.

462. Volkovitsh, M.G., Two new species of Acmaeodera Eschscholtz, 1829 from the Arabian Peninsula and Maurita nia (Coleoptera: Buprestidae: Polycestinae: Acmaeoderini). *Zootaxa*, 2019. **4648**(3): p. zootaxa.
463. Hussain, S., et al., Notes on the genus Micropsalliota (Agaricales, Basidiomycota) and the description of a new species from Southern Oman. *Phytotaxa*, 2022. **543**(2): pp. 113–126.
464. Li, M.-Y., et al., The genome sequence of celery (Apium graveolens L.), an important leaf vegetable crop rich in apigenin in the Apiaceae family. *Horticulture Research*, 2020. **7**.
465. Lin, L.-Z., S. Lu, and J.M. Harnly, Detection and quantification of glycosylated flavonoid malonates in celery, Chinese celery, and celery seed by LC-DAD-ESI/MS. *Journal of Agricultural and Food Chemistry*, 2007. **55**(4): pp. 1321–1326.
466. Hertog, M.G., P.C. Hollman, and D.P. Venema, Optimization of a quantitative HPLC determination of potentially anticarcinogenic flavonoids in vegetables and fruits. *Journal of Agricultural and Food Chemistry*, 1992. **40**(9): pp. 1591–1598.
467. Hedayati, N., et al., Beneficial effects of celery (Apium graveolens) on metabolic syndrome: a review of the existing evidences. *Phytotherapy Research*, 2019. **33**(12): pp. 3040–3053.
468. *Treating Kidney Stones with Tibb-e-Nabawi*. Available from: https://www.tibbenabawi.org/index.php?option=com_content&%20view=article&%20id=352&%20Itemid=394%20
469. Mencherini, T., et al., An extract of Apium graveolens var. dulce leaves: structure of the major constituent, apiin, and its anti-inflammatory properties. *Journal of Pharmacy and Pharmacology*, 2007. **59**(6): pp. 891–897.
470. Ovodova, R.G., et al., Chemical composition and anti-inflammatory activity of pectic polysaccharide isolated from celery stalks. *Food Chemistry*, 2009. **114**(2): pp. 610–615.
471. Tang, D., et al., Pharmacokinetic properties and drug interactions of apigenin, a natural flavone. *Expert Opinion on Drug Metabolism & Toxicology*, 2017. **13**(3): pp. 323–330.
472. Kun, L., et al., The quantitation of flavonoids in leaf and stalk of different celery cultivars and the correlation with antioxidation activity. *Acta Horticulturae Sinica*, 2011. **38**(1): p. 69.
473. Gadermaier, G., et al., Molecular characterization of Api g 2, a novel allergenic member of the lipid-transfer protein 1 family from celery stalks. *Molecular Nutrition & Food Research*, 2011. **55**(4): pp. 568–577.
474. Kooti, W., et al., A review on medicinal plant of Apium graveolens. *Advanced Herbal Medicine*, 2015. **1**(1): pp. 48–59.
475. Tyagi, S., et al., Medical benefits of Apium graveolens (celery herb). *Journal of Drug Discovery and Therapeutics*, 2013. **1**(5): pp. 36–38.
476. Breiteneder, H., et al., Molecular characterization of Api g 1, the major allergen of celery (Apium graveolens), and its immumological and structural relationships to a group of 17-kDa tree pollen allergens. *European Journal of Biochemistry*, 1995. **233**(2): pp. 484–489.
477. Donovan, A., S. Isaac, and H. Collin, Inhibitory effects of essential oil components extracted from celery (Apium graveolens) on the growth of Septoria apiicola, causal agent of leaf spot disease. *Plant Pathology*, 1993. **42**(5): pp. 691–700.
478. Kaur, R. and N. Dogra, A review on traditional uses, chemical constituents and pharmacology of Ageratum conyzoides L. (Asteraceae). *International Journal of Pharmaceutical and Biological Archives*, 2014. **5**(5): pp. 33–45.
479. Dũng, N.n.X., et al., Chemical composition of the oil of Ageratum conyzoides L. from Vietnam. *Journal of Essential Oil Research*, 1989. **1**(3): pp. 135–136.
480. Yadav, N., et al., Phytochemical constituents and ethnopharmacological properties of Ageratum conyzoides L. *Phytotherapy Research*, 2019. **33**(9): pp. 2163–2178.
481. Kamboj, A. and A.K. Saluja, Isolation of stigmasterol and β-sitosterol from petroleum ether extract of aerial parts of Ageratum conyzoides (Asteraceae). *nternational Journal of Pharmaceutical Sciences and Research*, 2011. **3**(1): pp. 94–96.
482. Bosi, C.F., et al., Pyrrolizidine alkaloids in medicinal tea of Ageratum conyzoides. *Revista Brasileira de Farmacognosia*, 2013. **23**(3): pp. 425–432.
483. Kotta, J.C., et al., Medicinal effect, in silico bioactivity prediction, and pharmaceutical formulation of Ageratum conyzoides L.: a review. *Scientifica*, 2020. **2020**: pp. 1–12.
484. de Mello, S.V.G.V., et al., Beneficial effect of Ageratum conyzoides Linn (Asteraceae) upon inflammatory response induced by carrageenan into the mice pleural cavity. *Journal of Ethnopharmacology*, 2016. **194**: pp. 337–347.
485. Destaa, T., et al., Isolation and structural elucidation of toxic pyrrolizidine alkaloids from Ageratum conyzoides collected from Vod disease affected communities. *International Journal of Pharmaceutical Technology*, 2014. **6**: pp. 6281–6290.
486. Diallo, A., et al., Acute and sub-chronic (28-day) oral toxicity studies of hydroalcohol leaf extract of Ageratum conyzoides L (Asteraceae). *Tropical Journal of Pharmaceutical Research*, 2010. **9**(5): pp. 463–467.
487. Saunders, K., et al., A unique virus complex causes Ageratum yellow vein disease. Proceedings of the National Academy of Sciences, 2000. **97**(12): pp. 6890–6895.
488. Saunders, K. and J. Stanley, A nanovirus-like DNA component associated with yellow vein disease of Ageratum conyzoides: evidence for interfamilial recombination between plant DNA viruses. *Virology*, 1999. **264**(1): pp. 142–152.
489. Sunaina, V., V. Kishore, and G. Shekhawat, Latent survival of Pseudomonas solanacearum in potato tubers and weeds/ Latentes Überleben von Pseudomonas solanacearum in Kartoffelknollen und Unkräutern. *Zeitschrift für Pflanzenkrankheiten und Pflanzenschutz/Journal of Plant Diseases and Protection*, 1989. **96**(4): pp. 361–364.
490. Zem, A. and L.G.E. Lordello, Studies on hosts of Radopholus similis and Helicotylenchus multicinctus. Trabalhos apresentados à VII Reunião Brasileira de Nematologia, Brasília, DF, 21-25 de fevereiro de 1983. *Publicação*1983. 7: pp. 175–187.
491. *Geographical distribution of Arnebia decumbens*. Available from: https://www.cabi.org/isc/datasheet/112432
492. Eddine, T.N., et al., Antioxidant and antimicrobial activity of flavonoids fraction extract from Arnebia decumbens (Vent) growing in South East Algeria. *International Journal of Current Pharmaceutical Review and Research*, 2016. **7**: pp. 110–116.
493. Al-Mussawi, A.A., Isolation and identification of shikonin from Arnebia decumbens L. and its antibacterial activity. *Journal of Applied Sciences Research*, 2010. **6**(9): pp. 1452–1456.
494. Taghizadeh, S.F., et al., Cytotoxic activity of cis-(E)-and trans-(Z)-spiroethers isolated from various Arnebia species. *South African Journal of Botany*, 2021. **142**: pp. 114–123.

495. Singh, B., et al., Anti-inflammatory activity of shikonin derivatives from Arnebia hispidissima. *Phytomedicine*, 2003. **10**(5): pp. 375–380.
496. He, J.-M., et al., Isovalerylshikonin, a new resistance-modifying agent from Arnebia euchroma, supresses antimicrobial resistance of drug-resistant Staphylococcus aureus. *International Journal of Antimicrobial Agents*, 2019. **53**(1): pp. 70–73.
497. Ahmad, L., et al., Toxic pyrrolizidine alkaloids provide a warning sign to overuse of the ethnomedicine Arnebia benthamii. *Journal of Ethnopharmacology*, 2018. **210**: pp. 88–94.
498. Smyrska-Wieleba, N., K.K. Wojtanowski, and T. Mroczek, Comparative HILIC/ESI-QTOF-MS and HPTLC studies of pyrrolizidine alkaloids in flowers of Tussilago farfara and roots of Arnebia euchroma. *Phytochemistry Letters*, 2017. **20**: pp. 339–349.
499. Yuzbasioglu Baran, M., et al., Cytotoxic naphthoquinones from Arnebia densiflora (Nordm.) Ledeb and determining new apoptosis inducers. *Natural Product Research*, 2020. **34**(12): pp. 1669–1677.
500. Hosseini, A., et al., The traditional medicine aspects, biological activity and phytochemistry of Arnebia spp. *Medicinski Glasnik*, 2018. **15**(1): pp. 1–9.
501. Xu, H., et al., Authentication of three source spices of Arnebiae Radix using DNA barcoding and HPLC. *Frontiers in Pharmacology*, 2021. **12**: p. 677014.
502. Sardari, S., M.A. Shokrgozar, and G. Ghavami, Cheminformatics based selection and cytotoxic effects of herbal extracts. *Toxicology in Vitro*, 2009. **23**(7): pp. 1412–1421.
503. Roeder, E. and B. Rengel, *Pyrrolizidine Alkaloids from Lithospermum Erythrorhizon*, Phytochemistry, 1990.
504. Singh, B. and R.A. Sharma, Yield enhancement of phytochemicals by Azotobacter chroococcum biotization in hairy roots of Arnebia hispidissima. *Industrial Crops and Products*, 2016. **81**: pp. 169–175.
505. Boulos, L., *Flora of Egypt. Volume 2, Geraniaceae – Boraginaceae*. Cairo, Egypt: Al Hadara Pub., 2000.
506. Sharma, R., et al., Ethnomedicinal, pharmacological properties and chemistry of some medicinal plants of Boraginaceae in India. *Journal of Medicinal Plants Research*, 2009. **3**(13): pp. 1153–1175.
507. Phulwaria, M. and N. Shekhawat, An efficient in vitro shoot regeneration from immature inflorescence and ex vitro rooting of Arnebia hispidissima (Lehm). DC.-A red dye (Alkannin) yielding plant. *Physiology and Molecular Biology of Plants*, 2013. **19**: pp. 435–441.
508. Shukla, Y.N., et al., New naphthoquinones from Arnebia hispidissima roots. *Planta Medica*, 2001. **67**(06): pp. 575–577.
509. Hamdard, M., et al., Revised phytochemical study of Arnebia hispidissima. *Pakistan Journal of Pharmaceutical Sciences*, 1988. **1**(1): pp. 19–20.
510. Singh, B., et al., Estimation of naphthaquinones from Arnebia hispidissima (Lehm.) DC. in vivo and in vitro. I. anti-inflammatory screening. *Phytotherapy Research: An International Journal Devoted to Pharmacological and Toxicological Evaluation of Natural Product Derivatives*, 2004. **18**(2): pp. 154–159.
511. Ahmed, A.F., H.-E.A. Saad, and E.M. Abd El-Karim, Two polycyclic geranylhydroquinone-derived metabolites from roots of Arnebia hispidissima (Lehm.) DC. *Molecules*, 2014. **19**(5): pp. 5940–5951.
512. Jain, A. and S. Mathur, A chemical study of the pigment of Arnebia. *Bulletin of the National Institute of Sciences of India*, 1965. **28**: pp. 52–56.
513. Shukla, Y., et al., Chemical constituents of the antibiotic fraction of Arnebia nobilis. *Experientia*, 1969. **25**: pp. 357–358.
514. Tandon, V. K., et al., Synthesis and evaluation of novel 1,4-naphthoquinone derivatives as antiviral, antifungal and anticancer agents. *Bioorganic & Medicinal Chemistry Letters*, 2024. **14**(11): pp. 2901–2904. https://doi.org/10.1016/j.bmcl.2004.03.047
515. Khan, H.A., I. Chandrasekharan, and A. Ghanim, Naphthazarins from Arnebia hispidissima. *Phytochemistry*, 1983. **22**(2): pp. 614–615.
516. Uses *of Arnebia hispidissima (Lehm.) DC*. Available from: https://niscpr.res.in/includes/images/wealthofindia/wealth-of-indiaFolder2010.pdf
517. Sunil, C., et al., Antioxidant, free radical scavenging and liver protective effects of friedelin isolated from Azima tetracantha Lam. leaves. *Food Chemistry*, 2013. **139**(1-4): pp. 860–865.
518. Antonisamy, P., V. Duraipandiyan, and S. Ignacimuthu, Anti-inflammatory, analgesic and antipyretic effects of friedelin isolated from Azima tetracantha Lam. in mouse and rat models. *Journal of Pharmacy and Pharmacology*, 2011. **63**(8): pp. 1070–1077.
519. Duraipandiyan, V., M. Gnanasekar, and S. Ignacimuthu, Antifungal activity of triterpenoid isolated from Azima tetracantha leaves. *Folia Histochemica et Cytobiologica*, 2010. **48**(2): pp. 311–313.
520. Ismail, T.S., et al., Anti-inflammatory activity of Salacia oblonga Wall. and Azima tetracantha Lam. *Journal of Ethnopharmacology*, 1997. **56**(2): pp. 145–152.
521. Hepsibha, B.T., et al., In vitro studies on antioxidant and free radical scavenging activities of Azima tetracantha Lam leaf extracts. *Indian Journal of Science and Technology*, 2010. **3**(5): pp. 571–577.
522. Begum, T.N., et al., Effects of ethanolic leaf extract of Azim tetracantha Lam. on Ehrlich ascites carcinoma tumour bearing mice. *Research Journal of Medicine and Medical Sciences*, 2009. **4**(2): pp. 351–354.
523. Arthika, S., et al., Hepatoprotective activity of the ethanolic extract of Azima tetracantha against paracetamol-induced hepatotoxicity in wistar albino rats. *Journal of Advances in Pharmacy and Healthcare Research*, 2011. **1**(2): pp. 14–21.
524. Konda, V.R., et al., Nephroprotective effect of ethanolic extract of Azima tetracantha root in glycerol induced acute renal failure in Wistar albino rats. *Journal of Traditional and Complementary Medicine*, 2016. **6**(4): pp. 347–354.
525. Janardhan, B., et al., In vitro screening and evaluation of antivenom phytochemicals from Azima tetracantha Lam. leaves against Bungarus caeruleus and Vipera russelli. *Journal of Venomous Animals and Toxins Including Tropical Diseases*, 2014. **20**: pp. 1–8.
526. Antonisamy, P., et al., Protective effects of friedelin isolated from Azima tetracantha Lam. against ethanol-induced gastric ulcer in rats and possible underlying mechanisms. *European Journal of Pharmacology*, 2015. **750**: pp. 167–175.
527. Al-Fatimi, M., et al., Antioxidant, antimicrobial and cytotoxic activities of selected medicinal plants from Yemen. *Journal of Ethnopharmacology*, 2007. **111**(3): pp. 657–666.
528. Elgorban, A.M., et al., Natural products of Alternaria sp., an endophytic fungus isolated from Salvadora persica from Saudi Arabia. *Saudi Journal of Biological Sciences*, 2019. **26**(5): pp. 1068–1077.
529. Samia, H., K. Elmalik, and H. Khalid, Therapeutic effect of Aristolochia bractealata extract against experimental Trypanosoma evansi infection. *International Journal of Tropical Medicine*, 2006. **1**(4): pp. 170–172.

530. Abdelgadir, A.A., E.M. Ahmed, and M.S. Eltohami, Isolation, characterization and quantity determination of aristolochic acids, toxic compounds in Aristolochia bracteolata L. *Environmental Health Insights*, 2011. **5**: p. EHI. S6292.
531. Kennedy, D.O. and E.L. Wightman, Herbal extracts and phytochemicals: plant secondary metabolites and the enhancement of human brain function. *Advances in Nutrition*, 2011. **2**(1): pp. 32–50.
532. Ahmed, E.-H.M., et al., Antiplasmodial activity of some medicinal plants used in Sudanese folk-medicine. *Environmental Health Insights*, 2010. **4**: p. EHI. S4108.
533. Elango, G., et al., Screening for feeding deterrent activity of herbal extracts against the larvae of malaria vector Anopheles subpictus Grassi. *Parasitology Research*, 2011. **109**: pp. 715–726.
534. Burkill, H., The Useful Plants of West Tropical. New York, USA: Royal Botanic Gardens, 1985.
535. Thirumal, M., et al., Aristolochia bracteolata: an overview on pharmacognostical, phytochemical and pharmacological properties. *Critical review in pharmaceutical sciences*, 2012. **1**(1): pp. 70–82.
536. El Tahir, A., G.M. Satti, and S.A. Khalid, Antiplasmodial activity of selected Sudanese medicinal plants with emphasis on Maytenus senegalensis (Lam.) Exell. *Journal of ethnopharmacology*, 1999. **64**(3): pp. 227–233.
537. Li, Y.-p., et al., Aristolochia bracteolate Retz. attenuates hyperuricemia in a metabolic arthritis rat model. *African journal of traditional, complementary and alternative medicines*, 2017. **14**(4): pp. 180–187.
538. Mathew, L.S., et al., Ethno medical knowledge and traditional use of Aristolochia bracteolata Lam. for malaria among local communities in Jubek State of South Sudan: a cross-sectional survey. *Journal of Ethnopharmacology*, 2021. **279**: p. 114314.
539. Gbadamosi, I. and A. Egunyomi, In–vitro propagation and antimycotic potential of extracts and essential oil of roots of Aristolochia Bracteolata Linn. (Aristolochiaceae). *African Journal of Traditional, Complementary and Alternative Medicines*, 2012. **9**(1): pp. 50–55.
540. Thanh, N.C., et al., Silver nanoparticles (AgNPs) fabricating potential of aqueous shoot extract of Aristolochia bracteolata and assessed their antioxidant efficiency. *Environmental Research*, 2022. **208**: p. 112683.
541. Chitme, H.R., et al., Antiallergic activity of Aristolochia bracteolata Lank in animal model. *Indian Journal of Experimental Biology*, 2010. **48**(1): pp. 46–52.
542. Kalachaveedu, M., et al., Fabrication and characterization of herbal drug enriched Guar galactomannan based nanofibrous mats seeded with GMSC's for wound healing applications. *International Journal of Biological Macromolecules*, 2020. **148**: pp. 737–749.
543. Sakthivel, G., et al., In vitro and in vivo evaluation of polyherbal formulation against Russell's viper and cobra venom and screening of bioactive components by docking studies. *Evidence-Based Complementary and Alternative Medicine*, 2013. **2013**.
544. Suliman Mohamed, M., et al., Activity of Aristolochia bracteolata against Moraxella catarrhalis. *International Journal of Bacteriology*, 2014. **2014**: pp. 481686.doi: 10.1155/2014/481686.
545. Shirwaikar, A., et al., Wound healing studies of Aristolochia bracteolata Lam. with supportive action of antioxidant enzymes. *Phytomedicine*, 2003. **10**(6–7): pp. 558–562.
546. Murugan Girija, D., et al., Osteogenic differentiation of human gingival mesenchymal stem cells by Aristolochia bracteolata supplementation through enhanced Runx2 expression. *Journal of Cellular Physiology*, 2017. **232**(7): pp. 1591–1595.
547. Heinrich, M., et al., Local uses of Aristolochia species and content of nephrotoxic aristolochic acid 1 and 2 – A global assessment based on bibliographic sources. *Journal of Ethnopharmacology*, 2009. **125**(1): pp. 108–144.
548. Debelle, F.D., J.-L. Vanherweghem, and J.L. Nortier, Aristolochic acid nephropathy: a worldwide problem. *Kidney International*, 2008. **74**(2): pp. 158–169.
549. Gillerot, G., et al., Aristolochic acid nephropathy in a Chinese patient: time to abandon the term "Chinese herbs nephropathy"? *American Journal of Kidney Diseases*, 2001. **38**(5): p. E26.
550. Hoang, M.L., et al., Mutational signature of aristolochic acid exposure as revealed by whole-exome sequencing. *Science Translational Medicine*, 2013. **5**(197): p. 197ra102.
551. Poon, S.L., et al., Genome-wide mutational signatures of aristolochic acid and its application as a screening tool. *Science Translational Medicine*, 2013. **5**(197): p. 197ra101.
552. *Carcinogens toxicity*. Available from: https://ntp.niehs.nih.gov/whatwestudy/assessments/cancer/roc
553. Thongkhao, K., C. Tungphatthong, and S. Sukrong, A PCR-lateral flow immunochromatographic assay (PCR-LFA) for detecting Aristolochia species, the plants responsible for aristolochic acid nephropathy. *Scientific Reports*, 2022. **12**(1): p. 12188.
554. Guevara-Araya, M., et al., Differences in community composition of endophytic fungi between above-and below-ground tissues of Aristolochia chilensis in an arid ecosystem. *Revista Chilena de Historia Natural*, 2020. **93**(1): pp. 1–9.
555. Moufid, A. and M. Eddouks, Artemisia herba alba: a popular plant with potential medicinal properties. *Pakistan Journal of Biological Sciences: PJBS*, 2012. **15**(24): pp. 1152–1159.
556. Mahboubi, M., Artemisia sieberi Besser essential oil and treatment of fungal infections. *Biomedicine & Pharmacotherapy*, 2017. **89**: pp. 1422–1430.
557. Ghasemi, G., et al., Phytochemical properties of essential oil from Artemisia sieberi Besser (Iranian accession) and its antioxidant and antifungal activities. *Natural Product Research*, 2021. **35**(21): pp. 4154–4158.
558. Mohamed, T.A., et al., Antimicrobial sesquiterpene lactones from Artemisia sieberi. *Journal of Asian Natural Products Research*, 2017. **19**(11): pp. 1093–1101.
559. Guetat, A., F.A. Al-Ghamdi, and A.K. Osman, The genus Artemisia L. in the northern region of Saudi Arabia: essential oil variability and antibacterial activities. *Natural Product Research*, 2017. **31**(5): pp. 598–603.
560. Bagheri, R., et al., Grazing affects essential oil compositions of Artemisia sieberi Besser. *Pakistan Journal of Biological Sciences: PJBS*, 2007. **10**(5): pp. 810–813.
561. Zomorodian, K., et al., Antimicrobial activity of seven essential oils from Iranian aromatic plants against common causes of oral infections. *Jundishapur Journal of Microbiology*, 2015. **8**(2): pp. e17766. https://doi.org/10.5812/jjm.17766.
562. Vakili, Z., et al., In vitro effects of Artemisia sieberi on Echinococcus granulosus protoscolices. *Experimental Parasitology*, 2019. **197**: pp. 65–67.
563. Naseri, B., et al., Fumigant toxicity and sublethal effects of Artemisia khorassanica and Artemisia sieberi on Sitotroga

cerealella (Lepidoptera: Gelechiidae). *Journal of Insect Science*, 2017. **17**(5): p. 100.
564. Mahboubi, M. and N. Kazempour, The antifungal activity of Artemisia sieberi essential oil from different localities of Iran against dermatophyte fungi. *Journal de Mycologie Medicale*, 2015. **25**(2): pp. e65–e71.
565. Ali, E.M. and B.M. Abdallah, Effective inhibition of invasive pulmonary aspergillosis by silver nanoparticles biosynthesized with Artemisia sieberi leaf extract. *Nanomaterials*, 2021. **12**(1): p. 51.
566. Akhtari, A., et al., Antifungal activities of pure and ZnO-encapsulated essential oil of Zataria multiflora on Alternaria solani as the pathogenic agent of tomato early blight disease. *Frontiers in Plant Science*, 2022. **13**: p. 932475.
567. Irshaid, F., K. Mansi, and T. Aburjai, Antidiabetic effect of essential oil from Artemisia sieberi growing in Jordan in normal and alloxan induced diabetic rats. *Pakistan Journal of Biological Sciences: PJBS*, 2010. **13**(9): pp. 423–430.
568. Irshaid, F., et al., Hepatoprotetive, cardioprotective and nephroprotective actions of essential oil extract of Artemisia sieberi in alloxan induced diabetic rats. *Iranian Journal of Pharmaceutical Research: IJPR*, 2012. **11**(4): p. 1227.
569. Tabari, M.A., M.R. Youssefi, and G. Benelli, Eco-friendly control of the poultry red mite, Dermanyssus gallinae (Dermanyssidae), using the α-thujone-rich essential oil of Artemisia sieberi (Asteraceae): toxic and repellent potential. *Parasitology Research*, 2017. **116**: pp. 1545–1551.
570. Khalaji, S., et al., Black cumin seeds, Artemisia leaves (Artemisia sieberi), and Camellia L. plant extract as phytogenic products in broiler diets and their effects on performance, blood constituents, immunity, and cecal microbial population. *Poultry Science*, 2011. **90**(11): pp. 2500–2510.
571. Towhidi, A., T. Saberifar, and E. Dirandeh, Nutritive value of some herbage for dromedary camels in the central arid zone of Iran. *Tropical Animal Health and Production*, 2011. **43**: pp. 617–622.
572. Heydari, F.E., et al., Comparison between in vitro effects of aqueous extract of Artemisia seiberi and artemisinin on Leishmania major. *Jundishapur Journal of Natural Pharmaceutical Products*, 2013. **8**(2): p. 70.
573. Ammar, N.M., et al., Gastro-protective effect of Artemisia sieberi essential oil against ethanol-induced ulcer in rats as revealed via biochemical, histopathological and metabolomics analysis. *Biomarkers*, 2022. **27**(3): pp. 247–257.
574. Zamani, S., et al., Sesquiterpene fractions of Artemisia plants as potent inhibitors of inducible nitric oxide synthase and cyclooxygenase-2 expression. *Iranian Journal of Basic Medical Sciences*, 2019. **22**(7): p. 774.
575. Farzaneh, M., et al., Chemical composition and antifungal activity of the essential oils of three species of Artemisia on some soil-borne phytopathogens. *Communications in AgriculturalandAppliedBiologicalSciences*,2006.**71**(3PtB): pp. 1327–1333.
576. Arab, H., et al., Determination of artemisinin in Artemisia sieberi and anticoccidial effects of the plant extract in broiler chickens. *Tropical Animal Health and Production*, 2006. **38**: pp. 497–503.
577. Nahrevanian, H., et al., Antimalarial effects of Iranian flora Artemisia sieberi on Plasmodium berghei in vivo in mice and phytochemistry analysis of its herbal extracts. *Malaria Research and Treatment*, 2012. **2012**: pp. 727032.doi: 10.1155/2012/727032.
578. Rad, F., et al., Randomized comparative clinical trial of Artemisia sieberi 5% lotion and clotrimazole 1% lotion for the treatment of pityriasis versicolor. *Indian Journal of Dermatology*, 2008. **53**(3): p. 115.
579. Kheirabadi, K.P., et al., Comparison of the anticoccidial effect of granulated extract of Artemisia sieberi with monensin in experimental coccidiosis in broiler chickens. *Experimental Parasitology*, 2014. **141**: pp. 129–133.
580. Kaboutari, J., et al., Prophylactic and therapeutic effects of a novel granulated formulation of Artemisia extract on broiler coccidiosis. *Tropical Animal Health and Production*, 2014. **46**: pp. 43–48.
581. Yarahmadi, F., et al., Investigating contact toxicity of Geranium and Artemisia essential oils on Bemisia tabaci Gen. *Avicenna Journal of Phytomedicine*, 2013. **3**(2): p. 106.
582. Hemmati, C. and M. Nikooei, Phytoplasma infection could affect chemical composition of Artemisia sieberi. *The Plant Pathology Journal*, 2019. **35**(3): p. 274.
583. Renault, S., et al., Effects of NaCl and Na2SO4 on red-osier dogwood (Cornus stolonifera Michx) seedlings. *Plant and Soil*, 2001. **233**: pp. 261–268.
584. Kaya, H., et al., FASCIATA genes for chromatin assembly factor-1 in Arabidopsis maintain the cellular organization of apical meristems. *Cell*, 2001. **104**(1): pp. 131–142.
585. Parida, A.K. and A.B. Das, Salt tolerance and salinity effects on plants: a review. *Ecotoxicology and Environmental Safety*, 2005. **60**(3): pp. 324–349.
586. Murakeözy, E., et al., Phylogenetic relationships and genetic diversity of the Salicornieae (Chenopodiaceae) native to the Atlantic coasts of France. *Plant Systematics and Evolution*, 2007. **264**: pp. 217–237.
587. Rodrigues, M.J., et al., Maritime halophyte species from southern Portugal as sources of bioactive molecules. *Marine Drugs*, 2014. **12**(4): pp. 2228–2244.
588. El-Wahab, R., et al., Diversity and distribution of medicinal plants in North Sinai, Egypt. *African Journal of Environmental Science and Technology*, 2008. **2**(7): pp. 157–171.
589. Gairola, S., A. Bhatt, and A. El-Keblawy, A perspective on potential use of halophytes for reclamation of salt-affected lands. *Wulfenia*, 2015. **22**(1): pp. 88–97.
590. Jurado-Mañogil, C., et al., Efficient in vitro platform for multiplication, acclimatization, and deliver of high-NaCl-tolerant clones of the Halophyte Arthrocaulon macrostachyum. *Journal of Plant Growth Regulation*, 2024: **43**: pp. 1–11
591. Sánchez-Gavilán, I., E. Ramírez Chueca, and V. de la Fuente García, Bioactive compounds in Sarcocornia and Arthrocnemum, two wild halophilic genera from the Iberian Peninsula. *Plants*, 2021. **10**(10): p. 2218.
592. Custódio, L., et al., The marine halophytes Carpobrotus edulis L. and Arthrocnemum macrostachyum L. are potential sources of nutritionally important PUFAs and metabolites with antioxidant, metal chelating and anticholinesterase inhibitory activities. *Botanica Marina*, 2012. **55**(3): pp. 281–288.
593. Zabka, M., R. Pavela, and L. Gabrielova-Slezakova, Promising antifungal effect of some Euro-Asiatic plants against dangerous pathogenic and toxinogenic fungi. *Journal of the Science of Food and Agriculture*, 2011. **91**(3): pp. 492–497.
594. Agoramoorthy, G., et al., Evaluation of antioxidant polyphenols from selected mangrove plants of India. *Asian Journal of Chemistry*, 2008. **20**(2): p. 1311.
595. Boulaaba, M., et al., In vitro antiproliferative effect of Arthrocnemum indicum extracts on Caco-2 cancer cells through cell cycle control and related phenol LC-TOF-MS

identification. *Evidence-Based Complementary and Alternative Medicine*, 2013. **2013**.

596. Ksouri, R., et al., Medicinal halophytes: potent source of health promoting biomolecules with medical, nutraceutical and food applications. *Critical Reviews in Biotechnology*, 2012. **32**(4): pp. 289–326.
597. Sekii, Y., et al., Two isorhamnetin glycosides from Arthrocnemum glaucum that inhibit adipogenesis in 3T3-L1 adipocytes. *Chemistry of Natural Compounds*, 2015. **51**: pp. 338–340.
598. Al-Tohamy, R., et al., Phytochemical analysis and assessment of antioxidant and antimicrobial activities of some medicinal plant species from Egyptian flora. *Journal of Applied Biomedicine*, 2018. **16**(4): pp. 289–300.
599. Barcia-Piedras, J.M., et al., Effect of prior salt experience on desalination capacity of the halophyte Arthrocnemum macrostachyum. *Desalination*, 2019. **463**: pp. 50–54.
600. Accogli, R., et al., Edible halophytes and halo-tolerant species in Apulia region (Southeastern Italy): biogeography, traditional food use and potential sustainable crops. *Plants*, 2023. **12**(3): p. 549.
601. ŁAZARSKI, G., Asperugo procumbens: a review of botany, traditional uses, phytochemistry and pharmacology. *Pakistan Journal of Medical and Health Sciences*, 2021. **15**(10).
602. Ahanjan, M., D. Mohana, and K. Raveesha, Antibacterial activity of Asperugo procumbens L. against some human pathogenic bacteria. *African Journal of Microbiology Research*, 2008. **2**(6): pp. 138–140.
603. Mirshafa, S.-A., M. Azadbakht, and N. Ahangar, Study of antidepressant and sedative-hypnotic activity of hydroalcoholic extract of Asperugo procumbens L. aerial parts in mice. *Iranian Journal of Pharmaceutical Research: IJPR*, 2013. **12**(3): p. 529.
604. Karimzadeh, S. and K.-B. Fotouhifar, Report of some fungi of Pleosporaceae family associated with leaf spot symptoms of plants in Chaharmahal and Bakhtiari province, Iran. *Journal of Crop Protection*, 2021. **10**(2): pp. 319–340.
605. Imperato, F., A flavanone glycoside from the fronds of Ceterach officinarum. *Phytochemistry*, 1983. **22**(1): pp. 312–313.
606. Froissard, D., et al., Asplenioideae species as a reservoir of volatile organic compounds with potential therapeutic properties. *Natural Product Communications*, 2015. **10**(6): p. 1934578X1501000671.
607. Živković, S., et al., Rehydration process in rustyback fern (Asplenium ceterach l.): profiling of volatile organic compounds. *Biology*, 2021. **10**(7): p. 574.
608. Tomou, E.-M. and H. Skaltsa, Phytochemical investigation of the fern Asplenium ceterach (Aspleniaceae). *Natural Product Communications*, 2018. **13**(7): p. 1934578X1801300715.
609. PEKGÖZ, A. and I. Çinbilgel, Phytochemical contents and antioxidant activities of fern, Asplenium ceterach L. In different altitudes. *Bangladesh Journal of Botany*, 2019. **48**(2).
610. Farràs, A., et al., Exploring wild Aspleniaceae ferns as safety sources of polyphenols: the case of Asplenium trichomanes L. and Ceterach officinarum Willd. *Frontiers in Nutrition*, 2022. **9**: p. 994215.
611. Petkov, V., et al., Selective anticancer properties, proapoptotic and antibacterial potential of three Asplenium species. *Plants*, 2021. **10**(6): p. 1053.
612. Karadeniz, A., et al., Antioxidant activity of some Turkish medicinal plants. *Natural Product Research*, 2015. **29**(24): pp. 2308–2312.
613. Lianza, M., et al., Screening of ninety herbal products of commercial interest as potential ingredients for phytocosmetics. *Journal of Enzyme Inhibition and Medicinal Chemistry*, 2020. **35**(1): pp. 1287–1291.
614. De Bellis, R., et al., In vitro effects on calcium oxalate crystallization kinetics and crystal morphology of an aqueous extract from Ceterach officinarum: analysis of a potential antilithiatic mechanism. *PLoS One*, 2019. **14**(6): p. e0218734.
615. Trewick, S., et al., Polyploidy, phylogeography and Pleistocene refugia of the rockfern Asplenium ceterach: evidence from chloroplast DNA. *Molecular Ecology*, 2002. **11**(10): pp. 2003–2012.
616. Passalacqua, N., P. Guarrera, and G. De Fine, Contribution to the knowledge of the folk plant medicine in Calabria region (Southern Italy). *Fitoterapia*, 2007. **78**(1): pp. 52–68.
617. Jones, D. and R. Baker, Introductions of non-native plant pathogens into Great Britain, 1970–2004. *Plant Pathology*, 2007. **56**(5): pp. 891–910.
618. Abd El-Fattah, H., Chemistry of Asphodelus fistulosus. *International Journal of Pharmacognosy*, 1997. **35**(4): pp. 274–277.
619. Fell, K., F. Hammouda, and A. Rizk, The constituents of the seeds of Asphodelus microcarpus Viviani and A. fistulosus L. *Journal of Pharmacy and Pharmacology*, 1968. **20**(8): pp. 646–649.
620. Al-Qudah, M., Antibacterial effect of Asphodelus fistulosus aqueous and ethanolic crude extracts on gram positive and gram negative bacteria. *Brazilian Journal of Biology*, 2022. **84**.
621. McGonigle, T.P., et al., A new method which gives an objective measure of colonization of roots by vesicular – arbuscular mycorrhizal fungi. *New Phytologist*, 1990. **115**(3): pp. 495–501.
622. Cavagnaro, T., et al., Quantitative development of Paris-type arbuscular mycorrhizas formed between Asphodelus fistulosus and Glomus coronatum. *New Phytologist*, 2001. **149**(1): pp. 105–113.
623. Solis-Hernández, A.P., et al., Effect of the natural establishment of two plant species on microbial activity, on the composition of the fungal community, and on the mitigation of potentially toxic elements in an abandoned mine tailing. *Science of the Total Environment*, 2022. **802**: p. 149788.
624. Dhar, P.P., A.A. Al-Qarawi, and M.A. Mridha, Arbuscular mycorrhizal fungal association in Asteraceae plants growing in the arid lands of Saudi Arabia. *Journal of Arid Land*, 2015. **7**: pp. 676–686.
625. Rios, J. and P. Waterman, A review of the pharmacology and toxicology of Astragalus. *Phytotherapy Research: An International Journal Devoted to Medical and Scientific Research on Plants and Plant Products*, 1997. **11**(6): pp. 411–418.
626. Pistelli, L., Secondary metabolites of genus Astragalus: Structure and biological activity. Studies in natural products chemistry. *Studies in Natural Products Chemistry*, 2002. **27**: pp. 443–545.
627. Nosrati, F., et al., Autecology of Astragalus fasciculifolius Bioss. in some natural habitats of south Sistan and Baluchestan province. *Iranian Journal of Medicinal and Aromatic Plants*, 2018. **34**(4).
628. Evans, W.C., Trease and *Evans' Pharmacognosy*. Elsevier Health Sciences, 2009.
629. Bratkov, V.M., et al., Flavonoids from the genus Astragalus: phytochemistry and biological activity. *Pharmacognosy Reviews*, 2016. **10**(19): p. 11.

630. Taherian, A., et al., The inhibition effects of two species of Astragalus extracts on mushroom tyrosinase activity. *Journal of Applied Biotechnology Reports*, 2014. **1**(2): pp. 63–66.
631. Shahrani, M., et al., Astragalus fasciculifolius manna; antinociceptive, antiinflammatoryand antioxidant properties in mice. *Immunopathologia Persa*, 2020. **7**(1): p. e02.
632. Ayubi-Rad, M., et al., Effects of Astragalus fasciculifolius gum on wound healing in streptozotocin-induced diabetic rats. *Journal of Herbmed Pharmacology*, 2020. **9**(4): pp. 328–332.
633. Xue, X.-F., H. Sadeghi, and A. Honarmand, Three eriophyoid mite species (Acari: Eriophyoidea: Eriophyidae) from Iran. *Zootaxa*, 2016. **4132**(3): pp. 403–412.
634. Leng, F., et al., Characterization of root microbial communities associated with Astragalus membranaceus and their correlation with soil environmental factors. *Rhizosphere*, 2023. **25**: p. 100656.
635. Chabani, S., et al., Flavonoid glycosides and triterpenoids from Atractylis flava. *Phytochemistry Letters*, 2013. **6**(1): pp. 9–13.
636. Chabani, S., et al., Three new oleanane-type triterpene saponins from Atractylis flava. *Phytochemistry Letters*, 2016. **15**: pp. 88–93.
637. Chabani, S., et al., Chemical composition of medicinal plant Atractylis serratuloides. *Industrial Crops and Products*, 2016. **88**: pp. 91–95.
638. Badaoui, M.I., et al., Pyrroloquinolone A, a new alkaloid and other phytochemicals from A tractylis cancellata L. with antioxidant and anticholinesterase activities. *Natural Product Research*, 2021. **35**(18): pp. 2997–3003.
639. Larrey, D. and G.P. Pageaux. Hepatotoxicity of herbal remedies and mushrooms. *Seminars in Liver Disease*, 1995. **15**(3): pp. 183–188. https://doi.org/10.1055/s-2007-1007276
640. El Rhaffari, L. and A. Zaid, Pratique de la phytothérapie dans le sud-est du Maroc (Tafilalet): un savoir empirique pour une pharmacopée rénovée. *Des Sources du Savoir Aux Médicaments du Futur*, 2002. **1**: pp. 293–318.
641. Daniele, C., et al., Atractylis gummifera L. poisoning: an ethnopharmacological review. *Journal of Ethnopharmacology*, 2005. **97**(2): pp. 175–181.
642. Akram, M.M., et al., Anti-inflammatory, anti-pyretic and acute toxicity effects of n-butanol extract of atractylis flava Desf in rats. *Pharmacognosy Journal*, 2018. **10**(4), pp. 763–767.
643. Bammou, M., et al., Ethnobotanical survey of Asteraceae family used in Meknes-Tafilalet region (Morocco). *International Journal of Innovation and Applied Studies*, 2015. **13**(4): pp. 789–815.
644. HAMOUDA, C., A review of acute poisoning from Atractylis gummifera L. *The Journal of Emergency Actors*, 2004. **3**: pp. 23–26.
645. Le Houérou, H., The role of saltbushes (Atriplex spp.) in arid land rehabilitation in the Mediterranean Basin: a review. *Agroforestry Systems*, 1992. **18**: pp. 107–148.
646. *Distribution of Atriplex farinosa Forssk.* Available from: https://powo.science.kew.org/taxon/urn:lsid:ipni.org:names:163791-1#:~:text=The%20native%20range%20of%20this,desert%20or%20dry%20shrubland%20biome.
647. Parvez, M.K., et al., Protective effect of Atriplex suberecta extract against oxidative and apoptotic hepatotoxicity. *Experimental and Therapeutic Medicine*, 2018. **15**(4): pp. 3883–3891.
648. Zeghib, K. and D.A. Boutlelis, Food additive (sodium benzoate)-induced damage on renal function and glomerular cells in rats; modulating effect of aqueous extract of Atriplex halimus L. I*ranian Journal of Pharmaceutical Research: IJPR*, 2021. **20**(1): p. 296.
649. Delli Compagni, E., A. Pardossi, and S. Pecchia, Fungal and Fungal-like Diseases of Halophytes in the Mediterranean Basin: a state-of-the-art review. *Horticulturae*, 2024. **10**(4): p. 313.
650. Vickerman, D., et al., Evaluation of Atriplex lines for selenium accumulation, salt tolerance and suitability for a key agricultural insect pest. *Environmental Pollution*, 2002. **120**(2): pp. 463–473.
651. Kumar, S. and A.K. Pandey, Chemistry and biological activities of flavonoids: an overview. *The Scientific World Journal*, 2013 (1).
652. Rastogi, S. and D.K. Kulshreshtha, Bacoside A 2 – a triterpenoid saponin from Bacopa monniera. *Phytochemistry*, 1999. **36**(1): pp. 133–137. https://doi.org/10.1016/s0031-9422(00)97026-2
653. Rastogi, S., R. Pal, and D.K. Kulshreshtha, Bacoside A3 a triterpenoid saponin from Bacopa monniera. *Phytochemistry*, 1994. **36**(1): pp. 133–137.
654. Basu, N., Chemical examination of Bacopa monniera, Wettst: part III Bacoside B. *Indian Journal of Chemistry*, 1967. **5**: pp. 84–86.
655. Garai, S., et al., Dammarane-type triterpenoid saponins from Bacopa monniera. *Phytochemistry*, 1996. **42**(3): pp. 815–820.
656. Mahato, S.B., S. Garai, and A.K. Chakravarty, Bacopasaponins E and F: two jujubogenin bisdesmosides from Bacopa monniera. *Phytochemistry*, 2000. **53**(6): pp. 711–714.
657. Chakravarty, A.K., et al., Bacopasides III – V: three new triterpenoid glycosides from Bacopa monniera. *Chemical and Pharmaceutical Bulletin*, 2003. **51**(2): pp. 215–217.
658. Pal, R. and J.P. Sarin, Quantitative determination of bacosides by UV-spectrophotometry. *Indian Journal of Pharmaceutical Sciences*, 1992. **54**(1): p. 17.
659. Devishree, R., S. Kumar, and A. Jain, Short term effect of Bacopa monnieri on memory – A brief review. *Journal of Pharmaceutical Research*, 2017. **11**: pp. 1447–1450.
660. Fatima, U., et al., Pharmacological attributes of Bacopa monnieri extract: current updates and clinical manifestation. *Frontiers in Nutrition*, 2022. **9**: p. 972379.
661. *Uses of Bacopa monnieri (L.) Wettst.* Available from: https://www.ayurveda.hu/api/API-Vol-3.pdf
662. Gohil, K.J. and J.A. Patel, A review on Bacopa monniera: current research and future prospects. *International Journal of Green Pharmacy (IJGP)*, 2010. **4**(1).
663. Jyoti, A. and D. Sharma, Neuroprotective role of Bacopa monniera extract against aluminium-induced oxidative stress in the hippocampus of rat brain. *Neurotoxicology*, 2006. **27**(4): pp. 451–457.
664. Calabrese, C., et al., Effects of a standardized Bacopa monnieri extract on cognitive performance, anxiety, and depression in the elderly: a randomized, double-blind, placebo-controlled trial. *The Journal of Alternative and Complementary Medicine*, 2008. **14**(6): pp. 707–713.
665. Kamkaew, N., et al., Bacopa monnieri increases cerebral blood flow in rat independent of blood pressure. *Phytotherapy Research*, 2013. **27**(1): pp. 135–138.
666. Russo, A. and F. Borrelli, Bacopa monniera, a reputed nootropic plant: an overview. *Phytomedicine*, 2005. **12**(4): pp. 305–317.
667. Sinha, S. and R. Saxena, Effect of iron on lipid peroxidation, and enzymatic and non-enzymatic antioxidants and bacoside-A content in medicinal plant Bacopa monnieri L. *Chemosphere*, 2006. **62**(8): pp. 1340–1350.

668. Ramasamy, S., et al., In silico and in vitro analysis of bacoside A aglycones and its derivatives as the constituents responsible for the cognitive effects of Bacopa monnieri. *PLoS One*, 2015. **10**(5): p. e0126565.
669. Singh, H. and B. Dhawan, Neuropsychopharmacological effects of the Ayurvedic nootropic Bacopa monniera Linn. (Brahmi). *Indian Journal of Pharmacology*, 1997. **29**(5): p. 359.
670. Charles, P.D., et al., Bacopa monniera leaf extract upregulates tryptophan hydroxylase (TPH2) and serotonin transporter (SERT) expression: implications in memory formation. *Journal of Ethnopharmacology*, 2011. **134**(1): pp. 55–61.
671. Rauf, K., F. Subhan, and R.D. Sewell, A bacoside containing Bacopa monnieri extract reduces both morphine hyperactivity plus the elevated striatal dopamine and serotonin turnover. *Phytotherapy Research*, 2012. **26**(5): pp. 758–763.
672. Sireeratawong, S., et al., Acute and chronic toxicities of Bacopa monnieri extract in Sprague-Dawley rats. *BMC Complementary and Alternative Medicine*, 2016. **16**(1): pp. 1–10.
673. Roodenrys, S., et al., Chronic effects of Brahmi (Bacopa monnieri) on human memory. *Neuropsychopharmacology*, 2002. **27**(2): pp. 279–281.
674. Morgan, A. and J. Stevens, Does Bacopa monnieri improve memory performance in older persons? Results of a randomized, placebo-controlled, double-blind trial. *The Journal of Alternative and Complementary medicine*, 2010. **16**(7): pp. 753–759.
675. Pase, M.P., et al., The cognitive-enhancing effects of Bacopa monnieri: a systematic review of randomized, controlled human clinical trials. *The Journal of Alternative and Complementary Medicine*, 2012. **18**(7): pp. 647–652.
676. Teschke, R. and R. Bahre, Severe hepatotoxicity by Indian Ayurvedic herbal products: a structured causality assessment. *Annals of Hepatology*, 2009. **8**(3): pp. 258–266.
677. Mishra, S., et al., Phytochelatin synthesis and response of antioxidants during cadmium stress in Bacopa monnieri L◊. *Plant Physiology and Biochemistry*, 2006. **44**(1): pp. 25–37.
678. Katoch, M., A. Salgotra, and G. Singh, Endophytic fungi found in association with Bacopa monnieri as potential producers of industrial enzymes and antimicrobial bioactive compounds. *Brazilian Archives of Biology and Technology*, 2014. **57**: pp. 714–722.
679. Katoch, M., et al., Cytotoxic and antimicrobial activities of endophytic fungi isolated from Bacopa monnieri (L.) Pennell (Scrophulariaceae). *BMC Complementary and Alternative Medicine*, 2014. **14**(1): pp. 1–8.
680. Shukla, N., et al., Synergism of endophytic Bacillus subtilis and Klebsiella aerogenes modulates plant growth and bacoside biosynthesis in Bacopa monnieri. *Frontiers in Plant Science*, 2022. **13**: p. 896856.
681. Tripathi, P., et al., Differential response of two endophytic bacterial strains inoculation on biochemical and physiological parameters of Bacopa monnieri L. under arsenic stress conditions. *Journal of Hazardous Materials Advances*, 2022. **6**: p. 100055.
682. Katoch, M., et al., Trichoderma lixii (IIIM-B4), an endophyte of Bacopa monnieri L. producing peptaibols. *BMC Microbiology*, 2019. **19**(1): pp. 1–10.
683. Chothani, D.L. and H. Vaghasiya, A review on Balanites aegyptiaca Del (desert date): phytochemical constituents, traditional uses, and pharmacological activity. *Pharmacognosy Reviews*, 2011. **5**(9): p. 55.
684. Orwa, C., et al., Agroforestree Database: a tree reference and selection guide. Version 4. Agroforestree Database: a tree reference and selection guide. *Version*, 2009. 4.
685. Bagul, R., Folk Ethnotherapies on scorpion sting among Tribals of east Khandesh region of Satpuda Forest, Maharashtra, India. *International Journal of Scientific Research*, 2016. **5**(10): pp. 1494–1496.
686. Murthy, H.N., et al., Phytochemicals and biological activity of desert date (Balanites aegyptiaca (L.) Delile). *Plants*, 2020. **10**(1): p. 32.
687. Hassan, L.E.A., et al., Chemotherapeutic potentials of the stem bark of Balanite aegyptiaca (L.) Delile: an antiangiogenic, antitumor and antioxidant agent. *BMC Complementary and Alternative Medicine*, 2016. **16**: pp. 1–13.
688. Anani, K., et al., Effects of hydroethanolic extracts of Balanites aegyptiaca (L.) Delile (Balanitaceae) on some resistant pathogens bacteria isolated from wounds. *Journal of Ethnopharmacology*, 2015. **164**: pp. 16–21.
689. Thirupathi, K., et al., Hepatoprotective effect of leaves of Balanites roxburghii against carbon tetrachloride-induced hepatic damage in rats. *Current Trends in Biotechnology and Pharmacy*, 2009. **3**(2): pp. 219–224.
690. Ahmed, M.M. and M.M. Eid, Evaluation of Balanites aegyptiaca oil as untraditional source of oil and its antiinflammatory activity. *Journal of Drug Research of Egypt*, 2015. **36**(1): pp. 1–11.
691. Kamel, M.S., et al., Studies on Balanites aegyptiaca fruits, an antidiabetic Egyptian folk medicine. *Chemical and Pharmaceutical Bulletin*, 1991. **39**(5): pp. 1229–1233.
692. Anto, F., et al., The relative susceptibilities of juvenile and adult Bulinus globosus and Bulinus truncatus to the molluscicidal activities in the fruit of Ghanaian Blighia sapida, Blighia unijugata and Balanites aegyptiaca. *Annals of Tropical Medicine & Parasitology*, 2005. **99**(2): pp. 211–217.
693. Gnoula, C., et al., Balanitin-6 and-7: diosgenyl saponins isolated from Balanites aegyptiaca Del. display significant anti-tumor activity in vitro and in vivo. *International Journal of Oncology*, 2008. **32**(1): pp. 5–15.
694. Kim, H., et al., Effects of naturally-occurring flavonoids and biflavonoids on epidermal cyclooxygenase and lipoxygenase from guinea-pigs. *Prostaglandins, Leukotrienes and Essential Fatty Acids*, 1998. **58**(1): pp. 17–24.
695. Okwuosa, V., B. Molta, and S. Ebele, Toxicity of aqueous bark extract of the tree Balanites aegyptiaca on the fish Oreochromis niloticus. *Applied Parasitology*, 1993. **34**(2): pp. 89–94.
696. Mohamed, A., et al., Some pharmacological and toxicological studies on Balanites aegyptiaca bark. *Phytotherapy Research*, 1999. **13**(5): pp. 439–441.
697. Zijlstra, E. and A. El-Hassan, Leishmaniasis in Sudan. 3. Visceral leishmaniasis. *Transactions of the Royal Society of Tropical Medicine and Hygiene*, 2001. **95**(Supplement_1): pp. S27–S58.
698. *Pathology of B. aegyptiaca*. Available from: https://www.google.com/url?sa=t&source=web&rct=j&opi=89978449&url=https://seahipaj.org/journals-ci/dec-2021/IJIABR/full/IJIABR-D-2-2021.pdf&ved=2ahUKEwicz92o9rmGAxX9efEDHWG4AsAQFnoECBMQAQ&usg=AOvVaw0kVD62Qx4neeptYm8_BJvu.
699. Yineger, H., et al., Ethnoveterinary medicinal plants at bale mountains national park, Ethiopia. *Journal of Ethnopharmacology*, 2007. **112**(1): pp. 55–70.

700. *Uses of Basananthe berberoides*. Available from: https://uses.plantnet-project.org/e/index.php?title=Basananthe_sandersonii.
701. Pervez, S., et al., Nephroprotective effect of Berberis baluchistanica against gentamicin-induced nephrotoxicity in rabbit. *Bangladesh Journal of Pharmacology*, 2018. **13**(3): pp. 222–230.
702. Gul, Z., A. Akbar, and S.K. Leghari, Elucidating therapeutic and biological potential of Berberis baluchistanica Ahrendt bark, leaf, and root extracts. *Frontiers in Microbiology*, 2022. **13**: p. 823673.
703. Shamma, M., et al., Pakistanine and pakistanamine, two new dimeric isoquinoline alkaloids. *Journal of the American Chemical Society*, 1973. **95**(17): pp. 5742–5747.
704. Miana, G., et al., Baluchistine, a new bisbenzylisoquinoline alkaloid. *Experientia*, 1979. **35**: pp. 1137–1138.
705. Pervez, S., et al., Antimicrobial and antioxidant potential of berberisinol, a new flavone from Berberis baluchistanica. *Chemistry of Natural Compounds*, 2019. **55**: pp. 247–251.
706. Tunalier, Z., et al., Antioxidant properties and phenolic composition of Sideritis species. *Chemistry of Natural Compounds*, 2004. **40**: pp. 206–210.
707. Karimov, A., et al., Berberis alkaloids. XXIX. an investigation of the alkaloids of Berberis sibirica. *Chemistry of Natural Compounds*, 1993. **29**(3): pp. 361–364.
708. Genkina, G., et al., Spectrophotometry of glycosides of oleanolic acid and hederagenin in concentrated sulfuric acid. *Chemistry of Natural Compounds*, 1977. **13**: pp. 188–193.
709. Kovganko, N. and Z.N. Kashkan, Advances in the chemical transformation of β-sitosterol. *Chemistry of Natural Compounds*, 1994. **30**(5): pp. 533–546.
710. Tadzhibaev, M., et al., Isolation of berberine. *Chemistry of Natural Compounds*, 1974. **10**(1): pp. 42–43.
711. Baloch, N., et al., In vitro antileishmanial, cytotoxic, antioxidant activities and phytochemical analysis of Berberis baluchistanica roots extracts and its fractions. *The Journal of Phytopharmacology*, 2013. **4**: pp. 282–287.
712. Kakar, S.A., et al., Screening of antibacterial activity of four medicinal plants of Balochistan-Pakistan. *Pakistan Journal of Botany*, 2012. **44**(2010): pp. 245–250.
713. Abbasi, M., et al., Berberis baluchistanica: assessment of natural antioxidants to reprieve from oxidative stress. *International Research Journal of Pharmacy*, 2013. **4**(5): pp. 101–105.
714. Yin, J., J. Ye, and W. Jia, Effects and mechanisms of berberine in diabetes treatment. *Acta Pharmaceutica Sinica B*, 2012. **2**(4): pp. 327–334.
715. Gul, Z., et al., High throughput screening for bioactive components of Berberis baluchistanica ahrendt root and their functional potential assessment. *BioMed Research International*, 2022. **2022**: pp. 1–11.
716. Rad, S.Z.K., M. Rameshrad, and H. Hosseinzadeh, Toxicology effects of Berberis vulgaris (barberry) and its active constituent, berberine: a review. *Iranian Journal of Basic Medical Sciences*, 2017. **20**(5): p. 516.
717. Bibi, T., et al., The endemic medicinal plants of Northern Balochistan, Pakistan and their uses in traditional medicine. *Journal of Ethnopharmacology*, 2015. **173**: pp. 1–10.
718. Sharma, S., et al., Diversity and bioactive potential of culturable fungal endophytes of medicinal shrub Berberis aristata DC.: a first report. *Mycobiology*, 2018. **46**(4): pp. 370–381.
719. Davis, C.C. and M.W. Chase, Elatinaceae are sister to Malpighiaceae; Peridiscaceae belong to Saxifragales. *American Journal of Botany*, 2004. **91**(2): pp. 262–273.
720. Anandjiwala, S., H. Srinivasa, and M. Rajani, Isolation and TLC densitometric quantification of gallicin, gallic acid, lupeol and β-sitosterol from Bergia suffruticosa, a hitherto unexplored plant. *Chromatographia*, 2007. **66**: pp. 725–734.
721. Ezzat, S.M., M.A. Choucry, and Z.A. Kandil, Antibacterial, antioxidant, and topical anti-inflammatory activities of Bergia ammannioides: a wound-healing plant. *Pharmaceutical Biology*, 2016. **54**(2): pp. 215–224.
722. Anandjiwala, S., et al., Free-radical scavenging activity of Bergia suffruticosa (Delile) Fenzl. *Journal of Natural Medicines*, 2007. **61**: pp. 59–62.
723. Siddiqa, A., et al., Bergia ammannioides: phytochemical screening, antioxidant activity and radical scavenging effects of its various fractions. *Asian Journal of Chemistry*, 2013. **25**(14): pp. 7921–7926.
724. Roux, J. and M. Coetzee, First report of pink disease on native trees in South Africa and phylogenetic placement of Erythricium salmonicolor in the homobasidiomycetes. *Plant Disease*, 2005. **89**(11): pp. 1158–1163.
725. *Distribution of Bidens biternata L.* Available from: https://powo.science.kew.org/taxon/urn:lsid:ipni.org:names:184771-1#synonyms
726. Xuan, T.D. and T.D. Khanh, Chemistry and pharmacology of Bidens pilosa: an overview. *Journal of Pharmaceutical Investigation*, 2016. **46**: pp. 91–132.
727. Lima Silva, F., et al., Compilation of secondary metabolites from Bidens pilosa L. *Molecules*, 2011. **16**(2): pp. 1070–1102.
728. Maarfia, S. and A. Zellagui, Study of essential oils and phenolic compounds their changes and anticancer activity in some species belonging to Asteraceae and Lamiaceae families. 2019.
729. Matos, M.J., et al., Coumarins – An important class of phytochemicals. Phytochemicals-isolation, characterisation and role in human health.. *InTech*, 2015. **25**: pp. 533–538.
730. Zhu, Y.J., et al., Two new polyacetylene glucosides and a new caffeoyl derivative with angiogenic activity from Bidens parviflora Willd. *Phytochemistry Letters*, 2021. **42**: pp. 82–86.
731. Kil, J.-S., et al., Okanin, a chalcone found in the genus Bidens, and 3-penten-2-one inhibit inducible nitric oxide synthase expression via heme oxygenase-1 induction in RAW264. 7 macrophages activated with lipopolysaccharide. *Journal of Clinical Biochemistry and Nutrition*, 2011. **50**(1): pp. 53–58.
732. Ma, T.-T., et al., Analysis of fingerprint and bioactive components of Bidens biternata by HPLC. *Zhong yao cai= Zhongyaocai= Journal of Chinese Medicinal Materials*, 2012. **35**(6): pp. 892–896.
733. Bhatt, J., et al., *Invasive Alien Plants an Ecological Appraisal for the Indian Subcontinent*, vol. 1. Cabi, 2011.
734. Panda, H., *Medicinal Plants Cultivation & Their Uses*. Asia Pacific Business Press Inc., 2002.
735. Wu, Z., P. Raven, and D. Hong, *Flora of China Volume 20–21 (Asteraceae)*. Science Press, Beijing & Missouri Botanical Garden Press, St. Louis, 2011.
736. *China Flora*. Available from: https://chineseherbinfo.com/chineseherbinfo.html
737. De Padua, L., Bunyapraphatsara and Lemmens, RHMJ. *Plant Resources of South East Asia*, 1999. **12**(1): pp. 210–218.
738. Zahara, K., et al., Isolation and Identification of Bioactive Compounds from Bidens spp. Using HPLC-DAD and GC-MS Analysis and Their Biological Activity as Anticancer Molecules. *Molecules*, 2022. **27**(6): pp. 1927.

739. Almosnid, N.M., et al., Evaluation of extracts prepared from 16 plants used in Yao ethnomedicine as potential anticancer agents. *Journal of Ethnopharmacology*, 2018. **211**: pp. 224–234.
740. Sukumaran, P., et al., Phytochemical investigation of Bidens biternata (Lour.) Merr. and Sheriff. – A nutrient-rich leafy vegetable from Western Ghats of India. *Applied Biochemistry and Biotechnology*, 2012. **167**: pp. 1795–1801.
741. Kinuthia, D.G., A.W. Muriithi, and P.W. Mwangi, Freeze dried extracts of Bidens biternata (Lour.) Merr. and Sheriff. show significant antidiarrheal activity in in-vivo models of diarrhea. *Journal of Ethnopharmacology*, 2016. **193**: pp. 416–422.
742. *WHO (World Health Organization) Tropical Disease Research Division*. Geneva: WHO, 2000.
743. Klayman, D.L., Qinghaosu (artemisinin): an antimalarial drug from China. *Science*, 1985. **228**(4703): pp. 1049–1055.
744. *Toxicity of Bidens biternata L.* Available from: https://raintree.com/picaopreto.htm
745. Tsai, L.-C., et al., Bidens identification using the noncoding regions of chloroplast genome and nuclear ribosomal DNA. *Forensic Science International: Genetics*, 2008. **2**(1): pp. 35–40.
746. Chen, P., P.-X. Lan, and F. Li, First report of tobacco mosaic virus infecting Bidens pilosa in China. *Plant Disease*, 2024. **108**(2): p. 538. PDIS-06-23-1178-PDN.
747. Mukhtar, I., et al., First Report of Powdery Mildew Caused by Golovinomyces ambrosiae on Bidens pilosa in China. *Plant Disease*, 2022. **106**(1): pp. 323.
748. Deng, Q., et al., Effects of intercropping with Bidens species plants on the growth and cadmium accumulation of Ziziphus acidojujuba seedlings. *Environmental Monitoring and Assessment*, 2019. **191**: pp. 1–8.
749. Dirar, A.I., et al., Genus Blepharis (Acanthaceae): a review of ethnomedicinally used species, and their phytochemistry and pharmacological activities. *Journal of Ethnopharmacology*, 2021. **265**: pp. 113255.
750. Mohamed, G.A., et al., Blepharisides A and B, new flavonol glycosides from Blepharis ciliaris growing in Saudi Arabia. *Phytochemistry Letters*, 2015. **11**: pp. 177–182.
751. Ashour, M.A.-G., New diacyl flavonoid derivatives from the Egyptian plant Blepharis edulis (Forssk.) Pers. *Bulletin of Faculty of Pharmacy*, 2015. **53**(1): pp. 11–17.
752. Afifi, M., A novel 4′-O-diglycoside of decarboxyrosmarinic acid from Blepharis ciliaris. *Pharmaceutical Biology*, 2003. **41**(7): pp. 487–490.
753. El-Shanawany, M.A., et al., A new isoflavone from Blepharis ciliaris of an Egyptian origin. *Medicinal Chemistry Research*, 2013. **22**: pp. 2346–2350.
754. Deshpande DJ. *A Handbook of Medicinal Herbs*. Jodhpur: Agrobios, 2006, pp. 342–343.
755. Tackholm, V., *Students Flora of Egypt'*, 2nd edn. Publ Cairo University, 1974.
756. Kamal, A. and G. Abdul, Blepharis ciliaris (L.) BL Burtt in notes Roy. *Bot Gard Edinb*, 1956. **22**: pp. 94.
757. Zeng, L., et al., Inhibitory mechanism of apigenin on α-glucosidase and synergy analysis of flavonoids. *Journal of Agricultural and Food Chemistry*, 2016. **64**(37): pp. 6939–6949.
758. Rajasekaran, A., V. Sivakumar, and S. Darlinquine, Role of Blepharis maderaspatensis and Ammannia baccifera plant extracts on in vitro oxygen radical scavenging, secretion of gastric fluid and gastroprotection on ulcer induced rats. *Pharmaceutical Biology*, 2012. **50**(9): pp. 1085–1095.
759. Al Nasr, I., In vitro anti-leishmanial assessment of some medicinal plants collected from Al Qassim, Saudi Arabia. *Acta Parasitologica*, 2020. **65**: pp. 696–703.
760. Alrajhei, K., I. Saleh, and M.H. Abu-Dieyeh, Biodiversity of arbuscular mycorrhizal fungi in plant roots and rhizosphere soil from different arid land environment of Qatar. *Plant Direct*, 2022. **6**(1): p. e369.
761. Vijayalakshmi, S. and K. Kripa, Therapeutic uses of plants of genus Blepharis-a systematic review. *International Journal of Pharma and Bio Sciences*, 2018. 7: B236–B243.
762. El-Shanawany, M., et al., Stigmasterol tetracosanoate, a new stigmasterol ester from the Egyptian Blepharis ciliaris. *Drug Research*, 2014: pp. 347–353.
763. Dirar, A.I., et al., Phenolic compounds from the aerial parts of Blepharis linariifolia Pers. and their free radical scavenging and enzyme inhibitory activities. *Medicines*, 2019. **6**(4): pp. 113.
764. El Ghazali, G., et al., *Medicinal Plants of Sudan Part II. Medicinal Plants of Eastern Nuba Mountains*. Khartoum: National Council for Research, 1987.
765. El-Kamali, H.H., Ethnopharmacology of medicinal plants used in North Kordofan (Western Sudan). *Ethnobotanical Leaflets*, 2009. **2009**(1): p. 24.
766. Doka, I. and S. Yagi, Ethnobotanical survey of medicinal plants in West Kordofan (Western Sudan). *Ethnobotanical Leaflets*, 2009. **2009**(11): p. 8.
767. Sowemimo, A., et al., Studies on the anti-inflammatory and anti-nociceptive properties of Blepharis maderaspatensis leaves. *Revista Brasileira de Farmacognosia*, 2013. **23**(5): pp. 830–835.
768. Rajasekaran, A., V. Sivakumar, and S. Darlinquine, Evaluation of wound healing activity of Ammannia baccifera and Blepharis maderaspatensis leaf extracts on rats. *Revista Brasileira de Farmacognosia*, 2012. **22**: pp. 418–427.
769. Neelambika, H. and S. Leelavathi, Comparative antioxidant activity of whole plant of Blepharis maderaspatensis (L.) Heyne ex Roth. and Blepharis molluginifolia Pers. of Mysore district by DPPH method. *Indo American Journal of Pharmacy Research*, 2015. **5**(3): pp. 1191–1196.
770. Devarajan, N., S. Ramalingam, and S.M. Subramaniam, Gas chromatography mass spectroscopy chromatogram and antimicrobial activity of leaf extracts of Blepharis maderaspatensis and Maesa indica. *Journal of Herbs, Spices & Medicinal Plants*, 2015. **21**(3): pp. 267–282.
771. Baskar, A.-A., et al., In vitro antioxidant and antiproliferative potential of medicinal plants used in traditional Indian medicine to treat cancer. *Redox Report*, 2012. **17**(4): pp. 145–156.
772. Saqib, F., et al., Ethnopharmacological studies on antispasmodic, bronchodilator and antiplatelet aggregation activities of Blepharis edulis Pers. *Asian Journal of Natural & Applied Sciences*, 2012. **1**: pp. 33–45.
773. Adikini, S., et al., Development of a specific molecular tool for detecting Xanthomonas campestris pv. musacearum. *Plant Pathology*, 2011. **60**(3): pp. 443–452.
774. Bohlmann, F., C. Zdero, and A. Nair, A new carvotanacetone derivative from Blumea wightiana. *Phytochemistry*, 1979. **18**(6): p. 1062.
775. Sivanandhan, S., et al., Effect of monoterpene ester from Blumea axillaris (Lam.) DC and its acetyl derivative against plant pathogenic fungi and their in silico molecular docking. *Natural Product Research*, 2021. **35**(24): pp. 5744–5751.

776. Owolabi, M.S., et al., Essential oil composition and insecticidal activity of Blumea perrottetiana growing in Southwestern Nigeria. *Natural Product Communications*, 2010. **5**(7): p. 1934578X1000500733.
777. Roy, B., et al., Leaf extracts of Shiyalmutra (Blumea lacera Dc.) as botanical pesticides against lesser grain borer and rice weevil. *Journal of Biological Sciences*, 2005. **5**(2): pp. 201–204.
778. Uphof, J.C.T., *Dictionary of Economic Plants*. 1959.
779. Sreelekha, K., et al., Pharmaco-chemical characterization of leaves of Blumea mollis (D. Don) merr. from Western Ghats of wayanad region of Kerala, India. *Journal of Pharmacognosy and Phytochemistry*, 2017. **6**(4): pp. 319–323.
780. Sivanandhan, S., et al., Activity of some medicinal plants against phytopathogenic fungi. *International Journal of Scientific Research in Biological Sciences*, 2018. **5**: p. 5.
781. Senthilkumar, A., K. Kannathasan, and V. Venkatesalu, Chemical constituents and larvicidal property of the essential oil of Blumea mollis (D. Don) Merr. against Culex quinquefasciatus. *Parasitology Research*, 2008. **103**: pp. 959–962.
782. Dash, J.P., L. Mani, and S.K. Nayak, Antibacterial activity of Blumea axillaris synthesized selenium nanoparticles against multidrug resistant pathogens of aquatic origin. *Egyptian Journal of Basic and Applied Sciences*, 2022. **9**(1): pp. 65–76.
783. Zahan, M.N., et al., Blumea lacera Plant Poisoning in Cattle; Epidemiology and Management. *Turkish Journal of Agriculture-Food Science and Technology*, 2015. 3(8): pp. 635–638.
784. Lavanya, N., et al., Endophytic fungal isolation from Blumea axillaris: identification and biological activity of secondary metabolites. *Notulae Scientia Biologicae*, 2021. **13**(2): pp. 10953–10953.
785. Joshi, R.K., et al., Identification of potentially bioactive compounds from Blumea lacera essential oil by gas chromatography-mass spectroscopy and molecular docking studies for targeting inflammatory bowel disease. *Natural Product Research*, 2023. **37**(9): pp. 1589–1593.
786. Satyal, P., et al., Chemical composition of Blumea lacera essential oil from Nepal. Biological activities of the essential oil and (Z)-lachnophyllum ester. *Natural Product Communications*, 2015. **10**(10): p. 1934578X1501001028.
787. Akter, R., et al., A new cytotoxic steroidal glycoalkaloid from the methanol extract of Blumea lacera leaves. *Journal of Pharmacy & Pharmaceutical Sciences*, 2015. **18**(4): pp. 616–633.
788. Ashrafi, S., et al., Chemico-biological profiling of Blumea lacera (Burm. f.) DC.(Family: Asteraceae) Provides new insights as a potential source of antioxidant, cytotoxic, antimicrobial, and antidiarrheal agents. *Evidence-Based Complementary and Alternative Medicine*, 2022. **2022**. https://doi.org/10.1155/2022/2293415
789. Mishra, P., R. Irchhiaya, and S.K. Mishra, Phytochemical investigation and spectral studies of isolated flavonoid from ethanolic extract of whole plant Blumea lacera DC. *Journal of Pharmacognosy and Phytochemistry*, 2015. **4**(2): pp. 1–4.
790. Khair, M.A., et al., Preliminary phytochemical screenings and Pharmacological activities of Blumea lacera (Burn. f.) DC. *Dhaka University Journal of Pharmaceutical Sciences*, 2014. **13**(1): pp. 69–73.
791. Hossen, M.A., et al., Bioactive metabolites of Blumea lacera attenuate anxiety and depression in rodents and computer-aided model. *Food Science & Nutrition*, 2021. **9**(7): pp. 3836–3851.
792. Shariare, M.H., et al., Liposomal drug delivery of Blumea lacera leaf extract: in-vivo hepatoprotective effects. *Nanomaterials*, 2022. **12**(13): p. 2262.
793. Ghani, A., *Medicinal Plants of Bangladesh: Chemical Constituents and Uses*. Asiatic Society of Bangladesh, 1998.
794. Akter, R., et al., A new cytotoxic diterpenoid glycoside from the leaves of Blumea lacera and its effects on apoptosis and cell cycle. *Natural Product Research*, 2016. **30**(23): pp. 2688–2693.
795. Chiang, L.-C., et al., In vitro anti-leukemic and antiviral activities of traditionally used medicinal plants in Taiwan. *The American Journal of Chinese Medicine*, 2004. **32**(5): pp. 695–704.
796. Uddin, S.J., I.D. Grice, and E. Tiralongo, Cytotoxic effects of Bangladeshi medicinal plant extracts. *Evidence-Based Complementary and Alternative Medicine*, 2011. **2011**.
797. Zhao, L., et al., Pathogenicity and diversity of large-spored Alternaria associated with three solanaceous vegetables (Solanum tuberosum, S. lycopersicum and S. melongena) in China. *Plant Pathology*, 2023. **72**(2): pp. 376–391.
798. Rehman, N.U., et al., A new anticancer Bisflavan-3-ol from Boerhavia elegans. *Chemistry of Natural Compounds*, 2020. **56**: pp. 235–238.
799. Sadeghi, Z., et al., Antioxidant activity and total phenolic content of Boerhavia elegans (choisy) grown in Baluchestan, Iran. *Avicenna Journal of Phytomedicine*, 2015. **5**(1): p. 1.
800. Ramazani, A., et al., In vitro and in vivo anti-malarial activity of Boerhavia elegans and Solanum surattense. *Malaria Journal*, 2010. **9**: pp. 1–8.
801. Najam, A., A.K. Singh, and H. Verma, Ancient and modern medicinal potential of Boerhaavia diffusa and Clerodendrum aculeatum. *Research in Environment and Life sciences*, 2008. **1**(1): pp. 1–4.
802. Sandhu, P., et al., Potential herbs used in ocular diseases. *Journal of Pharmaceutical Sciences and Research*, 2011. **3**(4): p. 1127.
803. Aniszewski, T., *Definition, Typology, and Occurrence of Alkaloids*, 2nd edn. Alkaloids. Elsevier BV, 2015.
804. Algfri, S.K., A.A. Qaid, and S.A. Mahde, Preliminary phytochemical analysis and in vitro antioxidant activity of Boscia arabica leaves. *International Journal of Pharmaceutical Sciences Review and Research*, 2021. **68**(1): pp. 127–134.
805. Myo, E.M., et al., Characterization of bacterial endophytes from Myanmar medicinal plants for antimicrobial activity against human and plant pathogens. *Brazilian Journal of Pharmaceutical Sciences*, 2020. **56**.
806. Kreck, C. and R. Saller, Indischer Weihrauch und seine Zubereitungen einschliesslich H15 als traditionelles und modernes Therapeutikum. *Internist Prax*, 1998. **38**: pp. 857–872.
807. Shen, T. and H.X. Lou, Bioactive constituents of myrrh and frankincense, two simultaneously prescribed gum resins in Chinese traditional medicine. *Chemistry & Biodiversity*, 2008. **5**(4): pp. 540–553.
808. Shah, B.A., G.N. Qazi, and S.C. Taneja, Boswellic acids: a group of medicinally important compounds. *Natural Product Reports*, 2009. **26**(1): pp. 72–89.
809. Roy, N.K., et al., The potential role of boswellic acids in cancer prevention and treatment. *Cancer Letters*, 2016. **377**(1): pp. 74–86.
810. Hussain, H., et al., Therapeutic potential of boswellic acids: a patent review (1990–2015). *Expert Opinion on Therapeutic Patents*, 2017. **27**(1): pp. 81–90.

811. Al-Yahya, A.A., et al., Repeat oral dose safety study of standardized methanolic extract of Boswellia sacra oleo gum resin in rats. *Saudi Journal of Biological Sciences*, 2020. **27**(1): pp. 117–123.
812. Al-Harrasi, A., et al., Chemical, molecular and structural studies of Boswellia species: β-Boswellic Aldehyde and 3-epi-11β-Dihydroxy BA as precursors in biosynthesis of boswellic acids. *PLoS One*, 2018. **13**(6): p. e0198666.
813. Halliwell, B. and J.M. Gutteridge, *Free Radicals in Biology and Medicine*. USA: Oxford University Press, 2015.
814. Miyamoto, S., et al., Singlet molecular oxygen generated by biological hydroperoxides. *Journal of Photochemistry and Photobiology B: Biology*, 2014. **139**: pp. 24–33.
815. Khan, A.L., et al., Genome structure and evolutionary history of frankincense producing Boswellia sacra. *Iscience*, 2022. **25**(7).
816. Asif, M., et al., Diuretic activity of Boswellia serrata Roxb. oleo gum extract in albino rats. *Pakistan Journal of Pharmaceutical Sciences*, 2014. **27**(6): pp. 1811–1817.
817. Ni, X., et al., Frankincense essential oil prepared from hydrodistillation of Boswellia sacra gum resins induces human pancreatic cancer cell death in cultures and in a xenograft murine model. *BMC Complementary and Alternative Medicine*, 2012. **12**(1): pp. 1–14.
818. Pan, Y.-N., et al., Comparative studies of pharmacokinetics and anticoagulatory effect in rats after oral administration of Frankincense and its processed products. *Journal of Ethnopharmacology*, 2015. **172**: pp. 118–123.
819. Al-Harrasi, A., et al., Analgesic effects of crude extracts and fractions of Omani frankincense obtained from traditional medicinal plant Boswellia sacra on animal models. *Asian Pacific Journal of Tropical Medicine*, 2014. **7**: pp. S485–S490.
820. Asad, M. and M. Alhumoud, Hepatoprotective effect and GC-MS analysis of traditionally used Boswellia sacra oleo gum resin (Frankincense) extract in. *African Journal of Traditional, Complementary and Alternative Medicines*, 2015. **12**(2): pp. 1–5.
821. Al-Harrasi, A., et al., Antiglycation and antioxidant activities and HPTLC analysis of Boswellia sacra Oleogum resin: the sacred frankincense. *Tropical Journal of Pharmaceutical Research*, 2013. **12**(4): pp. 597–602.
822. Yassin, N.A., et al., Effect of Boswellia serrata on Alzheimer′ s disease induced in rats. *Journal of the Arab Society for Medical Research*, 2013. **8**(1): pp. 1–11.
823. Al-Harrasi, A., et al., 11α-Ethoxy-β-boswellic Acid and Nizwanone, a New Boswellic Acid Derivative and a New Triterpene, Respectively, from Boswellia sacra. *Chemistry & Biodiversity*, 2013. **10**(8): pp. 1501–1506.
824. Wagner, A.E., et al., Ascorbic acid partly antagonizes resveratrol mediated heme oxygenase-1 but not paraoxonase-1 induction in cultured hepatocytes-role of the redox-regulated transcription factor Nrf2. *BMC Complementary and Alternative Medicine*, 2011. **11**: pp. 1–8.
825. Banno, N., et al., Anti-inflammatory activities of the triterpene acids from the resin of Boswellia carteri. *Journal of Ethnopharmacology*, 2006. **107**(2): pp. 249–253.
826. Zaki, A.A., et al., Cardioprotective and antioxidant effects of oleogum resin "Olibanum" from Bos Boswellia carteri Birdw.(Bursearceae). *Chinese journal of natural medicines*, 2014. **12**(5): pp. 345–350.
827. Asad, M. and M. Alhomoud, Proulcerogenic effect of water extract of Boswellia sacra oleo gum resin in rats. *Pharmaceutical Biology*, 2016. **54**(2): pp. 225–230.
828. Tolera, M., et al., Frankincense tree recruitment failed over the past half century. *Forest Ecology and Management*, 2013. **304**: pp. 65–72.
829. Suhail, M.M., et al., Boswellia sacra essential oil induces tumor cell-specific apoptosis and suppresses tumor aggressiveness in cultured human breast cancer cells. *BMC Complementary and Alternative Medicine*, 2011. **11**: pp. 1–14.
830. Ilyas, K., et al., Boswellia sacra extract-loaded mesoporous bioactive glass nano particles: synthesis and biological effects. *Pharmaceutics*, 2022. **14**(1): p. 126.
831. Aldahlawi, A.M., A.T. Alzahrani, and M.F. Elshal, Evaluation of immunomodulatory effects of Boswellia sacra essential oil on T-cells and dendritic cells. *BMC Complementary Medicine and Therapies*, 2020. **20**(1): pp. 1–14.
832. Alyahya, A.A., et al., Haematotoxic effects of methanolic extract of Boswellia sacra oleo gum resin (frankinccense) in rats. *Science Progress*, 2021. **104**(2): p. 00368504211015102.
833. Khan, A.L., et al., Endophytic fungi from frankincense tree improves host growth and produces extracellular enzymes and indole acetic acid. *PloS One*, 2016. **11**(6): p. e0158207.
834. Ahmad, V.U., et al., Two new N-acetylated spermidine alkaloids from Capparis decidua. *Journal of Natural Products*, 1992. **55**(10): pp. 1509–1512.
835. Fu, X.P., et al., New spermidine alkaloids from Capparis spinosa roots. *Phytochemistry Letters*, 2008. **1**(1): pp. 59–62.
836. Ahmad, V.U., et al., Cadabicine and Cadabicine Diacetate from Crataeva nurvala and Cadaba farinosa. *Journal of Natural Products*, 1987. **50**(6): p. 1186.
837. Telrandhe, U.B. and V. Uplanchiwar, Phyto-pharmacological perspective of Cadaba farinosa forsk. *American Journal of Phytomedicine and Clinical Therapeutics*, 2013. **1**: pp. 11–22.
838. Terna, G., I. Jideani, and I. Nkama, Nutrient and sensory qualities of kunun zaki from different saccharification agents. *International Journal of Food Sciences and Nutrition*, 2002. **53**(2): pp. 109–115.
839. Mengistu, G., et al., The in vitro anthelmintic properties of browse plant species against Haemonchus contortus is determined by the polyphenol content and composition. *Veterinary Parasitology*, 2017. **237**: pp. 110–116.
840. Awoke, A., A. Ambelu, and W. Legesse, Treatment of raw water using forssk farinosa shrub, a traditional water clarifier by "Tseme Bena" people. *Ethiopian Journal of Health Sciences*, 2006. **16**(1): pp. 121–129.
841. Tagele Haligamo, D., A. Ejeso, and E. Mengistie Beyene, Bacterial removal efficiency of Moringa stenopetala and Cadaba farinosa from surface water: laboratory-based cross-sectional study. *Environmental Health Insights*, 2022. **16**: p. 11786302221111842.
842. *Chemical composition of Cadaba farinosa*. Available from: https://uses.plantnet-project.org/en/Cadaba_farinosa_(PROTA).
843. Halilu, E.M., et al., Phytochemical and antioxidant activity of Cadaba farinosa Forssk stem bark extracts. *Physical Sciences Reviews*, 2021. **8**(5): pp. 703–714.
844. Viqar Uddin, A., et al., Cadabacine and cadabacine diacetate from Crataeva nurvala and Cadaba farinosa. *Journal of Natural Products*, 1987. **50**: p. 1186.
845. Aziz-Ur-Rehman, *Studies in Chemical Constituents of Cadaba fruticosa*. Pakistan Research Repository, PhD thesis. Karachi: University of Karachi, 1990.
846. Abdulaziz Al-Hamoud, G., et al., Four new flavonoids isolated from the aerial parts of Cadaba rotundifolia Forssk. (Qadab). *Molecules*, 2019. **24**(11): p. 2167.

847. Mohamed, G.A., et al., New anti-inflammatory flavonoids from Cadaba glandulosa Forssk. *Archives of Pharmacal Research*, 2014. **37**: pp. 459–466.
848. Gohar, A.A., Flavonol glycosides from Cadaba glandulosa. *Zeitschrift für Naturforschung C*, 2002. **57**(3–4): pp. 216–220.
849. Mythreyi, R., et al., Antipyretic activity of leaves of Cadaba fruticosa (L.) Druce. *Pharmacol Online*, 2008. **3**: pp. 136–142.
850. Amudha, M. and S. Rani, Assessing the bioactive constituents of Cadaba fruticosa (L.) Druce through GC-MS. *International Journal of Pharmaceutics*, 2014. **6**: pp. 383–385.
851. Shashikanth, J., C. Mohan, and P. Ramachandra, A potent folklore medicinal plant: Cadaba fruticosa (L.) Druce. *Research & Reviews: Journal of Botanical Sciences*, 2014. 3: pp. 30–33.
852. Mythreyi, R., et al., Antibacterial activity of leaves of Cadaba trifoliata (Roxb.) Wt. & Arn. *Indian Journal of Pharmaceutical Sciences*, 2009. **71**(2): p. 115.
853. El Dirdiri, N., S. Barakat, and S. Adam, The combined toxicity of Aristolochia bracteata and Cadaba rotundifolia to goats. *Veterinary and Human Toxicology*, 1987. **29**(2): pp. 133–137.
854. Lack, H.W., From Ethiopia to Fiesole, Kew and Paris: the discovery, naming and typification of Cadia purpurea (Fabaceae). *Flora Mediterranea*, 2021. **31**: pp. 9–21.
855. Zweekhorst-Van Laer, A. and T. Nelen, Flavanoids from Cadia purpurea (picc.) Ait. *Pharmaceutisch Weekblad*, 1976. **51**(111): pp. 1289–1293.
856. van Eijk, J. and M. Radema, Alkaloids from Cadia purpurea (Picc.) Ait. *Pharmaceutisch Weekblad*, 1972. **107**(2): pp. 13–20.
857. Van Eijk, J., M. Radema, and C. Versluis, New quinolizidine alkaloids from Cadia purpurea. *Tetrahedron Letters*, 1976. **17**(24): pp. 2053–2054.
858. van Laer, A. and O. Uffelie, Apigenine-7-O-neohesperiodoside in Cadia purpurea (Picc.) Ait. *Pharmaceutisch Weekblad*, 1971. **106**(16): pp. 337–339.
859. Van Eijk, J. and M. Radema, Lupanin alkaloids and other compounds from Cadia purpurea. *Planta Medica*, 1975. **28**(06): pp. 139–142.
860. Radema, M., Ethoxylupanin, a new alkaloid from Cadia purpurea. *Planta Medica*, 1975. **28**(6): p. 143.
861. Eijk, J.v.L. and M. Radema, Some alkaloids of ethiopian calpurnea aurea and cadia purpurea. *Planta Medica*, 1977. **32**(07): pp. 275–279.
862. Kiros, T., et al., Compounds with antibacterial and antioxidant activities from Cadia purpurea. *Natural Product Research*, 2023. **37**(16): pp. 2672–2680.
863. van Laer, A. and O. Uffelie, Luteoline in Cadia purpurea (picc.) ait. *Pharmaceutisch Weekblad*, 1971. **106**(48): pp. 890–892.
864. Edwards, S., S. Demissew, and I. Hedberg, *Flora of Ethiopia and Eritrea: Hydrocharitaceae to Arecaceae*. The National Herbarium, Addis Ababa University, 1997.
865. de Fitoquímica, S.L., Boletín Latinoamericano y del Caribe de Plantas Medicinales y Aromáticas. *BLACPMA*, 2004. **3**(6): p. 107.
866. Yemane, B., et al. Survey of some common medicinal plants used in Eritrean folk medicine. American Journal of Ethnomedicine, 2018. **112**: pp. 865–876.
867. Belayneh, A., et al., Medicinal plants potential and use by pastoral and agro-pastoral communities in Erer Valley of Babile Wereda, Eastern Ethiopia. *Journal of Ethnobiology and Ethnomedicine*, 2012. **8**(1): pp. 1–11.
868. Citerne, H.L., R.T. Pennington, and Q.C. Cronk, An apparent reversal in floral symmetry in the legume Cadia is a homeotic transformation. *Proceedings of the National Academy of Sciences*, 2006. **103**(32): pp. 12017–12020.
869. Soliman, S., et al., Mechanical and phytochemical protection mechanisms of Calligonum comosum in arid deserts. *PloS One*, 2018. **13**(2): p. e0192576.
870. Alzahrani, A.J., Potent antioxidant and anticancer activities of the methanolic extract of Calligonum comosum (L'Her) fruit hairs against human hepatocarcinoma cells. *Saudi Journal of Biological Sciences*, 2021. **28**(9): pp. 5283–5289.
871. Soliman, S.S., et al., Critical discovery and synthesis of novel antibacterial and resistance-modifying agents inspired by plant phytochemical defense mechanisms. *Chemico-Biological Interactions*, 2021. **333**: p. 109318.
872. Alzarah, M.I., Macro and micro-elements concentrations in Calligonum comosum wild grazing plant through its growth period. *Saudi Journal of Biological Sciences*, 2021. **28**(12): pp. 6992–6999.
873. Shalabi, M., et al., Anticancer activity of Aloe vera and Calligonum comosum extracts separetely on hepatocellular carcinoma cells. *Asian Pacific Journal of Tropical Biomedicine*, 2015. **5**(5): pp. 375–381.
874. Jahromi, M.A., et al., Evaluating the effects of Escanbil (Calligonum) extract on the expression level of Catsper gene variants and sperm motility in aging male mice. *Iranian Journal of Reproductive Medicine*, 2014. **12**(7): p. 459.
875. Abdel-Sattar, E.A., et al., Protective effect of Calligonum comosum on haloperidol-induced oxidative stress in rat. *Toxicology and Industrial Health*, 2014. **30**(2): pp. 147–153.
876. Ekrami, H., et al., The effects of calligonum extract and low-intensity ultrasound on motility, viability, and DNA fragmentation of human frozen-thawed semen samples. *International Journal of Fertility & Sterility*, 2020. **14**(2): p. 84.
877. Alehaideb, Z., et al., Anti-proliferative and pro-apoptotic effects of Calligonum comosum (L'Her.) methanolic extract in human triple-negative MDA-MB-231 breast cancer cells. *Journal of Evidence-Based Integrative Medicine*, 2020. **25**: p. 2515690X20978391.
878. Riadh, H., et al., Detection and extraction of anti-listeral compounds from Calligonum comosum, a medical plant from arid regions of Tunisia. *African Journal of Traditional, Complementary and Alternative Medicines*, 2011. **8**(3): pp. 322–327.
879. Badria, F.A., M. Ameen, and M.R. Akl, Evaluation of cytotoxic compounds from Calligonum comosum L. growing in Egypt. *Zeitschrift für Naturforschung C*, 2007. **62**(9–10): pp. 656–660.
880. Liu, X., et al., Anti-inflammatory and anti-ulcer activity of Calligonum comosum in rats. *Fitoterapia*, 2001. **72**(5): pp. 487–491.
881. Kiani, K., et al., Calligonum comosum (Escanbil) extract exerts anti-angiogenic, anti-proliferative and anti-inflammatory effects on endometriotic lesions. *Journal of Ethnopharmacology*, 2019. **239**: p. 111918.
882. Abdo, W., et al., Calligonum comosum extract inhibits diethylnitrosamine-induced hepatocarcinogenesis in rats. *Oncology Letters*, 2015. **10**(2): pp. 716–722.
883. Kumar, A., et al., Apoptosis-mediated anti-proliferative activity of Calligonum comosum against human breast cancer cells, and molecular docking of its major polyphenolics to Caspase-3. *Frontiers in Cell and Developmental Biology*, 2022. **10**: p. 972111.

884. Soliman, S.S., et al., Calli essential oils synergize with lawsone against multidrug resistant pathogens. *Molecules*, 2017. **22**(12): p. 2223.
885. Alwakeel, S.S., The effect of mycotoxins found in some herbal plants on biochemical parameters. *Pakistan Journal of Biological Sciences*, 2009. **12**(8): pp. 637–642.
886. Aly, H.A., et al., Lipoic acid and Calligonum comosumon attenuate aroclor 1260-induced testicular toxicity in adult rats. *Environmental Toxicology*, 2017. **32**(4): pp. 1147–1157.
887. Anak, H., M.F. Dönmez, and İ. Çoruh, Biological Control of Rhizoctonia Solani Kühn. with Rhizobacteria Isolated from Different Soiland Calligonum Polygonoides L. Subsp. Comosum (L'hér.). *Journal of Agriculture*, 2021. **4**(2): pp. 92–107.
888. Purohit, C. and R. Kumar, A review on genus Calligonum L.(Polygonaceae) from India and report Calligonum crinitum an addition for Flora of India. *Journal of Asia-Pacific Biodiversity*, 2020. **13**(2): pp. 319–324.
889. Samejo, M.Q., et al., Essential oil constituents in fruit and stem of Calligonum polygonoides. *Industrial Crops and Products*, 2013. **45**: pp. 293–295.
890. Soliman, S., et al., The scientific basis of the antibacterial traditional use of Calligonum comosum in UAE. *Journal of Herbal Medicine*, 2021. **27**: p. 100361.
891. Paolini, J., et al., Chemical composition, intraspecies variation and seasonal variation in essential oils of Calendula arvensis L. *Biochemical Systematics and Ecology*, 2010. **38**(5): pp. 865–874.
892. Ercetin, T., et al., Comparative assessment of antioxidant and cholinesterase inhibitory properties of the marigold extracts from Calendula arvensis L. and Calendula officinalis L. *Industrial Crops and Products*, 2012. **36**(1): pp. 203–208.
893. Maccioni, S., et al., Ricerche etnobotaniche in Liguria. La riviera spezzina (Liguria orientale). *Atti Della Società Toscana di Scienze Naturali Residente in Pisa. Memorie. Serie B*, 2008. **115**: pp. 77–82.
894. Addis, R., et al., Fibroblast proliferation and migration in wound healing by phytochemicals: evidence for a novel synergic outcome. *International Journal of Medical Sciences*, 2020. **17**(8): p. 1030.
895. Abudunia, A.-M., et al., Anticandidal, antibacterial, cytotoxic and antioxidant activities of Calendula arvensis flowers. *Journal de Mycologie Médicale*, 2017. **27**(1): pp. 90–97.
896. Abutaha, N., et al., Calendula arvensis L. as an anti-cancer agent against breast cancer cell lines. *Molecular Biology Reports*, 2019. **46**: pp. 2187–2196.
897. Fiorentino, M., et al., Calendula arvensis (Vaill.) L.: a systematic plant analysis of the polar extracts from its organs by UHPLC-HRMS. *Foods*, 2022. **11**(3): p. 247.
898. *Mechanism of action of Calendula arvensis*. Available from: https://my.clevelandclinic.org/health/treatments/24512-cardiac-glycosides
899. *Toxicity of Calendula arvensis*. Available from: https://www.aphis.usda.gov/plant_health/plant_pest_info/weeds/downloads/wra/Calendula-arvensis.pdf
900. Serjeant, R., Plants of Dhofar, the southern region of Oman, traditional, economic and medicinal uses. By Anthony G. Miller and Miranda Morris, illustrated by Susanna Stuart-Smith with a foreword by Sultan Qaboos Bin Saif and preface by RE Schultes. pp. xxvii, 361, front., 148 col. drawings, 10 col. photographs, 2 figs., 3 maps (2 on end papers). Oman, Office of the Adviser for Conservation of the Environment, Diwan of Royal Court, Sultanate of Oman, 1988.£ 35.00. *Journal of the Royal Asiatic Society*, 1989. **121**(2): pp. 338–340.
901. Rajkovic, J., et al., An updated pharmacological insight into calotropin as a potential therapeutic agent in cancer. *Frontiers in Pharmacology*, 2023. **14**: p. 1160616.
902. Mossa, J., et al., Pharmacological studies on aerial parts of Calotropis procera. *The American Journal of Chinese Medicine*, 1991. **19**(03n04): pp. 223–231.
903. Moustafa, A., et al., Extraction and phytochemical investigation of Calotropis procera: effect of plant extracts on the activity of diverse muscles. *Pharmaceutical Biology*, 2010. **48**(10): pp. 1080–1190.
904. Al-Rowaily, S.L., et al., Essential oil of Calotropis procera: comparative chemical profiles, antimicrobial activity, and allelopathic potential on weeds. *Molecules*, 2020. **25**(21): p. 5203.
905. Pattnaik, P.K., et al., Chemometric profile & antimicrobial activities of leaf extract of Calotropis procera and Calotropis gigantea. *Natural Product Research*, 2017. **31**(16): pp. 1954–1957.
906. Al Sulaibi, M.A., C. Thiemann, and T. Thiemann, Chemical constituents and uses of Calotropis procera and Calotropis gigantea–a review (Part I–the plants as material and energy resources). *Open Chemistry Journal*, 2020. **7**(1).
907. Salem, W., et al., Antibacterial activity of silver and zinc nanoparticles against Vibrio cholerae and enterotoxic Escherichia coli. *International Journal of Medical Microbiology*, 2015. **305**(1): pp. 85–95.
908. Priya, M., et al., First report of phytoplasma 'Candidatus Phytoplasma trifolii'(16SrVI) group associated with leaf yellows of Calotropis gigantea in India. *New Disease Reports*, 2010. **22**(29): p. 2044-0588.2010.
909. Alsharif, B., et al., Volatiles of Capparis cartilaginea Decne. from Saudi Arabia. *Plants*, 2022. **11**(19): p. 2518.
910. Tlili, N., et al., The caper (Capparis L.): ethnopharmacology, phytochemical and pharmacological properties. *Fitoterapia*, 2011. **82**(2): pp. 93–101.
911. S. Al Hatmi, 2024. The Natural Heritage of Musandam, Ministry of Heritage and Tourism. OMAN.
912. Moharram, B.A., et al., Phytochemical, anti-inflammatory, antioxidant, cytotoxic and antibacterial study of Capparis cartilaginea decne fromyemen. *International Journal of Pharmacy and Pharmaceutical Sciences*, 2018. **10**: pp. 38–44.
913. Wagner, A.E., A.M. Terschluesen, and G. Rimbach, Health promoting effects of brassica-derived phytochemicals: from chemopreventive and anti-inflammatory activities to epigenetic regulation. *Oxidative Medicine and Cellular Longevity*. 2013. pp. 1–12.
914. Cavaiuolo, M. and A. Ferrante, Nitrates and glucosinolates as strong determinants of the nutritional quality in rocket leafy salads. *Nutrients*, 2014. **6**(4): pp. 1519–1538.
915. Tebay, L.E., et al., Mechanisms of activation of the transcription factor Nrf2 by redox stressors, nutrient cues, and energy status and the pathways through which it attenuates degenerative disease. *Free Radical Biology and Medicine*, 2015. **88**: pp. 108–146.
916. Abutaha, N. and A. Al-Mekhlafi, Evaluation of the safe use of the larvicidal fraction of Capparis cartilaginea Decne. against Aedes caspius (Pallas)(Diptera: Culicidae) larvae. *African Entomology*, 2014. **22**(4): pp. 838–846.
917. Edmund, E., M.M. LK, and W. Eric, Pharmacognostic studies of the leaves, stem and root of Capparis erythrocarpos Isert (Capparaceae). *Pharmacognosy Journal*, 2019. **11**(1): pp. 112–118.

918. Hamed, A.R., et al., Chemical investigation of some Capparis species growing in Egypt and their antioxidant activity. *Evidence-Based Complementary and Alternative Medicine*, 2007. **4**: pp. 25–28.
919. Infantino, A., et al., Capparis spinosa-a new host for Sclerotium rolfsii. *Plant Pathology*, 2006. **55**(4): pp. 580–580.
920. Verma, P., et al., Pharmacognostical potential of Capparis deciduas Edgew. *Journal of Applied Pharmaceutical Science*, 2011(Issue): pp. 6–11.
921. Anjum, S., et al., Biological and phytochemical studies on Capparis decidua (Forssk) Edgew from Cholistan desert. *Natural Product Research*, 2020. **34**(16): pp. 2315–2318.
922. Chopra, R., S. Nayar, and I. Chopra, *Glossary of Indian Medicinal Plants, National Institute of Science Communication and Information Resources*, vol. 203. New Delhi: CSIR, 2006.
923. Matsuura, H.N. and A.G. Fett-Neto, Plant alkaloids: main features, toxicity, and mechanisms of action. *Plant Toxins*, 2015. **2**(7): pp. 1–15.
924. Upadhyay, R., Insecticidal and oviposition inhibition efficacy of Capparis decidua to Sitophilus oryzae Linn. (Coleoptera: Curculionidae). *International Journal of Chemical and Biochemical Sciences*, 2012. **2**(1): pp. 14–23.
925. Singh, R.P. and P.N. Jha, The multifarious PGPR Serratia marcescens CDP-13 augments induced systemic resistance and enhanced salinity tolerance of wheat (Triticum aestivum L.). *PLos One*, 2016. **11**(6): p. e0155026.
926. Kavak, H., Epidemic outbreaks of powdery mildew caused by Leveillula taurica on Capparis spinosa in Turkey. *Plant Pathology*, 2004. **53**(6).
927. Memariani, Z., et al., *Traditional uses, in Seyed Mohammad* Nabavi, Ipek Suntar, Davide Barreca, Haroon Khan, *Phytonutrients in Food*. 2020, Elsevier. p. 23–66.
928. Fici, S., A taxonomic revision of the Capparis spinosa group (Capparaceae) from the Mediterranean to Central Asia. *Phytotaxa*, 2014. **174**(1): p. 1–24.
929. Shahrajabian, M.H., W. Sun, and Q. Cheng, Plant of the Millennium, Caper (Capparis spinosa L.), chemical composition and medicinal uses. *Bulletin of the National Research Centre*, 2021. **45**: pp. 1–9.
930. Hacİseferoğullar İ, H., M. Özcan, and E. Duman, Biochemical and technological properties of seeds and oils of Capparis spinosa and Capparis ovata plants growing wild in Turkey. *Journal of Food Processing and Technology*, 2011. **2**(6).
931. Zhang, H. and Z.F. Ma, Phytochemical and pharmacological properties of Capparis spinosa as a medicinal plant. *Nutrients*, 2018. **10**(2): p. 116.
932. Rahnavard, R. and N. Razavi, A review on the medical effects of Capparis spinosa L. *Advanced Herbal Medicine*, 2017. **3**(1): pp. 44–53.
933. Safe, S., et al., Flavonoids: structure–function and mechanisms of action and opportunities for drug development. *Toxicological Research*, 2021. **37**: pp. 147–162.
934. Marketed product of *Calotropis*. Available from: https://kamal.pk/Product/3/Mother-Tinctures/461/CALOTROPIS/
935. Vahid, H., H. Rakhshandeh, and A. Ghorbani, Antidiabetic properties of Capparis spinosa L. and its components. *Biomedicine & Pharmacotherapy*, 2017. **92**: pp. 293–302.
936. Ciferri, R., Rassegna di parassiti e malattie del cappero (Capparis spinosa L.) in Italia. *Notiziario sulle Malattie delle Piante*, 1949. **3**: pp. 33–35.
937. Bruyns, P. and H. Jonkers, The genus Caralluma R. Br.(Asclepiadaceae) in Oman. *Bradleya*, 1993. **1993**(11): pp. 51–69.
938. Soobrattee, M.A., et al., Phenolics as potential antioxidant therapeutic agents: mechanism and actions. *Mutation Research/Fundamental and Molecular Mechanisms of Mutagenesis*, 2005. **579**(1–2): pp. 200–213.
939. Raees, M.A., et al., Desmiflavasides A and B: two new bioactive pregnane glycosides from the sap of Desmidorchis flava. *Phytochemistry Letters*, 2015. **12**: pp. 153–157.
940. Tatiya, A., et al., Antioxidant and hypolipidemic effect of Caralluma adscendens Roxb. in alloxanized diabetic rats. *IJP-International Journal of Pharmacology*, 2010. **6**(4): pp. 400–406.
941. Bader, A., et al., Further constituents from Caralluma negevensis. *Phytochemistry*, 2003. **62**(8): pp. 1277–1281.
942. Abdel-Sattar, E. and D.E. Ali, Russelioside B: a pregnane glycoside with pharmacological potential. *Revista Brasileira de Farmacognosia*, 2022. **32**(2): pp. 188–200.
943. Dutt, H.C., et al., Pharmacological review of Caralluma R. Br. with special reference to appetite suppression and anti-obesity. *Journal of Medicinal Food*, 2012. **15**(2): pp. 108–119.
944. Mothana, R.A., et al., Evaluation of the in vitro antiplasmodial, antileishmanial, and antitrypanosomal activity of medicinal plants used in Saudi and Yemeni traditional medicine. *Evidence-Based Complementary and Alternative Medicine*, 2014. **2014**.
945. Issiki, Z., et al., Toxicological evaluation of the aqueous extract of Caralluma europaea and its immunomodulatory and inflammatory activities. *Pharmacognosy Research*, 2017. **9**(4): p. 390.
946. Wadood, A., N. Wadood, and S. Shah, Effects of Acacia arabica and Caralluma edulis on blood glucose levels of normal and alloxan diabetic rabbits. JPMA. *The Journal of the Pakistan Medical Association*, 1989. **39**(8): pp. 208–212.
947. Ali, S., et al., Endophytic fungi from Caralluma acutangula can secrete plant growth promoting enzymes. *Fresenius Environmental Bulletin*, 2019. **28**: pp. 2688–2696.
948. Raees, M.A., A phytopharmacological review on an Arabian medicinal plant: Caralluma flava NE Br. *International Journal of Phytomedicine*, 2018. **10**: pp. 148–152.
949. Patzelt, A., *Oman Plant: Red Data Book*. Oman botanic Garden, 2015.
950. Raees, M.A., *Phytochemical Investigation on the Chemical Constituents of Caralluma flava*. Federal Urdu University of Arts, Sciences & Technology, 2017.
951. Al-Naqeb, G., Acute toxicity and anti-ulcerative potential of Caralluma flava NE Br methanolic extract against ethanol-induced gastric ulcers in rats. *Journal of Medicinal Plants*, 2017. **5**(6): pp. 21–25.
952. Hussain, H., et al., Nizwaside: a new anticancer pregnane glycoside from the sap of Desmidorchis flava. *Archives of Pharmacal Research*, 2015. **38**: pp. 2137–2142.
953. *Caralluma penicillata 2005*. Available from: https://www.llifle.com/Encyclopedia/SUCCULENTS/Family/Asclepiadaceae/33656/Caralluma_penicillata
954. Abdel-Sattar, E., et al., Penicillosides A– C, C-15 oxypregnane glycosides from Caralluma penicillata. *Phytochemistry*, 2001. **57**(8): pp. 1213–1217.
955. Al-Yahya, M., et al., Toxicity studies on Caralluma penicillata – a poisonous plant from Saudi Arabia. *Toxicon*, 1983. **21**: pp. 9–12.
956. Shareef, H., et al., In vitro antimicrobial and phytochemical analysis of Cardiospermum halicacabum L. *Pakistan Journal of Botany*, 2012. **44**(5): pp. 1677–1680.
957. Sheeba, M. and V. Asha, Cardiospermum halicacabum ethanol extract inhibits LPS induced COX-2, TNF-α and

iNOS expression, which is mediated by NF-κB regulation, in RAW264. 7 cells. *Journal of Ethnopharmacology*, 2009. **124**(1): pp. 39–44.

958. Raza, S.A., et al., Review of beneficial and remedial aspects of Cardiospermum halicacabum L. *African Journal of Pharmacy and Pharmacology*, 2013. **7**(48): pp. 3026–3033.
959. Veerappan, L. and N. Pradheepa, Evaluation of antimicrobial potentials of Cardiospermum halicacabum Linn. *Indonesian Journal of Pharmacy*, 2013. **24**(1): pp. 61–64.
960. Viji, M. and S. Murugesan, Phytochemical Analysis and Antibacterial Activity of Medicinal Plant Cardiospermum halicacabum Linn. *Journal of Phytology*, 2010. **2**(1).
961. Duke, J.A. and E.S. Ayensu, *Medicinal Plants of China*. 1985. https://www.abebooks.com/9780917256202/Medicinal-Plants-China-Duke-James-0917256204/plp
962. Subramanyam, R., et al., Exploring ethnobiological classifications for novel alternative medicine: a case study of Cardiospermum halicacabum L. Modakathon, Balloon Vine) as a traditional herb for treating rheumatoid arthritis. *Ethnobotany*, 2007. **19**(1): pp. 1–16.
963. Samarakoon, S., Phosphodiesterase-5 (PDE5) Inhibitory Potential of Major Phytochemicals of Withania somnifera and Cardiospermum halicacabum: an in silico comparison with approved PDE5 inhibitors. *Authorea Preprints*, 2023.
964. Linh, N.T.T., et al., Flavonoid from the aerial parts of Cardiospermum halicacabum L. *Vietnam Journal of Chemistry*, 2023. **61**(1): pp. 80–83.
965. Sadique, J., et al., Biochemical modes of action of Cassia occidentalis and Cardiospermum halicacabum in inflammation. *Journal of ethnopharmacology*, 1987. **19**(2): pp. 201–212.
966. Rao, N.V., K.C. Prakash, and S.S. Kumar, Pharmacological investigation of Cardiospermum halicacabum (Linn) in different animal models of diarrhoea. *Indian Journal of Pharmacology*, 2006. **38**(5): p. 346.
967. Huang, M.-H., et al., Antioxidant and anti-inflammatory properties of Cardiospermum halicacabum and its reference compounds ex vivo and in vivo. *Journal of Ethnopharmacology*, 2011. **133**(2): pp. 743–750.
968. Fai, D., et al., Cardiospermum halicacabum in atopic dermatitis: clinical evidence based on phytotherapic approach. *Dermatologic Therapy*, 2020. **33**(6): p. e14519.
969. *Toxicity of Cardiospermum helicacabum*. Available from: https://www.ema.europa.eu/en/documents/mrl-report/cardiospermum-halicacabum-summary-report-committee-veterinary-medicinal-products_en.pdf.
970. Waako, P., et al., The in vitro and in vivo antimalarial activity of Cardiospermum halicacabum L. and Momordica foetida Schumch. Et Thonn. *Journal of Ethnopharmacology*, 2005. **99**(1): pp. 137–143.
971. Ahmad, B., et al., Pharmacological and biological investigations of Chenopodium murale Linn. *Asian Journal of Plant Sciences*, 2003. **2**(15–16): pp. 1107–1111.
972. Raza, A., et al., Bio-ecology and the management of Chenopodium murale L.: a problematic weed in Asia. *Crop Protection*, 2023: p. 106332.
973. *Macro and Micro characteristics of Chenopodium murale*. Available from: https://plantnet.rbgsyd.nsw.gov.au/cgi-bin/NSWfl.pl?page=nswfl&lvl=sp&name=Chenopodium~murale#:~:text=Chenopodium%20murale%20L.&text=Description%3A%20Erect%20annual%20to%201.5,toothed%2C%20even%20on%20uppermost%20leaves.
974. *Description of Chenopodium murale*. Available from: https://en.wikipedia.org/wiki/Chenopodiastrum_murale.
975. El-Sayed, N., et al., A flavonol triglycoside from Chenopodium murale. *Phytochemistry*, 1999. **51**(4): pp. 591–593.
976. Abdel-Aziz, M.S., et al., Antioxidant and antibacterial activity of silver nanoparticles biosynthesized using Chenopodium murale leaf extract. *Journal of Saudi Chemical Society*, 2014. **18**(4): pp. 356–363.
977. El-Newary, S.A., et al., Chenopodium murale Juice Shows Anti-Fungal Efficacy in Experimental Oral Candidiasis in Immunosuppressed Rats in Relation to Its Chemical Profile. *Molecules*, 2023. **28**(11): p. 4304.
978. Said, A., et al., Mechanisms underlying the wound healing and tissue regeneration properties of Chenopodium album. *3 Biotech*, 2020. **10**: pp. 1–10.
979. *Toxicity of Chenopodium murale*. Available from: https://pfaf.org/User/Plant.aspx?LatinName=Chenopodium+murale#:~:text=Chenopodium%20murale%20%2D%20L.&text=The%20leaves%20and%20seeds%20of,small%20to%20do%20any%20harm
980. da Silva, M.G.C., et al., Acute and sub-chronic toxicity of aqueous extracts of Chenopodium ambrosioides leaves in rats. *Journal of Medicinal Food*, 2014. **17**(9): pp. 979–984.
981. Ali, N.A., et al., Screening of Yemeni medicinal plants for antibacterial and cytotoxic activities. *Journal of Ethnopharmacology*, 2001. **74**(2): pp. 173–179.
982. Capinera, J., *Handbook of Vegetable Pests*. Academic Press, 2020.
983. Hull, R., *Chapter 17-Variation, Evolution and Origins of Plant Viruses*. Matthews' Plant Virology, 4th edn. London: Academic Press, 2002, pp. 743–812.
984. Poonia, A. and A. Upadhayay, Chenopodium album Linn: review of nutritive value and biological properties. *Journal of Food Science and Technology*, 2015. **52**: pp. 3977–3985.
985. *Description of Chenopodium album*. Available from: https://en.wikipedia.org/wiki/Chenopodium_album
986. Krak, K., et al., Allopolyploid origin of Chenopodium album s. str.(Chenopodiaceae): a molecular and cytogenetic insight. *PLoS One*, 2016. **11**(8): p. e0161063.
987. Usman, L., et al., Chemical constituents and anti-inflammatory activity of leaf essential oil of Nigerian grown Chenopodium album L. *EXCLI Journal*, 2010. **9**: p. 181.
988. Yılmaz, P.K., et al., Chemical compositions by LC-MS/MS and GC-MS and biological activities of Chenopodium album subsp. album var. microphyllum. *Industrial Crops and Products*, 2019. **141**: p. 111755.
989. Prajapati, N.D., et al., A handbook of medicinal plants: a complete source book, in Prajapati, Narayan Das, S.S. Purohit, Arun K. Sharma & Tarun Kumar (eds.), *A Handbook of Medicinal Plants: A Complete Source Book*. 2003. p. 554. Agrobios, India.
990. Sarma, H., A. Sarma, and C. Sarma, Traditional knowledge of weeds: a study of herbal medicines and vegetables used by the Assamese people [India]. *Herba Polonica*, 2008. **2**(54): pp. 80–88.
991. Sharma, P., S. Tripathi, and R. Chandra, Phytoremediation potential of heavy metal accumulator plants for waste management in the pulp and paper industry. *Heliyon*, 2020. **6**(7): pp. e04559.
992. Warner, S., *Wild Edibles of the West Coast: Foraging Wild Edible Plants in California and the Pacific Northwest-2 Book Bundle*. Rowan's Publishing, LLC, 2023.
993. Fleischer, A., M.A. O'Neill, and R. Ehwald, The pore size of non-graminaceous plant cell walls is rapidly decreased by borate ester cross-linking of the pectic polysaccharide rhamnogalacturonan II. *Plant Physiology*, 1999. **121**(3): pp. 829–838.

994. Nerva, L., et al., Biological and molecular characterization of Chenopodium quinoa mitovirus 1 reveals a distinct small RNA response compared to those of cytoplasmic RNA viruses. *Journal of Virology*, 2019. **93**(7): p. 10.1128/jvi.01998-18.
995. Mandaville, J.P., *Flora of Eastern Saudi Arabia*. Routledge, 2013.
996. Al-Shammari, L.A., W.H. Hassan, and H.M. Al-Youssef, Chemical composition and antimicrobial activity of the essential oil and lipid content of Carduus pycnocephalus L. growing in Saudi Arabia. *Journal of Chemical and Pharmaceutical Research*, 2012. **4**(2): pp. 1281–1287.
997. El-Lakany, A., et al., A new flavone glycoside with antimicrobial activity from Carduus pycnocephalus L. *Die Pharmazie*, 1997. **52**(1): pp. 78–79.
998. Al-Shammari a, W.H.B.H.a.b., Hanan M. Al-Youssef, Phytochemical and biological studies of Carduus pycnocephalus L. Author links open overlay panelLatifa A. *Journal of Saudi Chemical Society*, 2015. 4(4 July 2015): pp. 410–416.
999. Esmaeili, A., et al., Volatile constituents of Centaurea depressa MB and Carduus pycnocephalus L. two compositae herbs growing wild in Iran. *Journal of essential oil research*, 2005. **17**(5): pp. 539–541.
1000. Goeden, R.D., Comparative survey of the phytophagous insect faunas of Italian thistle, Carduus pycnocephalus, in southern California and southern Europe relative to biological weed control. *Environmental Entomology*, 1974. **3**(3): pp. 464–474.
1001. Vogelsang, K.M. and J.D. Bever, Mycorrhizal densities decline in association with nonnative plants and contribute to plant invasion. *Ecology*, 2009. **90**(2): pp. 399–407.
1002. Mejri, D., T. Souissi, and D. Berner, First report of rust caused by Puccinia carduorum on Italian thistle in Tunisia. *Plant Disease*, 2008. **92**(1): p. 174.
1003. *Carissa edulisc (SANBI)*. Available from: https://pza.sanbi.org/carissa-edulis
1004. Giday, M., et al., An ethnobotanical study of medicinal plants used by the Zay people in Ethiopia. *Journal of Ethnopharmacology*, 2003. **85**(1): pp. 43–52.
1005. Bentley, M.D., S.R. Brackett, and A. Chapya, 2-Hydroxyacetophenone: principal root volatile of the East African medicinal plant, Carissa edulis. *Journal of Natural Products*, 1984. **47**(6): pp. 1056–1057.
1006. Ahmed, I.S. and D.E. Mustafa, Psychoactive chemical constituents of Carissa edulis (Forssk.) vahl roots and its safety limits in folk treatment: benzodiazepines analogous. *Journal of Chemical Sciences*, 2020. **14**: pp. 113–118.
1007. Fanta Yadang, S.A., et al., Quantification of bioactive compounds and evaluation of the antioxidant activity of Carissa edulis Valh (Apocynaceae) leaves. *The Scientific World Journal*, 2019. **2019**: pp. 7549620.
1008. Al-Youssef, H.M. and W.H. Hassan, Chemical constituents of Carissa edulis Vahl. *Arabian Journal of Chemistry*, 2017. **10**(1): pp. 109–113.
1009. Houngbè, U., et al., Carissa edulis Vahl (Apocynaceae) extract, a medicinal plant of Benin pharmacopoeia, induces potent endothelium-dependent relaxation of coronary artery rings involving nitric oxide. *Phytomedicine*, 2022. **105**: p. 154370.
1010. Kaczmarek, B., Tannic acid with antiviral and antibacterial activity as a promising component of biomaterials – A minireview. *Materials*, 2020. **13**(14): p. 3224.
1011. Allahverdiyev, A.M., et al., Development of new antiherpetic drugs based on plant compounds, in Mahendra Kumar Rai, Kateryna Volodymyrivna Kon (eds.), *Fighting Multidrug Resistance with Herbal Extracts, Essential Oils and Their Components*. 2013, Elsevier. pp. 245–259.
1012. Selvaraj, D., et al., Evaluation of DNA barcode candidates for the discrimination of the large plant family Apocynaceae. *Plant Systematics and Evolution*, 2015. **301**: pp. 1263–1273.
1013. Tolo, F.M., et al., Anti-viral activity of the extracts of a Kenyan medicinal plant Carissa edulis against herpes simplex virus. *Journal of Ethnopharmacology*, 2006. **104**(1-2): pp. 92–99.
1014. *Pathology of Carissa*. Available from: https://ipm.ucanr.edu/PMG/GARDEN/PLANTS/DISEASES/dampingoff.html
1015. *Rush Milkweed*. Available from: https://www.flowersofindia.net/catalog/slides/Rush%20Milkweed.html
1016. Ansari, B., et al., Caralluma edulis (Apocynaceae): a comprehensive review on its traditional uses, phytochemical profile and pharmacological effects. *Current Topics in Medicinal Chemistry*, 2022. **22**(18): pp. 1501–1514.
1017. Singh,V.andN.Nimbkar,Safflower(CarthamustinctoriusL.). *Chapter*, 2006. **6**: pp. 167–194.
1018. Zohary, D., M. Hopf, and E. Weiss, *Domestication of Plants in the Old World: The Origin and Spread of Domesticated Plants in Southwest Asia, Europe, and the Mediterranean Basin*. Oxford University Press, 2012.
1019. Asgarpanah, J. and N. Kazemivash, Phytochemistry, pharmacology and medicinal properties of Carthamus tinctorius L. *Chinese Journal of Integrative Medicine*, 2013. **19**: pp. 153–159.
1020. Zhou, F.-R., M.-B. Zhao, and P.-F. Tu, Simultaneous determination of four nucleosides in Carthamus tinctorius L. and Safflower injection using highperformance liquid chromatography. *Journal of Chinese Pharmaceutical Sciences*, 2009. **18**: pp. 326–330.
1021. Asgary, S., et al., Antidiabetic effect of hydroalcoholic extract of Carthamus tinctorius L. in alloxan-induced diabetic rats. *Journal of Research in Medical Sciences: the Official Journal of Isfahan University of Medical Sciences*, 2012. **17**(4): p. 386.
1022. Tu, Y., et al., Carthami flos: a review of its ethnopharmacology, pharmacology and clinical applications. *Revista Brasileira de Farmacognosia*, 2015. **25**: pp. 553–566.
1023. Mirhoseini, M., M. Mohamadpour, and L. Khorsandi, Toxic effects of Carthamus tinctorius L.(Safflower) extract on mouse spermatogenesis. *Journal of Assisted Reproduction and Genetics*, 2012. **29**: pp. 457–461.
1024. Nobakht, M., et al., A study on the teratogenic and cytotoxic effects of safflower extract. *Journal of Ethnopharmacology*, 2000. **73**(3): pp. 453–459.
1025. Holz, N., et al., Point-of-care suitable identification of the adulterants Carthamus tinctorius and Curcuma longa in Crocus sativus based on loop-mediated isothermal amplification (LAMP) and lateral-flow-assay (LFA). *Food Control*, 2023. **148**: p. 109637.
1026. Dai, Y., et al., Natural deep eutectic solvents as a new extraction media for phenolic metabolites in Carthamus tinctorius L. *Analytical Chemistry*, 2013. **85**(13): pp. 6272–6278.
1027. Northern, G.S., *Grains Research and Development Corporation*. Australia: GRDC, 2017.
1028. Carev, I., et al., Centaurea triumfetii essential oil chemical composition, comparative analysis, and antimicrobial activity of selected compounds. *Scientific Reports*, 2023. **13**(1): p. 7475.
1029. Alamri, F.B., et al., Bioactivity and molecular docking of lactones isolated from Centaurea pseudosinaica Czerep. *Saudi Pharmaceutical Journal*, 2023. **31**(6): pp. 773–782.

1030. Yang, W., et al., Advances in pharmacological activities of terpenoids. *Natural Product Communications*, 2020. **15**(3): p. 1934578X20903555.
1031. *Toxicity of Centaurea*. Available from: https://www.gardenersworld.com/plants/centaurea-cyanus/#:~:text=Is%20Centaurea%20poisonous%3F,has%20no%20toxic%20effects%20reported
1032. Mothana, R.A., et al., Antimicrobial, antioxidant and cytotoxic activities and phytochemical screening of some yemeni medicinal plants. *Evidence-based Complementary and Alternative Medicine*, 2010. **7**: pp. 323–330.
1033. Brown, E., Observations on a race of Ditylenchus dipsaci attacking annual aster and sweet sultan. *Plant Pathology*, 1958. **7**(4): pp. 150–151.
1034. *Lesser Centaury*. Available from: https://luontoportti.com/en/t/2287/lesser-centaury
1035. Van der Sluis, W., Chemotaxonomical investigations of the genera Blackstonia and Centaurium (Gentianaceae). *Plant Systematics and Evolution*, 1985. **149**: pp. 253–286.
1036. Bibi, H., et al., Phytochemical studies and antibacterial activity of Centaurium pulchellum Druce. *Natural Product Research*, 2006. **20**(10): pp. 896–901.
1037. *Extinction. 2023*. Available from: https://www.nhm.ac.uk/our-science/data/uk-species/species/centaurium_scilloides.html
1038. Šiler, B., et al., Antibacterial and antifungal screening of Centaurium pulchellum crude extracts and main secoiridoid compounds. *Natural Product Communications*, 2010. **5**(10): p. 1934578X1000501001.
1039. *Centaurium pulchellum toxicity*. Available from: https://www.maltawildplants.com/GENT/Centaurium_pulchellum.php
1040. Hillcoat, D., G. Lewis, and B. Verdcourt, A new species of Ceratonia (Leguminosae-Caesalpinioideae) from Arabia and the Somali Republic. *Kew Bulletin*, 1980: pp. 261–271.
1041. Dahmani, W., et al., Exploring Carob (Ceratonia siliqua L.): a comprehensive assessment of its characteristics, ethnomedicinal uses, *Phytochemical Aspects, and Pharmacological Activities. Plants*, 2023. **12**(18): p. 3303.
1042. Basharat, Z., et al., Nutritional and functional profile of carob bean (Ceratonia siliqua): a comprehensive review. *International Journal of Food Properties*, 2023. **26**(1): pp. 389–413.
1043. Rathod, N.B., et al., Recent developments in polyphenol applications on human health: a review with current knowledge. *Plants*, 2023. **12**(6): p. 1217.
1044. *Ceratonia siliqua (carob)*. Available from: https://foodallergytest.co.uk/product/carob-allergy-test/
1045. El-Gali, Z.I., The causative of leaf spot on carob: isolation and identification. *European Journal of Botany Plant Sciences and Pathology*, 2015. **2**(2): pp. 11–18.
1046. *Vernacular Name*. Available from: https://species.wikimedia.org/wiki/Chlorophytum_laxum
1047. Books, S., *Sunset Western Garden Book*. 1975. TI Inc. Books.
1048. *Spider Plant*. Available from: https://hort.extension.wisc.edu/articles/spider-plant-chlorophytum-comosum/
1049. Katoch, M., et al., Identification of Chlorophytum species (C. borivilianum, C. arundinaceum, C. laxum, C. capense and C. comosum) using molecular markers. *Industrial Crops and Products*, 2010. **32**(3): pp. 389–393.
1050. Abubakar, A., et al., Antihyperglycaemic activity of ethylacetate extract of Chlorophytum alismifolium in type 2 diabetes: the involvement of peroxisome proliferator-activated receptor-γ and dipeptidyl peptidase-4. *Journal of Integrative Medicine*, 2021. **19**(1): pp. 78–84.
1051. Gao, Y.H., et al., Chlorophytoside A, a new labdane diterpene glycoside from Chlorophytum laxum. *Chinese Chemical Letters*, 2005. **16**(7): p. 925.
1052. Auwal, M.S., et al. Preliminary phytochemical and elemental analysis of aqueous and fractionated pod extracts of Acacia nilotica (Thorn mimosa). *Veterinary Research Forum: an International Quarterly Journal*. 2014. **5**(2): 95–100.
1053. Niranja, M.H., Chlorophytum laxum, R. Br. (Liliaceae): a threatened medicinal herb of soliga tribe from B.R.Hills, Karnataka, India. *Journal of Applied Pharmaceutical Science* 2011.
1054. Adhami, S., et al., Chemical profiling of Chlorophytum comosum (Thunb.) jaques by GC-MS/LC-ESIMS and its antiproliferative effects on human carcinoma cell lines. *Anti-Cancer Agents in Medicinal Chemistry (Formerly Current Medicinal Chemistry-Anti-Cancer Agents)*, 2021. **21**(13): pp. 1697–1707.
1055. *Toxicity of Spider Plants*. Available from: https://www.kansaspoisoncenter.com/poisons/plants-and-mushrooms/spider-plant
1056. Chuaphasuk, C. and B. Prapagdee, Effects of biochar-immobilized bacteria on phytoremediation of cadmium-polluted soil. *Environmental Science and Pollution Research*, 2019. **26**: pp. 23679–23688.
1057. Sattar, A., et al., First report of leaf blight of Chlorophytum borivilianum caused by Colletotrichum capsici in northern India. *Plant Pathology*, 2006. **55**(2): p. 301.
1058. Kamel, M.R., et al., Ent-trachylobane-3 β-hydroperoxide, a new diterpene from the root bark of Chrozophora oblongifolia (Fam.; Euphorbiaceae). *Natural Product Research*, 2021. **35**(19): pp. 3177–3184.
1059. Ahmed, M., et al., Review: biochemical composition and medicinal uses of Chrozophora genus. *International Journal of Pharmaceutical Sciences and Research*, 2014. **4**: pp. 227–232.
1060. Abdallah, H.M., et al., Phenolics from Chrozophora oblongifolia aerial parts as inhibitors of α-glucosidases and advanced glycation end products: in-vitro assessment, molecular docking and dynamics studies. *Biology*, 2022. **11**(5): p. 762.
1061. Kamel, M.R., et al., Phytochemical and biological study of the aerial parts of Chrozophora oblongifolia (Delile) Spreng.(Euphorbiaceae). *Journal of Pharmacognosy and Phytochemistry*, 2016. **5**(4): pp. 17–24.
1062. Dipankar, C., S. Murugan, and P.U. Devi, Review on medicinal and pharmacological properties of Iresine herbstii, Chrozophora rottleri and Ecbolium linneanum. *African Journal of Traditional, Complementary and Alternative Medicines*, 2011. **8**(5S): pp. 124–129.
1063. *Toxicity of Chrozophora*. Available from: https://uses.plantnet-project.org/en/Chrozophora_plicata_(PROTA).
1064. Sher, A.A., et al., Prokinetic and laxative effects of Chrozophora tinctoria whole plant extract. *Molecules*, 2022. **27**(7): p. 2143.
1065. *Adulteration of Chrozophora oblongifolia*. Available from: https://www.inaturalist.org/taxa/489423-Chrozophora-oblongifolia
1066. Mothana, R.A., et al., Assessment of selected Yemeni medicinal plants for their in vitro antimicrobial, anticancer, and antioxidant activities. *Pharmaceutical Biology*, 2011. **49**(2): pp. 200–210.

1067. *Pathology of Chrozophora oblongifolia.* Available from: https://www.botanikks.com/plants/chrozophora-oblongifolia-delile-ajuss-ex-spreng/637245/1
1068. *Britannica Chicory.* Available from: https://www.britannica.com/plant/chicory
1069. Nørbæk, R., K. Nielsen, and T. Kondo, Anthocyanins from flowers of Cichorium intybus. *Phytochemistry*, 2002. **60**(4): pp. 357–359.
1070. Varotto, S., M. Lucchin, and P. Parrini, Immature embryos culture in Italian red chicory. *Plant Cell, Tissue and Organ Culture*, 2000. **62**: pp. 75–77.
1071. Nwafor, I.C., K. Shale, and M.C. Achilonu, Chemical composition and nutritive benefits of chicory (Cichorium intybus) as an ideal complementary and/or alternative livestock feed supplement. *The Scientific World Journal*, 2017. **2017**: pp. 11.
1072. Judžentienė, A. and J. Būdienė, Volatile constituents from aerial parts and roots of Cichorium intybus L.(chicory) grown in Lithuania. *Chemija*, 2008. **19**(2), 25–28.
1073. Conforti, F., et al., Antiproliferative activity against human tumor cell lines and toxicity test on Mediterranean dietary plants. *Food and Chemical Toxicology*, 2008. **46**(10): pp. 3325–3332.
1074. *Substitution of Cichorium intybus.* Available from: https://www.gardenia.net/guide/native-plant-alternatives-to-cichorium-intybus
1075. Baxter, N.T., et al., Dynamics of human gut microbiota and short-chain fatty acids in response to dietary interventions with three fermentable fibers. *MBio*, 2019. **10**(1): p. 10.1128/mbio.02566-18.
1076. *CABI.* Available from: https://www.cabidigitallibrary.org/doi/10.1079/cabicompendium.13409
1077. Mishra, G., S. Srivastava, and B. Nagori, Pharmacological and therapeutic activity of Cissus quadrangularis: an overview. *International Journal of Pharmtech Research*, 2010. **2**(2): pp. 1298–1310.
1078. Sundaran, J., et al., A short review on pharmacological activity of Cissus quadrangularis. *Bioinformation*, 2020. **16**(8): p. 579.
1079. Stohs, S.J. and S.D. Ray, A review and evaluation of the efficacy and safety of Cissus quadrangularis extracts. *Phytotherapy Research*, 2013. **27**(8): pp. 1107–1114.
1080. Hasani-Ranjbar, S., et al., A systematic review of the efficacy and safety of herbal medicines used in the treatment of obesity. *World Journal of Gastroenterology: WJG*, 2009. **15**(25): p. 3073.
1081. HUMPHREYS- JONES, D., Leaf Blotch (Glomerella cingulata (Stonem.) Spanld, & Schrenk) on Cissus antarctica Vent. *Plant Pathology*, 1976. **25**(2): p. 115.
1082. Al-Snafi, A.E., Bioactive metabolites and pharmacology of Cistanche tubulosa-A review. *IOSR Journal of Pharmacy*, 2020. **10**(1): pp. 37–46.
1083. Gu, C., X. Yang, and L. Huang, Cistanches Herba: a neuropharmacology review. *Frontiers in Pharmacology*, 2016. **7**: p. 289.
1084. Bao, X., et al., Effects of the Cistanche tubulosa aqueous extract on the gut microbiota of mice with intestinal disorders. *Evidence-Based Complementary and Alternative Medicine*, 2021. **2021**: pp. 1–11.
1085. Liu, J., et al., In silico-based screen synergistic drug combinations from herb medicines: a case using Cistanche tubulosa. *Scientific Reports*, 2017. **7**(1): p. 16364.
1086. Wu, C.-R., H.-C. Lin, and M.-H. Su, Reversal by aqueous extracts of Cistanche tubulosa from behavioral deficits in Alzheimer's disease-like rat model: relevance for amyloid deposition and central neurotransmitter function. *BMC Complementary and Alternative Medicine*, 2014. **14**(1): pp. 1–11.
1087. Geng, X., et al., Neuroprotective effects of echinacoside in the mouse MPTP model of Parkinson's disease. *European Journal of Pharmacology*, 2007. **564**(1–3): pp. 66–74.
1088. Shi, H.M., et al., Identification of Cistanche species by chemical and inter-simple sequence repeat fingerprinting. *Biological and Pharmaceutical Bulletin*, 2009. **32**(1): pp. 142–146.
1089. Chen, H., et al., Echinacoside prevents the striatal extracellular levels of monoamine neurotransmitters from diminution in 6-hydroxydopamine lesion rats. *Journal of Ethnopharmacology*, 2007. **114**(3): pp. 285–289.
1090. Liu, X.-M., et al., Chemical constituents from Cistanche sinensis (Orobanchaceae). *Biochemical Systematics and Ecology*, 2013. **47**: pp. 21–24.
1091. Shen, C.Y., et al., Anti-ageing active ingredients from herbs and nutraceuticals used in traditional Chinese medicine: pharmacological mechanisms and implications for drug discovery. *British Journal of Pharmacology*, 2017. **174**(11): pp. 1395–1425.
1092. Pravin, B., et al., Review on Citrullus colocynthis. *International Journal of Current Pharmaceutical Research*, 2013. **3**(1): pp. 46–53.
1093. Najafi, S., et al., Phytochemical screening and antibacterial activity of Citrullus colocynthis (Linn.) Schrad against Staphylococcus aureus. *Journal of Medicinal Plants Research*, 2010. **4**(22): pp. 2321–2325.
1094. Zheng, M.-S., et al., Research progress on chemical constituents of Citrullus colocynthis and their pharmacological effects. Zhongguo Zhong yao za zhi= Zhongguo zhongyao zazhi= *China Journal of Chinese Materia Medica*, 2020. **45**(4): pp. 816–824.
1095. Lavie, D., D. Willner, and Z. Merenlender, Constituents of Citrullus colocynthis (L.) Schrad. *Phytochemistry*, 1964. **3**(1): pp. 51–56.
1096. Masayuki, Y., et al., Bioactive Saponins and Glycosides (27) Structures or New Cucurbitane-Type Triterpene Glycosides and Antiallergic Constituents from Citrullus colocynthis. *Chemical & Pharmaceutical Bulletin*, 2007. **55**(3): pp. 428–434.
1097. Chen, J.C., et al., Cucurbitacins and cucurbitane glycosides: structures and biological activities. *Natural Product Reports*, 2005. **22**(3): pp. 386–399.
1098. Li, Q.-Y., et al., Citrullus colocynthis (L.) Schrad (Bitter Apple Fruit): promising traditional uses, pharmacological effects, aspects, and potential applications. *Frontiers in Pharmacology*, 2022. **12**: p. 791049.
1099. Kaushik, U., V. Aeri, and S.R. Mir, Cucurbitacins–an insight into medicinal leads from nature. *Pharmacognosy Reviews*, 2015. **9**(17): p. 12.
1100. Kirtikar, K. and B. Basu, *Indian Medicinal Plants*, vol. 3. International Book Distributors, 1999.
1101. Cai, Y., et al., Cucurbitacins: a systematic review of the phytochemistry and anticancer activity. *The American Journal of Chinese Medicine*, 2015. **43**(7): pp. 1331–1350.
1102. Faraj, S.A., Haemorrhagic colitis induced by Citrullus colocynthis. *Annals of Tropical Medicine & Parasitology*, 1995. **89**(6): pp. 695–696.
1103. Dr. Anjali Rani, P.M.L., and Dr. Ramanand, Controversy on Indravaruni. *World Journal of Pharmaceutical, and Medical Research*, 2023. **9**(6, 2023): p. 119.

1104. Patel, D., et al., An overview on antidiabetic medicinal plants having insulin mimetic property. *Asian Pacific Journal of Tropical Biomedicine*, 2012. **2**(4): pp. 320–330.
1105. Al-Musa, A., Severe mosaic caused by zucchini yellow mosaic virus in cucurbits from Jordan. *Plant Pathology*, 1989. **38**(4): pp. 541–546.
1106. *Pathology of Citrullus colocynthis*. Available from: https://en.wikipedia.org/wiki/Zucchini_yellow_mosaic_virus
1107. Su, H.-J., et al., Complete chloroplast genome sequence of Omani lime (Citrus aurantiifolia) and comparative analysis within the rosids. *Plos One*, 2014. **9**(11): p. e113049.
1108. Gorgus, E., et al., Limettin and furocoumarins in beverages containing citrus juices or extracts. *Food and Chemical Toxicology*, 2010. **48**(1): pp. 93–98.
1109. Indriyani, N.N., et al., Bioactive Components and Their Activities from Different Parts of Citrus aurantifolia (Christm.) Swingle for Food Development. *Foods*, 2023. **12**(10): p. 2036.
1110. Spadaro, F., et al., Volatile composition and biological activity of key lime Citrus aurantifolia essential oil. *Natural Product Communications*, 2012. **7**(11): p. 1934578X1200701128.
1111. Adokoh, C.K., et al., Chemical profile and in vivo toxicity evaluation of unripe Citrus aurantifolia essential oil. *Toxicology Reports*, 2019. **6**: pp. 692–702.
1112. Vand, S.H. and T.L. Abdullah, Identification and introduction of thornless lime (Citrus aurantifolia) in Hormozgan, Iran. *Indian Journal of Science and Technology*, 2012. **5**(11): pp. 3670–3673.
1113. Bipat, R., From rat poison to medicine: medical applications of coumarin derivatives, in *Phytochemicals in Human Health*. 2019, IntechOpen. pp. 91–104.
1114. Meiyanto, E., A. Hermawan, and A. Anindyajati, Natural products for cancer-targeted therapy: citrus flavonoids as potent chemopreventive agents. *Asian Pacific Journal of Cancer Prevention*, 2012. **13**(2): pp. 427–436.
1115. Bar-Joseph, M. and R. Lee, Citrus tristeza virus. *AAB Descriptions of Plant Viruses*, 1989. **353**(7).
1116. Pappu, H., et al., Nucleotide sequence and organization of eight 3′ open reading frames of the citrus tristeza closterovirus genome. *Virology*, 1994. **199**(1): pp. 35–46.
1117. Hasan, U.H., Antiarthritic efficacy of Clematis orientalis. *Bangladesh Journal of Pharmacology*, 2018. **13**(2): pp. 142–148.
1118. *Description of Clematis orientalis*. Available from: https://en.wikipedia.org/wiki/Clematis_orientalis
1119. Karimi, E., et al., Antioxidant potential assessment of phenolic and flavonoid rich fractions of Clematis orientalis and Clematis ispahanica (Ranunculaceae). *Natural Product Research*, 2018. **32**(16): pp. 1991–1995.
1120. Tabatadze, N., et al., Triterpene glycosides from Clematis orientalis. *Chemistry of Natural Compounds*, 2007. **43**: pp. 363–363.
1121. Li, Y., et al., Chemical constituents from Clematis delavayi var. spinescens. *Molecules*, 2009. **14**(11): pp. 4433–4439.
1122. Chawla, R., S. Kumar, and A. Sharma, The genus Clematis (Ranunculaceae): chemical and pharmacological perspectives. *Journal of Ethnopharmacology*, 2012. **143**(1): pp. 116–150.
1123. Hasan, U.H., et al., Inhibitory effects of Clematis orientalis aqueous ethanol extract and fractions on inflammatory markers in complete Freund's adjuvant-induced arthritis in Sprague–Dawley rats. *Inflammopharmacology*, 2019. **27**: pp. 781–797.
1124. Li, R.W., et al., Inhibition of COXs and 5-LOX and activation of PPARs by Australian Clematis species (Ranunculaceae). *Journal of Ethnopharmacology*, 2006. **104**(1–2): pp. 138–143.
1125. *Mechanism of action of Clematis orientalis*. Available from: https://www.drugs.com/npp/clematis.html
1126. *Toxicity of Clematis orientalis*. Available from: https://plants.ces.ncsu.edu/plants/clematis-tangutica/#:~:text=Clematis%20can%20cause%20severe%20mouth,confusion%2C%20fainting%2C%20and%20convulsions
1127. Phukhamsakda, C., et al., Microfungi associated with Clematis (Ranunculaceae) with an integrated approach to delimiting species boundaries. *Fungal Diversity*, 2020. **102**: pp. 1–203.
1128. Molino, P., *A guide to Medicinal Plants in North Africa*. ISBN, Malaga, 2005.
1129. Khlifi, A., et al., Gas chromatography-mass spectrometry (GM-MS) analysis and biological activities of the aerial part of Cleome amblyocarpa Barr. and Murb. *Environmental Science and Pollution Research*, 2020. **27**: pp. 22670–22679.
1130. Abdel-Monem, A.R., A new alkaloid and a new diterpene from Cleome paradoxa B. Br.(Cleomaceae). *Natural Product Research*, 2012. **26**(3): pp. 264–269.
1131. Abd-ElGawad, A.M., et al., Chemical composition, allelopathic, antioxidant, and anti-inflammatory activities of sesquiterpenes rich essential oil of Cleome amblyocarpa Barratte & Murb. *Plants*, 2021. **10**(7): p. 1294.
1132. Archi, F.F., et al., Potential evaluation of central nervous system anti-depressant activity of Cleome rutidosperma in mice. *Biomedical Research and Therapy*, 2016. **3**: pp. 1–13.
1133. Le Floc'h, E., *Contribution à une étude Ethnobotanique de la flore Tunisienne*. Ministère de l'Enseignement Supériur et de la Recherche Scientifique, 1983.
1134. Hashem, M., Antifungal properties of crude extracts of five Egyptian medicinal plants against dermatophytes and emerging fungi. *Mycopathologia*, 2011. **172**: pp. 37–46.
1135. Abdel-Motaal, F.F., et al., Leaf spot disease of Hyoscyamus muticus (Egyptian henbane) caused by Cladosporium herbarum. *Journal of General Plant Pathology*, 2009. **75**: pp. 437–439.
1136. Aparadh, V.T., R.J. Mahamuni, and B. Karadge, Taxonomy and physiological studies in spider flower (Cleome species): a critical review. *Plant Science Feeds*, 2012. **2**(3): pp. 25–46.
1137. Rassouli, E., et al., The essential oil composition of Cleome brachycarpa Vahl ex DC. *Journal of Essential Oil Bearing Plants*, 2014. **17**(1): pp. 158–163.
1138. Ahmad, V.U., K.A. Alvi, and M.A. Khan, The molecular structure and absolute configuration of Brachycarpone, A new trinortriterpenoid dilactone from Cleome brachycarpa. *Journal of Natural Products*, 1986. **49**(2): pp. 249–252.
1139. Naeem, H., et al., Anti-inflammatory and Analgesic Activity of Cleome brachycarpa ethanolic extract in lower mammals and effects on blood and liver enzymes. *Jundishapur Journal of Natural Pharmaceutical Products*, 2023 (In Press), **18**: pp. 1–10.
1140. Sumra, A.A., et al., Biogenic synthesis, characterization, and in vitro biological evaluation of silver nanoparticles using Cleome brachycarpa. *Plants*, 2023. **12**(7): p. 1578.
1141. Naeem, H., et al., Cleome brachycarpa: a review on ethnobotany, phytochemistry, and pharmacology. *RADS Journal of Pharmacy and Pharmaceutical Sciences*, 2019. **7**(2): pp. 107–111.

1142. Malik, A., et al., In silico and in vivo characterization of cabralealactone, solasodin and salvadorin in a rat model: potential anti-inflammatory agents. *Drug Design, Development and Therapy*, 2018: pp. 1431–1443.
1143. Singh, H., A. Mishra, and A.K. Mishra, The chemistry and pharmacology of Cleome genus: a review. *Biomedicine & Pharmacotherapy*, 2018. **101**: pp. 37–48.
1144. Khuntia, A., et al., Theoretical evaluation of Cleome species' bioactive compounds and therapeutic potential: a literature review. *Biomedicine & Pharmacotherapy*, 2022. **151**: p. 113161.
1145. Boulos, L., *Flora of Egypt*. Nordic Journal of Botany. 1999. **19**: pp. 328–328. https://doi.org/10.1111/j.1756-1051.1999.tb01119.x
1146. Saad, B., et al., Evaluation of medicinal plant hepatotoxicity in co-cultures of hepatocytes and monocytes. *Evidence-Based Complementary and Alternative Medicine*, 2006. **3**: pp. 93–98.
1147. Choi, Y.-J., S.-B. Hong, and H.-D. Shin, Genetic diversity within the Albugo candida complex (Peronosporales, Oomycota) inferred from phylogenetic analysis of ITS rDNA and COX2 mtDNA sequences. *Molecular Phylogenetics and Evolution*, 2006. **40**(2): pp. 400–409.
1148. *Description of Cleome gynandra*. Available from: https://pza.sanbi.org/cleome-gynandra-0#:~:text=Description-,Description,%2C%20with%203%2D5%20leaflets
1149. Van Wyk, B.-E. and N. Gericke, *People's Plants: A Guide to Useful Plants of Southern Africa*. Briza Publications, 2000.
1150. Anbazhagi, T., et al., Studies on the pharmacognostical and in vitro antioxidant potential of Cleome gynandra Linn. leaves. *Natural Product Radiance*. 2009, **8**(2): pp. 151–157.
1151. Moyo, M. and A.O. Aremu, Nutritional, phytochemical and diverse health-promoting qualities of Cleome gynandra. *Critical Reviews in Food Science and Nutrition*, 2022. **62**(13): pp. 3535–3552.
1152. Pieroni, A., G. Prance, and M. Nesbitt (eds.), *The Cultural History of Plants*. Routledge, 2005.
1153. Moyo, M., et al., Determination of mineral constituents, phytochemicals and antioxidant qualities of Cleome gynandra, compared to Brassica oleracea and Beta vulgaris. *Frontiers in Chemistry*, 2018. **5**: p. 128.
1154. Narendhirakannan, R., S. Subramanian, and M. Kandaswamy, Anti-inflammatory and lysosomal stability actions of Cleome gynandra L. studied in adjuvant induced arthritic rats. *Food and Chemical Toxicology*, 2007. **45**(6): pp. 1001–1012.
1155. Ahouansinkpo, E., et al., Ethnobotany, phytochemical screening and toxicity risk of Cleome gynandra and Cleome viscosa, two traditional leafy vegetables consumed in Benin. *International Journal of Current Microbiology and Applied Sciences*, 2016. **5**(2): pp. 813–829.
1156. Bräutigam, A., et al., An mRNA blueprint for C4 photosynthesis derived from comparative transcriptomics of closely related C3 and C4 species. *Plant Physiology*, 2011. **155**(1): pp. 142–156.
1157. Mashamaite, C.V., A. Manyevere, and E. Chakauya, Cleome gynandra: a wonder climate-smart plant for nutritional security for millions in semi-arid areas. *Frontiers in Plant Science*, 2022. **13**: p. 1003080.
1158. Al-Humaidia, J.Y., et al., Antioxidant activity and chemical composition of essential oils of selected Cleome species growing in Saudi Arabia. *Jordan Journal of Chemistry (JJC)*, 2019. **14**(1): pp. 29–37.
1159. Al-Rehaily, A.J., et al., Chemical constituents of Cleome rupicola growing in Saudi Arabia. *Chemistry of Natural Compounds*, 2017. **53**: pp. 670–673.
1160. Al-Mekhlafi, F.A., et al., Larvicidal activity of selected xerophytic plants against Culex pipiens and Aedes caspius (Diptera: Culicidae). *Pakistan Journal of Zoology*, 2013. **45**(1): pp. 241–246.
1161. *Toxicity of Cleome rupicola*. Available from: https://plantura.garden/uk/flowers-perennials/spider-flower/spider-flower-overview#:~:text=Yes%2C%20spider%20flowers%20contain%20toxic,or%20choose%20an%20inaccessible%20location
1162. Shaheen, N., et al., Analgesic, anti-inflammatory and anti-emetic activities of Cleome scaposa DC. *Phytopharmacology*, 2013. **4**(1): pp. 106–113.
1163. Abbasi, W.M., et al., Preliminary phytochemical analysis and in vivo evaluation of antipyretic effects of hydro-methanolic extract of Cleome scaposa leaves. *Journal of Traditional and Complementary Medicine*, 2018. **8**(1): pp. 147–149.
1164. Calapai, G. and A.P. Caputi, Herbal medicines: can we do without pharmacologist? *Evidence-Based Complementary and Alternative Medicine*, 2007. **4**(S1): pp. 41–43.
1165. Mali, R.G., Cleome viscosa (wild mustard): a review on ethnobotany, phytochemistry, and pharmacology. *Pharmaceutical Biology*, 2010. **48**(1): pp. 105–112.
1166. Singh, H., A. Mishra, and A.K. Mishra, Cleome viscosa linn (Capparaceae): a review. *Pharmacognosy Journal*, 2015. **7**(6): pp. 326–329.
1167. Gupta, P.C. and C.V. Rao, *Pharmacognostical Studies of Cleome viscosa Linn*. Indian Journal of Natural Products and Resources. 2012: pp. 527–534.
1168. *Cultivation Cleome viscosa*. Available from: https://housing.com/news/cleome-viscosa/.
1169. Yaniv, Z., et al., Plants used for the treatment of diabetes in Israel. *Journal of Ethnopharmacology*, 1987. **19**(2): pp. 145–151.
1170. Wicker, E., et al., Epidemiological evidence for the emergence of a new pathogenic variant of Ralstonia solanacearum in Martinique (French West Indies). *Plant Pathology*, 2009. **58**(5): pp. 853–861.
1171. Arora, S. and M. Saini, Gas chromatography mass spectrometry profiling in methanolic and ethyl-acetate root and stem extract of Corbichonia decumbens (Forssk.) exell from Thar Desert of Rajasthan, India. *Pharmacognosy Research*, 2017. **9**(Suppl 1): p. S48.
1172. Arora, S. and M. Saini, Morphological studies on medicinally important plant of Gisekia pharnaceoides Linn. and Corbichonia decumbens (Forssk.) Exell of Molluginaceae from Thar Desert of Rajasthan, India. *Biolife*, 2016. **4**(2): pp. 327–332.
1173. *Description of Corbichonia decumbens*. Available from: https://www.botswanaflora.com/speciesdata/species.php?species_id=122960#:~:text=Procumbent%20or%20prostrate%20annual%20or,oblanceolate%2C%20glaucous%3B%20petiole%20winged
1174. Uma, G., V. Balasubramaniam, and K. Jagathes, Assessment of anti-ulcer activity of Corbichonia decumbens Forssk. Methanolic extract by aspirin plus pyloric ligation model. *International Journal of Pharmaceutical Sciences and Research*, 2014. **3**: pp. 757–61.
1175. Arora, S. and M. Saini, Biochemical screening of leaf extract of Corbichonia decumbens (Forssk.) Exell of Molluginaceae from Thar Desert of Rajasthan, India. *Advances in Plant Sciences*, 2016. **29**(11): pp. 275–278.
1176. Al-Yahya, M.A., *Saudi Plants: A Phytochemical and Biological Approach*. King Abdulaziz City for Science and Technology, 1990.

1177. Dudai, N., et al., Chemical composition and monoterpenoid enantiomeric distribution of the essential oils from Apharsemon (Commiphora gileadensis). *Medicines*, 2017. **4**(3): p. 66.
1178. Shadid, K.A., et al., Exploring the chemical constituents, antioxidant, xanthine oxidase and COX inhibitory activity of Commiphora gileadensis commonly grown wild in Saudi Arabia. *Molecules*, 2023. **28**(5): p. 2321.
1179. Groom, N., Frankincense and Myrrh: A Study of the Arabian Incense Trade. London. 1981.
1180. Wineman, E., et al., Commiphora gileadensis sap extract induces cell cycle-dependent death in immortalized keratinocytes and human dermoid carcinoma cells. *Journal of Herbal Medicine*, 2015. **5**(4): pp. 199–206.
1181. Al-Sieni, A.I., The antibacterial activity of traditionally used Salvadora persica L.(miswak) and Commiphora gileadensis (palsam) in Saudi Arabia. *African Journal of Traditional, Complementary and Alternative Medicines*, 2014. **11**(1): pp. 23–27.
1182. Amiel, E., et al., β-Caryophyllene, a compound isolated from the biblical balm of gilead (Commiphora gileadensis), is a selective apoptosis inducer for tumor cell lines. *Evidence-Based Complementary and Alternative Medicine*, 2012. **2012**.
1183. Verma, N., et al., Insect-pest and diseases disturbances associated with an endangered plant Commiphora wightii (Arnott) Bhandari in Western Region, India. *Ecology, Environment, and Conservation*, 2023. **29** (April Suppl. Issue): pp. S75–S79.
1184. *Botanical Description Commiphora foliacea*. Available from: https://bihrmann.com/caudiciforms/subs/com-fol-sub.asp#:~:text=The%20stem%20can%20grow%20to,name%20foliacea%20means%20'leafy'
1185. *Geographical distribution of Commiphora foliacea*. Available from: https://powo.science.kew.org/taxon/urn:lsid:ipni.org:names:127654-1
1186. Shah, M., et al., Chemical composition and biomedical effects of essential oil from Commiphora foliacea Sprague stem. *Journal of Essential Oil Bearing Plants*, 2023. **26**(5): pp. 1319–1337.
1187. Khan, A., et al., First complete chloroplast genomics and comparative phylogenetic analysis of Commiphora gileadensis and C. foliacea: Myrrh producing trees. *PloS One*, 2019. **14**(1): p. e0208511.
1188. de Melo Alcântara, L.F., et al., Toxicological safety, antioxidant activity and phytochemical characterization of leaf and bark aqueous extracts of Commiphora leptophloeos (Mart.) JB Gillett. *Journal of Toxicology and Environmental Health, Part A*, 2023, **86**(16): pp. 557–574.
1189. *Botanical description of Commiphora kua*. Available from: https://plants.jstor.org/compilation/commiphora.kua
1190. Manguro, L.O.A., I. Ugi, and P. Lemmen, Further bisabolenes and dammarane triterpenes of Commiphora kua resin. *Chemical and Pharmaceutical Bulletin*, 2003. **51**(5): pp. 479–482.
1191. Dekebo, A., et al., Two octanordammarane triterpenes from Commiphora kua. *Phytochemistry*, 2002. **59**(4): pp. 399–403.
1192. Ullah, H., et al., Lophenol and lathosterol from resin of Commiphora kua possess hepatoprotective effects in vivo. *Journal of Ethnopharmacology*, 2020. **252**: p. 112558.
1193. Omer, S. and S. Adam, Toxicity of Commiphora myrrha to goats. *Veterinary and Human Toxicology*, 1999. **41**(5): pp. 299–301.
1194. *Overview of Commiphora myrrha*. Available from: https://www.worldagroforestry.org/treedb/AFTPDFS/Commiphora_myrrha.PDF
1195. Hanuš, L.O., et al., Myrrh-commiphora chemistry. *Biomedical Papers*, 2005. **149**(1): pp. 3–28.
1196. Shen, T., et al., The genus Commiphora: a review of its traditional uses, phytochemistry and pharmacology. *Journal of Ethnopharmacology*, 2012. **142**(2): pp. 319–330.
1197. Batiha, G.E.-S., et al., Commiphora myrrh: a phytochemical and pharmacological update. *Naunyn-Schmiedeberg's Archives of Pharmacology*, 2023. **396**(3): pp. 405–420.
1198. Tao, S. and L. Hong-xiang, Chemical constituents from resin of Commiphora species and their biological activities. *Natural Product Research & Development*, 2008. **20**(2): p. 20.
1199. Alahmari, A.S., et al., Effect of natural Commiphora myrrha extract against hepatotoxicity induced by alcohol intake in rat model. *Toxics*, 2022. **10**(12): p. 729.
1200. Al-Romaiyan, A., et al., Commiphora myrrha stimulates insulin secretion from β-cells through activation of atypical protein kinase C and mitogen-activated protein kinase. *Journal of Ethnopharmacology*, 2023. **302**: p. 115937.
1201. Mekonnen, Y., A. Dekebo, and E. Dagne, Toxicity study in mice of resins of three Commiphora species. *SINET: Ethiopian Journal of Science*, 2003. **26**(2): pp. 151–153.
1202. Dekebo, A., E. Dagne, and O. Sterner, Furanosesquiterpenes from Commiphora sphaerocarpa and related adulterants of true myrrh. *Fitoterapia*, 2002. **73**(1): pp. 48–55.
1203. Bhatia, A., et al., Metabolic profiling of Commiphora wightii (guggul) reveals a potential source for pharmaceuticals and nutraceuticals. *Phytochemistry*, 2015. **110**: pp. 29–36.
1204. *Botanical Description of Commiphora wightii*. Available from: https://en.wikipedia.org/wiki/Commiphora_wightii
1205. Bhatt, G. and R. Dixit, A preliminary study on extensive cultivation of guggul at Mangliawas Hebal Farm, Ajmer, Rajasthan. *Journal of Research Indian Medicine*, 1974. **9**: pp. 51–58.
1206. Rani, R. and S. Mishra, Phytochemistry of guggul (Commiphora wightii): a review. *Asian Journal of Research in Chemistry*, 2013. **6**(4): pp. 415–426.
1207. Sairkar, P.K., A. Sharma, and N. Shukla, Estimation of guggulsterone E and Z in the guggul-based commercial formulations using high-performance thin-layer chromatography. *Journal of Pharmacy & Bioallied Sciences*, 2017. **9**(1): p. 1.
1208. Urizar, N.L. and D.D. Moore, GUGULIPID: a natural cholesterol-lowering agent. *Annual Review of Nutrition*, 2003. **23**(1): pp. 303–313.
1209. Kunnumakkara, A.B., et al., Googling the Guggul (Commiphora and Boswellia) for prevention of chronic diseases. *Frontiers in Pharmacology*, 2018. **9**: p. 686.
1210. Liu, F.-G., et al., Z-guggulsterone produces antidepressant-like effects in mice through activation of the BDNF signaling pathway. *International Journal of Neuropsychopharmacology*, 2017. **20**(6): pp. 485–497.
1211. Wanjari, M.M., et al., Oral toxicity evaluation of gokshuradi guggulu, an ayurvedic formulation. *Drug and Chemical Toxicology*, 2022. **45**(5): pp. 1986–1994.
1212. Ahmed, R., et al., Chemical characterization of a commercial Commiphora wightii resin sample and chemical profiling to assess for authenticity. *Planta Medica*, 2011: pp. 945–950.
1213. Urizar, N.L., et al., A natural product that lowers cholesterol as an antagonist ligand for FXR. *Science*, 2002. **296**(5573): pp. 1703–1706.

1214. *Botanical description of Convolvulus arvensis*. Available from: https://keys.lucidcentral.org/keys/v3/eafrinet/weeds/key/weeds/Media/Html/Convolvulus_arvensis_(Field_Bindweed).htm
1215. Distribution of Convolvulus arvensis. Available from: https://www.first-nature.com/flowers/convolvulus-arvensis.php#:~:text=clockwise%20twining%20climbers.-,Distribution,Wales%20and%20in%20northern%20France
1216. Kaur, M. and A. Kalia, Pharmacognostic parameters and phytochemical screening of Convolvulus arvensis Linn. *International Research Journal of Pharmacy*, 2012. **3**(10): pp. 162–163.
1217. Al-Snafi, A.E., The chemical constituents and pharmacological effects of Convolvulus arvensis and Convolvulus scammonia-A review. *IOSR Journal of Pharmacy*, 2016. **6**(6): pp. 64–75.
1218. Salamatullah, A.M., Convolvulus arvensis: antioxidant, antibacterial, and antifungal properties of chemically profiled essential oils: an approach against nosocomial infections. *Life*, 2022. **12**(12): p. 2138.
1219. Abdul Khaliq, H., et al., Dereplication and quantification of major compounds of Convolvulus arvensis L. extracts and assessment of their effect on LPS-activated J774 macrophages. *Molecules*, 2022. **27**(3): p. 963.
1220. Riordan, N., et al., Anti-angiogenic, anti-tumor and immunostimulatory effects of a nontoxic plant extract (PMG). *Allergy Research Group Focus Newsletter*, 2001.
1221. Leporatti, M.L. and S. Ivancheva, Preliminary comparative analysis of medicinal plants used in the traditional medicine of Bulgaria and Italy. *Journal of Ethnopharmacology*, 2003. **87**(2–3): pp. 123–142.
1222. Schultheiss, P., et al., Toxicity of field bindweed (Convolvulus arvensis) to mice. *Veterinary and Human Toxicology*, 1995. **37**(5): pp. 452–454.
1223. Zouhar, K., *Convolvulus arvensis*. US Department of Agriculture, Forest Service, Rocky Mountain Research, 2004.
1224. *Adulteration of Convolvulus arvensis*. Available from: https://www.inaturalist.org/taxa/53021-Convolvulus-arvensis
1225. Tepfer, D., Transformation of several species of higher plants by Agrobacterium rhizogenes: sexual transmission of the transformed genotype and phenotype. *Cell*, 1984. **37**(3): pp. 959–967.
1226. Kosovac, A., et al., Molecular tracing of the transmission routes of bois noir in Mediterranean vineyards of Montenegro and experimental evidence for the epidemiological role of Vitex agnus-castus (Lamiaceae) and associated Hyalesthes obsoletus (Cixiidae). *Plant Pathology*, 2016. **65**(2): pp. 285–298.
1227. Sotelo-Cerón, N.D., et al., Isolation, selection, and identification of phytopathogenic fungi with bioherbicide potential for the control of field bindweed (Convolvulus arvensis L.). *Weed Biology and Management,* **23**(3-4): pp. 99–109.
1228. Mathur, M. and S. Sundaramoorthy, Mineral composition in Corchorus depressus at heterogeneous environmental conditions and their relationships with bottom-up, top-down, and plant metabolite factors. *Communications in soil science and plant analysis*, 2009. **40**(13–14): pp. 2028–2043.
1229. Pareek, A., A. Godavarthi, and B.P. Nagori, In vitro hepatoprotective activity of Corchorus depressus L. against CCl4 induced toxicity in HepG2 cell line. *Pharmacognosy Journal*, 2013. **5**(4): pp. 191–195.
1230. Zahid, M., et al., New cycloartane and flavonol glycosides from Corchorus depressus. *Helvetica Chimica Acta*, 2002. **85**(2): pp. 689–697.
1231. Ahmad, V.U., et al., Cycloartane triterpene glucosides from Corchorus depressus. *Phytochemistry*, 1998. **49**(3): pp. 829–834.
1232. Kataria, S., et al., In vitro and in vivo aphrodisiac properties of Corchorus depressus Linn. on rabbit corpus cavernosum smooth muscle relaxation and sexual behavior of normal male rats. *Journal of Ethnopharmacology*, 2013. **148**(1): pp. 210–217.
1233. Duke, J.A., and Wain, K.K., Computer index with more than 85,000 entries. *Medicinal Plants of the World*, 1981. **3**.
1234. Afzal, S., et al., Antioxidant, acetylcholinesterase, butyrylcholinesterase, and α-glucosidase inhibitory activities of Corchorus depressus. *Pharmacognosy Magazine*, 2017. **13**(52): p. 647.
1235. *Adulteration of Corchorus depressus*. Available from: https://books.google.com.om/books?id=StInDwAAQBAJ&pg=PT618&lpg=PT618&dq=Corchorus+depressus+adulterants&source=bl&ots=xI8KIHPCy8&sig=ACfU3U2x30DP2la7iYaxPoBPn1NzRDfoUg&hl=en&sa=X&ved=2ahUKEwjzwtC3yIGCAxXzXvEDHeG9AJ8Q6AF6BAgdEAM
1236. Ikram, M., S.G. Khattak, and S.N. Gilani, Antipyretic studies on some indigenous Pakistani medicinal plants: II. *Journal of Ethnopharmacology*, 1987. **19**(2): pp. 185–192.
1237. Kataria, S., Kaur, D., Rao, S.K. and Khajuria, R.K., 2013. In vitro and in vivo aphrodisiac properties of Corchorus depressus Linn. on rabbit corpus cavernosum smooth muscle relaxation and sexual behavior of normal male rats. *Journal of Ethnopharmacology*, *148*(1), pp.210-217.
1238. Das, A., et al., An endophytic Basidiomycete, Grammothele lineata, isolated from Corchorus olitorius, produces paclitaxel that shows cytotoxicity. *PLoS One*, 2017. **12**(6): p. e0178612.
1239. Mbaye, M., et al., Elements de precision sur la systematique d'especes adventices dugenre corchorus l. (tiliaceae) au Senegal. *African Journal of Science and Technology*, 2001. **2**(1), pp. 51–64.
1240. Ali, S. and E. Nasik, Feroz sons. *Flora of Pakistan, Karachim Pakistan*, 1974. **75**: p. 20.
1241. Gupta, M., et al., Anti-steroidogenic activity of methanolic extract of Cuscuta reflexa Roxb. stem and Corchorus olitorius Linn. seed in mouse ovary. Indian Journal of Experimental Biology, 01 Jun 2003, 41(6):641–644.
1242. N'danikou, S., *Corchorus aestuans L. Record from PROTA4U. Brink, M. & Achigan-Dako, EG PROTA (Plant Resources of Tropical Africa/Ressources végétales de l'Afrique tropicale)*, Wageningen, Netherlands, 2011.
1243. Khare, C.P., *Indian Medicinal Plants: An Illustrated Dictionary*. Springer Science & Business Media, 2008.
1244. Ramadevi, D. and S. Ganapaty, Phytochemical examination of Corchorus aestuans (Tiliaceae) capsule. *International Journal of Pharmacognosy*, 2012. **4**: pp. 54–58.
1245. Dhanalakshmi, R. and R. Manavalan, Determination of bioactive constituents of leaves of Corchorus aestuans (L.) by GC–MS Analysis. *International Journal of Pharmacy and Pharmaceutical Sciences*, 2014. **6**(9): pp. 248–251.
1246. Mahato, S.B. and B.C. Pal, Triterpenoid glycosides of Corchorus acutangulus Lam. *Journal of the Chemical Society, Perkin Transactions* 1, 1987: pp. 629–634.
1247. Pancho, J.V. and S.R. Obien, *Manual of Ricefield Weeds in the Philippines,* AGRIS - International System for Agricultural Science and Technology. 1995.
1248. Chen, J.-C., et al., Saikosaponin-A induces apoptotic mechanism in human breast MDA-MB-231 and MCF-7 cancer cells. *The American Journal of Chinese Medicine*, 2003. **31**(3): pp. 363–377.

1249. Mallick, S., et al., Corchorusin-D, a saikosaponin-like compound isolated from Corchorus acutangulus Lam., targets mitochondrial apoptotic pathways in leukemic cell lines (HL-60 and U937). *Cancer Chemotherapy and Pharmacology*, 2010. **66**: pp. 709–719.
1250. *Adulteration of Corchorus.* Available from: https://uses.plantnet-project.org/e/index.php?mobileaction=toggle_view_mobile&title=Corchorus_trilocularis_%28PROTA%29
1251. Mathieu, G. and D. Meissa, Traditional leafy vegetables in Senegal: diversity and medicinal uses. *African Journal of Traditional, Complementary and Alternative Medicines*, 2007. **4**(4): pp. 469–475.
1252. Nuwangburuka, C. and O. Denton, Heritability, character association and genetic advance in six agronomic and yield related characters in leaf Corchorus olitorius. Departement of Agriculture and Industrial Technology, Babcock University, Ilishan-Remo, PMB 21244, Ikeja, Lagos State, Nigeria. *International Journal of Agriculture Innovations and Research*, 2012. **7**: pp. 367–375.
1253. Loumerem, M. and A. Alercia, Descriptors for jute (Corchorus olitorius L.). *Genetic Resources and Crop Evolution*, 2016. **63**: pp. 1103–1111.
1254. Biswas, A., et al., A comprehensive review of C. Capsularis and C. olitorius: a source of nutrition, essential phytoconstituents and pharmacological activities. *Antioxidants*, 2022. **11**(7): p. 1358.
1255. Zakaria, Z., et al., The in vitro antibacterial activity of Corchorus olitorius extracts. *International Journal of Pharmacology*, 2006. **2**(2): pp. 213–215.
1256. Oyedele, D., C. Asonugho, and O. Awotoye, Heavy metals in soil and accumulation by edible vegetables after phosphate fertilizer application. *Electronic Journal of Environmental, Agricultural and Food chemistry*, 2006. **5**(4): pp. 1446–1453.
1257. Al Batran, R., et al., Gastroprotective effects of C orchorus olitorius leaf extract against ethanol-induced gastric mucosal hemorrhagic lesions in rats. *Journal of Gastroenterology and Hepatology*, 2013. **28**(8): pp. 1321–1329.
1258. Nakaziba, R., et al., Phytochemical Analysis, Acute Toxicity, as well as Antihyperglycemic and Antidiabetic Activities of Corchorus olitorius L. Leaf Extracts. *The Scientific World Journal*, 2022. **2022**.
1259. Orieke, D., et al., Identification of phytocomponents and acute toxicity evaluation of Corchorus olitorius leaf extract. *European Journal of Medicinal Plants*, 2018. **23**(1): pp. 1–16.
1260. Mohammed, A., et al., Evaluation of the effect of aqueous leaf extract of jute mallow Corchorus olitorius on some biochemical parameters in alloxan-induced diabetic rats. *European Journal Pharmaceutical and Medical Research*, 2019. **6**(10): pp. 652–658.
1261. Parvin, S., et al., Preliminary phytochemical screening, antihyperglycemic, analgesic and toxicity studies on methanolic extract of aerial parts of Corchorus olitorius L. *Journal of Applied Pharmaceutical Science*, 2015. **5**(9): pp. 68–71.
1262. Khan, M. S. Y., Bano, S., Javed, K., & Asad Mueed, M. (2006). A comprehensive review on the chemistry and pharmacology of Corchorus species: a source of cardiac glycosides, triterpenoids, ionones, flavonoids, coumarins, steroids and some other compounds. Journal of Scientific & industrial research, 65(4), 283-298.
1263. *Adulteration of Corchorus olitorius.* Available from: https://www.sciencedirect.com/topics/engineering/olitorius#:~:text=Jute%20fibers%20are%20obtained%20from,jute%20(from%20Abutilon%20theophrasti)
1264. Yoshikawa, M., et al., Medicinal foodstuffs. V. Moroheiya. (1): Absolute stereostructures of corchoionosides A, B, and C, histamine release inhibitors from the leaves of Vietnamese Corchorus olitorius L.(Tiliaceae). *Chemical and Pharmaceutical Bulletin*, 1997. **45**(3): pp. 464–469.
1265. Ahmed, A.M., et al., Evaluation of the anti-infective potential of the seed endophytic fungi of Corchorus olitorius through metabolomics and molecular docking approach. *BMC Microbiology*, 2023. **23**(1): p. 355.
1266. *Botanical Description of Corchorus trilocularis L.* Available from: https://indiabiodiversity.org/species/show/229261
1267. Dhanalakshmi, R. and R. Manavalan, Bioactive compounds in leaves of Corchorus trilocularis L. by GC-MS analysis. *International Journal of PharmTech Research*, 2014. **6**(7): pp. 1991–1998.
1268. Ahirrao, R., L. Borsel, and S. Desai, Evaluation of anti-inflammatory activity of Corchorus trilocularis linn. Seed oil, *Indian Science Abstracts*, 2009. **17512**(45): p. 18.
1269. Senthilkumar, M., P. Gurumoorthi, and K. Janardhanan, *Some Medicinal Plants Used by Irular, the Tribal People of Marudhamalai hills*, Coimbatore, Tamil Nadu, 2006.
1270. Siringo, C.G., et al., Gastro-protective and anti-acidic effects of Corchorus trilocularis Linn against diclofenac-induced gastric ulcers. *Discovery Phytomedicine - Journal of Natural Products Research and Ethnopharmacology*, 2020. **7**(1): p. 33.
1271. Kumar, P. and P.S. Gupta, Antidiabetic evaluation of various polar & non-polar extracts of Corchorus trilocularis linn. in streptozotocin induced diabetic animals. *International Journal of Pharmacy and Biological Sciences*, 2017. **7**(3): pp. 222–228.
1272. *Corchorus trilocularis*. Available from: https://uses.plantnet-project.org/en/Corchorus_trilocularis_(PROTA)
1273. Sharmin, S., et al., Xyloglucan endotransglycosylase/hydrolase genes from a susceptible and resistant jute species show opposite expression pattern following Macrophomina phaseolina infection. *Communicative & Integrative Biology*, 2012. **5**(6): pp. 598–606.
1274. Al-Snafi, A.E., The Pharmacological and therapeutic importance of Cordia myxa-A review. *IOSR Journal of Pharmacy*, 2016. **6**(6): pp. 47–57.
1275. Rashed, K., et al., Evaluation of anti-HIV-1 activity of Cordia myxa L. and phytochemical profile. *Banat's Journal of Biotechnology*, 2014. **5**(9): p. 51.
1276. Matcheme, M., et al., Chemical constituents from Cordia myxa L. (Boraginaceae) and their antibacterial activity. *Natural Product Research*, 2023: pp. 1–9.
1277. Al-Musawi, M.H., K.M. Ibrahim, and S. Albukhaty, In vitro study of antioxidant, antibacterial, and cytotoxicity properties of Cordia myxa fruit extract. *Iranian Journal of Microbiology*, 2022. **14**(1): p. 97.
1278. Abdel-Aleem, E.R., et al., Total phenolic and flavonoid contents and antioxidant, anti-inflammatory, analgesic, antipyretic and antidiabetic activities of Cordia myxa L. leaves. *Clinical Phytoscience*, 2019. **5**: pp. 1–9.
1279. Nowicke, J.W. and J.E. Ridgway, Pollen studies in the genus Cordia (Boraginaceae). *American Journal of Botany*, 1973. **60**(6): pp. 584–591.
1280. Keshani-Dokht, S., et al., Extraction, chemical composition, rheological behavior, antioxidant activity and functional properties of Cordia myxa mucilage. *International Journal of Biological Macromolecules*, 2018. **118**: pp. 485–493.
1281. Choudhary, S. and R. Ghasolia, Activation of defense mechanism in Lehsua (Cordia myxa) through systemic acquired resistance inducers against Alternaria leaf spot disease. *Journal of Plant Biochemistry and Biotechnology*, 2021. **30**(1): pp. 205–207.

1282. *Description of Dorstenia foetida*. Available from: https://en.wikipedia.org/wiki/Dorstenia_foetida

1283. Oza, M.J. and Y.A. Kulkarni, Traditional uses, phytochemistry and pharmacology of the medicinal species of the genus Cordia (Boraginaceae). *Journal of Pharmacy and Pharmacology*, 2017. **69**(7): pp. 755–789.

1284. Alshegaihi, R.M., et al., The first complete chloroplast genome of Cordia monoica: structure and comparative analysis. *Genes*, 2023. **14**(5): p. 976.

1285. Glover, P., J. Stewart, and M. Gwynne, Masai and Kipsigis notes on East African plants: part III – –medicinal uses of plants. *East African Agricultural and Forestry Journal*, 1966. **32**(2): pp. 200–207.

1286. Pradheeps, M. and G. Poyyamoli, Ethnobotany and utilization of plant resources in Irula villages (Sigur plateau, Nilgiri Biosphere Reserve, India). *Journal of Medicinal Plants Research*, 2013. **7**(6): pp. 267–276.

1287. Ruffo, C.K., A. Birnie, and B. Tengnäs, *Edible Wild Plants of Tanzania*. AGRIS - International System for Agricultural Science and Technology, 2002.

1288. Tahir, M., et al., The traditional use of wild edible plants in pastoral and agro-pastoral communities of Mieso District, eastern Ethiopia. *Tropical Medicine and Health*, 2023. **51**(1): p. 10.

1289. Dev, R., et al., Collection, characterization, conservation and utilization of Cordia sinensis Lam.: an underexploited multipurpose fruit species of hot arid regions. *Plant Genetic Resources*, 2020. **18**(6): pp. 427–436.

1290. Ahmed, A. and M.A. Kordofani, Morphological and anatomical studies of Cordia abyssinica R. Br. and Cordia sinensis Lam.(Boraginaceae) in Sudan. *Journal of Chemical and Pharmaceutical Research*, 2012. **4**(10): pp. 4582–4588.

1291. Choi, W.-H., et al., Phenolic compounds from the stem bark of Cornus walteri Wanger. *Korean Journal of Pharmacognosy*, 1998. **29**(3): pp. 217–224.

1292. Ayinde, B., D. Onwukaeme, and E. Omogbai, Isolation and characterization of two phenolic compounds from the stem bark of Musanga. *Biointerface*, 2007. **24**: pp. 297–308.

1293. DÜRÜST, N., et al., The isolation of carboxylic acids from the flowers of Delphinium formosum. *Turkish Journal of Chemistry*, 2001. **25**(1): pp. 93–97.

1294. Woo, E.-R. and M.S. Piao, Antioxidative constituents from Lycopus lucidus. *Archives of Pharmacal Research*, 2004. **27**: pp. 173–176.

1295. Lu, L., et al., Studies on the homolytic and heterolytic cleavage of kaempferol and kaempferide glycosides using electrospray ionization tandem mass spectrometry. *Rapid Communications in Mass Spectrometry*, 2010. **24**(1): pp. 169–172.

1296. Youssef Moustafa, A.M., A.I. Khodair, and M.A. Saleh, Isolation, structural elucidation of flavonoid constituents from Leptadenia pyrotechnica and evaluation of their toxicity and antitumor activity. *Pharmaceutical Biology*, 2009. **47**(6): pp. 539–552.

1297. Kindt R., et al. Documentation of the Agroforestry Species Switchboard web database (version 1.3). World Agroforestry Centre (ICRAF), 2016. Nairobi.

1298. Eltayeib, A.A. and W. Ishag, Phytochemical screening, antimicrobial, antioxidant and cytotoxicity activities of bark's crude extracts of Cordia sinensis. *Advancement in Medicinal Plant Research*, 2015. **2**: pp. 39–45.

1299. Al-Musayeib, N., et al., Antioxidant, anti-glycation and anti-inflammatory activities of phenolic constituents from Cordia sinensis. *Molecules*, 2011. **16**(12): pp. 10214–10226.

1300. Yeshi, K. and P. Wangchuk, Essential oils and their bioactive molecules in healthcare, in Subhash C. Mandal, Amit Kumar Nayak, Amal Kumar Dhara (eds.), *Herbal Biomolecules in Healthcare Applications*. 2022, Elsevier. pp. 215–237.

1301. Lans, C., Ethnomedicines used in Trinidad and Tobago for reproductive problems. *Journal of Ethnobiology and Ethnomedicine*, 2007. **3**: pp. 1–12.

1302. *Description of Cornulaca monacantha*. Available from: https://en.wikipedia.org/wiki/Cornulaca_monacantha

1303. Kumar, A., et al., Photo-degradation of noxious pollutants from water system using Cornulaca monacantha stem supported ZnFe2O4 magnetic bio-nanocomposite. *Sustainable Chemistry and Pharmacy*, 2020. **18**: pp. 100290.

1304. Kandil, F.E. and M.H. Grace, Polyphenols from Cornulaca monacantha. *Phytochemistry*, 2001. **58**(4): pp. 611–613.

1305. Mhiri, R., et al., New isoflavones with antioxidant activity isolated from Cornulaca monacantha. *Chemistry & Biodiversity*, 2020. **17**(12): p. e2000758.

1306. Salem, S.H., et al., The antimicrobial, antibiofilm, and wound healing properties of ethyl acetate crude extract of an endophytic fungus Paecilomyces sp.(AUMC 15510) in earthworm model. *Scientific Reports*, 2022. **12**(1): p. 19239.

1307. Pandey, C., et al., *Medicinal Plants of Gujarat*, vol. 410. Gujarat Ecological Education and Research (GEER) Foundation, Gandhinagar, 2005.

1308. Del Río, J.A., et al., Furanocoumarins: biomolecules of therapeutic interest. *Studies in Natural Products Chemistry*, 2014. **43**: pp. 145–195.

1309. Aiswarya, S.U.D., et al., Cucurbitacin B, purified and characterized from the rhizome of Corallocarpus epigaeus exhibits anti-melanoma potential. *Frontiers in Oncology*, 2022. **12**: p. 903832.

1310. Gupta, J., et al., The Occurrence of Ishwarane and Ishwarone in the Roof Oil of Corallocarpus epigaeus Benth. ex Hook. f. *Journal of Essential Oil Research*, 1997. **9**(6): pp. 667–672.

1311. Ishnava, K.B. and P.S. Konar, In vitro anthelmintic activity and phytochemical characterization of Corallocarpus epigaeus (Rottler) Hook. f. tuber from ethyl acetate extracts. *Bulletin of the National Research Centre*, 2020. **44**: pp. 1–10.

1312. Ahmed, M.S., L.C. Kopel, and F.T. Halaweish, Structural optimization and biological screening of a steroidal scaffold possessing cucurbitacin-like functionalities as B-raf inhibitors. *ChemMedChem*, 2014. **9**(7): pp. 1361–1367.

1313. Gnananath, K., et al., Evaluation of antidiabetic activity in Corallocarpus epigaeus rhizomes. *International Current Pharmaceutical Journal*, 2013. **2**(3): pp. 53–6.

1314. Nayar, M., *Meaning of Indian Flowering Plant Names*. 1985. https://www.banglajol.info/index.php/ICPJ/article/view/13581/0

1315. Adibatti, N., et al., A pyridine alkaloid from Ceropegia juncea. *Phytochemistry*, 1991. **30**(7): pp. 2449–2450.

1316. Phulwaria, M., et al., An efficient in vitro regeneration and ex vitro rooting of Ceropegia bulbosa Roxb. – a threatened and pharmaceutical important plant of Indian Thar Desert. *Industrial Crops and Products*, 2013. **42**: pp. 25–29.

1317. Hussain, G., et al., Role of plant derived alkaloids and their mechanism in neurodegenerative disorders. *International Journal of Biological Sciences*, 2018. **14**(3): p. 341.

1318. *Pathology of Ceropegia*. Available from: https://greg.app/bugs-on-ceropegia-ampliata/

1319. Ganeshaiah, K., R. Vasudeva, and R.U. Shaanker, In search of Sanjeevani. *Current Science*, 2009: pp. 484–489.

1320. Shahat, A., et al., Flavonoids from Cressa cretica. *Pharmaceutical Biology*, 2004. **42**(4-5): pp. 349–352.
1321. Rizk, A. and G.A. El-Ghazaly, *Medicinal and Poisonous Plants of Qatar*. University of Qatar, 1995.
1322. Chopra, R.N., *Glossary of Indian Medicinal Plants*. 1956. Council of Scientific & Industrial Research.
1323. Macdonald, H., *A dictionary of natural products*, vol. 7. Medford: NJ, 1997, p. 187.
1324. Priyashree, S., S. Jha, and S.P. Pattanayak, Bronchodilatory and mast cell stabilising activity of Cressa cretica L.: evaluation through in vivo and in vitro experimental models. *Asian Pacific Journal of Tropical Medicine*, 2012. **5**(3): pp. 180–186.
1325. Kumari, B.L., et al., Experimental evidences of methanolic extraction of Cressa cretica Linn. on alloxan induced hyperglycemic Wistar rats. *Perspectives in Science*, 2016. **8**: pp. 179–182.
1326. Shah, S., et al., Prospecting for Cressa cretica to treat COVID-19 via in silico molecular docking models of the SARS-CoV-2. *Journal of Biomolecular Structure and Dynamics*, 2022. **40**(12): pp. 5643–5652.
1327. Gashgari, R., et al., Molecular characterization and analysis of antimicrobial activity of endophytic fungi from medicinal plants in Saudi Arabia. *Jundishapur Journal of Microbiology*, 2016. **9**(1): pp. e26157.
1328. *Botanical description of Crotalaria retusa*. Available from: https://handwiki.org/wiki/Biology%3ACrotalaria_retusa.
1329. Le Roux, M.M. and B.-E. Van Wyk, The systematic value of flower structure in Crotalaria and related genera of the tribe Crotalarieae (Fabaceae). Flora-Morphology, Distribution, *Functional Ecology of Plants*, 2012. **207**(6): pp. 414–426.
1330. Aragão, D.P., et al., The anti-inflammatory and antinociceptive activity of albumins from Crotalaria retusa seeds. *Biomedicine & Pharmacotherapy*, 2017. **93**: pp. 536–542.
1331. Culvenor, C. and L. Smith, The alkaloids of Crotalaria retusa L. *Australian Journal of Chemistry*, 1957. **10**(4): pp. 464–473.
1332. Rouamba, A., et al. (2018). Ethno-medicinal use of Crotalaria retusa L.(Fabaceae), a pyrrolizidine alkaloid toxic plant. *International Journal of Biochemistry Research & Review,,* 2018. 23(2): pp. 1–6.
1333. Rouamba, A., et al., Ethno-medicinal use of Crotalaria retusa L.(Fabaceae), a pyrrolizidine alkaloid toxic plant. *International Journal of Biochemistry Research & Review*, 2018. **23**(2): pp. 1–6.
1334. Adams, R. and E. Rogers, The structure of monocrotaline, the alkaloid in Crotalaria spectabilis and Crotalaria retusa. I. *Journal of the American Chemical Society*, 1939. **61**(10): pp. 2815–2819.
1335. Solofomalala, A.H.D., et al., Pyrrolizidine-derived alkaloids: highly toxic components in the seeds of Crotalaria cleomifolia used in popular beverages in Madagascar. *Molecules*, 2021. **26**(11): p. 3464.
1336. Soto-Estrada, C., & Cruz-Durán, R. (2023). Crotalaria heteromorpha (Leguminosae, Papilionoideae), a new species of the trifoliolate crotalarias group from the state of Guerrero, Mexico. *Brittonia*, 75(3), 240-248.
1337. Pitanga, B.P., et al., Assessment of neurotoxicity of monocrotaline, an alkaloid extracted from Crotalaria retusa in astrocyte/neuron co-culture system. *Neurotoxicology*, 2011. **32**(6): pp. 776–784.
1338. Spence, N. and D. Walkey, Variation for pathogenicity among isolates of bean common mosaic virus in Africa and a reinterpretation of the genetic relationship between cultivars of Phaseolus vulgaris and pathotypes of BCMV. *Plant Pathology*, 1995. **44**(3): pp. 527–546.
1339. Salatino, A., M.L.F. Salatino, and G. Negri, Traditional uses, chemistry and pharmacology of Croton species (Euphorbiaceae). *Journal of the Brazilian chemical society*, 2007. **18**: pp. 11–33.
1340. Block, S., et al., Diterpenes from the leaves of Croton zambesicus. *Phytochemistry*, 2004. **65**(8): pp. 1165–1171.
1341. Milanowski, D.J., et al., Geographic Distribution of Three Alkaloid Chemotypes of Croton l echleri. *Journal of Natural Products*, 2002. **65**(6): pp. 814–819.
1342. Xu, W.-H., W.-Y. Liu, and Q. Liang, Chemical constituents from Croton species and their biological activities. *Molecules*, 2018. **23**(9): p. 2333.
1343. Espinoza-Hernández, F.A., A.D. Moreno-Vargas, and A. Andrade-Cetto, Diabetes-related mechanisms of action involved in the therapeutic effect of croton species: a systematic review. *Plants*, 2023. **12**(10): p. 2014.
1344. Salatino, A., et al. Traditional uses, chemistry and pharmacology of Croton species (Euphorbiaceae). *Journal of the Brazilian Chemical Society,* 2007. **18**: pp. 11–33.
1345. Bertaccini, A. and M. Bellardi, A rhabdovirus inducing vein yellowing in croton. *Plant Pathology*, 1992. **41**(1): pp. 79–82.
1346. Sharma, S., K.L. Tiwari, and S.K. Jadhav, Diversity of fungal endophytes in Typha latifolia (L.) and their lead biosorption activity. *Euro-Mediterranean Journal for Environmental Integration*, 2018. **3**: pp. 1–8.
1347. *Cucumis canoxyi details and information*. Available from: https://www.selinawamucii.com/plants/cucurbitaceae/cucumis-canoxyi/
1348. Insanu, M., et al., Chemical compounds and pharmacological activities of cucumis genus. *Biointerface Research in Applied Chemistry*, 2022. **12**(1): pp. 1324–1334.
1349. Mukherjee, P.K., et al., Phytochemical and therapeutic potential of cucumber. *Fitoterapia*, 2013. **84**: pp. 227–236.
1350. Wu, X.-Y., et al., Chemical constituents contained in fatty oil from seeds of Cucumis sativus. Zhongguo Zhong yao za zhi= Zhongguo Zhongyao Zazhi= *China Journal of Chinese Materia Medica*, 2012. **37**(21): pp. 3252–3255.
1351. Swamy, K. R. M. Origin, distribution, taxonomy, botanical description, genetics, genetic diversity and breeding of cucumber (Cucumis sativus L.). *International Journal of Development Research, 2023. 13*(02): pp. 61542–61559.
1352. Ghebretinsae, A.G., M. Thulin, and J.C. Barber, Relationships of cucumbers and melons unraveled: molecular phylogenetics of Cucumis and related genera (Benincaseae, Cucurbitaceae). *American Journal of Botany*, 2007. **94**(7): pp. 1256–1266.
1353. Huang, S., et al., The genome of the cucumber, Cucumis sativus L. *Nature Genetics*, 2009. **41**(12): pp. 1275–1281.
1354. Liang, Y., et al., Effects of foliar-and root-applied silicon on the enhancement of induced resistance to powdery mildew in Cucumis sativus. *Plant Pathology*, 2005. **54**(5): pp. 678–685.
1355. *Wikipedia of Cucumis prophetarum.* Available from: https://www.flowersofindia.net/catalog/slides/Wild%20Gourd.html.
1356. Afifi, M., et al., Cucurbitacins of Cucumis prophetarum and Cucumis prophetarum. *Journal of Chemical Ecology*, 1999. **25**: pp. 847–859.
1357. Alsayari, A., et al., Isolation of anticancer constituents from Cucumis prophetarum var. prophetarum through bioassay-guided fractionation. *BMC Complementary and Alternative Medicine*, 2018. **18**: pp. 1–12.
1358. Assefa, T., et al., Chemical composition and antibacterial activity of essential oils from selected species of the genus

Cucumis in Ethiopia. *Bulletin of the Chemical Society of Ethiopia*, 2023. **37**(3): pp. 703–715.
1359. Aljohani, O.S., Phytochemical evaluation of Cucumis prophetarum: protective effects against carrageenan-induced prostatitis in rats. *Drug and Chemical Toxicology*, 2022. **45**(4): pp. 1461–1469.
1360. Chung, W. Y., Jadhav, S., Hsu, P. K., & Kuan, C. M. (2022). Evaluation of acute and sub-chronic toxicity of bitter melon seed extract in Wistar rats. *Toxicology Reports*, *9*, 1024-1034.
1361. Hemlata, et al., Biosynthesis of silver nanoparticles using Cucumis prophetarum aqueous leaf extract and their antibacterial and antiproliferative activity against cancer cell lines. *ACS Omega*, 2020. **5**(10): pp. 5520–5528.
1362. Huang, L.-Q., et al., The potential of endophytic fungi isolated from cucurbit plants for biocontrol of soilborne fungal diseases of cucumber. *Microbiological Research*, 2020. **231**: p. 126369.
1363. *Botanical Description of Cucumis melo*. Available from: https://www.missouribotanicalgarden.org/PlantFinder/PlantFinderDetails.aspx?taxonid=292722&isprofile=0&#:~:text=Noteworthy%20Characteristics,the%20species%20when%20cut%20open
1364. Mallek-Ayadi, S., et al., Bioactive compounds from Cucumis melo L. fruits as potential nutraceutical food ingredients and juice processing using membrane technology. *Food Science & Nutrition*, 2022. **10**(9): pp. 2922–2934.
1365. Mallek-Ayadi, S., N. Bahloul, and N. Kechaou, Characterization, phenolic compounds and functional properties of Cucumis melo L. peels. *Food Chemistry*, 2017. **221**: pp. 1691–1697.
1366. Rabadán, A., et al., From by-product to the food chain: melon (Cucumis melo L.) seeds as potential source for oils. *Foods*, 2020. **9**(10): p. 1341.
1367. Asif, H.M., et al., Medicinal Properties of Cucumis melo Linn. *RADS Journal of Pharmacy and Pharmaceutical Sciences*, 2014. **2**(1): pp. 58–62.
1368. Wahid, M., et al., Integrated Mechanisms of polarity–based extracts of Cucumis melo L. Seed kernels for airway smooth muscle relaxation via key signaling pathways based on WGCNA, in vivo, and in vitro analyses. *Pharmaceuticals*, 2022. **15**(12): p. 1522.
1369. Wahid, S., et al., Analgesic, anti-inflammatory and toxic effects of ethanol extracts of Cucumis melo and Citrullus lanatus seeds. *Pakistan Journal of Pharmaceutical Sciences*, 2020. **33**(3): 1049–1055.
1370. Kim, K., et al., Effective Phytosanitary Treatment for Export of Oriental Melons (Cucumis melo var L.) Using Ethyl Formate and Modified Atmosphere Packaging to Control Trialeurodes vaporariorum (Hemiptera: Aleyrodidae). *Insects*, 2023. **14**(5): p. 442.
1371. Meshram, V., et al., Endophytic Fusarium clavum confers growth and salt tolerance in Cucumis melo. *Environmental and Experimental Botany*, 2023. **206**: p. 105153.
1372. Yuncker, T.G., The genus cuscuta. *Memoirs of the Torrey Botanical Club*, 1932. **18**(2): pp. 109–331.
1373. Chabra, A., et al., Ethnopharmacology of Cuscuta epithymum: a comprehensive review on ethnobotany, phytochemistry, pharmacology and toxicity. *Journal of Ethnopharmacology*, 2019. **231**: pp. 555–569.
1374. Khan, S., K.J. Mirza, and M.Z. Abdin, Development of RAPD markers for authentication of medicinal plant Cuscuta reflexa. *Eurasian Journal of Biosciences*, 2010. **4:** pp. 1–7.
1375. Firoozabadi, A., et al., Effectiveness of Cuscuta planiflora Ten. and Nepeta menthoides Boiss. & Buhse in major depression: a triple-blind randomized controlled trial study. *Journal of Evidence-Based Complementary & Alternative Medicine*, 2015. **20**(2): pp. 94–97.
1376. Shah, G., et al., Scientific basis for the therapeutic use of Cymbopogon citratus, stapf (Lemon grass). *Journal of Advanced Pharmaceutical Technology & Research*, 2011. **2**(1): pp. 3–8.
1377. Ganjewala, D., Cymbopogon essential oils: chemical compositions and bioactivities. *International Journal of Essential Oil Therapeutics*, 2009. **3**(2–3): pp. 56–65.
1378. Seibert, J.B., et al., Seasonality study of essential oil from leaves of Cymbopogon densiflorus and nanoemulsion development with antioxidant activity. *Flavour and Fragrance Journal*, 2019. **34**(1): pp. 5–14.
1379. Mukhtar, M.H., et al., Cymbopogon citratus and Citral Overcome Doxorubicin Resistance in Cancer Cells via Modulating the Drug's Metabolism, *Toxicity, and Multidrug Transporters. Molecules*, 2023. **28**(8): p. 3415.
1380. Akande, I., et al., Comparative proximate analysis of ethanolic and water extracts of Cymbopogon citratus (lemon grass) and four tea brands. *Plant Science Research*, 2011. **3**(4): pp. 29–35.
1381. Takaisi-Kikuni, N., D. Tshilanda, and B. Babady, Antibacterial activity of the essential oil of Cymbopogon densiflorus. *Fitoterapia*, 2000. **71**(1): pp. 69–71.
1382. Lulekal, E., et al., Phytochemical analysis and evaluation of skin irritation, acute and sub-acute toxicity of Cymbopogon citratus essential oil in mice and rabbits. *Toxicology Reports*, 2019. **6**: pp. 1289–1294.
1383. Bidinotto, L.T., et al., Protective effects of lemongrass (Cymbopogon citratus STAPF) essential oil on DNA damage and carcinogenesis in female Balb/C mice. *Journal of Applied Toxicology*, 2011. **31**(6): pp. 536–544.
1384. Tibenda, J.J., et al., Review of phytomedicine, phytochemistry, ethnopharmacology, toxicology, and pharmacological activities of Cymbopogon genus. *Frontiers in Pharmacology*, 2022. **13**: p. 997918.
1385. *Substitution of Lemongrass*. Available from: https://thestonesoup.com/lemongrass-substitutes/
1386. Muyembe, D.K., et al., Molecular, chemical and microscopic analysis of medicinal plant (Cymbopogon distans) from its adulterants. *Journal of Diseases and Medicinal Plants*, 2018. **4**(2): pp. 35–47.
1387. Soltanzadeh, M., et al., Chitosan nanoparticles encapsulating lemongrass (Cymbopogon commutatus) essential oil: physicochemical, structural, antimicrobial and in-vitro release properties. *International Journal of Biological Macromolecules*, 2021. **192**: pp. 1084–1097.
1388. Alam, M., et al., Collar rot and wilt: a new disease of Java citronella (Cymbopogon winterianus) caused by Fusarium moniliforme Sheldon. *Plant Pathology*, 1994. **43**(6): pp. 1057–1061.
1389. Fais, M.M., Y. Yurnaliza, and L.D.S. Hastuti, Isolation and antibacterial activity of endophytic fungi from citronella grass (*Cymbopogon nardus). Biodiversitas Journal of Biological Diversity*, 2022. **23**(12): pp. 6565–6569.
1390. Prasad, C., et al., Cymbopogon jwarancusa-An important medicinal plant: a review. *The Pharma Innovation*, 2014. **3**(6, Part A): p. 13.
1391. Dutta, S., et al., A short review on chemical composition, therapeutic use and enzyme inhibition activities of Cymbopogon species. *Indian Journal of Science and Technology*, 2016. **9**(46): pp. 1–9.

1392. Barfroshan, Y., et al., Comparative study of bioactivities and chemical constituents of Cymbopogon jwarancusa subsp. olivieri (Boiss.) Soenarko harvested in spring and winter. *Journal of Essential Oil Bearing Plants*, 2018. **21**(4): pp. 1107–1118.
1393. Singh, R. and M. Pathak, Variability in herb yield and volatile constituents of Cymbopogon jwarancusa (Jones) Schult. cultivars. *Industrial Crops and Products*, 1994. **2**(3): pp. 197–199.
1394. Khan, S.J., S. Afroz, and R.A. Khan, Diuretic effect of Cymbopogon jwarancusa after single and multiple doses in rats. *Pharmacology & Pharmacy*, 2018. **9**(7): pp. 250–256.
1395. Sumaiya, S., et al., Isolation and characterization of bioactive components from hydroalcoholic extract of Cymbopogon jwarancusa (Jones) Schult. to evaluate its hepatoprotective activity. *Journal of Ethnopharmacology*, 2024. **319**: p. 117185.
1396. Fandohan, P., et al., Toxicity and gastric tolerance of essential oils from Cymbopogon citratus, Ocimum gratissimum and Ocimum basilicum in Wistar rats. *Food and Chemical Toxicology*, 2008. **46**(7): pp. 2493–2497.
1397. Cerceau, C.I., et al., 1H-NMR and GC for detection of adulteration in commercial essential oils of Cymbopogon ssp. *Phytochemical Analysis*, 2020. **31**(1): pp. 88–97.
1398. Ullah, S.A., et al., Microscopic implication and evaluation of herbaceous melliferous plants of southern Khyber Pakhtunkhwa-Pakistan using light and scanning electron microscope. *Microscopy Research and Technique*, 2021. **84**(8): pp. 1750–1764.
1399. Zhang, Q., et al., Diseases of Cymbopogon citratus (Poaceae) in China: Curvularia nanningensis sp. nov. *MycoKeys*, 2020. **63**: p. 49.
1400. Bayala, B., et al., Chemical composition of essential oil of cymbopogon schoenanthus (L.) Spreng from Burkina Faso, and effects against prostate and cervical cancer cell lines. *Molecules*, 2023. **28**(11): p. 4561.
1401. Malti, C.E.W., et al., Composition, chemical variability and biological activity of Cymbopogon schoenanthus essential oil from Central Algeria. *Chemistry & Biodiversity*, 2020. **17**(6): p. e2000138.
1402. Bellik, F.-Z., et al., Chemical composition, kinetic study and antimicrobial activity of essential oils from Cymbopogon schoenanthus L. Spreng extracted by conventional and microwave-assisted techniques using cryogenic grinding. *Industrial Crops and Products*, 2019. **139**: p. 111505.
1403. Ketoh, G.K., et al., Comparative effects of Cymbopogon schoenanthus essential oil and piperitone on Callosobruchus maculatus development. *Fitoterapia*, 2006. **77**(7-8): pp. 506–510.
1404. Hashim, G.M., et al., Biological activity of Cymbopogon schoenanthus essential oil. *Saudi Journal of Biological Sciences*, 2017. **24**(7): pp. 1458–1464.
1405. Sawadogo, I., et al., Antifungal and antiaflatoxinogenic effects of Cymbopogon citratus, Cymbopogon nardus, and Cymbopogon schoenanthus essential oils alone and in combination. *Journal of Fungi*, 2022. **8**(2): p. 117.
1406. Fathifar, E., T. Rastegar, and J. Asgarpanah, Histopathological and biochemical toxicity of Cymbopogon schoenanthus essential oil in female mice. *Research Journal of Pharmacognosy*, 2021. **8**(1): pp. 53–62.
1407. de Souza Chagas, A.C., et al., In vitro efficacy of plant extracts and synthesized substances on Rhipicephalus (Boophilus) microplus (Acari: Ixodidae). *Parasitology Research*, 2012. **110**: pp. 295–303.
1408. Zito, P., et al., Essential oil of Cyphostemma juttae (Vitaceae): chemical composition and antitumor mechanism in triple negative breast cancer cells. *Plos One*, 2019. **14**(3): p. e0214594.
1409. Bello, O.M., et al., Phytochemistry, pharmacology and perceived health uses of non-cultivated vegetable Cyphostemma adenocaule (Steud. ex A. Rich.) Desc. ex Wild and RB Drumm: a review. *Scientific African*, 2019. **2**: p. e00053.
1410. Bello, O.M., et al., Bioactive compounds of Swahili [Cyphostemma Adenocaule (Steud. ex A. Rich.) Desc. ex Wild and RB Drumm.], in *Bioactive Compounds in Underutilized Vegetables and Legumes*, 2020, pp. 1–14.
1411. *Toxicity of Cyphostemma*. Available from: https://tula.house/blogs/tulas-plant-library/cyphostemma-juttae
1412. Teklehaymanot, T. and M. Giday, Ethnobotanical study of medicinal plants used by people in Zegie Peninsula, *Northwestern Ethiopia. Journal of ethnobiology and Ethnomedicine*, 2007. **3**(1): pp. 1–11.
1413. Väre, H. and I. Kukkonen. Seven new species of Cyperus (Cyperaceae) section Arenarii and one new combination and typification. in *Annales Botanici Fennici*. 2005. JSTOR.
1414. Hu, Q.-P., et al., Chemical composition, antioxidant, DNA damage protective, cytotoxic and antibacterial activities of Cyperus rotundus rhizomes essential oil against foodborne pathogens. *Scientific Reports*, 2017. **7**(1): pp. 45231.
1415. Feizbakhsh, A. and A. Naeemy, Chemical composition of the essential oil of Cyperus conglomeratus Rottb. From Iran. *Journal of Chemistry*, 2011. **8**: pp. S293–S296.
1416. Abu-Serag, N., et al. Analysis of bioactive phytochemical compound of (Cyperus aucheri Jaub.) by using gas chromatography–mass spectrometry. in *IOP Conference Series: Earth and Environmental Science*. 2019. IOP Publishing, 388, 012063.
1417. Al-Hazmi, G.H., et al., Anticandidal activity of the extract and compounds isolated from Cyperus conglomertus Rottb. *Saudi Pharmaceutical Journal*, 2018. **26**(6): pp. 891–895.
1418. Zhang, S., et al., Cyperus (Cyperus esculentus L.): a review of its compositions, medical efficacy, antibacterial activity and allelopathic potentials. *Plants*, 2022. **11**(9): p. 1127.
1419. Duarte, M.C.T., et al., Anti-Candida activity of Brazilian medicinal plants. *Journal of Ethnopharmacology*, 2005. **97**(2): pp. 305–311.
1420. García-Jiménez, J., et al., A tuber rot of Cyperus esculentus caused by Rosellinia necatrix. *Plant Disease*, 1998. **82**(11): pp. 1281–1281.
1421. Elshamy, A.I., et al., UPLC-qTOF-MS phytochemical profile and antiulcer potential of Cyperus conglomeratus Rottb. alcoholic extract. *Molecules*, 2020. **25**(18): p. 4234.
1422. Abdel-Razik, A.F., et al., New prenylflavans from *Cyperus conglomeratus*. *Fitoterapia*, 2005. **76**(7–8): pp. 762–764.
1423. Khojaste, M., et al., Cell toxicity and inhibitory effects of *Cyperus rotundus* extract on *Streptococcus mutans, Aggregatibacter actinomycetemcomitans* and *Candida albicans*. *European Journal of Translational Myology*, 2018. **28**(4): pp. 7917.
1424. Shang, N., et al., Biodegradation of malachite green by an endophytic bacterium Klebsiella aerogenes S27 involving a novel oxidoreductase. *Applied Microbiology and Biotechnology*, 2019. **103**: pp. 2141–2153.
1425. Mifsud, S., *Cyperus longus Subsp. badius (Compact Galingale): MaltaWildPlants*. com-the online Flora of the Maltese Islands, 2002.
1426. Memariani, T., et al., Evaluation of the cytotoxic effects of Cyperus longus extract, fractions and its essential oil on the PC3 and MCF7 cancer cell lines. *Oncology Letters*, 2016. **11**(2): pp. 1353–1360.

1427. Ait-Ouazzou, A., et al., Evaluation of the chemical composition and antimicrobial activity of Mentha pulegium, Juniperus phoenicea, and Cyperus longus essential oils from Morocco. *Food Research International*, 2012. **45**(1): pp. 313–319.
1428. Xu, F., et al., Structures of New Sesquiterpenes and Hepatoprotective Constituents from the Egyptian Herbal Medicine Cyperus l ongus. *Journal of Natural Products*, 2004. **67**(4): pp. 569–576.
1429. Dávid, C.Z., J. Hohmann, and A. Vasas, Chemistry and pharmacology of Cyperaceae stilbenoids: a review. *Molecules*, 2021. **26**(9): p. 2794.
1430. *IUCN report*. Available from: https://www.iucnredlist.org/search?query=Cyperus%20longus&searchType=species
1431. Ngoc, Q.N. and T.N. Minh. Cyperus rotundus cyperaceae: a study of phytochemistry, total polyphenol content, flavonoid content, and antioxidant activity. in *E3S Web of Conferences*. 2021. EDP Sciences. 332, 2021
1432. Ju, Y. and B. Xiao, Chemical constituents of Cyperus rotundus L. and their inhibitory effects on uterine fibroids. *African Health Sciences*, 2016. **16**(4): pp. 1000–1006.
1433. Peerzada, A.M., et al., Cyperus rotundus L.: traditional uses, phytochemistry, and pharmacological activities. *Journal of Ethnopharmacology*, 2015. **174**: pp. 540–560.
1434. Ahmad, M., et al., Toxicological and biochemical evaluation of ethanolic crude extract of Cyperus rotundus. *International Journal of Pharmacy and Pharmaceutical Sciences*, 2013. **5**: pp. 538–544.
1435. Chang, K.-S., et al., Contact and fumigant toxicity of Cyperus rotundus steam distillate constituents and related compounds to insecticide-susceptible and-resistant Blattella germanica. *Journal of Medical Entomology*, 2012. **49**(3): pp. 631–639.
1436. Thanabhorn, S., et al., Acute and subacute toxicities of the ethanol extract from the rhizomes of Cyperus rotundus Linn. *Mahidol University Joural of Pharmaceutical Sciences*, 2005. **32**: pp. 15–22.
1437. *Adulterant of Cyperus rotundus*. Available from: https://www.thscenter.com/pdf/Cyperus%20rotundus.pdf
1438. Delfosse, P., et al., Dynamics of Polymyxa graminis and Indian peanut clump virus (IPCV) infection on various monocotyledonous crops and groundnut during the rainy season. *Plant Pathology*, 2002. **51**(5): pp. 546–560.
1439. Kusmiyati, N., S.T. Wicaksono, and D. Maknuna, Identification of endophytic fungi in nutgrass (Cyperus rotundus L.) as solubilizing phosphate and indole-3-acetic acid producers. *Biogenesis Jurnal Ilmiah Biologi*, 2021. **9**(1): pp. 93–101.
1440. Zega, A. and D. Suryanto. An ability of endophytic bacteria from nutgrass (Cyperus rotundus) from lafau beach of north nias in producing indole acetic acid and in solubilizing phosphate. in *IOP Conference Series: Earth and Environmental Science*. 2018. IOP Publishing130: 012007.
1441. Sheikh, M., A quick guide to useful nitrogen fixing trees from around the world. *NFT Highlights, NFTA*, 1989. **89**(7).
1442. Al-Snafi, A.E., Chemical constituents and pharmacological effects of Dalbergia sissoo-A review. *IOSR Journal of Pharmacy*, 2017. **7**(2): pp. 59–71.
1443. Gnanaraja, R., et al., Qualitative and quantitative phytochemicals analysis of selected fabaceae medicinal plants from Allahabad region. *The Pharma Innovation Journal*, 2014. **3**(7): pp. 53–56.
1444. Dixit, P., et al., Constituents of Dalbergia sissoo Roxb. leaves with osteogenic activity. *Bioorganic & Medicinal Chemistry Letters*, 2012. **22**(2): pp. 890–897.
1445. Mannan, M.A., A. Khatun, and M.F.H. Khan, Antinociceptive effect of methanol extract of Dalbergia sissoo leaves in mice. *BMC Complementary and Alternative Medicine*, 2017. **17**(1): pp. 1–13.
1446. Chakravarty, B., Herbal medicines. Safety and Efficacy Guidelines. *The Regulatory Affairs Journal*, 1993. **4**: pp. 699–701.
1447. Naik, H.N., et al., Dalbergia sissoo phytochemicals as EGFR inhibitors: an in vitro and in silico approach. *Journal of Biomolecular Structure and Dynamics*, 2023: pp. 1–13.
1448. Asif, M. and A. Kumar, Phytochemical investigation and evaluation of antinociceptive activity of ethanolic extract of Dalbergia sissoo (Roxb.) bark. *Journal of Natural Science, Biology, and Medicine*, 2011. **2**(1): p. 76.
1449. Khan, A.G., Relationships between chromium biomagnification ratio, accumulation factor, and mycorrhizae in plants growing on tannery effluent-polluted soil. *Environment International*, 2001. **26**(5–6): pp. 417–423.
1450. Ghazali, H.M.Z.U., et al., Fungi species causing dieback and wilt diseases in shisham [Dalbergia sissoo (Roxb)] and impact of various fungicides on their management. *Journal of King Saud University-Science*, 2022. **34**(4): p. 101970.
1451. Bakshi, B., Wilt of shisham (Dalbergia sissoo Roxb.) due to Fusarium solani sensu Snyder and Hansen. *Nature*, 1954. **174**(4423): pp. 278–279.
1452. *Description of Dalechampia scandens*. Available from: https://www.flowersofindia.net/catalog/slides/Gujarat%20Spurge-Creeper.html#:~:text=Male%20bracts%20are%20ovate%20in,heart%2Dshaped%20at%20the%20base
1453. Bolstad, G.H., et al., Genetic constraints predict evolutionary divergence in Dalechampia blossoms. *Philosophical Transactions of the Royal Society B: Biological Sciences*, 2014. **369**(1649): p. 20130255.
1454. Srivastava, S., D. Panchani, and N. Modi, Phytochemical analysis and antioxidant activity of selected plant. *Molecules*, 2023. **28**(6), p. 2867.
1455. *Uses of Dalechampia scandens*. Available from: https://powo.science.kew.org/taxon/319177-2#:~:text=The%20native%20range%20of%20this,a%20poison%20and%20a%20medicine
1456. Hansen, T.F., et al., Evolvability and genetic constraint in Dalechampia blossoms: components of variance and measures of evolvability. *Journal of Evolutionary biology*, 2003. **16**(4): pp. 754–766.
1457. *Macroscopic and microscopic characteristics of Daphne mucronata*. Available from: http://www.efloras.org/florataxon.aspx?flora_id=5&taxon_id=250062927.
1458. Khodadadian, Z., et al., Analgesic and anti-inflammatory potential of aerial parts of the Daphne mucronata Royle extract in mice: opioid-independent action. *Asian Pacific Journal of Tropical Biomedicine*, 2016. **6**(3): pp. 198–201.
1459. Zaidi, A., et al., Ethnobotanical, phytochemical and pharmacological aspects of Daphne mucronata (Thymeleaceae). *Tropical Journal of Pharmaceutical Research*, 2015. **14**(8): pp. 1517–1523.
1460. Ashraf, I., et al., Chemical composition, antioxidant and antimicrobial potential of essential oils from different parts of Daphne mucronata Royle. *Chemistry Central Journal*, 2018. **12**(1): pp. 1–8.
1461. Nazir, N., et al., Phytochemical profiling and antioxidant potential of Daphne mucronata Royle and action against paracetamol-induced hepatotoxicity and nephrotoxicity in rabbits. *Saudi Journal of Biological Sciences*, 2021. **28**(9): pp. 5290–5301.

1462. Ghorbani, L., et al., Phytochemicals, essential oils composition and antioxidant activity of Astragalus spp., Phlomis olivieri and Daphne mucronata in habitats of Central Iran. *Chemistry & Biodiversity*, 2023. **20**(9): p. e202300811.

1463. Murad, W., et al., Indigenous knowledge and folk use of medicinal plants by the tribal communities of Hazar Nao Forest, Malakand District, North Pakistan. *Journal of Medicinal Plants Research*, 2011. **5**(7): pp. 1072–1086.

1464. Bader, A., et al., The anti-angiogenic effect of polyphenols from the roots of Daphne mucronata Royle subsp. linearifolia (Hart) Halda (Thymelaeaceae). *European Journal of Integrative Medicine*, 2022. **53**: p. 102151.

1465. Javidnia, K., et al., A preliminary study on the biological activity of Daphne mucronata Royle. *DARU*, 2003. **11**(1): pp. 28.

1466. Mehrabi-Koushki, M. and S. Artand, First report of Coniothyrium ferrarisianum from Iran. *Rostaniha*, 2022. **23**(2): pp. 306–308.

1467. Brusotti, G., et al., Isolation and characterization of bioactive compounds from plant resources: the role of analysis in the ethnopharmacological approach. *Journal of Pharmaceutical and Biomedical Analysis*, 2014. **87**: pp. 218–228.

1468. *Wikipedia of Datura innoxia*. Available from: https://en.wikipedia.org/wiki/Datura_innoxia

1469. Li, Q., et al., Molecular cloning and functional analysis of hyoscyamine 6β-hydroxylase (H6H) in the poisonous and medicinal plant Datura innoxia mill. *Plant Physiology and Biochemistry*, 2020. **153**: pp. 11–19.

1470. Al-Snafi, A.E., Medical importance of Datura fastuosa (syn: Datura metel) and Datura stramonium-A review. *IOSR Journal of Pharmacy*, 2017. **7**(2): pp. 43–58.

1471. Ali, K., M. Sagheer, and A. Rashid, Impact of extracts of Azadirachta indica and Datura inoxia on the esterases and phosphatases of three stored grains insect pests of economic importance. *Pakistan Journal of Agricultural Sciences*, 2017. **54**(1): pp. 71 –81.

1472. Ayuba, V., T. Ojobe, and S. Ayuba, Phytochemical and proximate composition of Datura innoxia leaf, seed, stem, pod and root. *Journal of Medicinal Plants Research*, 2011. **5**(14): pp. 2952–2955.

1473. *Mechanism of Tropane alkoloids*. Available from: https://efsa.onlinelibrary.wiley.com/doi/pdf/10.2903/j.efsa.2008.691#:~:text=The%20mechanism%20of%20action%20of,preventing%20the%20binding%20of%20acetylcholine

1474. Jaafar, F.R., S.J. Ajeena, and S.S. Mehdy, Anti-inflammatory impacts and analgesiac activity of aqueous extract Datura innoxia leaves against induced pain and inflammation in mice. *Journal of Entomology and Zoology Studies*, 2018. **6**: pp. 1894–1899.

1475. Sharma, M., et al., Phytochemistry, pharmacology, and toxicology of Datura species – A review. *Antioxidants*, 2021. **10**(8): p. 1291.

1476. Fatima, H., et al., Extraction optimization of medicinally important metabolites from Datura innoxia Mill.: an in vitro biological and phytochemical investigation. *BMC Complementary and Alternative Medicine*, 2015. **15**: pp. 1–18.

1477. Al-Zharani, M., et al., In vitro cytotoxic evaluation and apoptotic effects of Datura innoxia grown in Saudi Arabia and phytochemical analysis. *Applied Sciences*, 2021. **11**(6): p. 2864.

1478. Romancer, M.L., C. Kerlan, and M. Nedellec, Biological characterisation of various geographical isolates of potato virus Y inducing superficial necrosis on potato tubers. *Plant Pathology*, 1994. **43**(1): pp. 138–144.

1479. *Distribution of Datura metel*. Available from: https://prota.prota4u.org/protav8.asp?g=pe&p=Datura+metel

1480. Sharma, G., Studies on antimycotic properties of Datura metel. *Journal of Ethnopharmacology*, 2002. **80**(2–3): pp. 193–197.

1481. Islam, T., et al., Ethnobotanical uses and phytochemical, biological, and toxicological profiles of Datura metel L.: a review. *Current Research in Toxicology*, 2023: p. 100106.

1482. Alam, W., et al., Datura metel: a review on chemical constituents, traditional uses and pharmacological activities. *Current Pharmaceutical Design*, 2021. **27**(22): pp. 2545–2557.

1483. Yang, B.-Y., et al., Chemical constituents from roots of Datura metel. *Zhongguo Zhong yao za zhi= Zhongguo Zhongyao Zazhi= China Journal of Chinese Materia Medica*, 2018. **43**(8): pp. 1654–1661.

1484. Kuang, H.-x., et al., Chemical constituents from the flower of Datura metel L. *Archives of Pharmacal Research*, 2008. **31**: pp. 1094–1097.

1485. Han, X., et al., Study on chemical constituents in seeds of Datura metel from Xinjiang. *Zhong yao cai= Zhongyaocai= Journal of Chinese Medicinal Materials*, 2015. **38**(8): pp. 1646–1648.

1486. Xue, J., et al., Chemical composition and cytotoxicity of the essential oil from different parts of Datura metel L. *Natural Product Research*, 2016. **30**(17): pp. 1938–1940.

1487. Mutebi, R.R., et al., Large outbreak of Jimsonweed (Datura stramonium) poisoning due to consumption of contaminated humanitarian relief food: Uganda, March–April 2019. *BMC Public Health*, 2022. **22**(1): pp. 1–10.

1488. Trancă, S.D., R. Szabo, and M. Cociş, Acute poisoning due to ingestion of Datura stramonium–a case report. *Romanian Journal of Anaesthesia and Intensive Care*, 2017. **24**(1): p. 65.

1489. Al-Snafi, A. E. Medical importance of Datura fastuosa (syn: Datura metel) and Datura stramonium-A review. *IOSR Journal of Pharmacy*, 2017. **7**(2): pp: 43–58.

1490. Ayyanar, M. and S. Ignacimuthu, Ethnobotanical survey of medicinal plants commonly used by Kani tribals in Tirunelveli hills of Western Ghats, India. *Journal of Ethnopharmacology*, 2011. **134**(3): pp. 851–864.

1491. Nisbet, C., et al., Characterization of Potato rough dwarf virus and Potato virus P: distinct strains of the same viral species in the genus Carlavirus. *Plant Pathology*, 2006. **55**(6): pp. 803–812.

1492. Singh, S. and S.N. Kumar, A review: introduction to genus Delonix. *World Journal of Pharmacy and Pharmaceutical Sciences*, 2014. **3**(6): pp. 2042–2055.

1493. Hegazi, G., In vitro studies on Delonix elata L., an endangered medicinal plant. *World Applied Sciences Journal*, 2011. **14**(5): pp. 679–686.

1494. Amabye, T., A. Bezabh, and F. Mekonen, Phytochemical constituents and antioxidant activity of Delonix elata L. flower extract. *Journal of Analytical & Pharmaceutical Research*, 2016. **2**(1): p. 00006.

1495. Krishnappa, P., et al., Antioxidant and prophylactic effects of Delonix elata L., stem bark extracts, and flavonoid isolated quercetin against carbon tetrachloride-induced hepatotoxicity in rats. *BioMed Research International*, 2014. **2014**: pp. 507851

1496. Al-Taweel, A.M., et al., Isolation of flavonoids from Delonix data and determination of its rutin content using capillary electrophoresis. *Pakistan Journal of Pharmaceutical Sciences*, 2015. **28:** pp. 1897–1903.

1497. Krishnappa, P., et al., Wound healing activity of Delonix elata stem bark extract and its isolated constituent

quercetin-3-rhamnopyranosyl-(1-6) glucopyranoside in rats. *Journal of Pharmaceutical Analysis*, 2016. **6**(6): pp. 389–395.
1498. Vasugi, C., S. Kamalakannan, and K. Murugan, Toxicity effect of Delonix elata (Yellow Gulmohr) and predatory efficiency of Copepod, Mesocyclops aspericornis for the control of dengue vector, Aedes aegypti. *Asian Pacific Journal of Tropical Disease*, 2013. **3**(2): pp. 119–126.
1499. Lin, J.-Y., et al., Trypsin inhibitor from the seeds of Acacia confusa. *The Journal of Biochemistry*, 1991. **110**(6): pp. 879–883.
1500. Al-Bahry, S., et al., First report of Ganoderma colossum on Ficus altissima and Delonix regia in Oman. *Plant Pathology*, 2005. **54**(2): pp. 245–245.
1501. *Dionysia mira*. Available from: https://www.selinawamucii.com/plants/primulaceae/dionysia-mira/
1502. Mamadalieva, N., et al., Chemical composition of essential oil from Dionysia hissarica. *Chemistry of Natural Compounds*, 2018. **54**: pp. 593–594.
1503. Theodosis, D.T., D.A. Poulain, and S.H. Oliet, Activity-dependent structural and functional plasticity of astrocyte-neuron interactions. *Physiological Reviews*, 2008. **88**(3): pp. 983–1008.
1504. El-Shabrawy, M.O., et al., Flavonoid constituents of Dipcadi erythraeum Webb. & Berthel. *Asian Pacific Journal of Tropical Disease*, 2016. **6**(5): pp. 404–405.
1505. Marzouk, M.M., et al., C-glycosyl flavonoids-rich extract of Dipcadi erythraeum Webb & Berthel. bulbs: phytochemical and anticancer evaluations. *Journal of Applied Pharmaceutical Science*, 2019. **9**(6): pp. 94–98.
1506. Lin, M.-T., et al., Chemical Constituents from Drypetes l ittoralis. *Journal of Natural Products*, 2001. **64**(6): pp. 707–709.
1507. Aswal, S., et al., Drimia indica: a plant used in traditional medicine and its potential for clinical uses. *Medicina*, 2019. **55**(6): p. 255.
1508. Samanta, T., et al., Comparative cytogenetics and fluorescent chromosome banding in five indian species of Dipcadi Medik. *Plants*, 2023. **12**(13): p. 2534.
1509. Park, R.F., et al., Leaf rust of cultivated barley: pathology and control. *Annual Review of Phytopathology*, 2015. **53**(1): pp. 565–589.
1510. *Description of Dipcadi viride*. Available from: https://www.zimbabweflora.co.zw/speciesdata/species.php?species_id=114390#:~:text=Description%3A,bulb%20sometimes%20loose%20and%20scaly
1511. Al Farraj, D.A., et al., In-vitro antimicrobial activities of organic solvent extracts obtained from Dipcadi viride (L.) Moench. *Journal of King Saud University-Science*, 2020. **32**(3): pp. 1965–1968.
1512. Adly, F., et al., Phytochemical screening and biological study of ethanol extractives of Dipcadi serotinum (L.) Medik. *European Journal of Advanced Research in Biological and Life Sciences* Vol, 2015. **3**(3).
1513. *Toxicity of Dipcadi viride*. Available from: https://repository.up.ac.za/handle/2263/8458#:~:text=Toxicity%20variable%2C%20depends%20on%20growth,Aimless%20wandering%3B%20pushing%20against%20objects
1514. Park, R.F., et al., Leaf rust of cultivated barley: pathology and control. *Annual Review of Phytopathology*, 2015. **53**: pp. 565–589.
1515. Ahmed, A.F., et al., A major diplotaxis harra-derived bioflavonoid glycoside as a protective agent against chemically induced neurotoxicity and Parkinson's models; In silico target prediction; and biphasic HPTLC-based quantification. *Plants*, 2022. **11**(5): p. 648.
1516. Hashem, F. and M. Saleh, Antimicrobial components of some cruciferae plants (Diplotaxis harra Forsk. and Erucaria microcarpa Boiss.). *Phytotherapy Research: An International Journal Devoted to Pharmacological and Toxicological Evaluation of Natural Product Derivatives*, 1999. **13**(4): pp. 329–332.
1517. Kassem, M.E., et al., Two new flavonol glycosides and biological activities of Diplotaxis harra (Forssk.) Boiss. *Natural Product Research*, 2013. **27**(24): pp. 2272–2280.
1518. Melrose, J. The glucosinolates: a sulphur glucoside family of mustard anti-tumour and antimicrobial phytochemicals of potential therapeutic application. *Biomedicines*, 2019. **7**(3): pp. 62.
1519. Nasri, I., et al., Anti-inflammatory and anticancer effects of flavonol glycosides from Diplotaxis harra through GSK3β regulation in intestinal cells. *Pharmaceutical Biology*, 2017. **55**(1): pp. 124–131.
1520. Atta, E.M., et al., Phytochemical studies on Diplotaxis harra growing in Sinai. *European Journal of Chemistry*, 2011. **2**(4): pp. 535–538.
1521. Nicoletti, R., F. Raimo, and G. Miccio, First report of Rhizoctonia solani on Diplotaxis tenuifolia in Italy. *Plant pathology*, 2004. **53**(6): pp. 811–811.
1522. *Botanical Description of Dipterygium glaucum*. Available from: https://plants.jstor.org/compilation/dipterygium.glaucum
1523. Shaheen, U., et al., Flavonoidal constituents, antioxidant, antimicrobial, and cytotoxic activities of Dipterygium glaucum grown in Kingdom of Saudi Arabia. *Pharmacognosy Magazine*, 2017. **13**(Suppl 3): p. S484.
1524. Altwaty, N.H., et al., Molecular and cytogenetic assessment of Dipterygium glaucum genotoxicity. *Anais da Academia Brasileira de Ciências*, 2016. **88**: pp. 623–634.
1525. Choudhary, D., J. Shekhawat, and V. Kataria, GC-MS analysis of bioactive phytochemicals in methanol extract of aerial part and callus of Dipterygium glaucum decne. *Pharmacognosy Journal*, 2019. **11**(5): pp. 1055–1063.
1526. Shahid, A., et al., Chemical profiling and biological activities of Dipterygium glaucum Decne.: an in-vivo, in-vitro and in-silico evaluation. *South African Journal of Botany*, 2023. **160**: pp. 715–730.
1527. Sherwani, N. and S.A. Farooq, Impact of habitat heterogeneity on growth dynamics and physiological responses of Dipterygium glaucum. *Asian Journal of Plant Sciences*, 2019. **18**(2): pp. 75–84.
1528. Manjulatha, K., *A Comparative Study of Different Parts of Flavonoid Rich Plant, Dodonaea viscosa for Antimicrobial and Antioxidant Potential.* LAP Lambert Academic Publishing, 2012. https://www.lap publishingcom/catalog/details//store/gb/book/978-3-659-28793-0/dodonaea-viscosa-flavonoid-rich-plant-and-its-biological-potential.
1529. Al-Snafi, A.E., A review on Dodonaea viscosa: a potential medicinal plant. *IOSR Journal of Pharmacy*, 2017. **7**(2): pp. 10–21.
1530. Tong, Z.-W., et al., Determination of in vivo biological activities of Dodonaea viscosa flowers against CCL4 toxicity in albino mice with bioactive compound detection. *Scientific Reports*, 2021. **11**(1): p. 13336.
1531. Al Bimani, B.M.H. and M.A. Hossain, A new antimicrobial compound from the leaves of Dodonaea viscosa for infectious diseases. *Bioactive Materials*, 2020. **5**(3): pp. 602–610.
1532. Muhammad, A., et al., Phytochemicals from Dodonaea viscosa and their antioxidant and anticholinesterase activities with structure–activity relationships. *Pharmaceutical Biology*, 2016. **54**(9): pp. 1649–1655.

1533. Herrera-Calderon, O., et al., Dodonaea viscosa Jacq. induces cytotoxicity, antiproliferative activity, and cell death in colorectal cancer cells via regulation of caspase 3 and p53. *Frontiers in Pharmacology*, 2023. **14**.

1534. Meenu, J., S. Sunil, and K. Manoj, Evaluation of antihyperglycemic activity of Dodonaea viscosa leaves in normal and STZ-diabetic rats. *International Journal of Pharmacy and Pharmaceutical Sciences*, 2011. **3**(1): pp. 69–74.

1535. Yu, H., et al., Antifeedant Mechanism of Dodonaea viscosa Saponin A Isolated from the Seeds of Dodonaea viscosa. *Molecules*, 2022. **27**(14): pp. 4464.

1536. Malik, M.N., et al., Bioprospecting Dodonaea viscosa Jacq.; a traditional medicinal plant for antioxidant, cytotoxic, antidiabetic and antimicrobial potential. *Arabian Journal of Chemistry*, 2022. **15**(3): p. 103688.

1537. *Pathology of Dodonaea viscosa*. Available from: https://prota.prota4u.org/protav8.asp?g=pe&p=Dodonaea+viscosa

1538. Heinke, R., et al., Furanocoumarins from Dorstenia foetida. *Phytochemistry*, 2011. **72**(9): pp. 929–934.

1539. Algabr, M., et al., Wound healing activity and phytochemical screening of latex extract of Dorstenia foetida plant. *Advances in Bioresearch*. 2022. **13**(6): pp. 161–166.

1540. *Toxicity of Dorstenia foetida*. Available from: https://www.llifle.com/Encyclopedia/SUCCULENTS/Family/Moraceae/11246/Dorstenia_sp

1541. *Pathology of Dorstenia foetida*. Available from: https://www.indoor-plant-care.com/plant-list/dorstenia-foetida/

1542. Vahalík, P., et al., The conservation status and population mapping of the endangered Dracaena serrulata in the Dhofar Mountains, Oman. *Forests*, 2020. **11**(3): p. 322.

1543. Vaníčková, L., et al., Terpenoid profiles of resin in the genus Dracaena are species specific. *Phytochemistry*, 2020. **170**: p. 112197.

1544. Lu, P.-L. and C.W. Morden, Phylogenetic relationships among Dracaenoid genera (Asparagaceae: Nolinoideae) inferred from chloroplast DNA loci. *Systematic Botany*, 2014. **39**(1): pp. 90–104.

1545. Ying, L., et al., Chemical constituents from dragon's blood of Dracaena cambodiana. *Chinese Journal of Natural Medicines*, 2011. **9**(2): pp. 112–114.

1546. Al-Awthan, Y.S. and O.S. Bahattab, Phytochemistry and pharmacological activities of Dracaena cinnabari resin. *BioMed Research International*, 2021. **2021**: pp. 1–7.

1547. *Uses of Dracaena serrulata*. Available from: https://en.wikipedia.org/wiki/Dracaena_serrulata

1548. Yannasithinon, S., et al., Dolichandrone serrulata flower extract ameliorates male reproductive damages in type 2 diabetic rats. *Andrologia*, 2021. **53**(2): p. e13911.

1549. Al-Afifi, N.A., et al., Acute and sub-acute oral toxicity of Dracaena cinnabari resin methanol extract in rats. *BMC Complementary and Alternative Medicine*, 2018. **18**: pp. 1–14.

1550. Chaimontri, C. and S. Iamsaard, Avaliação da toxicidade do extrato da flor de Dolichandrone serrulata em órgãos vitais e reprodutivos em ratos machos adultos. *Brazilian Journal of Biology*, 2021. **84**: p. e251224.

1551. Durán, I., et al., Iconic, threatened, but largely unknown: biogeography of the Macaronesian dragon trees (Dracaena spp.) as inferred from plastid DNA markers. *Taxon*, 2020. **69**(2): pp. 217–233.

1552. Ahmad, W., et al., Complete chloroplast genome sequencing and comparative analysis of threatened dragon trees Dracaena serrulata and Dracaena cinnabari. *Scientific Reports*, 2022. **12**(1): p. 16787.

1553. Bowman, C. and B. PW, Panonychus citri (McGregor) (Acari: Tetranychidae) infesting imported Citrus mitis in England. *Plant Pathology*, 1978 **27**: pp. 201–201.

1554. Gong, L. and S. Guo, Endophytic fungi from Dracaena cambodiana and Aquilaria sinensis and their antimicrobial activity. *African Journal of Biotechnology*, 2009. **8**(5).

1555. Mottaghipisheh, J., et al., Ducrosia spp., rare plants with promising phytochemical and pharmacological characteristics: an updated review. *Pharmaceuticals*, 2020. **13**(8): p. 175.

1556. Hajhashemi, V., et al., Evaluation of antianxiety and sedative effects of essential oil of Ducrosia anethifolia in mice. *Clinics*, 2010. **65**: pp. 1037–1042.

1557. Shalaby, N.M., et al., Preliminary *in vitro* and *in vivo* evaluation of antidiabetic activity of Ducrosia anethifolia Boiss. and its linear furanocoumarins. *BioMed Research International*, 2014. 480545.

1558. Mothana, R.A., et al., Ducrosia ismaelis Asch. essential oil: chemical composition profile and anticancer, antimicrobial and antioxidant potential assessment. *Open Chemistry*, 2020. **18**(1): pp. 175–184.

1559. Hung, W.-L., J.H. Suh, and Y. Wang, Chemistry and health effects of furanocoumarins in grapefruit. *Journal of Food and Drug Analysis*, 2017. **25**(1): pp. 71–83.

1560. Zamyad, M., et al., Os efeitos anticonvulsivantes do óleo essencial de Ducrosia anethifolia (Boiss) são realizados pelo seu principal componente alfa-pineno em ratos. *Arquivos de Neuro-Psiquiatria*, 2019. **77**: pp. 106–114.

1561. Zamyad, M., et al., The anticonvulsant effects of Ducrosia anethifolia (Boiss) essential oil are produced by its main component alpha-pinene in rats. *Arquivos de Neuro-Psiquiatria*, 2019. **77**: pp. 106–114.

1562. *Botanical Description of Dyerophytum indicum*. Available from:https://www.flowersofindia.net/catalog/slides/Indian%20Desert%20Statice.html

1563. Chernyshev, A. and K. Melghit, Characterization of a mineral coating of the plant Dyerophytum indicum. *Zeitschrift für Naturforschung C*, 2015. **70**(3-4): pp. 59–64.

1564. Rastogi, A., Biotic elicitors used to enhance plumbagin production in Plumbago Zeylanica and assesment of antioxidant and antibacterial activity. *Plant Cell Tissue and Organ Culture* 2020. **151**(3): pp. 1–11.

1565. Ji, T., et al., Non-palm plant volatile α-pinene is detected by antenna-biased expressed odorant receptor 6 in the Rhynchophorus ferrugineus (Olivier) (Coleoptera: Curculionidae). *Frontiers in Physiology*, 2021. **12**: p. 701545.

1566. Tyagi, R., E. Menghani, and G. Sharma, An endangered medicinal plant, Plumbago zeylanica L. and its substitute, Dyerophytum indicum (Gibbs ex Wight) Kuntze: a comparative HPTLC, HPLC study, and screening of antioxidant activity. *JPC–Journal of Planar Chromatography–Modern TLC*, 2017. **30**: pp. 17–24.

1567. Migahid, A. and M. Hammouda, *Flora of Saudi Arabia*. Riyadh:, King Saud University Press, 1996.

1568. Salehi, B., et al., Therapeutic potential of α-and β-pinene: a miracle gift of nature. *Biomolecules*, 2019. **9**(11): p. 738.

Index